Distributed Cooperative Laboratories: Networking, Instrumentation, and Measurements

T0137965

Distributed Cooperative Laboratories: Networking, Instrumentation, and Measurements

Edited by:

Franco Davoli
DIST - University of Genoa

Sergio Palazzo
DIIT – University of Catania

Sandro Zappatore
DIST - University of Genoa

UNIONE EUROPEA

 Springer

Prof. Franco Davoli
University of Genoa - DIST
Via Opera Pia,13
16145 GENOVA
ITALY

Prof. Sergio Palazzo
Universitá di Catania
Dip. di Ingegneria Informatica e delle Telecomunicazioni
Viale A. Doria,6
95125 CATANIA
ITALY

Prof. Sandro Zappatore
University of Genoa - DIST
Via Opera Pia,13
16145 GENOVA
ITALY

Distributed Cooperative Laboratories: Networking, Instrumentation, and Measurements

e-ISBN 0-387-30394-4

ISBN 978-1-4419-4002-5

e-ISBN 978-0-387-30394-9

Printed on acid-free paper.

©2006 Springer Science+Business Media, Inc.
Softcover reprint of the hardcover 1st edition 2006

9 8 7 6 5 4 3 2 1

springer.com

Table of Contents

Chapter III. Data Acquisition And Aggregation In Sensor Networks

Chapter IV. GRID Structures For Distributed Cooperative Laboratories

Chapter VI. Virtual Immersive Communications And Distance Learning

Preface

Progress in Telecommunications and Networking is fostering the development of high-speed and ubiquitous networks, both wired and wireless, characterized by an unprecedented degree of transport capacity and flexibility. At the same time, laboratory equipment for measurement and experimental evaluation of devices and systems is available, with a wide range of sophistication and complexity. Activities such as distance learning, performance monitoring and testing may now receive a support capable of making them truly distributed and highly cooperative.

Yet another aspect strictly related with the increased cooperation and coordination of distributed laboratory equipment is that of Grid computing. Some of the main characteristics of a Grid structure, namely, distributed and coordinated resource usage, standardized middleware and quality of service delivery, play a very important role in the interconnection of measurement devices of various kinds. On the other hand, the extension of the Grid platforms beyond the computing world to such environments poses new challenges in dealing with aspects such as real-time constraints.

The visualization of experiment results and the introduction of the usage of laboratory equipment in distance learning brings other components to this already complex picture, namely, those related to multimedia and immersive communications, virtual and augmented reality.

Based on the recent developments in these multidisciplinary fields, research effort is being dedicated worldwide to the investigation of the main issues related to the sustainable realization of tele-laboratories, where real and virtual instrumentation can be shared and used in a collaborative environment. Such issues are related, but not limited to, multimedia communications and networking, sensor networks, Grid technology, Quality of Service (QoS) provisioning and control, network management, measurement instrumentation and methodology, architecture of measurement systems.

This book is dedicated to highlighting some state-of-the-art research aspects in this multi-faceted scenario. All papers in the book were presented at the 2005 Tyrrhenian Workshop on Digital Communications. The workshop, besides being a forum of discussion among internationally known experts in the various related fields, represented also the closing event of the CNIT (the Italian National Consortium for

Telecommunications) project "Technological Network for Telecommunication Measurement Instrumentation", funded by the Italian Ministry of Education, University and Research (MIUR), in the framework of the European Union (EU). The additional equipment acquired by the CNIT National Laboratory for Multimedia Communications in Naples, Italy, within this project, set the basis for the participation in two other currently active ones, namely, GRIDCC (Grid-Enabled Remote Instrumentation with Distributed Control and Computation), funded by the EU, within the 6th Framework Program, and CRIMSON (Cooperative Remote Measurement Systems Over Networks), also funded by MIUR. Quite a few researchers operating in these projects took part in the Workshop.

According to the structure of the Workshop, the book is organized in six chapters.

Chapter I, *Technologies for Real-Time Interactive Multimedia Communications*, addresses issues regarding content presentation and interaction over the network, resource allocation and scalability.

Chapter II, *Monitoring, Management and Configuration of Networks and Networking Devices*, is dedicated to measurements over network traffic and router performance, both in terms of tools and experimental analysis.

Chapter III, *Data Acquisition and Aggregation in Sensor Networks*, touches the aspects related with data acquisition, particularly from wireless sensors, with respect to processing, detection, routing, energy consumption, and the construction of test beds.

Chapter IV, *Grid Structures for Distributed Cooperative Laboratories*, opens a window on the Grid computing world and investigates its relation with the management of instrumentation and measurement devices. Here the major issues are the architecture of the collaborative environment, the presence of real-time aspects, and the allocation of resources.

Chapter V, *Architectures and Techniques for Tele-Measurements*, deals with the essence of the measurement systems and methodologies needed to realize truly distributed laboratory spaces. Besides the methodological aspects, the specific environment of telecommunications is addressed, regarding the management of both physical layer and networking distributed measurement platforms.

Chapter VI, *Virtual Immersive Communications and Distance Learning*, is devoted to context-awareness and to the construction of virtual and immersive environments. Most of the material in this chapter comes from the results of two national CNIT projects, namely, VICOM (Virtual Immersive Communications) and Teledoc2.

Overall, the book addresses complementary and strictly related aspects of a highly multidisciplinary framework, which characterizes the status and the evolution of distributed cooperative laboratories. As such, it can be viewed as a source of reference for those interested in this challenging field.

FRANCO DAVOLI
SERGIO PALAZZO
SANDRO ZAPPATORE

Acknowledgments

The realization of this book would have not been possible without the help of many people and organizations that contributed to the success of the Tyrrhenian Workshop on Digital Communications, held in Sorrento, Italy, in July 2005, and focused on "Distributed Cooperative Laboratories: Issues in Networking, Instrumentation, and Measurements". The editors wish to thank the organizers of the technical sessions, whose cooperation was invaluable: Magda El Zarki, from the University of California at Irvine, USA; Mario Gerla, from the University of California at Los Angeles, USA; Luigino Benetazzo, from the University of Padova, Italy; Christophe Diot, from Intel Research, Cambridge, UK; Oreste Andrisano and Davide Dardari, from the University of Bologna, Italy; Franco Vatalaro, from the University of Rome Tor Vergata, Italy. The sincerest appreciation goes to the keynote speakers, Ian Akyildiz, from Georgia Institute of Technology, USA, Geoffrey Fox, from Indiana University, USA, and Hui Zhang, from Carnegie Mellon University, Pittsburgh, USA, for their enlightening and insightful talks. The editors also thank all the Program Committee members, and the authors for their fine contributions. A special thank goes to Stefano Vignola, for his help in the organization of the workshop and in the preparation of the book, and to the staff of the CNIT National Laboratory for Multimedia Communications in Naples, Italy.

We gratefully acknowledge the support of the Italian National Consortium for Telecommunications (CNIT), of the European Union, of the Italian Ministry of Education, University and Research (MIUR) and of the following companies, whose precious sponsorship was greatly appreciated.

AERSat	**Agilent Tecnologies**
CISCO Systems	**Eutelsat**
ITNet	**Light Comm**
SELEX Communications	**MBI**
SIR Soluzioni in Rete	**WIND**

Chapter I

Technologies For Real-Time Interactive Multimedia Communications

Team Collaboration Mixing Immersive Video Conferencing With Shared Virtual 3D Objects

Ralf Tanger [1], Peter Kauff [1], Oliver Schreer [1],
Dominique Pavy [2], Stephane Louis Dit Picard [2], Grégory Saugis [2]

[1] Fraunhofer Institute for Telecommunications – Heinrich Hertz Institut, Einsteinufer 37
10587 Berlin, Germany
{Ralf.Tanger, Peter.Kauff, Oliver.Schreer}@fhg.hhi.de
[2] France Telecom Division R&D – TECH/IRIS/VIA, 2 Avenue Pierre Marzin
22307 Lannion Cedex, France
{ dominique.pavy, stephane.louisditpicard, gregory.saugis }@ francetelecom.com

(Invited Paper)

Abstract. This paper presents a novel approach on a joint team collaboration system. It combines two fields of applications that have been developed separately in the past: collaborative virtual environments and video-based tele-conferencing. For this purpose technologies from both areas have been integrated to a common test platform. After a specific description of the initial technology branches, the paper mainly focuses on the conceptual work for merging them to a novel virtual team user environment.

Keywords. Immersive tele-collaboration, immersive telecommunication, synchronous team collaboration, video conferencing

1 Introduction

The migration of immersive media to telecommunication continues to advance and to become cheaper through digital representation. The ability to evoke a state of "being there" and/or of "being immersed" will no longer remain the domain of the flight simulators, CAVE systems, cyberspace applications, theme parks or IMAX theatres. It will arise in offices, venues and homes and it has the potential to enhance the workflow of management, training, manufacturing and marketing in general. First steps in this direction could already been observed in two application areas during the last few years.

On one hand, today's high-end videoconferencing systems offer tele-presence capabilities to achieve most natural communication conditions. Thanks to this evolution video conferencing is going to become more and more attractive for various lines of business and it will definitively benefit from the advent of tele-immersion in the same manner therefore.

On other hand, the market of synchronous team collaboration systems is growing drastically to satisfy demands of increasing competition in costs, innovation, productivity and development cycles. In addition, emerging tele-immersion systems, like collaborative virtual environments, will provide further improvements including intuitive interaction and communication capabilities.

This paper aims at the enormous potential of merging these two technologies. Especially the embedding of video conferencing functionalities into shared virtual environments paves the road for a new era in telecommunications. It combines two fields of applications that have been developed separately in the past: the VR-based functionality of shared applications and the realism of video-conferencing.

Therefore Fraunhofer Institute for Telecom-munications/Heinrich Hertz-Institut (HHI) and France Telecom Research & Development (FTR&D) have initiated a joint research activity with the objective to integrate technologies from both areas into a new test platform for immersive tele-collaboration. This work is based on two systems that have been developed separately so far at the two institutions: the immersive videoconference system im.point (HHI) and the synchronous team collaboration software Spin3D (FTR&D). After a specific description of im.point and Spin3D in section 2 and 3, respectively, section 4 will focus on the conceptual work for merging these technologies to a novel virtual team user environment.

2 Immersive Meeting Point (im.point)

As shown in Fig. 1, the im.point represents a semi-immersive videoconferencing system that follows the concept of a round table conversation. The idea is to place suitable video reproductions of all participants at predefined positions in a shared virtual environment (SVE). For a typical three-party conference these positions are located at the corners of an equilateral triangle enclosing the round table.

Fig. 1. Multi-party conference system im.point

Thanks to this configuration, it can be ensured that each conferee sees his interlocutors under an individual viewpoint [1]. As a consequence, those communication modalities that are particularly important for multi-party conferences, such as eye-contact, gaze, postures, gestures or directive sound, can be reproduced under right 3D geometry. Moreover, at each side the virtual scene is rendered from the local user's viewpoint to enhance the illusion of a seamless transition between real and virtual world across the screen. To enforce this effect, elements in the real room like the table are mirrored and virtual objects are aligned with their real counterparts. Thus, participants get the impression of sitting around a common table as they know it from a real conference situation.

One main objective of the im.point development was to bridge the gap between research systems on one hand, which usually need expensive hardware and are dedicated to a particular academic topic, and com-mercial products on other hand, which still do not meet the requirements of human-centered communication system, especially not in the case of multi-party applications.

For this purpose the im.point is based on a modular and extendable system architecture. In its basic version two cameras are placed as near as possible to the head of the remote participants which results in small but acceptable lack of eye contact. The difference between viewing direction and camera axis is less than 10 degrees. Non-verbal communication cues that contain directive information (who is pointing on what, who is looking in which direction, etc.) are supported to the greatest possible extent. This especially holds for the active line of sight (two conferees have direct eye-contact) as well as for the passive line of sight (a third conferee observes when two others have eye-contact).

To reduce the remaining lack of eye contact, the basic version can also be extended towards a 3D version. In this case, two cameras are used on each display side. Hence, disparities can be estimated from each camera pair. This 3D information is used to reconstruct the view of a virtual camera that is exactly located at the current line of sight between two related partners.

Fig. 2 shows the general system architecture of the im.point system. The basic modules are grey-shaded, additional modules of the 3D version are not shaded. All modules are entirely implemented in software and run on standard PCs. No special hardware is needed. Each terminal consists of a server and a client.

At the server all captured videos are first segmented to separate the person's silhouette from the background. This allows a seamless integration of the video objects into the virtual scene. In the 3D extension a "video plus depth" representation format is generated by rectification, disparity analysis and 3D view combining [2].

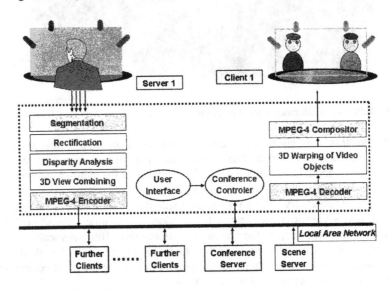

Fig. 2. Architecture of im.point system

Either the shaped video objects (basic version) or the "video plus depth" objects (3D version) are then encoded and transmitted using the MPEG-4 multimedia standard. The client decodes the incoming videos from the remote participants. In the 3D extension the additionally transmitted depth information is used for calculating the virtual view. The resulting videos are then integrated into the virtual scene using BIFS (Binary Format for Scenes) of MPEG-4.

3 Synchronous CVE-Platform Spin3D

FTR&D, in collaboration with INRIA Futurs, carries out studies on design and realization of a collaborative platform called Spin3D based on a 3D collaborative virtual environment (CVE) approach [3]. The aim of this work is to allow strong synchronous collaboration around virtual objects to a small group of users.

3.1 Interface and Interaction design

The first main objective is to design a user friendly 3D interface that privileges the easiness of use, the collaborative task interactions, and the quality of the communication between the participants.

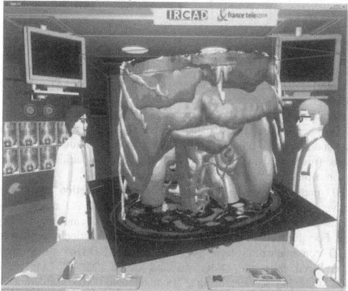

Fig. 3. The "meeting room" metaphor

Usually, collaboration is effective when only a small amount of persons are actively working together on a small number of objects, the inter-actors being situated in a common place, usually around a table, or at least inside a unique room. Moreover, in order to get a better understanding of the performed tasks, each user needs to be constantly aware of any object transformation, and thus no manipulation must be hidden. Based on those observations, the visual interface of the Spin-3D platform is built on a "meeting room" metaphor (Fig. 3), within which all virtual objects and users are displayed: a central table is a manipulation area on which the users put the objects they want to interact with at a given time; a rotative band around a table is a place holder for all the objects needed for a given task; finally, few virtual tools, set down on the table, allow users to manipulate or modify virtual objects. The mechanisms of the Spin3D interface can be changed using a MVC concept (Model-View-Controller).

Within the interface, the support of telepresence is enhanced by synthetic 3D avatars, which are realistic representation of remote users in terms of texture, morphology and animation [4]. The non-verbal part of the communication between users is represented through these avatars by analysis of the user inputs (eg. interaction actions, voice, video-tracking). Technologies such as facial animation, inverse kinematics, and acoustic spatial localization have been implemented for this purpose.

The user interacts with objects thanks to a bi-manual interaction mechanism: a first input device is used for designation/selection of objects in the interface, whereas a second input device is used for manipulation of objects (3D rotations and translations). To ease the 3D interaction and the understanding of the collaborative activity ("*who is doing what*"), the interface provides feedbacks (such as bounding boxes, shadows, or object ownership with color changes).

3.2 Technical platform

The second main objective is to develop a technical platform supporting the management of collaborative applications running on standard PC connected to the Internet through common access-points (e.g. xDSL).

The 3D visualization core of the collaborative platform is built around an extended VRML97 browser integrating extensions in terms of interaction (3D interaction and feedbacks) and sharing of objects [5]. In order to display various (i.e. non-VRML97) 3D formats (medical data, CAD models, etc.), the capabilities of the Spin3D visualization core can be extended by using plug-ins.

The Spin3D platform provides the underlying communication layers required for collaborative activities namely the replication of shared 3D objects amongst all workstations using a server-less architecture thus reducing the latency in the interaction. For group communication and state synchronization of replicated objects, the Spin3D communication layer [6] is based on CORBA using an enhanced protocol, inspired from the OMG standard called "MIOP", and an enhanced multimedia streaming service, inspired from another OMG standard called "A/V Streams service". The communication layer is kept independent of the underlying transport layer (multicast IP or IP bridge emulating multicast) in order to run the collaborative platform on various kinds of networks.

4 Joint Team Collaboration System

A comparison between Spin3D and im.point shows a lot of similarities with respect to the "meeting room" metaphor. Both systems use the concept of a shared virtual table as the connecting element between the real and the virtual world. However, Spin3D is more focused on a 3D CVE-based application allowing a strong interaction with 3D virtual objects where avatars represent the non-verbal part of the communication at an abstract level but with a limited amount of realism. In contrast, the im.point provides a high amount of realism and represents non-verbal communication cues in highest video quality, but it is limited in the use of joint application tools. Thus, to benefit from both system approaches, it seems to be straight-forward to combine them to a novel joint team collaboration system based on the concept of a shared virtual table.

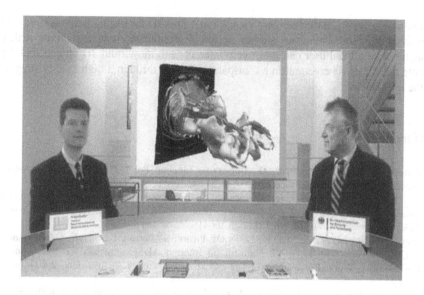

Fig. 4. Screen shot of joint team collaboration

As shown by the screen shot in Fig. 4, the im.point system runs in the foreground to display the virtual meeting room including the distant participants. The virtual Spin3D tools for manipulating or modifying shared objects are integrated into the im.point scene and placed on the conference table. As in Spin3D, these tools can be used through a bi-manual interaction mechanism, either for selecting or for manipulating objects.

The Spin3D application runs in the background and the results of joint interaction are depicted at a virtual screen additionally integrated in the im.point scene. Thus, as soon as a user selects a Spin3D tool and enters the active area of the virtual screen, he can use all functionalities of the Spin3D application. In addition, the user can zoom in to see details of the Spin3D visualization, and he can zoom out again to discuss the results with the distant partners in the im.point conference mode.

Both systems, im.point and Spin3D are connected using a custom protocol which allows a rapid integration of both approaches without affecting the original systems too much.

5 Conclusions

This paper concentrates on the description of an immersive tele-collaborative tool based on a pragmatic approach combining two separately developed application platforms (immersive videoconference im.point for the telepresence support and synchronous CVE-software Spin3D for the 3D interaction on shared virtual objects). To this end, the first common implementation step will focus on an ad-hoc, easy to develop prototyping system mixing im.point and Spin3D for testing purposes and will a be start for more ambitious studies on an ideal solution, that is a mixed reality system where a local user can intuitively interact with the virtual 3D objects and the remote users can observe that in a transparent manner through a high quality support of the telepresence.

References

1. Tanger, R., Kauff, P., Schreer, O.: Immersive Meeting Point – An Approach Towards Immersive Media Portals, Proceedings of PCM 04, Tokyo, December (2004) 89-96
2. N. Atzpadin, Kauff, P., Schreer, O.: Stereo Analysis by Hybrid Recursive Matching for Real-Time Immersive Video Conferencing, Trans. on CSVT, Special Issue on Immersive Telecommunication, Vol.14, No.3 (March 2004)
3. Pavy, D., Bouguet, A., Le Mer, P., Louis Dit Picard, S., Perron, L., Saugis, G., Degrande, S.,Chaillou, C.,: Spin3D: a VR-platform on Internet ADSL networks for Synchronous Collaborative Work, Proceedings of IST eChallenges, Vienna Austria (Oct. 2004) 1486-1493
4. Le Mer, P., Perron, L., Chaillou, C., Degrande, S, Saugis, G., Collaborating with Virtual Humans, Proceedings of HCI 2001, Springer, Lille France (Sept. 2001) 83-103
5. Louis Dit Picard, S., Degrande, S., Gransart, C., Saugis, G. Chaillou, C., VRML Data Sharing in the Spin3D CVE, Proceedings of Web3D'02, ACM, Tempe Arizona USA (Feb. 2002) 165-172
6. Louis Dit Picard, S., Degrande, S., Gransart, C., Saugis, G., Chaillou, C., A CORBA-based platform as communication support for synchronous collaborative environments, Proceedings of Multimedia Middleware Workshop (M3W), ACM, Ottawa Canada (Oct. 2001) 56-59

On The Stochastic Scalability
Of Information Sharing Platforms

Phuoc Tran-Gia and Andreas Binzenhöfer

Department of Distributed Systems
Institute of Computer Science
University of Würzburg, Am Hubland, 97074 Würzburg, Germany
{trangia,binzenhoefer}@informatik.uni-wuerzburg.de

(Invited Paper)

Abstract. Recently, centrally controlled information distribution systems are rapidly emerging to decentralized structures. This tendency can be observed in software distribution applications using BitTorrent or information sharing platforms based on distributed hash table structures like Chord or Kademlia. To ensure that the emerging platforms will function properly with a growing number of users and services the issue of *scalability* turned into one of the hottest research topics.
Traditionally, the term scalability often restricts to the functional scalability, which describes the scalability in terms of the system size. In this regard the basic structure is stationary, i.e. it does not fluctuate frequently. However, when the stochastic behavior of system components, the network structure, and user applications has to be taken into account, the *stochastic scalability* has to be investigated in the context of performance evaluation. In this paper we discuss the stochastic scalability of information sharing platforms. We give a classification of current information sharing platforms and define the terms functional and stochastic scalability in detail. A distributed phone book based on a Chord ring will be discussed as an example to motivate other areas of application and to show the potential of the evaluation of stochastic scalability.

Keywords: Information Sharing, Information Mediation, P2P, Stochastic Scalability, Performance Analysis

1 Information Mediation and Platforms

The scale of distributed systems has significantly changed with the growing size of the Internet in the last decade. Distributed applications have to serve thousands to millions of customers in parallel. Due to stochastic user behavior, the number of customers and their world wide locations, as well as the dynamic of current network architecture, such as overlay networks in file-sharing platforms, the underlying network structure and the usage can considerably change on different time scales. Together with the enormous growth of the size and the complexity of such systems, the need for information sharing platforms became immanently important. To guarantee the functionality of the system, investigations concerning the scalability of the application have to be carried out early during the conception and the dimensioning phases of the service deployment.

Concerning the architecture of information sharing platforms some current trends can be observed:

- *Transition to business cases*: Content distribution platforms and information sharing systems gain importance in the context of booming peer-to-peer (P2P) file sharing systems (e.g. Kaaza, eDonkey, ...). These applications are being transformed from a disruptive technology with rather gray-scale content (e.g. music downloads) to thoroughly designed business cases (e.g. distributed directory services, ...).
- *Information mediation*: In general, these systems can be categorized as information mediation platforms. The main task is similar to a telecommunication system: to efficiently mediate information from information providers to information consumers. Thus, the main trend is to move away from client-server based data centers or server farms to Internet oriented information storage and distribution services.
- *Distributed dynamic architecture*: Recently, centrally controlled information distribution systems are rapidly emerging to decentralized structures, which are usually implemented on a web-based platform. Examples are software distribution using BitTorrent or information sharing platforms based on distributed hash table structures like Chord [12] or Kademlia [9]. The structures of these systems are highly dynamic: during the system runtime, customers or network nodes can join or leave the system without notice. The system has to be designed to survive such so called "churns" with minimal service degradation.

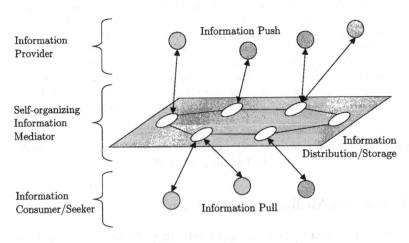

Fig. 1. Overview of a self-organizing information mediator

As illustrated in Figure 1 most of recent distributed information sharing platforms [5] are characterized by three main components: The information provider, the self-organizing mediator and the information consumer or seeker:

– *Information providing site*: Information providers mediate data and resources using the information distribution plane. The information can be the varying profile of users in a distributed phone book embedded in a Voice-over-IP (VoIP) application. The profile contains the nickname of the user, his current IP-address, charging information, etc.

– *Information mediation plane*: The self-organizing information mediator itself can be used to store the information in a reliable and consistent way to distribute the data to the information consumer at a later point in time. Thus, the mediator has two basic (control) functions, the mediation of resources and the coordination of the resource access and exchange. It must control, schedule and conduct the exchange of resources in a scalable and efficient way. The information mediation plane is often referred to as "overlay network", like in P2P systems. In our VoIP example above, the information mediation plane can be built by a number of mediation nodes, which are connected using a Kademlia structure or a Chord ring. Due to the use of hash functions, the location of a profile, which is a search object for the information consumer, is well defined. Mediation nodes can be embedded in the system and are thus identical to customers.

– *Information consuming site*: The information consumer must be able to search and locate a resource at any given time. In a running environment a participant can be both information provider and information consumer at the same time. In the VoIP example a calling subscriber searching for the current location of a nickname is an information consumer. Using the search algorithm he can find out in which mediation node the information is stored.

The advantage of the described architecture is its bandwidth efficiency in the distribution of information. It autonomically enables fast access, resilience and scalability. However, so far there are no guarantees for security or data consistency and completeness.

In the traditional client-server architecture the server has the role of the information mediator. Current information sharing platforms, however, tend to rely on P2P overlay networks or mediation planes. The P2P paradigm reflects a highly distributed and adaptive application architecture. P2P systems solve two basic functions resource mediation, i.e. search for and location of resources, as well as resource exchange. The underlying P2P algorithms are highly efficient, scalable, and robust.

As illustrated in Figure 2, a P2P network builds a virtual overlay topology on top of an already existing IP network, like e.g. the Internet itself. These P2P overlays are increasingly used as a self-organizing and scalable information mediator. The first wave of P2P systems (∼1999), Napster being the most popular representative, relied on direct peer-to-peer communication and central index servers. The second wave, including P2P applications like KaZaA (2000∼2002) made use of supernodes and introduced unique file IDs using a hash function. The current wave (2003∼), enables fast and scalable resource discovery using distributed hash tables. The corresponding P2P protocols are suitable to serve as the information mediator in an information sharing platform.

However, due to their highly distributed application architecture those P2P systems are too complex for large scale emulation. Even simulations on packet level proved to be rather intractable. To better understand the dynamics of such systems one has to approach

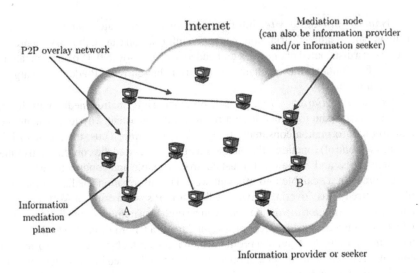

Fig. 2. A P2P overlay network used as information mediator

the problem on different levels of detail. While a detailed simulation can give valuable insights about the fundamental functionalities [7], a mathematical analysis of the main aspects of the problem often helps to investigate the scalability of the system itself [2, 13]. Due to the numerous stochastic processes involved in such highly distributed multi-user applications it is crucial to study not only the functional but also the stochastic scalability of the systems. In the following sections we define and further motivate the need for the evaluation of the stochastic scalability. Our goal is to better understand the dynamics of large scale information sharing platforms such as P2P systems. If we want to build reliable large scale information sharing platforms based on P2P mechanisms we need to master the complexity of such systems. Investigating the stochastic scalability, we will be able to get those systems under control and achieve carrier grade availability systems in a resource-efficient but also simple manner.

In this paper we refer to properties of the scalability of a system as stochastic scalability. One of this properties could be, to stay in our VoIP example, the quantile value, i.e. the time bound of search delays when we expect 99% (or even 99,99%) of searches to remain below this limit. Measures considering stochastic scalability can be used, e.g. to define and enforce Service Level Agreements (SLA) in communication systems and applications.

Traditionally, the term scalability often restricts to the functional scalability, which describes the scalability in terms of the system size, where the basic underlying application scenario and network structure are rather fixed or change only in a long-term time scale. More precisely, the question was: if a service or a solution, i.e. with a network carrying a target application, works properly for ten customers, will they also function accordingly for one thousand or for one million customers, following the potential growth of the market?

If the mid-term and shot-term stochastic behavior of system components and user applications has to be taken into account, the stochastic scalability has to be investigated in the context of classical performance evaluation. One possible question is: If a platform works properly (SLA is complied) in an environment with a network latency coefficient of variation $c_T = 0.5$, will it also support the same number of customers in a higher variation network with $c_T = 2$? This is crucial to ensure the network resilience in overload cases.

The remainder of this paper is structured as follows. In Section 2 we summarize the evolution of existing P2P-based information sharing platforms. Definitions of functional and stochastic scalability can be found in Section 3. A P2P-based VoIP solution is given as an example for the evaluation of the stochastic scalability of P2P-based information sharing platforms in Section 4. Section 5 finally summarizes and concludes the paper.

2   Evolution of P2P-based Information Sharing Platforms

2.1   The Traditional Client-Server Model

The client-server paradigm was the most prevalent model for information mediation in classical networks. It can be described as a service-oriented request-response protocol. A central server host runs a server process and provides access to a specific service such as web content or a centralized index. The client runs the corresponding client process and accesses the service offered by the central server host. Well known examples include but are not limited to protocols such as http or ftp. In the traditional client-server architecture the server had the role of the information mediator. It has become one of the central ideas of computer networks but severely suffers from two major drawbacks that come along with a centralized mediator:

– It represents a single point of failure. Once the central service provider, the server, fails, the offered service will be disrupted and no longer available to the customers, i.e. to the clients. The same problem could, e.g., be caused by a distributed denial of service attack which is targeted at a specific service. That is, the functionality of an entire business solution depends on the functionality of a single central unit.
– It hardly scales. The number of hosts that can be served at the same time is mainly restricted by two important properties of the central server: Its processing power and its available bandwidth. The latter is especially crucial in connection with the distribution of large files, such as software updates or multimedia content. The processing power of the central server might e.g. be the limiting factor when offering web services.

A new wave of highly distributed information mediation platforms emerged to cope with these problems and to provide reliable and scalable access to electronic information stored in a computer network. The distributed architecture of such platforms enables the offered service to be still available even if parts of the system crash or fail. The peer-to-peer paradigm plays an important role in this context and will be discussed in the next section.

2.2 P2P-based Information Sharing Platforms

A peer-to-peer (P2P) network can be described as a group of entities denoted as peers, with a common interest, that build a self-organizing overlay network on top of a mixture of already existing networks. That is, P2P is about the networked cooperation among equals. The main task is the discovery and sharing of pooled and exchangeable resources. An ever increasing number of companies discover the advantages of decentralized P2P networks. Skype [11], a P2P-based telephone directory, e.g., already attracts millions of users every day. P2P algorithms are also used to overcome the problems of distributed network management [3]. Due to the highly distributed application architecture companies using P2P mechanisms are no longer dependent on a single central unit nor do they have to invest in server farms to guarantee the scalability of their systems. Together with those new P2P systems, however, new challenges arise as well. In a business environment, they have to be able to guarantee efficient and, most important, scalable business solutions.

There are different P2P approaches trying to create information sharing platforms supporting tens and up to millions of entities to provide a highly scalable mediation platform. Those systems are designed to be highly dynamic, robust and resilient. In the ideal case any peer can be removed without resulting in any loss of service. P2P

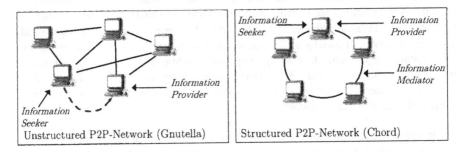

Fig. 3. Unstructured and structured P2P overlay topologies

mechanisms can roughly be divided into two main classes, those that build structured and those that build unstructured overlay topologies. Figure 3 gives a visual comparison between the two classes. Unstructured P2P networks, like e.g. Gnutella [6], build an arbitrary overlay topology. Peers are randomly connected to each other resulting in a fully decentralized use of the overlay network paradigm. There are no dedicated peers that store specific information. That is, resources are located at arbitrary peers. To cope with this indetermination of the desired information, searches in unstructured P2P networks are performed using an expanding ring principle. A peer searching for information simply floods the query to all its neighbors in the overlay network, who in turn forward the query to all of their overlay neighbors until the desired information is eventually found. To keep the involved overhead traffic within reasonable limits a Time To Live (TTL) counter is associated with the query. The TTL is decreased on every hop, while query packets with a TTL of zero are simply discarded. It is easy to see, that such unstructured overlay topologies suffer from two main drawbacks. First, there is no guarantee that a search returns a positive result even if the searched information is stored on a large number of peers. Second and even more important, unstructured P2P mechanisms do not scale to a large number of information consumers.

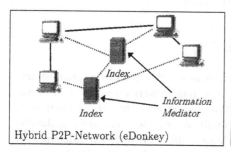 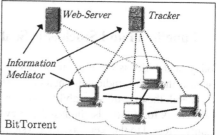

Hybrid P2P-Network (eDonkey)   BitTorrent

Fig. 4. Hybrid P2P overlay topologies

Structured overlay topologies on the other hand build a network structure, which specifies communication relationship among the participating peers. The resource mediation can then take advantage of the structure in the overlay. Local information of a peer is sufficient to globally search for information and find resources in an efficient way. In difference to unstructured networks the search and routing processes are scalable, deterministic and always return positive results if the resource is available in the overlay. The predefined overlay structure, which in a generic way specifies the communication relationship among the peers, is usually realized using a distributed hash table (DHT). The best known DHT algorithm is Chord [12] which arranges the participating peers on a ring topology. The position on this ring is chosen according to the hash value of a unique attribute of each peer. The basic idea is that each peers has a good knowledge about its overlay neighbors, i.e. its predecessors and successors on the Chord ring, while only maintaining a few connections to more distant peers. This way the mediation of the information stored in the distributed network can be done using only $O(\log_2(n))$ messages to other peers, where n is the current size of the overlay network.

Besides the (un)structured topologies there are different hybrid overlay mechanisms, which partly use a structured control or rely on distributed index servers. The eDonkey overlay [4] is a classical example of a hybrid P2P network shown in the left part of Figure 4. While the information exchange is realized using direct P2P communication, the signaling and resource location relies on central index servers, which reflect the functionality of a classic server farm. There are also some more exotic overlay topologies serving a special purpose. BitTorrent, shown in the right part of Figure 4, e.g., is used for the rapid distribution of one single file. The underlying mechanism slices the information into small parts and uses multiple source download to mediate the information as fast as possible. The meta-information has to be stored on a central unit like a web server and a centralized tracker takes control of the coordination of the download process. The mediation of the file itself is done by distributed information transfer.

The size of the system and the behavior of the customers significantly affect the performance and the functionality of all above mentioned topologies. While the number of peers in the overlay topology interferes with the logic of the system, the stochastic and dynamic behavior of the user pushes the system to its limits and might even cause it to fail entirely. These procedures can be associated with stochastic and functional scalability which will be discussed in the next section.

3   Functional and Stochastic Scalability

Today, scalability is the most important performance measure a carrier grade system has to withstand. It indicates whether a system is going to work on a large scale or not. In general, the question scalability asks, is: If a solution works for 10 customers, does it also work for hundreds, thousands, or even millions of customers? So far, scalability mainly referred to the mere size of a system. Most studies were intended to determine if a system at hand does work for growing customer clusters. We summarize this kind of analysis under the term functional scalability. It tells us whether the fundamental logic of a solution is scalable.

The mere size of a system, however, is not the only factor in terms of scalability a running application has to cope with. There are more and more system parameters having a stochastic character. Consider, e.g., the stochastic behavior of customers. There are numerous different random variables describing values like the inter-arrival time, the mean on-line time, and the query rate of customers of large scale systems. In P2P networks this stochastic behavior is defined as the autonomy of the participating peers, i.e., the peers may join or leave the system arbitrarily. This leads to the requirement to evaluate P2P algorithms with respect to the stochastic on-line behavior, which is summarized under the term "churn" [10]. This unpredictable stochastic behavior of the end user results in a highly dynamic evolution of the P2P network and thus has a significant impact on the functionality of the system [8]. The customer, however, is not the only variable introducing probabilistic properties into the system. A running system also faces stochastic network loads, probabilistic variations in traffic volumes and random transmission delays, to name just a few. Thus, in order to provide stochastic scalability, P2P networks with resilience requirements have to be able to survive in case of stochastic breakdowns. Stochastic scalability can be analyzed combining methods and techniques of both probability theory and performance analysis.

Figure 5 visualizes the difference between functional and stochastic scalability. The functional scalability verifies whether the interworking logic is extendable to larger crowds of customers. It mathematically analyzes whether the functionality of a system, like the search delay in the indicated Chord ring, also works for a large number of customers. Stochastic scalability on the other hand tries to verify whether a system can sustain the stochastic behavior of its components. It investigates whether a system can cope with the non-deterministic arrival, departure and query times of the participating customers. In respect of our Chord ring example stochastic scalability comprises the question whether a system, which can sustain minor churn rates, also works under extreme high churn rates? That is, we want to know how long the average customer has to stay on-line in order to guarantee the functionality of the running system.

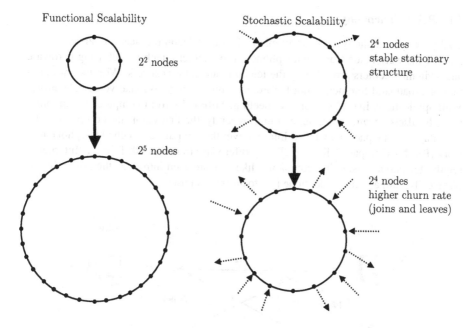

Fig. 5. On the definition of stochastic scalability

In the end a successful system most be scalable in both a functional and a stochastic way. Without functional scalability a system will collapse under its own size, without stochastic scalability a system will collapse under the random variations of its components.

4 An Example of Stochastic Scalability

Due to the increased bandwidth of the end user there is a growing demand for the mediation of information especially in multimedia applications. In this context, more and more companies are using P2P mechanisms to realize their business solutions. Such P2P systems are, e.g., used for content distribution, as index servers or even for distributed network monitoring. At the moment the most predominant structured P2P architecture in the research community is the ring based Chord algorithm. While its main functionality is to store and retrieve key-value pairs, it can be used for a broad variety of applications. In this section we will have a closer look at the stochastic scalability of Chord when used as a distributed phone book for voice-over-IP (VoIP) telephony. After a short description of IP telephony in general, we will explain how to use a P2P network in this context, define the problem areas, and show how to approach a performance analysis of the stochastic scalability of such a system.

4.1 P2P IP Telephony

Traditionally, telephony was the domain of large telecommunication carriers. Telephone calls were made from on telephone set to another telephone set using hardware and switching centers provided by the telecommunication carriers. With the introduction of broadband Internet, if not before, new possibilities to make voice calls arose. With applications like Microsoft Netmeeting, VoIP calls from computer to computer using headsets became possible. Not until recently, the first companies discovered the advantages of the packet switched Internet over the old public switched telephone network (PSTN). Companies like Net2Phone offer ways to make calls from the Internet to regular telephone sets. Other companies like Sipgate even introduce the possibility to place calls from a regular telephone line to an Internet user.

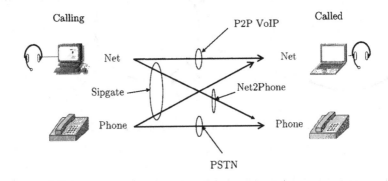

Fig. 6. Different approaches to transmit voice from the caller to the callee

The different ways of making telephone calls between PSTN and Internet users are summarized in Figure 6. Even so direct calls between Internet users are realized transmitting datagrams over IP networks, they still rely on a central unit, which is in charge of phone book lookups. In the meanwhile, however, highly distributed P2P-based VoIP solutions emerge, which do no longer rely on any central unit. The phone book is now realized in a distributed way, using a P2P-based information sharing platform. The most prominent example of a P2P-based VoIP telephony application is Skype. It offers free calls between Internet users, cheap calls from the Internet into the PSTN using the SkypeOut service and even calls from the PSTN to the Internet using the SkypeIn service. The advantages of a P2P-based VoIP solution are obvious. They are highly scalable, do not need any concentrated processing power, nor do they suffer from a single point of failure. In addition they are very inexpensive for the end user.

From a technical point of view, the main difference between a central and a P2P-based VoIP solution can be found in the call setup. In a P2P-based solution, the P2P overlay network is used to realize a distributed phone book, which in this case represents the information mediator. If a VoIP customer wants to publish his personal phone book entry, he becomes an information provider and stores his contact information in the mediation plane. When another VoIP customer wants to call this client at a later point in

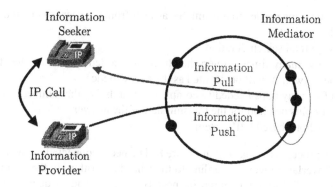

Fig. 7. Call setup using a Chord based information mediator

time, he takes the role of an information seeker and retrieves the corresponding contact information from the information mediator. So far, however, P2P-based IP telephony solutions come without guarantees for data consistency and security in the information plane. Furthermore, they involve signaling and information exchange overhead to maintain a consistent view of the stored resources. If a telecommunication carrier intends to build a large scale, P2P-based application, it must meet higher demands than a best effort service. The reachability of the customers has to be guaranteed. In this case, the functional scalability of the system alone does not suffice, since it only guarantees the scalability in terms of system size. The telecommunication operator, however, wants to be able to guarantee a certain quality of service. It should, e.g., be guaranteed that 99.99 percent of all call setups can be completed within a certain time limit. This, however, highly depends on a number of stochastic processes. The network transmission delay, e.g., can be regarded as a random variable. To be able to make any quality of service statements one needs to know the entire distribution function of the call setup delay. In the following example we show how to prove the stochastic scalability of the search delay in a Chord-based information sharing platform, calculating the quantiles of the search delay in such a system.

4.2 Performance Analysis of a VoIP Platform

In this section we show how to analyze the stochastic scalability of a Chord-based information sharing platform. The results can be used to realize a phone book for a P2P-based VoIP application. In particular, we analyze the time needed to complete a search in a Chord-based P2P system. Since the physical path delay strongly influences the performance of searches in such P2P systems, the stochastic impact of network delay variation is taken into consideration. The following random variables describe some of the stochastic processes, which are involved in a search for resources:

T_N: describes the delay of a query packet, which is transferred from one peer to another peer

T_A: represents the time needed to transmit the answer from the peer (having the answer) back to the originator

T: describes the total search duration

X: indicates how many times a query has to be forwarded until it reaches the peer having the answer. X will be denoted as the peer distance

H: number of overlay hops needed to complete a search, i.e. the number of forwards of the query plus one hop for the transmission of the answer

n: size of the Chord-based P2P system

The search process is visualized in Figure 8. The peers connected by the blue lines build the P2P overlay network according to the Chord algorithm. In the example peer A is searching for information stored on peer B. Peer A sends a query, which will recursively be forwarded until it finally reaches peer B. Each of the X overlay hops can be described by the random variable T_N.

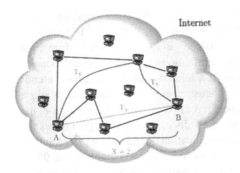
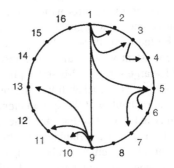

Fig. 8. Model of the search in a P2P network   **Fig. 9.** The model applied to Chord

Figure 9 shows the extended model of the search applied to the Chord algorithm. In this example the peer with $id_p = 1$ issues some queries for other peers. According to the Chord algorithm peers 2, 3, 5, and 9 can be reached using only $X = 1$ overlay hop. Furthermore, it takes $X = 2$ overlay hops to reach peers 4, 6, 7, 10, 11, 13. Extending this model, we are able to calculate the number of hops needed to reach a peer, that answers a specific query. We are thus able to derive the probability $p_i = P(X = i)$ that the searched peer is exactly i hops away from the searching peer. The detailed analysis can be found in [1].

Knowing the peer distance distribution X, we derive the length in hops of the path a particular search-query takes through the overlay network. Together with the probability p_i, that a search takes the corresponding path through the overlay, we can then compute the entire distribution of the search delay as a function of the stochastic network delay characteristics.

The phase diagram of the search delay is depicted in Figure 10. A particular path i is chosen with probability p_i where phase i consists of i network transmissions T_N to

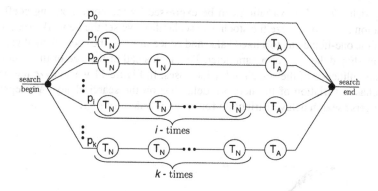

Fig. 10. Phase diagram for the information pull in a call setup

forward the query to the closest known finger and one network transmission T_A to send the answer back to the searching peer as illustrated in Figure 8. By means of the phase diagram, the generating function and the Laplace-Transform respectively can be derived to cope with the case of discrete-time or continuous-time network transfer delay:

$$X(z) = p_0 + \sum_{i=1}^{k} p_i \cdot X_A(z) \cdot X_N^i(z)$$

and the Laplace-Transform

$$\Phi(s) = p_0 + \sum_{i=1}^{k} p_i \cdot \Phi_A(s) \cdot \Phi_N^i(s).$$

The mean and the coefficient of variation of the search delay are such:

$$E[T] = \sum_{i=1}^{k} p_i \cdot (E[T_A] + i \cdot (E[T_N]))$$

$$E[T^2] = \sum_{i=1}^{k} p_i \cdot (VAR[T_A] + i \cdot VAR[T_N] + (E[T_A] + i \cdot E[T_N])^2)$$

and

$$c_T^2 = \frac{E[T^2] - E[T]^2}{E[T]^2}.$$

Other than the functional scalability the telecommunication carrier is now especially interested in the stochastic scalability of the system. In this context, the stochastic component with the most significant impact on the search delay is the variation of the one-hop delay T_N. This variation can be expressed by the corresponding coefficient of variation c_{T_N}. To analyze the stochastic scalability, we set the network size and the mean of the one-hop delay to a fixed value and concentrate on the coefficient of variation of the one-hop delay c_{T_N} as a parameter. That is, instead of the size of the system the stochastic influence of the search delay increases. In Figure 11 we analyze the impact the stochastic variation of the network delay has on the search delay. We depict the entire inverse search delay distribution for different values of c_{T_N}.

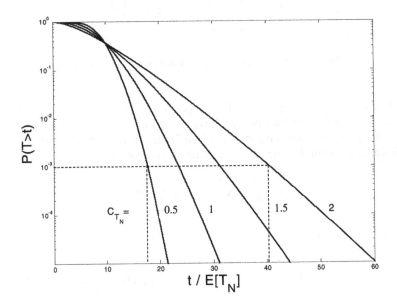

Fig. 11. Distribution of the search delay for different values of c_{T_N}

Note, that the search delay increases for larger values of c_{T_N}. The values $P(T > t)$ on the y-axis indicate how many percent of all searches will take longer than the corresponding time on the x-axis. Accordingly, $1 - P(T > t)$ percent of all lookups will take less time than the corresponding value on the x-axis. As indicated by the dotted lines, the value 10^{-3} on the y-axis, e.g., indicates that 99.9 percent of all phone book searches take less than roughly 18 overlay hops in the case of $c_{T_N} = 0.5$ and about 40 hops in the case of $c_{T_N} = 2$. Note, that since the results are independent of the mean value of the one-hop delay the values on the x-axis are normalized by $E[T_N]$. That is, in the case of $E[T_N] = 50$ms, e.g., 18 hops correspond to 900ms. In this scenario, it would therefore take 99.9 percent of all costumers less than 900ms to find their VoIP communication partner given that $c_{T_N} = 0.5$.

In a real business case, however, the operator of the system needs to assure both the functional and the stochastic scalability at the same time. In particular, he wants to know a search delay bound, which will be met by say 99.99 percent of all queries issued in the system independent of the current size of the system. Due to the mathematic analysis of the underlying stochastic processes it is possible to prove the functional and the stochastic scalability of the search delay. Figure 12 depicts the quantiles of the search delay T again normalized by $E[T_N]$.

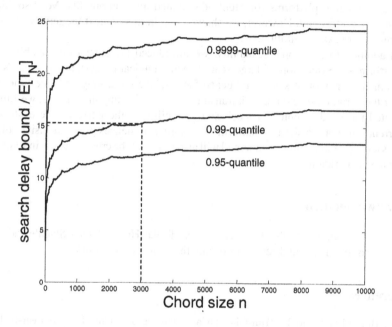

Fig. 12. Different quantiles for the search delay in a distributed phone book

Next to the mean delay, which shows the functional scalability, different quantiles for the search delay are taken as a parameter. The curve with the 99%-quantile, e.g., indicates that 99 percent of search durations lie below that curve. For a peer population of, e.g., $n = 3000$ in 99 percent of all cases the search delay is less then roughly 15 times the average network latency. It can be seen that the curves indicate stochastic bounds of the search delay. This can be used for dimensioning purposes, e.g. to know the quality of service in a search process with real-time constraints like looking at a phone directory, taking into account the patience of the users. Compared to the mean of the search delay the quantiles of the search delay are on a significantly higher level. Still the search delay scales in an analogous manner for the search delay quantiles. The above example shows the importance of stochastic scalability in the network planning process of a telecommunication carrier. Needless to say, that a comprehensive study of the stochastic scalability of a real system is more complex than the above example. The stochastic churn behavior of the participating peers, e.g., also has a great influence on the functionality of a running system. The more frequently customers enter and leave the system, the harder it is to maintain a stable overlay structure and the more timeouts will occur during a search process.

5 Conclusion

In this paper we gave a brief introduction to existing information sharing platforms. In particular, we provided a simple classification as well as a short description of P2P mechanisms. Current P2P algorithms are thought to be scalable and robust enough to serve as mediation platforms for highly distributed applications like VoIP solutions without any central unit. However, we showed that in this context the term scalability requires a more exact definition.

In addition to functional scalability, we introduced the stochastic scalability for the performance evaluation of large scale telecommunication systems. It regards the probabilistic behavior of system influence factors, like the average on-line time of a user or the variation of the network transmission delay. Using an example, we further motivated the need to consider stochastic scalability in the performance evaluation of current information sharing platforms. Stochastic influences play a decisive role in today's telecommunication systems and will also be one of the crucial factors in solutions of the next generation.

Acknowledgements

The authors would like to thank Robert Henjes, Tobias Hoßfeld, Dirk Staehle and Kurt Tutschku for the help and discussions during the course of this work.

References

1. A. Binzenhöfer and P. Tran-Gia. Delay Analysis of a Chord-based Peer-to-Peer File-Sharing System. In *ATNAC 2004*, Sydney, Australia, December 2004.
2. Andreas Binzenhöfer, Dirk Staehle, and Robert Henjes. On the stability of chord-based p2p systems. Technical Report 347, University of Würzburg, 11 2004.
3. Andreas Binzenhöfer, Kurt Tutschku, and Björn auf dem Graben. DNA – A P2P-based Framework for Distributed Network Management. In *Peer-to-Peer Systeme und -Anwendungen*, GI/ITG Work-In-Progress Workshop in Cooperation with KiVS 2005, Kaiserslautern, March 2005.
4. eDonkey. URL: http://www.edonkey2000.com/.
5. Patrick Th. Eugster, Pascal A. Felber, Rachid Guerraoui, and Anne marie Kermarrec. The many faces of publish/subscribe. In *ACM Computing Surveys*, 2003.
6. Gnutella website. http://www.gnutelliums.com.
7. Tobias Hoßfeld, Kurt Tutschku, Frank-Uwe Andersen, Hermann de Meer, and Jens Oberender. Simulative performance evaluation of a mobile peer-to-peer file-sharing system. In *Next Generation Internet Networks NGI2005*, Rome, Italy, April 2005.
8. D. Liben-Nowell, H. Balakrishnan, and D. Karger. Observations on the dynamic evolution of peer-to-peer networks. In *International Workshop on Peer-to-Peer Systems (IPTPS '02)*, Cambridge, MA, March 2002.
9. Petar Maymounkov and David Mazieres. Kademlia: A peer-to-peer information system based on the xor metric. In *IPTPS 2002*, MIT Faculty Club, Cambridge, MA, USA, March 2002.
10. Sean Rhea, Dennis Geels, Timothy Roscoe, and John Kubiatowicz. Handling Churn in a DHT. In *2004 USENIX Annual Technical Conference*, Boston, MA, June 2004.

11. Skype. URL: http://www.skype.com.
12. Ion Stoica, Robert Morris, David Karger, M. Frans. Kaashoek, and Hari Balakr-ishnan. Chord: A Scalable Peer-to-peer Lookup Service for Internet Applications. In *ACM SIGCOMM 2001*, San Diego, CA, August 2001.
13. Kurt Tutschku and Phuoc Tran-Gia. Traffic characteristics and performance eval-uation of peer-to-peer systems. In Klaus Wehrle Ralf Steinmetz, editor, *Peer-to-Peer-Systems and Applications*. Springer, 2005.

A Multimedia Adaptive-Quality Platform For Real-Time E-Learning Over IP

F. Licandro[1], D. Marchese[1], A. Lombardo[2], G. Morabito[2], C. Panarello[2] and G. Schembra[2]

[1] Research Department - LightComm srl
Via C. Vivante, 48 – 95123 Catania (Italy)
info@lightcomm.it
[2]Dip. di Ingegneria Informatica e delle Telecomunicazioni
University of Catania , Viale A. Doria 6 – 95125 – Catania (Italy)
{lombardo,gmorabito,cpana,schembra}@diit.unict.it

Abstract. The interest in creating multimedia support for remote learning is increasing explosively, thanks to the development of a variety of multimedia software packages for slide presentation, on the one hand, and the spread of the Internet in schools, Universities, research laboratories and the majority of student houses, on the other hand. However, most of the e-learning platforms used today are merely off-line. In other words, they consist of a multimedia site where teachers and students can exchange information and didactic material through file transfer and e-mail, or where the teacher can propose evaluation tests to the students. At the same time many videoconference hardware and software platforms have been deployed to support audio/video communications between two or more users. However, their main target is videoconference applications, and they are not suitable for teaching environments. The target of this paper is to provide an overview of the main aspects related to the development of multimedia e-learning tools which allow on-line remote learning, the main characteristic of which is close real-time interaction between teachers and students, and between students themselves. The authors conclude the paper by describing a relevant experience, the VIP-Teach tool, developed according to the principles illustrated in the paper.

Keywords. E-learning, Multimedia communications, multipoint-to-multipoint interaction, multicast, adaptive QoS.

1   Introduction

Distance education technologies have expanded at an extremely rapid rate in the last decade. The term e-learning has been applied by many different researchers to a great variety of programs, providers, audiences, and media [1]. Its hallmarks are the separation of teacher and learner in space and/or time [2], the volitional control of learning by the student rather than the distant instructor [3], and noncontiguous communication between student and teacher, mediated by print or some form of technology [4][5]. Remote learners have a wide variety of reasons for pursuing learning at a distance: constraints of time, distance, and finances, the opportunity to take courses or hear outside speakers who would otherwise be unavailable, and the ability to come into contact with other students from different social, cultural, economic, and experiential backgrounds [6]. As a result, they gain not only new knowledge but also new social skills, including the ability to communicate and collaborate with widely dispersed colleagues and peers whom they may never have seen.

Although the instructional needs of students are often considered as the only focus of e-learning programs, technology is an integral part of distance education and must be carefully taken into consideration if it is to be successful. In fact, technology for the deployment of distance learning systems must meet the important target of a caring, concerned teacher who is confident, experienced, at ease with the equipment, uses the media creatively, and maintains a high level of interactivity with the students [9]. This is the main factor for successful distance learning. We find a rich history as each form of instructional media evolved from print, to instructional television, to current interactive technologies. The earliest form of distance learning took place through correspondence courses in Europe, and evolved to audio/video-supported distance learning thanks to the application of radio and broadcast television. However, the major drawback of those means for instruction was the lack of a 2-way communication channel between teacher and student. In the last few years the interest in creating multimedia support for remote learning has increased explosively thanks to the development of a variety of multimedia software packages for slide presentation, on the one hand, and the spread of the Internet in schools, Universities, research laboratories and the majority of student houses, on the other hand. As increasingly sophisticated interactive communications technologies became available, they were adopted by distance educators. Currently, the most popular media are computer-based communications including electronic mail (E-mail), bulletin board systems (BBSs), and the World Wide Web (WWW), supported by 2-way audio telephone-based audio-conferencing, or 1-way video broadcasting.

However, most of the e-learning platforms used today are merely non-real-time. Successful distance education systems should, however, involve interactivity between teachers and students, between students and the learning environment, and between students themselves, as well as active learning in the classroom. McNabb [7] noted that, though students felt that the accessibility of distance learning courses far outweighs the lack of dialogue, there is still a considerable lack of dialogue in telecourses when compared to face-to-face classes. Millbank [8] studied the effectiveness of a mix of audio plus video in corporate training. When he introduced real-time interactivity, the retention rate of the trainees was raised from about 20 percent (using ordinary classroom methods) to about 75 percent.

The target of this paper is to give an overview of the main problems in realizing a multicast real-time e-learning platform, and to discuss some possible solutions.

More specifically, Section 2 considers audio/video communication aspects of e-learning. Section 3 considers data communication protocol to guarantee the data reliability. Section 4 reports a relevant experience, the VIP-Teach tool, developed according to the principles illustrated in this paper. Finally, the authors conclude the paper in Section 5.

2   Audio-video communications

Real time audio/video communications are a fundamental means to obtain an interactive e-learning platform supporting virtual classroom services. Unfortunately, real time media have strong delay and loss requirements, so at present expensive network connections providing QoS guarantees are used to support them. This discourages the deployment of e-learning and for this reason new transmission paradigms are necessary in order to work over any communications network, and in particular over networks characterized by variable connection bandwidths, for example the current Internet or wireless networks. *Adaptive multimedia systems* therefore represent a challenging paradigm in this field. In this perspective, two important features have to be considered:

1.   Bandwidth monitoring
2.   Output Rate adaptation

These features are the building blocks of a system where the audio/video encoders adapt their coding parameters to the available network resources while the minimal QoS is maintained.

2.1   Bandwidth monitoring

The first problem regards the monitoring of network bandwidth. This is a well-known and still open issue in telecommunications network research. The target is to achieve an estimation of the network bandwidth according to end-to-end measurements, with no knowledge of any internal characteristics of the network. Great effort has been devoted to this problem in the literature. Work in this context is mainly based on what are called TCP-friendly algorithms [27][28]. Although these algorithms were first proposed with the target of achieving UDP-based flows behaving like TCP, they were then introduced and modified as bandwidth monitors. However, their use presents the following main problems:

1   they increase the available bandwidth according to TCP principles, that is, they mainly react to the occurrence of loss, whereas real-time media are mainly delay sensitive.
2   they provide too variable a bandwidth estimation to be actually used for multimedia traffic;
3   they have not been defined in multipoint scenarios.

In this section we show how the TFRC algorithm can be modified to cope with these problems.

The TFRC protocol is defined in [27]; it implements an equation-based algorithm to calculate the available network bandwidth by using loss and delay information.

More specifically, the available bandwidth is calculated according to the following equation:

$$T = \frac{s}{RTT\sqrt{\frac{2p}{3}} + t_{RTO}\left(3\sqrt{\frac{3p}{8}}\right)p\left(1+32p^2\right)} \tag{1}$$

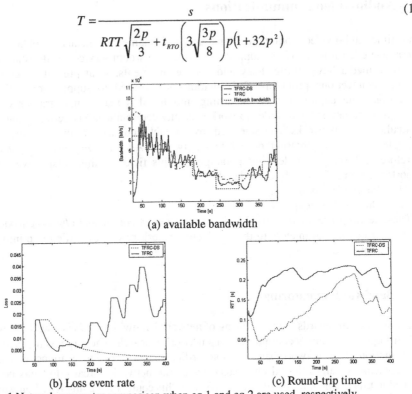

(a) available bandwidth

(b) Loss event rate   (c) Round-trip time

Fig. 1 Network parameter comparison when eq.1 and eq 2 are used, respectively.

where s is the packet size, RTT is the round-trip time, p is the steady-state loss-event rate, and t_{RTO} is the TCP retransmit timeout value.

In order to make TFRC more sensitive to an increase in RTT we exploit the fact that, due to the drop tail policy implemented in the IP routers, RTT increases faster than loss probability. So, as the loss probability p is the denominator in eq. 1, replacing p with RTT results in a smoothing of the available bandwidth. By using the formula:

$$T = \frac{s}{RTT\sqrt{\frac{2p}{3}} + t_{RTO}\left(3\sqrt{\frac{3p}{8}}\right)RTT\left(1+32p^2\right)} \tag{2}$$

we achieve the network parameter shape shown in Fig 1. In Fig. 1(a) we compare the available bandwidth achieved by using the original TFRC and the formula we introduce in this paper (named TFRC-DS in the figure) when a TFRC source estimates the available bandwidth on an ISDN access loaded by VBR traffic; likewise, in Figs. 1(b) and 1(c) we show a comparison between the measured RTT and loss in the same conditions.

In order to avoid excessive bandwidth variations, the bandwidth estimated by eq. 3 is then smoothed with an exponential weighted moving average (EWMA) filter defined as follows:

$$\hat{T}_s(t_n) = \alpha \cdot T_{MTFRC}(t_{n-1}) + (1-\alpha) \cdot \hat{T}_s(t_{n-1})$$ (3)

Where $\alpha \in [0,1]$ is the weighting parameter, $T(n)$ is the n-th sample of the bandwidth estimated by the TFRC, and $T_s(n-1)$ is the $(n-1)$-th sample of the smoothed bandwidth. Usually the choice of the parameter α value differs according to whether the monitoring algorithm determines a bandwidth increase or decrease, that is, higher α values are taken when the measured bandwidth decreases than when it increases.

The TFRC protocol was defined in a unicast environment. In this case the TFRC congestion control mechanism is distributed over three main protocol entities: the *TX* and *RX* entities in the sender part, and the *PR* entity in the receiver part.

The main target of the *PR* entity is to send feedback to the *RX* entity every time it receives a packet, in order to echo the sequence number from the most recent data packet. This will allow the *RX* to calculate the round-trip time and send it back to the *PR* entity. Another important target of the *PR* entity is to calculate the loss event rate. Every time the *PR* entity receives feedback, it uses the control equation in (1) to calculate a new value for the allowed sending rate and send it to the sender part.

E-learning systems are deployed in a multicast [9] or multipoint environment. To cope with the above problems, we have defined a new bandwidth monitoring algorithm, named Multicast TCP-Friendly Rate Control (MTFRC) by extending the TFRC protocol. The sender computes an $N-1$ step cycle where it calculates the RTT of one of the N-1 receivers and sends it to all the receivers at each step (multicast). When a receiver has to measure the available bandwidth in a multicast/multipoint session, the MTFRC on the receiving end computes the available bandwidth over the point-to-point path between it and the sender by using the following in the TFRC equation

1. the measured *RTT*, when the sender part sends an MTFRC packet which contains the *RTT* indication referring to it;
2. the *RTT* estimated according to the following formula in the remaining *N-2* steps, that is, when it receives an MTFRC packet which contains the *RTT* indication referring to another receiver:

$$estRTT = lastMeasuredRTT + Sender\_Receicer\_Jitter +$$ (4)
$$+ estReceiver\_Sender\_Jitter$$

In the above formula the *Sender_Receiver_Jitter* is computed by the receiver as $(\Delta_2 - \Delta_1)$ where Δ_1 is the difference between the local time on receipt of the packet and the time stamp in the received packet which contains the measured *RTT* referring to it (see step 1), and Δ_2 is the difference between the local time on receipt of the packet and the time stamp in the received packet which contains the measured *RTT* referring to it (see step 2).

As regards the estimated jitter between the receiver and the sender (*estSender_Receiver_Jitter*), the receiver uses the averaged value of the RTT jitter, which the sender calculates as before at the sender side, and sends it together with the measured RTT.

As an example, Figure 2 shows the error introduced by MTRFC when 40 users participate in a multipoint session.

In order to compute the *available bandwidth* over the multipoint session,

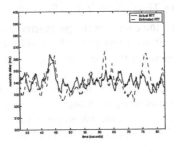

Fig. 2 Comparison of the actual RTT and estimated RTT.

the minimum bandwidth value computed over the point-to-point paths is usually taken. However, this may result in too strong a bandwidth decrease if a bottleneck in the network affects one of the point-to-point connections; for this reason, hierarchical coding schemes have to be used, where the lower-level coding corresponds to the minimum measured available bandwidth. In any case, MTFRC does not take into account computed bandwidth values lower than a given threshold corresponding to the minimal QoS required.

2.2 Output Rate adaptation

The second important feature for audio/video communications in this context regards rate adaptation according to the suggestion provided by the bandwidth monitoring process defined in the previous section. In this case, *MTFRC* feeds back the run-time computed available bandwidth to the upper-layer application so that it can adapt the encoding parameters used accordingly. For example, when MPEG video encoding is used, the application can exploit this information to reconfigure its quantizer scale and/or frame rate parameters with the purpose of matching the available bandwidth on the network. Let us stress that in this case the target we have to pursue is twofold: on the one hand, the output bit rate has to follow the bandwidth available in the network; on the other, it has to respect user requirements in terms of encoding quality and, in particular, it should protect the quality of the movie being encoded from oscillations due to rapid variations in the network bandwidth. With this in mind, the system we propose is shown in Fig. 3, where the three main component blocks of the system are represented: the *MTFRC*, the *Network bandwidth smoother* and the *Rate/Quality MPEG video source* (RQ-source), whose functions will be described in the following subsections.

The *Network bandwidth smoother* receives the available network bandwidth estimated by the *MTFRC*, $T_{MTFRC}(t_n)$, and has the aim of eliminating the high frequencies of this process. This is achieved by filtering the process by means of a low-pass filter with an Exponential Weighted Moving Average (EWMA) as in eq. 3.

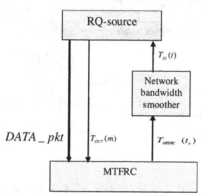

Fig. 3 : Rate/Quality controlled MPEG video transmission system.

Finally, the output of the *Network bandwidth smoother* is obtained as a continuous time step function derived from $\hat{T}_s(t_n)$ as follows:

$$T_{IN}(t) = \hat{T}_s(t_{n-1}) \qquad \forall t \in \left[t_{n-1}, t_n\right[\tag{5}$$

In this way the process $T_{IN}(t)$ represents the available bandwidth calculated by the *MTFRC*, and its variations can be regulated by the parameter α.

The *Rate/Quality MPEG video source* (RQ-source) is an MPEG video source whose emission rate is controlled through a timely choice of both the quantizer scale for each macroblock to be encoded and the frame rate, in such a way as to respect the network bandwidth calculated by the *MTFRC* block, and to keep the encoding quality as stable as possible. Quantizer scale variation only provides a fine tuning. Of course, given that frame rate variations are more disturbing for receivers, the system only has to apply frame rate variations when situations really require this drastic solution. A scheme of the RQ-source is represented in Fig. 4.

The video flow produced by the Video capture card is then passed to the *Rate controller*. The *Rate controller* has the task of modifying encoding parameters with the aim of adapting the output bit rate of the source to the available network bandwidth calculated by the *MTFRC*. It comprises two blocks: the *PSNR controller* and the *FR-controller*. At the beginning of the generic GoP *h*, the *PSNR controller* receives the budget of bits to be used in encoding the starting GoP. This budget is calculated as the difference between the budget derived by the *Sampler* output, and the amount of redundancy imposed by the *Redundancy controller*.

At the beginning of each GoP, the *Sampler* samples the smoothed network bandwidth provided by the Smoother.

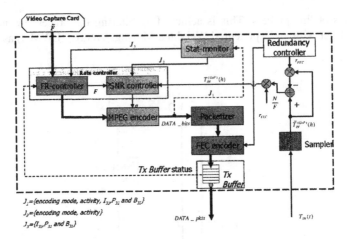

Fig. 4 RQ-source scheme.

This value is used to calculate the budget of a GoP by multiplying by N/F , which represents the number of GoPs per second. The *PSNR controller* calculates the *qsp q* very similarly to the TM-5, using the encoding mode of the frame being encoded, and the activity of the previous frame, received by the Stat-monitor. The only difference with respect to the TM-5 lies in the fact that the *PSNR controller* used here is memoryless, in order to be more reactive to network bandwidth behavior. The *Frame-Rate controller (FR-controller)* modifies the encoding frame rate when quantizer scale variations are not enough to follow the available bandwidth. It applies an algorithm which uses the value of the sampled bit rate, $T_{IN}^{(GoP)}(h)$, received by the *Sampler*, and the average values of the MINIMUM size of frames I, P and B, that is, when they are encoded with a *qsp q* of $q = 31$; let these values be I_{31}, P_{31}, and B_{31}, respectively. They are received by the *Stat-monitor*.

The *FR-controller* calculates the frame rate as follows:

$$\hat{F} = \frac{N}{\Delta} \tag{6}$$

where Δ represents the expected minimum size of the GoP. If we indicate the number of frames in one GoP encoded with encoding modes P and B as N_P and N_B, respectively, we have:

$$\Delta = \frac{I_{31} + N_P \cdot P_{31} + N_B \cdot B_{31} + R_b}{T_{IN}^{(GoP)}(h)} \tag{7}$$

The *FR-controller* checks whether to update the frame rate at the beginning of each GoP. The decision to update the frame rate is taken as follows: the *FR-controller* decreases the frame rate by 1 fps if some residue is present in the Transmission buffer (i.e. MTFRC was not able to transmit all the information generated by the video source in the previous GoP) and the estimated frame rate, $\hat{F}$, is less than the current frame rate, F, by a value greater than a given threshold, σ_{Decr}. On the other hand, the *FR-controller* increases the frame rate by 1 fps if no residue is present in the Transmission buffer, and the estimated frame rate, $\hat{F}$, is greater than the current frame rate, F, by a value greater than another given threshold, σ_{Incr}.

3   Data Communication

E-learning services have to provide reliable data transfer in order to support services such as slide shows, chat sessions, file transfer or application sharing. Unfortunately, UDP is generally used to support data transfer in a multicast environment, so reliable delivery is still missing in today's networks [10]-[26]. For this reason, we propose a novel protocol named Reliable Multicast Transport Protocol (RMTP), which provides a reliable service for a multicast session where UDP is used.

RMTP can be implemented in two different modes:

1   fully distributed
2   centralized

The former is suitable for multicast data transfer applications, such as file transfer or messaging; the latter is suitable for any multicast concurrent data sharing applications. RMTP in the centralized mode, however, supports multicast data transfer application as well; for this reason, only the RMTP centralized mode protocol will be described.

In the centralized mode, a participant in an RMTP session must be identified as responsible for managing concurrent handling of the shared data; we will refer to this user as the *server*. The other users are referred to as *clients*.

All clients joining an RMTP session open a point-to-point TCP connection with the server. A client that requires a reliable multicast data transport service transmits this information to the server through the TCP connection. Then, the server puts the information into RMTP packets, and transmits them to all other clients using IP multicast facilities. All RMTP packets sent by the server have a sequence number; therefore concurrent operations that the participating clients require on the shared data via RMTP, if any, are serialized by the server according to a FIFO discipline. Upon detecting one or several missing packets, a client sends the server a NAK using the TCP connection. The NAK contains information about the packets that have been lost so far. The server sends the lost packets to the client using the TCP connection.

The server periodically schedules the transmission of an ACK_REQUEST message containing the sequence number of the last packet transmitted to each client through the TCP connection. A client receiving such a message is expected to answer with a STATUS message which contains information about all the lost packets. In this way the server can update the list of packets that have been lost and can remove from a local memory the copy of packets that have been received by all clients.

If a multicast data transfer application with no concurrent data sharing is to be deployed, the distributed mode RMTP protocol can easily be implemented by assuming that each participant starting multicast data transfer takes the role of the server for that data transfer session.

4   An e-learning tool implementation: VIP-Teach

In this section we will analyze the performance of the MPEG 2 video transmission system described in Section 2, considering the following two Internet scenarios:

- SCENARIO 1: source and receiver are both in the same Ethernet LAN; this scenario will be analyzed in Section 4.1;
- SCENARIO 2: source and receiver are connected to each other through the Internet; the video source belongs to the University of Catania Campus, and has high-band ATM access with 155 Mbits/sec, while the video receiver is connected to its Internet Service Provider (ISP) through ADSL access; this will be analyzed in Section 4.2;

In order to compare the two different cases correctly, we captured the same 6-minute scene. In our analysis we used the six-frame GoP structure IPPPPP. The video sources used are equipped with an INTEL Pentium IV 1000 MHz processor, 256 MB of RAM, and a Pinnacle PC-TV capture card working with the PAL Video standard and a capture rate of $\tilde{F}$ =25 fps.

The operating system is Windows 2000 with DirectX 8.0 installed above. In all cases we considered a capture resolution of 320x240.

4.1   SCENARIO 1: LAN connection

In this section we perform two different performance analyses, both on the same scenario, where source and destination are connected to each other through an Ethernet local area network (LAN) with the configuration shown in Fig. 5, where the video stream generated by the video source is disturbed by the presence of background traffic generated by a traffic generator, and sent to a traffic recipient.

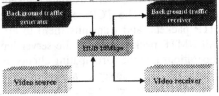

Fig. 5 LAN scenario.

The first analysis is a transient analysis obtained by varying the bandwidth of the background traffic as shown in Fig. 6(a). Fig. 6(b) shows a comparison between the rate processes at the input and output of the MPEG encoder.

Given that the input rate process $T_{IN}^{GoP}(h)$ is the number of bits to be used in one GoP, we define the process $T_{OUT}^{GoP}(h)$ similarly by averaging the number of bits per frame over all the frames belonging to the same GoP h. In this figure we can observe that in the first period corresponding to the first two background traffic values, the available bandwidth is so high that the video source does not use all of it, although it

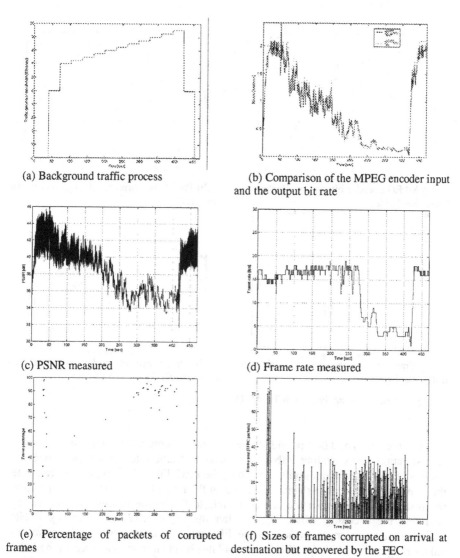

(a) Background traffic process

(b) Comparison of the MPEG encoder input and the output bit rate

(c) PSNR measured

(d) Frame rate measured

(e) Percentage of packets of corrupted frames

(f) Sizes of frames corrupted on arrival at destination but recovered by the FEC

Fig. 6 Transient analysis in SCENARIO 1.

is working at maximum quality. When the background traffic increases, both the available bandwidth estimated by the *MTFRC* and the bandwidth used by the video source decrease, and the latter fits the former well. Only when the background traffic reaches 9.5 Mbit/s is the available network bandwidth too low for the video source which, although at the lowest quality, is not able to respect $T_{IN}^{GoP}(h)$. When at the instant $t=423$ s the background traffic is reduced to 5 Mbit/s and then turned off, both the available bandwidth estimated by *MTFRC* and that used by the video source increase again as at the beginning of this experiment.

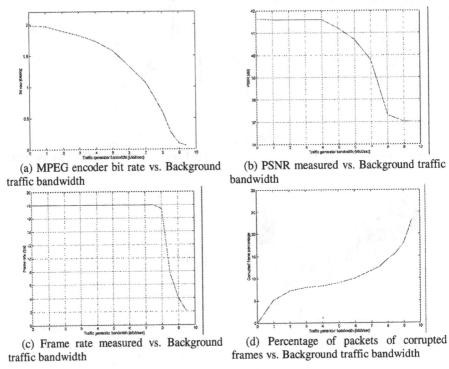

(a) MPEG encoder bit rate vs. Background traffic bandwidth

(b) PSNR measured vs. Background traffic bandwidth

(c) Frame rate measured vs. Background traffic bandwidth

(d) Percentage of packets of corrupted frames vs. Background traffic bandwidth

Fig. 7 Steady-state analysis in SCENARIO 1.

Figs. 6(c), 6(d) and 6(e) provide three elements representing the quality perceived at destination: the encoding PSNR, the encoding frame rate and the percentage of packets of corrupted frames not recoverd by the FEC. From these figures we can note that the frame rate remains almost constant in the range between 15 and 19 fps up to the instant t=276 s, when the source is able to follow the bandwidth decrease by changing the encoding PSNR. When, after the instant t=276 s, encoding PSNR variations are no longer sufficient, the FR-controller decreases the frame rate to values ranging around 6 fps. When the available bandwidth increases again, both the encoding PSNR and the frame rate increase return to the maximum values of the beginning of this experiment. As regards the percentage of packets of corrupted frames not recovered by the FEC shown in Fig. 6(e), we can note that we have losses of packets after 30 s, when the background traffic generator is turned on at 5 Mbit/s.

Fig. 6(f) presents the sizes of frames recovered by the FEC. In this figure we can appreciate the presence of FEC encoding which, in this scenario and in the presence of a large amount of background traffic, allows a very large number of frames to be recovered.

The second performance analysis regards the calculation of steady-state parameters against background traffic bandwidth. Figs. 7(a), 7(b), 7(c) and 7(d) show the average values of the process $T_{OUT}^{GoP}(h)$, the encoding PSNR, the frame rate and the percentage of corrupted frames not recoverd by the FEC, respectively. From Fig. 7(a) we can observe that the average video source output rate is not very sensitive to background traffic bandwidth variations as long as this bandwidth remains less than 4 Mbit/s. Then the average output rate rapidly decreases with futher background traffic increases. From Figs. 7(b), 7(c) we can observe that, as already discussed, low background traffic bandwidth values (ranging in the interval between 4 Mbit/s and 7.5 Mbit/s) cause no change in the frame rate, but only encoding PSNR degradation. With higher background traffic bandwidth values it is the frame rate variation which allows the source to adapt the available bandwidth.

In addition, let us note that, with the same high background traffic values, not only does the average frame rate rapidly degrade, but also more than 10% of the frames received at destination are corrupted and not recovered by the FEC, as we can see from Fig. 7(d).

4.2 SCENARIO 2: ADSL Internet access

In this section we present the results obtained when source and receiver are connected to each other through the Internet, as shown in Fig. 8; the video source belongs to the University of Catania Campus, and has highband ATM access with 155 Mbits/s, while the video receiver is connected to its Internet Service Provider (ISP) through ADSL access;

Fig. 8 ADSL scenario.

Fig. 9(a) shows a comparison between the rate processes at the input and output of the MPEG encoder, $T_{IN}^{GoP}(h)$ and $T_{OUT}^{GoP}(h)$, respectively. Figs. 9(b), 9(c) and 9(d) represent the quality perceived at destination, in terms of the encoding PSNR, the encoding frame rate and the percentage of packets of corrupted frames not recovered by the FEC.

From Fig. 9(a) we can observe that the video source output rate follows the available bandwidth process estimated by the *MTFRC* very well. Let us note that in the periods when MPEG violates the estimated available bandwidth, the video source reacts by reducing the frame rate, as can be seen in Fig. 9(c) (see for example the frame rate decrease at the instant $t=62.3$ s, corresponding to the violation period between the instants 54 and 79 seconds, or the frame rate decrease at the instant $t=173$ s, corresponding to the violation period between the instants 170 and 209 seconds). In any case, we can observe that the frame rate remains almost constant in time, while network bandwidth variations are compensated for by varying the encoding PSNR.

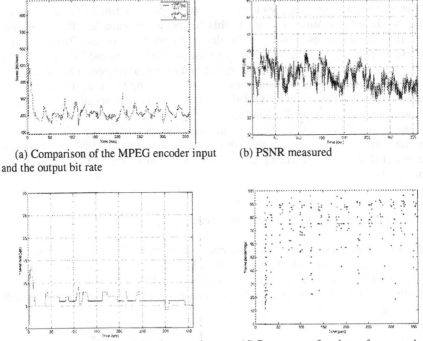

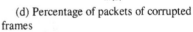

(a) Comparison of the MPEG encoder input and the output bit rate

(b) PSNR measured

(c) Frame rate measured in the transient analysis

(d) Percentage of packets of corrupted frames

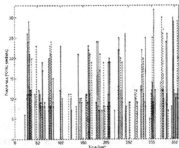

(e) Sizes of frames corrupted on arrival at destination but recovered by the FEC

Fig. 9 Transient analysis in SCENARIO 2.

Finally, Fig. 9(e) presents the sizes of frames recovered by the FEC. As in Scenario 1, we can appreciate the presence of FEC encoding, which allows a very high number of frames to be recovered.

5 Conclusions

In this paper we have provided an overview of the main aspects related to the development of multimedia e-learning tools which allow on-line remote learning, the main characteristic of which is close real-time interaction between teachers and students, and between students themselves.

Unfortunately, real time media have strong delay and loss requirements, so expensive network connections providing QoS guarantees are currently used to support them. In this paper a novel solution named *Adaptive Multimedia System* has been illustrated; moreover, its implementation in a real service scenario has been presented and the main performance indexes measured have been outlined.

The proposed solution consists of:

- A **congestion control** protocol (MTFRC): the objective of the protocol is to make the application TCP-friendly, i.e., the average transmission rate must not exceed the transmission rate of hypothetic TCP flows between each of the senders and each of the destinations.

- An **error control** protocol (RMTP): the objective of this protocol is to guarantee the level of data reliability required by any specific application.

The specific service scenario introduced to demonstrate the feasibility of both MTFRC and RMTP is a videoconference where the speaker transmits an audio/video stream, and a number of multicast data applications, such as multicast file transfer, multicast messaging and slide sharing, are available to support her/his talk.

Moreover, this service scenario has allowed us to demonstrate how the MTFRC protocol can be successfully used to support any adaptive-rate real-time multimedia applications over not guaranteed bandwidth networks such as best-effort Internet or wireless networks.

References

1. Poon, S.K., Reed, S. & Tang, C. (1997). Problem-based Learning in Distance Education. *Proceedings of the 5th International Conference on Modern Industrial Training*, Jinan, China. pp. 593-600.
2. H. Perraton, *A theory for distance education*. In D. Sewart, D. Keegan, & B. Holmberg (Ed.), Distance education: International perspectives (pp. 34-45). New York: Routledge - 1988.
3. M.A. Cambre, *The state of the art of instructional television*. In G.J. Anglin, (ed.), Instructional technology, past, present, and future (pp. 267-275). Englewood, CO: Libraries Unlimited.Jonassen - 1991
4. D. Keegan, *The foundations of distance education*. London: Croom Helm -1986.
5. D. R. Garrison, and D. Shale, *Mapping the boundaries of distance education: Problems in defining the field*. The American Journal of Distance Education, 1(1): 7-17. – 1987.
6. B. Willis, *Distance Education: A Particle Guide*. Educational Technology Publications, Englewood Cliffs: New Jersey -1993.
7. J. McNabb, Telecourse effectiveness: Findings in the current literature. Tech Trends, 39-40. -1994.
8. G. Millbank, *Writing multimedia training with integrated simulation*. Paper presented at the Writers' Retreat on Interactive Technology and Equipment. Vancouver, BC:

The University of British Columbia Continuing Studies. Retrieved June 18, 2002 from: http://www.nwlink.com/~donclark/hrd/elearning/proscons.html

9. (Nora, 2003) Characteristics of Successful Adult Distance Instructors for Adult Learners, by Nora N. Smith, from *Inquiry*, Volume 8, Number 1, Spring 2003, © Copyright 2003 Virginia Community College System.

10. C. Diot *et al.*, "Deployment Issues for the IP Multicast Services and Architecture," *IEEE Network*, Jan/Feb 2000.

11. H. Eriksson, "MBONE: The Multicast Backbone," *Communications of The ACM*, Aug 1994, pp. 54–60.

12. S. Paul, *Multicasting on the Internet and its Applications*, Kluwer Academic Publishers, 1998.

13. C. Diot, W. Dabbous, and J. Crowcroft, "Multipoint Communications: A Survey of Protocols, Functions, and Mechanisms," *IEEE JSAC*, Apr 1997, pp. 277–90.

14. J. Widmer, R. Denda, and M. Mauve, "A Survey of TCP-Friendly Congestion Control", *IEEE Network*, May/Jun 2001.

15. A. Basu and S. J. Golestani, "Architectural Issues for Multicast Congestion Control", *Proc. of NOSSDAV*, 1999.

16. K. Obraczka, "Multicast Transport Protocols: A Survey and Taxonomy," *IEEE Commun. Mag.*, Jan 1998.

17. S. J. Golestani and K. K. Sabnani, "Fundamental Observations on Multicast Congestion Control in the Internet," *Proc. of IEEE INFOCOM*, 1999.

18. D. Chiu and R. Jain, "Analysis of the Increase and Decrease Algorithms for Congestion Avoidance in Computer Networks," *Computer Networks and ISDN Systems*, 1989.

19. A. Matrawy and I. Lambadaris, "Real-Time Transport for Assured Forwarding: An Architecture for both Unicast and Multicast Applications," *Proc. of IEEE ICC*, 2003.

20. T. Turletti and C. Huitema, "Videoconferencing on the Internet", *IEEE/ACM Trans. on Networking*, vol. 4, no. 3, June 1996, pp. 340–51.

21. T. Kim and M. H. Ammar, "A Comparison of Layering and Stream Replication Video Multicast Schemes", *Proc. of NOSSDAV*, 2001.

22. L. Rizzo, "PGMCC: A TCP-Friendly Single-Rate Multicast Congestion Control Scheme", *Proc. of SIGCOMM*, 2000.

23. J. Widmer and M. Handley, "Extending Equation-Based Congestion Control to Multicast Applications", *Proc. of SIGCOMM*, 2001.

24. J. K. Shapiro, D. Towsley, and J. Kurose, "Optimization-Based Congestion Control for Multicast Communications", *IEEE Commun. Mag.*, Sep 2002.

25. K. Kar, S. Sarkar, and L. Tassiulas, "Optimization-Based Rate Control for Multirate Multicast Sessions", *IEEE Infocom*, 2001.

26. J. Byers, M. Luby, and M. Mitzenmacher, "Fine-Grained Layered Multicast", *Proc. of INFOCOM*, 2001.

27. M. Handley, S.Floyd, J. Pahdye and Widmer, "TCP Friendly Rate Control (TFRC): Protocol Specification", J. RFC 3448, Proposed Standard, January 2003.

28. R. Rejaie, M. Handley, and D. Estrin. "RAP: An end-to-end rate-based congestion control mechanism for realtime streams in the internet", Proc. IEEE Infocom, March 1999.

Real-Time Applications In 803.11 WLAN Using Feedback-Based Bandwidth Allocation

A. Barbuzzi, G. Binetti, G. Boggia, P. Camarda, L. A. Grieco, and S. Mascolo

DEE - Politecnico di Bari, Via E. Orabona, 4 - 70125 Bari Italy
{a.barbuzzi, g.binetti, g.boggia, camarda, a.grieco, mascolo}@poliba.it

Abstract. The Hybrid Coordination Function (HCF) has been recently proposed by the 802.11e working group in order to provide real-time services in Wireless Local Area Networks (WLANs). HCF is made of a contention-based channel access, known as Enhanced Distributed Coordination Access (EDCA), and of a HCF Controlled Channel Access (HCCA), which requires a centralized controller, called Hybrid Coordinator (HC). This paper proposes two feedback-based bandwidth allocation algorithms exploiting HCCA for dynamically assigning the WLAN channel bandwidth to mobile stations hosting real-time traffic streams. Proposed algorithms, which have been referred to as Feedback Based Dynamic Scheduler (FBDS) and Proportional Integral (PI)-FBDS, have been designed using classic discrete-time feedback control theory. Simulation results, obtained using the *ns*-2 simulator, have shown that, unlike the simple scheduler proposed by the 802.11e working group, both FBDS and PI-FBDS provide a real-time service regardless of the network load. Moreover, when the PI-FBDS is used, the best trade-off between one-way packet delays and network utilization is achieved.

Keywords: Quality of service, Wireless LAN, Dynamic bandwidth allocation, Feedback control.

1 Introduction

Although 802.11 Wireless Local Area Networks (WLANs) [1] are nowadays very broadly diffused, 802.11 Medium Access Control (MAC) is not well suited for providing real-time services [2]. Recently, to overcome this limitation, the 802.11e working group has proposed: the Hybrid Coordination Function (HCF), which is an enhanced access method; a Call Admission Control (CAC) algorithm; specific signaling messages for service request and Quality of Service (QoS) level negotiation; four Access Categories (ACs) with different priorities to map QoS users' requirements [3].

The HCF is made of a contention-based channel access, known as the Enhanced Distributed Coordination Access (EDCA), and of a HCF Controlled Channel Access (HCCA), which requires a centralized controller, called the Hybrid Coordinator (HC), which is generally located at the access point. EDCA operates as the basic DCF access method [1], but using different contention parameters per AC. In this way, a service differentiation among ACs is statistically pursued . EDCA parameters have to be properly set to provide prioritization of ACs. Tuning them in order to meet specific QoS needs is a current research topic [2]. In particular, regarding the goal of providing delay guarantees, several papers have pointed out that the EDCA can provide a real-time service to highest priority flows, but it starves flows with lower priority, especially at high network load [4, 5]. For that purpose, adaptive algorithms that dynamically tune EDCA parameters have been recently proposed in [2, 6]; however, the effectiveness of these heuristic schemes have been proved only using simulations and no theoretical bounds on their performance in a general scenario has been derived.

On the other hand, with the HCCA, the HC is responsible for assigning the right to transmit at nodes hosting applications with QoS requirements, i.e., to perform dynamic bandwidth allocation within the WLAN. However, the 802.11e draft does not specify an effective bandwidth allocation algorithm for providing the QoS required by real-time flows; it only suggests a simple scheduler that uses static values declared by data sources for providing a Constant Bit Rate (CBR) service. As a consequence, this scheduler is not well suited for bursty media flows [7]. An adaptive version of the simple scheduler, which is based on the Delay-Earliest Due-Date algorithm, has been proposed in [7]. However, this scheduler does not exploit any feedback information from mobile stations, but implements a trial and error procedure to discover the optimal amount of resources to assign to each AC. The Fair Scheduling scheme proposed in [8] allocates the WLAN channel bandwidth to wireless nodes in order to fully deplete transmit queues, which are estimated by taking into account the delayed feedbacks from the wireless nodes.

This paper proposes two feedback-based bandwidth allocation algorithms exploiting HCCA. Proposed algorithms, which have been referred to as Feedback Based Dynamic Scheduler (FBDS) and Proportional Integral (PI)-FBDS, have been designed using classic discrete-time feedback control theory. Simulation results, obtained using the *ns*-2 simulator [9], have shown that, unlike the simple scheduler proposed by the 802.11e working group, both FBDS and PI-FBDS provide a real-time service regardless of the network load. Moreover, when the PI-FBDS is used, the best trade-off between one-way packet delays and network utilization is achieved.

The rest of the paper is organized as follows: Section 2 gives an overview of the HCCA method; in Section 3 FBDS and PI-FBDS algorithms are proposed; Section 4 shows simulation results; finally, the last Section draws the conclusions.

2   Overview of the HCCA Method

The core of the 802.11e proposal is the HCF, which is responsible for assigning TXOPs (Transmission Opportunities) to each AC in order to satisfy its QoS needs. TXOP is defined as the time interval during which a station has the right to transmit and is characterized by a starting time and a maximum duration.

The contiguous time during which TXOPs are granted to the same station with QoS capabilities (i.e., a QoS station, QSTA) is called Service Period (SP). The interval T_{SI} between two successive SPs is called Service Interval [3].

HCCA method combines some EDCA characteristics with some features of the Point Coordination Function (PCF) scheme, which is an optional contention-free access method defined by the 802.11 standard [1]. The time is divided into repeated periods, called *SuperFrames* (SFs). Each superframe starts with a beacon frame after which, for legacy purpose, there could be a contention free period (CFP) for PCF access. The remaining part of the superframe forms the Contention Period (CP), during which QSTAs contend to access the radio channel using the EDCA mechanism (see Fig. 1). During the CP or the CFP, the HC can

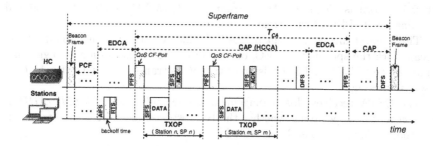

Fig. 1. Scheme of a superframe using the HCF controlled access method.

start a Contention Access Phase (CAP). During the CAP, only QSTAs, polled and granted with the *QoS CF-Poll frame*, are allowed to transmit during their TXOPs. Thus, the HC implements a prioritized medium access control.

The number of CAPs and their locations in each superframe are chosen by the HC in order to satisfy QoS needs of each station. Moreover, at least one CP interval, long enough to transmit a data frame with the maximum size at the minimum rate, must be contained in each superframe; this CP interval can be used for management tasks, such as associations of new stations, new traffic negotiations, and so on. CAP length cannot exceed the value of the system variable *dot11CAPLimit*, which is advertised by the HC in the Beacon frame when each superframe starts [3].

The simple scheduler designed in the draft [3] states that the $TXOP_i$ assigned to the ith queue should be computed as follows:

$$TXOP_i = max\left\{ \frac{N_i \cdot L_i}{C_i} + O, \frac{M}{C_i} + O \right\} \qquad (1)$$

where, with respect to the ith queue, L_i is the nominal size of MAC Service Data Units (MSDUs); C_i is the rate at which the data are transmitted over the WLAN; O is the protocol overhead; M is the maximum MSDU size; and $N_i = \left\lceil \frac{T_{SI} \cdot \rho_i}{L_i} \right\rceil$, where ρ_i is the Mean Data Rate associated with the queue and T_{SI} is the Service Interval.

According to IEEE 802.11e specifications, each QSTA can feed back queue length of each AC to the HC in the frames' headers. As will be shown in this paper, this information can be fruitfully exploited to design novel HCCA-based dynamic bandwidth allocation algorithms using feedback control theory [10].

2.1 QoS signalling

In the 802.11e proposal [3] each Traffic Stream (TS), i.e., a data flow with QoS needs, is described by a *Traffic SPECification* (TSPEC), which indicates the main characteristics of the stream [3]. Specific signalling has been introduced to manage new TS requests and QoS provisioning. In particular, when a new TS has to be started, the QSTA issues a setup phase by generating a message, which is known as *Mac Layer Management Entity (MLME)-ADDTS request* and contains the TSPEC of the stream.

This request message is sent to the HC which decides whether to admit the stream with the specified TSPEC, or to refuse it, or to not admit it suggesting an alternative TSPEC. The decision of the HC is transmitted with the *ADDTS response* message.

The QSTA receives this management frame and sends a *MLME-ADDTS confirm* message specifying whether the HC response meets its needs or not; if not, the whole process can be repeated [3].

2.2 Call Admission Control

In a IEEE 802.11 network the HC is used as admission control unit. Since the QoS facility supports two access mechanisms, there are two distinct admission control schemes: one for the contention-based access and the other for the controlled-access. Herein, we will focus on the latter mechanism. Details regarding the contention-based admission control can be found in [3].

Let m be the number of admitted flows. When in the presence of a new TS admission request, the admission control unit in the HC calculates the TXOP duration needed by the stream ($TXOP_{m+1}$) as imposed by the simple scheduler (see Eq. (1)). The stream is admitted if the following inequality is satisfied:

$$\frac{TXOP_{m+1}}{T_{SI}} + \sum_{i=1}^{m} \frac{TXOP_i}{T_{SI}} \leq \frac{T - T_{CP}}{T} \qquad (2)$$

where T indicates the superframe duration, and T_{CP} is the time during which EDCA is used for frame transmission in the superframe.

3   FBDS and PI-FBDS Algorithms

In this section, FBDS and PI-FBDS algorithms will be designed using feedback-based control theory. We will assume that both algorithms, running at the HC, allocate the WLAN channel bandwidth to wireless stations hosting real-time applications, using HCCA functionalities. These allow the HC to assign TXOPs to ACs by taking into account their specific time constraints and transmission queue levels [5].

We will refer to a WLAN system made of an Access Point and a set of quality of service enabled mobile stations (QSTAs). Each QSTA has up to 4 queues, one for each AC in the 802.11e proposal.

Let T_{CA} be the time interval between two successive CAPs (see Fig. 1). Every time interval T_{CA}, assumed constant, the HC has to allocate the bandwidth that will drain each queue during the next CAP. We assume that at the beginning of each CAP, the HC is aware of all the queue levels q_i, $i = 1, \ldots, M$ at the beginning of the previous CAP, where M is the total number of traffic queues in the WLAN system. [1]

The following discrete time linear model describes the dynamics of the ith queue:

$$q_i(n+1) = q_i(n) + d_i(n) \cdot T_{CA} + u_i(n) \cdot T_{CA}, \qquad i = 1, \ldots, M, \qquad (3)$$

where, with respect to the ith queue, $q_i(n) \geq 0$ is the queue level at the beginning of the nth CAP; $u_i(n) \leq 0$ is the average depletion rate (i.e., the bandwidth assigned to drain the queue); $d_i(n) = d_i^s(n) - d_i^{CP}(n)$ is the difference between $d_i^s(n) \geq 0$, which is the average input rate at the queue during the nth T_{CA} interval, and $d_i^{CP}(n) \geq 0$, which is the amount of data transmitted by the queue during the nth T_{CA} interval, using EDCA, divided by T_{CA}.

The signal $d_i(n)$ is unpredictable since it depends on the behavior of the source that feeds the ith queue and on the number of packets transmitted using EDCA. Thus, from a control theoretic perspective, $d_i(n)$ can be modelled as a disturbance. Without loss of generality, the following piece-wise constant model for the disturbance $d_i(n)$ can be assumed [11]:

$$d_i(n) = \sum_{j=0}^{+\infty} d_{0j} \cdot 1(n - t_j) \qquad (4)$$

where $1(n)$ is the unitary step function, $d_{0j} \in \mathbb{R}$, and t_j is a time lag.

Due to the assumption (4), the linearity of the system (3), and the superposition principle that holds for linear systems, we will design the feedback control law by considering a step disturbance: $d_i(n) = d_0 \cdot 1(n)$.

3.1 The closed loop control scheme

Our goal is to design a control law that drives the queuing delay τ_i, experienced by each frame going through the ith queue, to a desired target value τ_i^T that represents the QoS requirement of the AC associated to the queue.

---

[1] This is a worst case assumption, in fact, queue levels are fed back using frame headers as described in Sec. 2; as a consequence, if the ith queue length has been fed at the beginning of the previous CAP, then the feedback signal might be delayed up to T_{CA} seconds.

We will consider the closed loop control system shown in Fig. 2, where the set point q_i^T has been set equal to 0, i. e., we would ideally obtain empty queues. Regarding the controller transfer function $G_i(z)$, we will focus the attention on two possible kinds of controller: a proportional (P) controller, obtained by setting $G_i(z) = k_{p_i}$, and a proportional and integral (PI) controller, obtained as $G_i(z) = k_{p_i}\left(1 + \dfrac{z}{z-1} \cdot \dfrac{1}{T_{I_i}}\right)$. The corresponding bandwidth allocation algorithms will be referred to as Feedback Based Dynamic Scheduler (FBDS), and PI-FBDS.

Using a proportional controller.

By considering the control scheme in Fig. 2 where $G_i(z) = k_{p_i}$ it is straightforward to compute the $\mathcal{Z}$-transform of $q_i(n)$ and $u_i(n)$:

$$Q_i(z) = \frac{z \cdot T_{CA}}{z^2 - z + k_{pi} \cdot T_{CA}} \cdot D_i(z); \quad U_i(z) = -\frac{k_{pi} \cdot T_{CA}}{z^2 - z + k_{pi} \cdot T_{CA}} \cdot D_i(z) \quad (5)$$

with $D_i(z) = \mathcal{Z}[d_i(n)]$.

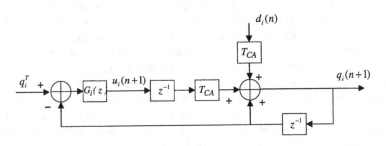

Fig. 2. Closed-loop control scheme based on HCCA.

From Eqs. (5) the system poles are $z_p = \dfrac{1 \pm \sqrt{1 - 4k_{pi} \cdot T_{CA}}}{2}$; thus, the system is asymptotically stable if and only if $|z_p| < 1$, that is:

$$0 < k_{pi} < 1/T_{CA}. \quad (6)$$

To investigate the steady-state behaviour of the control system, we apply the final value theorem to Eqs. (5).By considering that the $\mathcal{Z}$-transform of the step function $d_i(n) = d_0 \cdot 1(n)$ is $D_i(z) = d_0 \cdot \frac{z}{z-1}$, the following results turn out:

$$u_i(+\infty) = \lim_{n \to +\infty} u_i(n) = \lim_{z \to 1}(z-1)U_i(z) = -d_0; \quad q_i(+\infty) = d_0/k_{pi},$$

which implies that the steady state queueing delay is:

$$\tau_i(+\infty) = |q_i(+\infty)/u_i(+\infty)| = 1/k_{pi}. \quad (7)$$

It is worth to note that $q(+\infty) > 0$ even if $q_i^T = 0$, which means that the proportional controller is not able to fully reject the step disturbance $d_0 \cdot 1(n)$.

From Eq. (7), it turns out that the following inequality has to be satisfied in order to achieve a steady-state delay smaller than τ_i^T:

$$k_{pi} \geq 1/\tau_i^T. \tag{8}$$

By considering inequalities (6) and (8),the T_{CA} parameter has to fulfill the following constraint:

$$T_{CA} < \min_{i=1..M} \tau_i^T. \tag{9}$$

From these results we conclude that, when a proportional controller is used, the gain k_{pi} can vary in the range $\left[\frac{1}{\tau_i^T}, \frac{1}{T_{CA}}\right[$. We will set $k_{pi} = 1/\tau_i^T$, i.e., at its lowest admissible value. This choice allows, for a given feedback queue level $q_i(n)$, the lowest bandwidth assignment ensuring bounded delays, so that, a cautios usage of the WLAN channel bandwidth is achieved.

Using a PI controller.

Similarly to the case of the proportional controller, by considering Fig. 2 where $G_i(z) = k_{p_i}\left(1 + \dfrac{z}{z-1} \cdot \dfrac{1}{T_{I_i}}\right)$ and $q_i^T = 0$, after a little algebra, the following $\mathcal{Z}$-transforms of $q_i(n)$ and $u_i(n)$ can be obtained:

$$Q_i(z) = \frac{T_{CA}T_{I_i}z(z-1)}{T_{I_i}z^3 - 2T_{I_i}z^2 + (T_{I_i} + T_{CA}k_{p_i}T_{I_i} + T_{CA}k_{p_i})z - T_{CA}k_{p_i}T_{I_i}}D_i(z); \tag{10}$$

$$U_i(z) = \frac{T_{CA}k_{p_i}T_{I_i} - T_{CA}k_{p_i}T_{I_i}z - T_{CA}k_{p_i}z}{T_{I_i}z^3 - 2T_{I_i}z^2 + (T_{I_i} + T_{CA}k_{p_i}T_{I_i} + T_{CA}k_{p_i})z - T_{CA}k_{p_i}T_{I_i}}D_i(z). \tag{11}$$

By applying the Jury criterion [10], the system in Fig. 2 is asymptotically stable if the following inequalities are satisfied:

$$k_{p_i} < 1/T_{CA}; \qquad T_{I_i} > 1/\left(1 - T_{CA}k_{p_i}\right). \tag{12}$$

Also in this case, to investigate the steady-state behaviour of the control system, we apply the final value theorem to Eq. (10), thus obtaining: $q_i(+\infty) = 0$, which implies that the steady state queuing delay is zero.

This result is due to the integral action of the controller, which is able to fully reject the step disturbance at steady state. In this way, the parameter set of the PI regulator is only subject to the stability constraints (12). As a consequence we have more degrees of freedom in the choice of k_{p_i} and T_{I_i} with respect to the case of the proportional controller.

When the PI controller is used, it might happen that the depletion rate computed by the controlled $|u_i(n+1)|$ is larger than $q_i(n)/T_{CA}$, which is the amount of bandwidth required to fully deplete the ith queue during the $(n+1)^{th}$ CAP. This assignement would obviously waste WLAN resources. To overcome this drawback, we will employ the following shortcut:

$$u_i(n+1) \leftarrow max\left\{u_i(n+1), -q_i(n)/T_{CA}\right\}. \tag{13}$$

where, taking into account that $u_i \leq 0$, the term $-q_i(n)/T_{CA}$ is the depletion rate needed to fully empty the queue.

The rationale of eq. (13) is to correct the bandwidth allocation that would be performed by the PI regulator by not allowing a bandwidth assignment larger than the one than would fully deplete the ith queue.

3.2 TXOP assignment

We have seen in Sec. 2 that, every time interval T_{CA}, the HC allocates TXOPs to mobile stations in order to meet the QoS constraints. Herein, we shows how to transform the bandwidth u_i into a $TXOP_i$ assignment. In particular, if the ith queue is drained at rate C_i, the following relation holds:

$$TXOP_i(n) = \frac{|u_i(n) \cdot T_{CA}|}{C_i} + O \qquad (14)$$

where $TXOP_i(n)$ is the TXOP assigned to the ith queue during the nth CAP and O is the protocol time overhead. The extra quota of TXOP due to the overhead O depends on the number of MSDUs corresponding to the amount of data $|u_i(n) \cdot T_{CA}|$ to be transmitted. O could be estimated by assuming that all MSDUs have the same nominal size specified into the TSPEC. Moreover, when $|u_i(k) \cdot T_{CA}|$ does not correspond to a multiple of MSDUs, the TXOP assignment will be rounded in excess in order to guarantee a queuing delay equal or smaller than the target value τ_i^T.

3.3 Channel saturation

The above bandwidth allocation algorithm is based on the implicit assumption that the sum of the TXOPs assigned to each traffic stream is smaller than the maximum CAP duration, which is defined by the system variable *dot11CAPLimit*; this value can be violated when the network is saturated. In order to avoid heavy channel saturations, we will adopt a CAC scheme obtained by improving the one proposed by the 802.11e working group. However, since transient overloads cannot be avoided due to the burstiness of the multimedia flows, when for a given n_0 we have that: $\sum_{i=1}^{M} TXOP_i(n_0) > dot11CAPLimit$, each computed $TXOP_i(n_0)$ is decreased by an amount $\Delta TXOP_i(n_0)$, so that the following capacity constraints is satisfied:

$$\sum_{i=1}^{M} [TXOP_i(n_0) - \Delta TXOP_i(n_0)] = dot11CAPLimit . \qquad (15)$$

In particular, the generic amount $\Delta TXOP_i(n_0)$ is evaluated as a fraction of the total amount $\Delta = \sum_{j=1}^{M} TXOP_i(n_0) - dot11CAPLimit$, as follows:

$$\Delta TXOP_i(n) = \frac{TXOP_i(n)C_i}{\sum_{j=1}^{M} [TXOP_j(n)C_j]} \Delta. \qquad (16)$$

Notice that Eq. (16) provides a $\Delta TXOP_i(n_0)$, which is proportional to $TXOP_i(n_0)C_i$; in this way, connections transmitting at low rates are not too much penalized.

3.4 Call Admission Control

When the number of multimedia flows sharing the WLAN increases, the channel saturates and delay bounds cannot be guaranteed [6]. Under these conditions, a Call Admission Control scheme is required to provide QoS. Herein, we describe the proposed CAC scheme, which exploits the 802.11e CAC proposal (see Sec. 2).

In particular, starting from the TXOPs allocated to the active traffic streams in each CAP, a new flow request is admitted if

$$\frac{TXOP_{m+1}}{T_{CA}} + \sum_{i=1}^{m} \frac{TXOP_i}{T_{CA}} \leq \frac{T - T_{CP}}{T} \qquad (17)$$

where m is the number of admitted flows, T is the superframe duration, and T_{CP} is the time used by EDCA during the superframe.

Notice that the proposed CAC scheme given by eq. (17) has been obtained from the CAC suggested in the draft (see sec. 2.2), by replacing the constant TXOPs used by the simple scheduler with the time-varying ones allocated by the proposed bandwidth allocation algorithms. In this way, our suggested CAC takes into account the bandwidth actually used by the flows and not just the sum of the average source rates declared in the TSPECs.

4 Performance Evaluation

In order to assert the validity of the proposed bandwidth allocation algorithms in realistic scenarios, computer simulations involving voice, video and FTP data transfers have been run. We have used the *ns-2* simulator [9] considering a scenario where a 802.11a wireless channel with data rate of 54Mbps is shared by a mix of 3α voice flows encoded with the G.729 standard [12], α MPEG-4 encoded video flows [13], α H.263 video flows [14], and α FTP best effort flows. Each flow is hosted by a wireless node. Therefore, in such a scenario the traffic load is directly related to the parameter α.

For video flows, we have used traffic traces available from the video trace library [15]. For voice flows, we have modeled the G.729 sources using Markov ON/OFF sources [16]. The ON period is exponentially distributed with mean 3 s and the OFF period has a truncated exponential pdf with an upper limit of 6.9 s and an average value of 3 s [17]. During the ON period, the source sends packets of 20 bytes every 20 ms (i.e., the source data rate is 8 kbps and we are considering two G.729 frames combined into one packet [18]). By considering the overheads of the RTP/UDP/IP protocol stack, during the ON periods the total rate over the wireless channel becomes 24 kbps. During the OFF period the rate is approximated by zero assuming the presence of a Voice Activity Detector (VAD).

During CAPs, stations access the channel with HCCA method, otherwise they use EDCA. In simulations, EDCA parameters have been set as suggested in [3].

The target delay τ_i^T has been set equal to 30 ms for the voice flow and 40 ms for the video flows. According to the 802.11 standard, in our *ns-2* implementation the T_{CA} is expressed in Time Unit (TU), which is equal to 1024 μs [1]; we assume a T_{CA} of 29 TU in order to satisfy inequality (9). The value of the system variable *dot11CAPlimit* has been set in order to allow the transmission of at least 10 MSDUs of maximum size using EDCA, between two successive CAPs.

When FBDS is used, the proportional gain k_{p_i} is set equal to $1/\tau_i^T$ (see Sec. 2). With the PI-FBDS scheme, we consider the following parameter sets: $(k_{p_i} = 22s^{-1}, T_{I_i} = 10)$, $(k_{p_i} = 18s^{-1}, T_{I_i} = 7)$, and $(k_{p_i} = 14s^{-1}, T_{I_i} = 4)$.

The main characteristics of the considered multimedia flows are summarized in Table 1.

Table 1. Main features of the considered multimedia flows.

Type of flow	Nominal (Maximum) MSDU Size [Byte]	Mean (Maximum) Data Rate [Byte]	Inactivity Interval [s]
MPEG-4	1536 (2304)	770 (3300)	3
H.263 VBR	1536 (2304)	450 (3400)	3
G.729 VAD	60 (60)	12 (24)	10

Before starting data transmission, a multimedia source has to set up a new Traffic Stream as specified in Sec. 2.1. If the reply to the admission message for the new stream is not received within a Δ_{TO} timeout interval, the request is repeated up to a maximum number of times, N_{Adm}; in our simulations, we have chosen $N_{Adm} = 10$ and $\Delta_{TO} = 1.5$ s. If after the N_{Adm} admission tries no reply is received back, then the request is considered lost and a new admission procedure is initiated after an exponential distributed random time Δ_{defer} with average value equal to 1 min. The duration of video flows is deterministic and equal to 10 min, whereas voice flows durations are exponentially distributed with average value of 120 s. When a multimedia flow terminates, a new stream of the same type is generated after an exponentially distributed random time, with an average value equal to 1 min. Each terminated flow is withdrawn from the polling list by the HC after that no more packets from that flow are received for a time equal to the Inactivity Interval reported in Table 1. Each simulation lasts 1 hour.

In the sequel, we will first compare the performances of the PI-FBDS, the FBDS, and the Simple scheduler, then, we will investigate the impact of the parameters of the PI-FBDS (i.e., k_{p_i} and T_{I_i}) on the WLAN behavior.

Tab. 2 reports the ratio of admitted flows for various values of the network load parameter α when PI-FBDS or FBDS or the simple scheduler are used. By looking at this table, it is straightforward to note that FBDS admits the smallest ratio of flows. On the other hand, almost all the flows are admitted when either the PI-FBDS or the Simple scheduler are used. The reason is that the Simple scheduler allocates TXOPs by taking into account only the declared average source rates, which can be much smaller than the actual source rates; so that, the CAC test realized with eq. (2) is always satisfied. When FBDS or PI-FBDS are used, the CAC test given by eq. (17) takes into account the actual load. The substantial difference between FBDS and PI-FBDS is that, as discussed in Sec. 2, the PI regulator gives the opportunity to choice its parameter set in a broader space of values with respect to the case of a proportional controller. In other words, we can select the parameter set of the PI regulator to filter out high frequency components of the traffic load better than in the case of the P regulator. As a consequence, PI-FBDS allows the CAC test of eq. (17) to consider only the low frequency components of the network load, and to admit a larger number of flows. In other words, PI-FBDS is able to smooth the peaks of the traffic load better than FBDS. As a consequence, when PI-FBDS is used the probability that a new starting flow finds a fully occupied wireless channel is smaller.

Table 2. Ratios of admitted flows.

α	PI-FBDS ($k_{p_i} = 14s^{-1}$, $T_{I_i} = 4$)	FBDS	Simple Scheduler
5	100%	100%	100%
10	100%	89%	100%
12	99%	59%	100%

At this point, due to the smaller number of admitted traffic streams, we would expect lower one-way packet delays when FBDS is employed with respect to the one-way packet delay values obtained with the Simple scheduler or PI-FBDS. This is shown in Figs. 3, 4, and 5 where the average one-way packet delays experienced by, respectively, the MPEG, H.263, and G.729 flows are reported. By looking at those figures, we can note that FBDS obtains the smallest delays due the smallest quota of admitted flows. The surprising result is that, with high load (i.e., $\alpha = 12$) although the Simple scheduler and PI-FBDS allow almost the same ratio of admitted flows, delays obtained with PI-FBDS are one order of magnitude smaller than those obtained with the Simple scheduler. The latter observation highlights that if the scheduling strategy works fine, i.e., the network resources are carefully administered, a good trade-off between the number of admitted flows and the one-way packet delay can be achieved, also with high load.

Table 3 reports the average and peak superframe utilization in HCCA mode. It shows that the Simple scheduler requires the highest average quota of WLAN

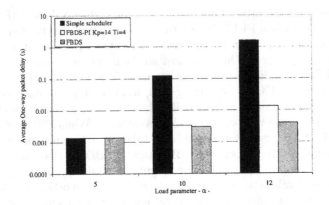

Fig. 3. Average one-way packet delay of the MPEG4 flows.

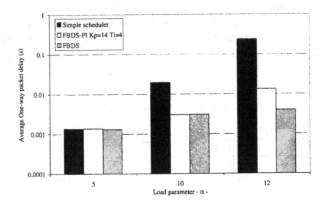

Fig. 4. Average one-way packet delay of the H.263 flows.

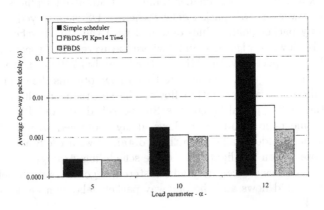

Fig. 5. Average one-way packet delay of the G.729 flows.

resources. In fact, it does not adapt the quota of allocated resources to the actual load because it provides a CBR service. For the same reason, the peak superframe utilizations achieved by the Simple scheduler for $\alpha = 10$ and $\alpha = 12$, i.e., at high traffic load, are smaller than those provided by FBDS and PI-FBDS. This result clearly highlights that the proposed control schemes enable a more proper usage of the bandwidth and allows the bandwidth requirements of the real-time flows to be tracked.

Table 3. Average (Peak) Superframe usage in HCCA mode.

α	PI-FBDS ($k_{p_i} = 14s^{-1}$, $T_{I_i} = 4$)	FBDS	Simple Scheduler
5	6.82(26)%	6.95(31)%	25.62(32.42)%
10	15.15(84)%	15.22(82)%	52.4(62.58)%
12	29.85(81)%	17.66(80)%	63.75(74.98)%

Table 4 reports the overall goodput achieved by the FTP flows, measured as the total amount of data received by FTP receivers over the simulation duration. It shows that FTP flows get the bandwidth left unused by the real-time flows. In fact, when FBDS is used, FTP flows get the highest goodput because, with FBDS, the smallest quota of real-time flows is admitted (see Tabs. 2 and 3).

Table 4. Overall goodput of the FTP flows (Mbps).

α	PI-FBDS ($k_{p_i} = 14s^{-1}$, $T_{I_i} = 4$)	FBDS	Simple Scheduler
5	4.57	4.58	4.57
10	4.32	4.38	4.27
12	2.47	4.09	2.32

In order to investigate the influence of the parameters k_{p_i} and T_{I_i} on system performance with PI-FBDS, we have considered the following parameter sets: $(k_{p_i} = 14s^{-1}, T_{I_i} = 4)$, $(k_{p_i} = 18s^{-1}, T_{I_i} = 7)$ and $(k_{p_i} = 22s^{-1}, T_{I_i} = 10)$. Tab. 5 shows the ratios of admitted flows obtained for various values of the load parameter α. The first observation that turns out by looking at these results

Table 5. Ratios of admitted flows using PI-FBDS with various parameter sets.

α	$k_{p_i} = 14s^{-1}$, $T_{I_i} = 4$	$k_{p_i} = 18s^{-1}$, $T_{I_i} = 7$	$k_{p_i} = 22s^{-1}$, $T_{I_i} = 10$
5	100%	100%	100%
10	100%	100%	99%
12	99%	98%	84%

is that, for increasing values of the two parameters, the ratio of admitted flows diminishes. This is due to the fact that as K_{pi} and T_{Ii} increase, the PI-FBDS becomes similar to FBDS. In fact, for increasing values of T_{Ii}, the integral action becomes less important, whereas, for increasing values of K_{pi}, the control-loop becomes faster.

Obviously, the one-way packet delays experienced by the flows diminish when the number of admitted streams decreases; this result is shown in Figs. 6 and 7, where the one way packet delays experienced by, respectively, the MPEG and G.729 flows are reported. Similar results, not shown here due to lack of space, have been obtained fot the H.263 flows.

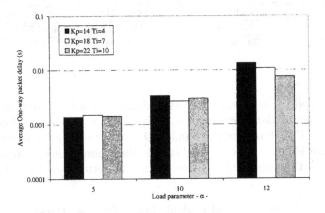

Fig. 6. Average one-way packet delay of the MPEG4 flows when using PI-FBDS with various parameter sets.

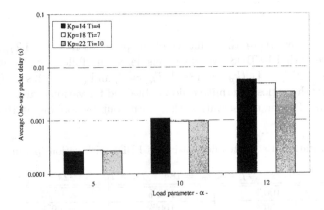

Fig. 7. Average one-way packet delay of the G.729 flows when using PI-FBDS with various parameter sets.

Tab. 6 reports the average and peak superframe utilization in HCCA mode using PI-FBDS with various parameter sets. Obviously, average and peak superframe usages follow the behaviour of the ratio of admitted flows (see Tab. 5). Finally, Table 7 reports the overall goodput achieved by the FTP flows. Also in

Table 6. Average (Peak) Superframe using PI-FBDS with various parameter sets.

α	$k_{p_i} = 14s^{-1}, T_{I_i} = 4$	$k_{p_i} = 18s^{-1}, T_{I_i} = 7$	$k_{p_i} = 22s^{-1}, T_{I_i} = 10$
5	6.82(26)%	6.91(30)%	6.95(29)%
10	15.15(84)%	15.03(82)%	15.34(73)%
12	29.85(81)%	27.69(81)%	24.24(81)%

this case, FTP flows get the highest goodput when the quota of real-time flows admitted is the smallest, i. e. for $k_{p_i} = 22s^{-1}$ and $T_{I_i} = 4$ (see Tab. 5).

Table 7. Overall goodput of FTP flows (Mbps) using PI-FBDS with various parameter sets.

α	$k_{p_i} = 14s^{-1}, T_{I_i} = 4$	$k_{p_i} = 18s^{-1}, T_{I_i} = 7$	$k_{p_i} = 22s^{-1}, T_{I_i} = 10$
5	4.57	4.57	4.58
10	4.32	4.32	4.38
12	2.47	2.47	4.09

5 Conclusion

In this paper a HCCA-based control theoretic framework for addressing the first hop bandwidth allocation issue using the 802.11e MAC has been proposed. This framework has been used to design two dynamic bandwidth allocation algorithms, which have been referred to as FBDS and PI-FBDS. Simulation results obtained using the *ns*-2 simulator have shown that, unlike the simple scheduler proposed by the 802.11e WG, both FBDS and PI-FBDS ensure bounded delays also at high network loads. Moreover, when the PI-FBDS is used, the best trade-off between one-way packet delays and network utilization is achieved.

References

1. IEEE 802.11: Information Technology - Telecommunications and Information Exchange between Systems Local and Metropolitan Area Networks Specific Requirements Part 11: Wireless LAN Medium Access Control (MAC) and Physical

Layer (PHY) Specifications. ANSI/IEEE Std. 802.11, ISO/IEC 8802-11. First edn. (1999)

2. Ni, Q., Romdhani, L., Turletti, T.: A survey of QoS enhancements for IEEE 802.11 wireless LAN. Jour. of Wirel. Commun. Mob. Comput. **4** (2004) 1–20

3. IEEE 802.11 WG: Draft Amendment to Standard for Information Technology - Telecommunications and Information Exchange between Systems - LAN/MAN Specific Requirements - Part 11: Wireless Medium Access Control (MAC) and Physical Layer (PHY) Specifications: Medium Access Control (MAC) Quality of Service (QoS) Enhancements. IEEE 802.11e/D10.0. (2004)

4. Kong, Z., Tsang, D.H.K., Bensaou, B., Gao, D.: Performance Analysis of IEEE 802.11e Contention-Based Channel Access. IEEE Journal of Selected Areas in Communications **22** (2004) 2095–2106

5. Boggia, G., Camarda, P., Grieco, L.A., Mascolo, S.: Feedback based bandwidth allocation with call admission control for providing delay guarantees in IEEE 802.11e networks. Computer Communications, Elsevier **28** (2005) 325–337

6. Xiao, Y., Li, H., Choi, S.: Protection and guarantee for voice and video traffic in IEEE 802.11e Wireless LANs. In: IEEE Infocom, Hong Kong (2004)

7. Grilo, A., Macedo, M., Nunes, M.: A scheduling algorithm for QoS support in IEEE 802.11e networks. IEEE Wireless Communications (2003) 36–43

8. Ansel, P., Ni, Q., Turletti, T.: Fhcf: A fair scheduling scheme for 802.11e wlan. Technical Report 4883, Institut National de Recherche en Informatique et en Automatique (INRIA) (2003)

9. Ns-2: Network simulator. available at http://www.isi.edu/nsnam/ns (2004)

10. Astrom, K.J., Wittenmark, B.: Computer controlled systems: theory and design. 3 edn. Prentice Hall, Englewood Cliffs (1995)

11. Mascolo, S.: Congestion control in high-speed communication networks using the smith principle. Automatica, Special Issue on Control methods for communication networks **35** (1999) 1921–1935

12. International Telecommunication Union (ITU): Coding of Speech at 8 kbit/s using Conjugate-Structure Algebraic-Code-Excited Linear Prediction (CS-ACELP). ITU-T Recommendation G.729. (1996)

13. MPEG-4 Video Group: Mpeg-4 overview. available at http://mpeg.telecomitalialab.com/ (2002)

14. International Telecommunication Union (ITU): Video coding for low bit rate communication. ITU-T Recommendation H.263. (1998)

15. Video Trace Library. (available at http://trace.eas.asu.edu/)

16. Chuah, C., Katz, R.H.: Characterizing Packet Audio Streams from Internet Multimedia Applications. In: Proc. of International Communications Conference (ICC 2002), New York, NY (2002)

17. Xu, Y., Guerin, R.: On evaluating loss performance deviation: a simple tool and its pratical implications. In: Quality of Service in multiservice IP networks (QoSIP2003), Milan, Italy (2003) 1–18

18. Wang, W., Liew, S.C., li, V.O.K.: Solutions to performance problems in VoIP over a 802.11 wireless LAN. IEEE Trans. Veh. Technol. **54** (2005) 366–384

Audio Rendering System For Multimedia Applications

Paola Pierleoni, Folco Fioretti, Giovanni Cancellieri, Tommaso Di Biase, Samuele Pasqualini, Fabrizio Nicolini*

Dipartimento di Elettronica, Intelligenza Artificiale e Telecomunicazioni
Università Politecnica delle Marche, Via Brecce Bianche
60131 Ancona, Italy
*Aethra S.p.A., Via Matteo Ricci 10, 60020 Ancona (AN), Italy
p.pierleoni@univpm.it

Abstract. An immersive audio environment for videoconferencing or virtual reality applications is presented. The proposed system allows the spatialisation of 3D sound fields around listener positioned in front of a screen (PC or TV monitor) with headphones or only two loudspeakers in front of him. Furthermore, the presented application is real-time customizable on the basis of the particular listener anthropometry. The source sound can be a real audio signal captured at the remote side of a videoconference system by a microphone or a synthetic one. The loudspeaker signals are derived by the knowledge of the spatial coordinate of the speaker at the remote end of a videoconference session and a subsequent appropriate filtering of the audio streams with head related transfer functions (HRTFs). Virtual sources can be positioned in azimuth, in elevation and in distance, too. However, the reproduction area around the listener and the possible source space are restricted to a defined area. Nevertheless, the proposed algorithm implementation into a commercial videoconference system demonstrates an high quality of the audio rendering performance. A special Visual C++ development tool has been established to carry out system parameters optimization.

Keywords: Immersive audio rendering, Teleconference, Virtual Reality, Head Related Transfer Functions (HRTFs).

1 Introduction

There is currently a great deal of interest in designing systems that can faithfully reproduce (or create) a full three dimensional auditory scene. These immersive audio systems can be used for applications such as videoconference, distance learning, virtual reality, home entertainment, displays for the visually impaired and pilot warning systems. Here, we propose signal audio processing that pertains to the acquisition and subsequent rendering of 3D sound fields over two loudspeakers or headphones. The proposed system is suitable for distance learning equally for virtual reality applications. In the first case, on the acquisition side, we have already developed a system of spatial filtering able to achieve steering in a given direction so localizing the active speaker and tracking the local television camera[1].The achieved spatial focusing in the direction of interest is a necessary element in immersive audio acquisition systems for teleconference. In fact, we need to transmit the acquired spatial information of the audio source at the opposite side of the teleconference connection (Fig. 1). Any application we consider, teleconference or virtual reality, the knowledge of real or synthetic source coordinates is required. On the rendering side, 3D audio signal processing methods are described that allow rendering of virtual sources around the listener using only two loudspeakers or headphones.

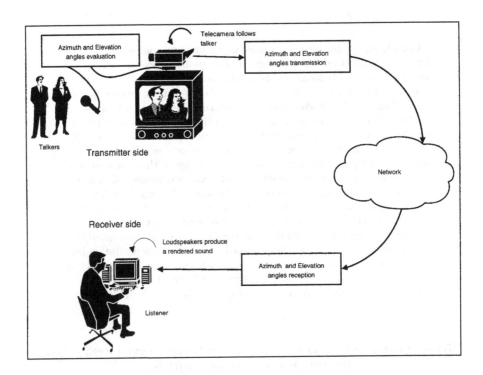

Fig. 1. A simple scheme of the proposed system for teleconference application

The basis of 3D audio is to control the acoustic signals at the receiving side and thereby reproduce the signals that the listeners would have received in the real audio environment that is being simulated. The way to do this is to deliver appropriate binaural signals through headphones or loudspeakers. Different 3D spatialization strategies are feasible. These strategies depend on whether the

---

[1] The active speaker localization tool is implemented in practice on Aethra Telecounicazioni S.p.A. videoconference commercial systems.

sound field is reproduced at the ears of the listeners (called transaural systems). Transaural systems can be realized by either using headphones or loudspeakers. In the transaural approach, human sound-source localization is based on the estimation of frequency-dependent differences in intensity and time of arrival at both ears for source localization in the horizontal (azimuth) plane. In the median plane time differences are constant. Localization is based on spectral filtering by the outer ear. The reflection and diffraction of sound waves from the head, torso, shoulders and pinnae, combined with resonances caused by the ear canal, form the physical basis for the head-related transfer function (HRTF) [1]. This system can be modelled as linear and time-invariant, and it is fully characterized by the HRTF in the frequency domain. Immersive audio rendering systems are based on digital implementation of such head-related transfer functions. The spectral information provided by the HRTF can be used to implement a set of filters that can process non-directional (monaural) sound to simulate a real HRTF. In principle, it is necessary the measurement of HRTF for each sound source direction and for each listeners. In fact, the magnitude and phase of these head-related transfer functions vary significantly not only for each sound direction, but also from person to person. Our attempt is focused on achieving good localization performance using HRTF derived through measurements realized at high spatial resolution over a set of more than 40 listeners plus a KEMAR mannequin with variable anthropometry [2]. Here, we have focused our discussion on delivering immersive audio through headphones or a pair of loudspeaker for two reasons: there is a large installed desktop computer with two loudspeakers on either side of the monitor and this loudspeakers configuration is widely used in videoconference systems too. At the same time, headphone is often an add-on component in both applications. While with headphones it is possible to deliver the appropriate sound field to each ear, with loudspeakers the drawback of unwanted channel crosstalk has to be reduced. This crosstalk arises because each loudspeaker sends sound to the same side ear, as well as undesired sound to the opposite site ear. A significant amount of methods have been proposed in literature to address crosstalk cancellation. This phenomenon will not be considered in this paper because widely treated in literature [3]. Regarding the distortion due to discrete early reflections, we have practically detected that in a typical desktop environment (two loudspeakers placed on either side of a video or a computer monitor and listener close to the loudspeakers) the direct sound is dominant with respect to the room reflection effects. These conclusions are supported by a test-bed of the audio rendering system over a videoconference environment.

2 Theory overview

Locating sound in 3D is a complex process for the human aural system. Even if it is not completely understood, the process is mainly based on the recognition of interaural arrive time differences (ITDs) and interaural intensity differences (IIDs) [4,5]. When not directly received from the front or from the back sounds do not arrive at both ears at the same time but with an interaural time difference as depicted in Fig. 2. Typically ITD ranges from 0 ms (direct front/back sound) to about 0.522 ms (sound received from straight left or right). The human brain is able to detect 3D sounds location in real-time for ITD value within a few μsecs.

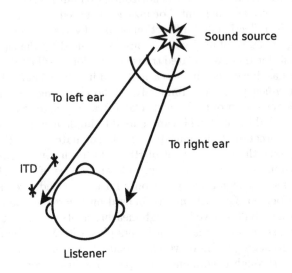

Fig. 2. Interaural arrive-Time Difference (ITD)

A simple approximated model to calculate ITD is presented by Kuhn [6]:

$$ITD \simeq \begin{cases} \frac{3a}{c} \sin \theta \text{ at low frequencies} \\ \frac{2a}{c} \sin \theta \text{ at low frequencies} \end{cases} \tag{1}$$

where a is the radius of the sphere used to model the head, c is the sound speed, and θ is the angle between the median plane and a ray passing from the centre of the head through the source position. The transition between the two frequency regions is located around 1 kHz. When sound travels from a source to a listeners ear it receives reflections at the human body such as head, shoulder, and outer ear. Those reflections and attenuations results in interaural intensity differences (IIDs) as outlined in Fig. 3.

An expression similar to (1) may be obtained for the IIDs, but they generally have more complicated variations with frequency. Because the influence of human anthropometry (the shapes of head, shoulders and outer ear) and its variability across listeners, simple approximation for the IID is not applicable. Furthermore, IID would be negligible for frequency below approximately 1 kHz because the wavelength is similar or larger than the distance between the listener ears. The combination of the two cues, ITD at high frequency and IID at low frequency, is known as the Rayleighs "duplex" theory of localization. Even if this theory is

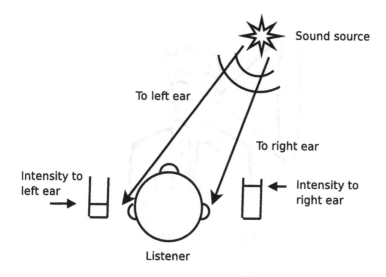

Fig. 3. Interaural Intensity Difference (IID)

still actual, and widely used in audio signal processing research, it is not able to entirely describe the human hearing system. The modelling based on ITD and IID fails into the so called cones of confusion. In fact, a locus of points exists for which ITD and IID measurements are constant. This surface is similar to a cone with axis collinear with interaural axis (Fig. 4). Each point over the cone of confusion is characterized by the same ITD and IID values so the duplex theory is not able to describe the overall human hearing phenomenon.

The previous model has to be integrated with the introduction of an appropriate acoustic transfer function taking into account the sound transmission from the source to the eardrum of the listener [7]. It is well known that the physical effects of the diffraction of sound waves by the human torso, shoulders, head and pinnae modify the spectrum of the sound that reaches the listener eardrums. Therefore, IIDs are often modelled by a set of so called Head-Related Transfer Functions (HRTFs) which not only vary in a complex way with azimuth, elevation, range, and frequency, but also vary significantly from listener to listener. Naturally, due to the symmetry of the listener head the HRTFs apply for both ears. The information in the HRTF is also contained in the temporal behaviour of the equivalent Head Related Impulse Response (HRIR). In real time synthesis, if the sound source moves relative to the head, the HRIR must be modified accordingly. This is typically done by computationally expensive interpolation. To produce convincing 3D effects, the HRTF or its equivalent HRIR must be measured for each condition, which is inconvenient and limits applications. Furthermore, serious perceptual distortion can occur when one listens to sound spatialized with a non-individualized HRTF.

Fig. 4. Cone of confusion

3   The proposed approach

To realize HRTF there are two major ways: measurements and model approaches. Although the determination of individualized HRTF can addressed in a lot of ways, we adopted the first strategy making use of a public domain database collected by CIPIC labs. This database collects head related transfer functions measured at high spatial resolution for 45 subjects (including KEMAR mannequin with two different ears conformation). Because sound source location is specified by the interaural polar coordinates (the azimuth angle θ and the elevation angle ϕ) this is the reference system used in this work. The azimuth angle θ is the angle between the vector to the sound source and the vertical median plane and varies from $-\pi/2$ (left) and $+\pi/2$ (right). The elevation angle ϕ is the angle of the horizontal plane with the projection of the source into the median plane and, starting directly below the subject, varies from $-\pi/2$ and $+3\pi/2$. The CIPIC measures involved subjects seated at the center of a hoop and their head aligned with interaural coordinates axes. Practically, the system is modelled by means of HRTFs (HRIRs) with a set of digital FIR filters whose coefficients are derived from the previous described measures. For a given number N of coefficients (taps), a FIR filter is defined as

$$u_T = \sum_{i=1}^{N} h_i v_{T-1-i} \qquad (2)$$

where u_T is the filter output at time T, v_i are the input values from previous time i, and h_i are the intensity of signal v_i at time i (i. e. the filter taps). We consider FIR filters with N=200 because the available CIPIC database dimension. This is, at the same time, an adequate compromise between acceptable run-time and approximation of the acoustic system. Fig. 5 shows the logical structure of CIPIC database: to evaluate filter coefficients we need the azimuth and elevation source coordinates to extract tap values. This structure is replicated for each subject involved in the CIPIC measurements and can be used to process values of signal u at the time T. The individual HRTF is so selected from database with respect to received sound direction.

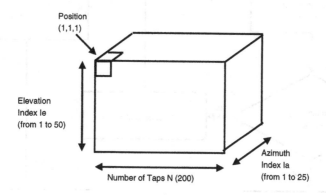

Fig. 5. Logical structure of CIPIC database

The involved parameters are evaluated by finite step measures: elevations were uniformly sampled in $360°/64 = 5.625°$ (from $-45°$ to $+230.625°$), while azimuth values were sampled at $-80°$, $-65°$, $-55°$, from $-45°$ to $45°$ with a step of $5°$, and $55°$, $65°$, $80°$. Fig. 6 describes the correspondence between a real spatial position captured by the remote camera and the discrete coordinate values we need to extract the right filter.

An azimuth range control is required, because an out of range angle is reasonable but there is not a corresponding coefficient in the database. Similarly, we have to determine elevation index from elevation angle. This procedure gives a sequence of elevation and azimuth coefficients. The elevation set belongs to the range from 1 to 25, while the azimuth set goes from 1 to 50.

4 Simulation and Listening Tests

To simulate and evaluate the three dimensional synthesized sound field a Simulink version of the described system has been developed. The off-line test procedure is summarized in the simple block diagram in Fig. 7.

This tool provides the loading of a monaural sound (in Wave format) among a set stored reference sounds and the choice of the azimuth and elevation source coordinates into the ranges $[-90°, 90°]$ and $[-45°, 90°]$ respectively. The acquisition of HRIR from CIPIC database and the conversion of coordinates (θ, ϕ) in index (I_e, I_a) gives the right filter taps for each channel (left and right). The binaural signal is then synthesized as a convolution of the input signal with

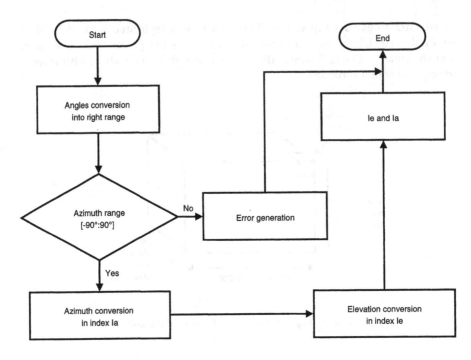

Fig. 6. Angle conversion

Head Related Impulse Response functions before its reproduction over a pair of loudspeaker. To establish a performance baseline, in an initial task the produced binaural sounds were listeners by each subject, so that a correct match with pseudo-individualized HRIR functions was possible. The Simulink block scheme is depicted in Fig. 8.

Once calibration was completed, several tests were performed to validate the localization model. In particular we are interested in verifying if fast variations in filter taps, due to sound source movement, could causes any distortion in reconstructed sounds. In systems like to videoconference there could be more than one active talker, or one talker can quickly move through the room. For this reason each stimulus was generated varying randomly the associated angular coordinates (θ, ϕ) and in a second time keeping constant a coordinate value and varying the other one. The coordinate changes occurred every 0.9 s: in real systems the remote camera tracking mechanism is characterized by a prearranged latency time to avoid unwanted training dues to noise or impulsive sounds. In our tests since the loudspeakers are placed in front of the listener only sources in front of the listener can be reproduced faithfully. However the position of sources can vary in azimuth, elevation and distance. The size of the reproduction area is restricted to a region around the listener depending on the distance between the two loudspeakers and that between the loudspeakers and the listener. Before

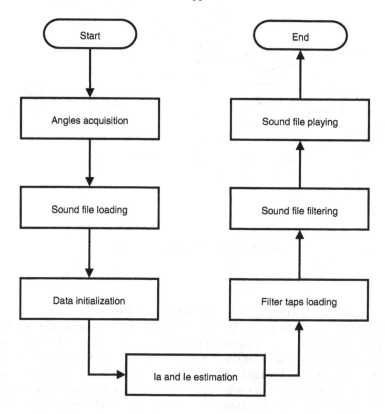

Fig. 7. Off-line filtering block diagram

implementing the corresponding C++ code in a real time application, a special Visual C++ development tool has been established to preset system optimal parameters and to evaluate the effect produced to listener by three dimensional sound synthesis. The features of this tool provide the simulation of 3D wave propagation at the output of the loudspeakers after HRTFs filtering. Fig. 9 shows the visual interface of the application tool we have developed. Its extensive possibilities for graphical evaluation can be used to examine the simulated system performance and to simplify the synthesis phase. The window shows the time and frequency responses (HRTFs, HRIRs, ITD) for a selected listener subject (and for each ear) as a function of source location. It is possible to choose among different subjects whose anthropometric characteristics are collected in the CIPIC database. Moreover we can change sampling frequency, spectral resolution in FFT calculation and spectral smoothing (related to the graphic representation of the data). We can also select among four playing mode called: *Play move, Play click, Play audio* and *Play cone of confusion.* In the first mode the tool, after a file wave acquisition, simulates the movement of the virtual source along a straight line in the frontal plane from left to right with ϕ=cost. The *Play click*

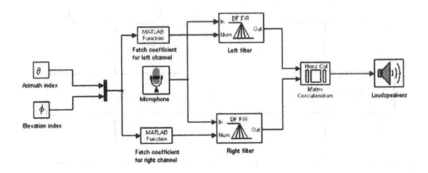

Fig. 8. Simulink implementation

mode produces three white noise bursts (100% amplitude modulation, f=30 Hz) at time difference of 150 ms each other. These sources are virtually set in the same azimuth angle θ, but in three different elevation angles ϕ. The *Play audio* mode allows the variation of all the input parameters. The last mode elaborates a sound source moving through all the allowed elevation position belonging to the cone of confusion. After choosing right parameters and desired playing mode the application gives several outputs: IDT as function of the source tilt; HRIRs, HRTFs as functions of time and frequency. The grey-scale images at the left side of Fig. 9 refer to HRTFs and HRIRs functions for all 50 allowed elevation values with a fixed tilt value (indicated by the vertical dashed line). The intersection among the two straight lines in the HRIRs representations is function of the azimuth value: therefore a simple observation of the two plots gives us the interaural delay.

These graphical representation gives some ideas of the HRTF and ITD variability for a single subject and the range of variability among subjects. Obviously it is also possible to play the synthesized binaural sounds. We observed that the effect of an immersive audio rendering is obtained if the listener is in a right position, that is listener must be in the middle between loudspeakers, obviously there are no problem using headphones. Another important issues is relative the area of audio rendering. Desired effects are given only for determinates movement of talkers, for example well not have the effect of a sound source behind listener. Finally the system was implemented on a distributed videoconference system, utilizing Aethra SpA standard commercial equipments, to test-real time performances (Fig. 10). Listening sessions with sound source moving at the remote side of a videoconference session showed that the system resulted in a good sound quality and localization, very similar to the sound perceived in a real environment.

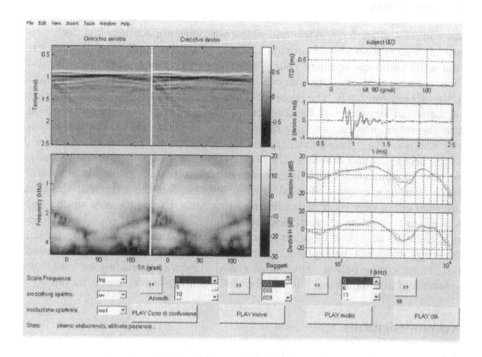

Fig. 9. Visual C++ interface of the developed tool

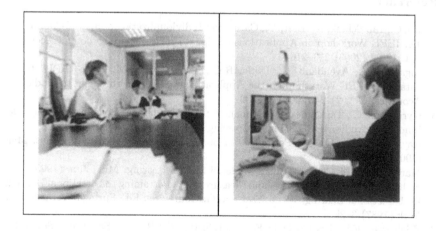

Fig. 10. Experimental set-up at the opposite side of videoconference connection

5 Conclusion

An audio interface for desktop and videoconference applications has been presented. The system should provide the user of screen application to immerse into a 3D audio environment using headphone or pair of loudspeaker. We also demonstrated that our approach can be taken for pre-processing sound in interactive 3D environments. The theoretical model approach is based on the filtering of the sound streams with head-related transfer functions (HRTFs). Based on this theoretical analysis we have shown that in common-use loudspeaker geometry and distance between listeners and loudspeakers, the spatial sound pattern of the original sound is sufficiently preserved in the reproduced sound. Finally, we have implemented a real-time version of this system and tested it in several videoconference sessions: informal listening tests indicated that the spatialization achieved offers very good 3D audio rendering sense through a very simple and low expensive system. Nevertheless, the quality of the spatial sound reproduction could be improved by adding crosstalk cancellation systems and informal listening tests must be replaced by a much better psychoacoustic localization test. However, we believe that the proposed system is promising for real-time synthesis of 3D sounds and the developed simulator is a very useful toolkit.

Acknowledgments

Part of this work was supported by the Aethra Telecommunication SpA Labs. We would also like to thanks Stefano Sarghini for his guidance on real-time testing procedure.

References

1. Jin, C., van Schaik, A., Best, V., Carlile, S.: Individualization in Spatial-Audio Coding. IEEE Workshop on Applications of Signal Processing to Audio and Acoustics, Oct. 19-22 (2003) 213–216
2. Algazi, V.R., Avendano, C., Duda, R.O., Thompson, D.M.: The CIPIC HRTF database. IEEE Workshop on the Applications of Signal Processing to Audio and Acoustics, Oct. 21-24 (2001) 99–102
3. Kyriakakis, C., Tsakalides, P., Holman, T.: Acquisition and Rendering Methods for Immersive Audio. IEEE Signal Processing Magazine, Jan. (1999) 55–66
4. Rayleigh, L. In: On our Perception of Sound Direction. Volume 13. Philos. Mag. (1907) 214–232
5. Martin, K.D.: A Computational Model of Spatial Hearing. MIT, Jun. (1995)
6. Kuhn, G.F.: Physical Acoustic and Measurements Pertaining to Directional Hearing. In: Directional Hearing. W. A. Yost and G. Gourevitch Ed., Springer-Verlang, New York (1987) 3–25
7. Wightman, F.L., Kistler, D.J.: Factors Effecting the Relative Salience of Sound Localization Cues. In: Binaural and Spatial Hearing in Real and Virtual Environments. Mahwah, New York (1997) 1–23

QoS Provisionig And Control In Real-Time Interactive Multimedia Communications Using Digital Watermarking

Francesco Benedetto[1], Gaetano Giunta[1], Alessandro Neri[1]

[1] Digital Signal Processing and Multimedia Communications Lab., Dept. of Applied Electronics, University of ROMA TRE, Via della Vasca Navale 84, 00146 Rome, Italy.

{emails: fbenedetto@ieee.org, giunta@ieee.org, neri@uniroma3.it}

Abstract. In the last decade, there has been an explosive growth in mobile computing technology and obviously an increase demand for platforms with multimedia application support. The desire to be connected *"any-time, any-where, and any-way"* has led to an increasing array of heterogeneous systems, multimedia applications, devices, and service providers. In response to this issue, quality of service (QoS) is designed to hide low-level application variation and provide necessary service guarantees. In this contribution, motivation and operating guidelines for QoS assessment in real-time interactive multimedia communications are discussed and investigated. In particular, we consider the end-to-end service quality methodologies proposing a QoS evaluation technique by means of an unconventional use of fragile watermarking. Thus, by knowing the end-to-end QoS, it is possible to adopt optimal pricing strategies in terms of the QoS profiling for all the active users involved in the communication.

Keywords. End-to-End Quality of Service (QoS), QoS profiling and pricing, multimedia communication, wireless/mobile networks, digital watermarking

1 Introduction

In recent years there has been an explosive growth in mobile computing technology and obviously an increase demand for platforms with multimedia application support. Progress in Telecommunications and Networking is fostering the development of high-speed and ubiquitous networks, both wired and wireless (i.e. heterogeneous configurations), characterized by an unprecedented degree of transport capacity and flexibility. The desire to be connected *"any-time, any-where, and any-way"* has led to an increasing array of heterogeneous systems, multimedia applications, devices, and service providers (see fig. 1). According to [1], this heterogeneity is unlikely to disappear in the foreseeable future for two reasons: one is that the variety of application requirements makes it difficult to find a single optimal and universal solution; the other is that in their eagerness to capture the market, competing organizations are releasing proprietary systems. As a result, the key to the success of next-generation mobile communications systems is the ability to provide seamless interactive real-time multimedia services in such a heterogeneous environment [2]-[3].

However, to provide users with satisfactory services, ubiquitous connectivity and corresponding best effort services are not enough. In such a wireless network environment, application performance could easily deteriorate for various reasons, and this performance fluctuation could be widespread. In response to this issue, quality of service (QoS) is designed to hide low-level application variation and provide necessary service guarantees. Dynamic QoS reconfiguration constitutes an important constraint that is yet to be addressed for the real-time interactive multimedia communications of 3GPP [4]-[10]. High-speed real-time multimedia applications such as Videoconferencing and HDTV have already become popular in many end-user communities. Developing better network protocols and systems that support such applications requires a sound understanding of the voice and video traffic characteristics and factors that ultimately affect end-user perception of audiovisual quality. Recently, there are a number of researches concerned with QoS guarantee for media data flow on the transport and network levels using QoS oriented protocols focusing on the end-to-end QoS control functions and mechanisms in accordance with different user's environment [11].

In this contribution, motivation and operating guidelines for QoS assessment in real-time interactive multimedia communications are discussed and investigated. In particular, we consider the end-to-end service quality methodologies proposing a QoS evaluation technique by means of an unconventional use of fragile watermarking [12]-[15]. Thus, by knowing the end-to-end QoS, it is possible to adopt optimal pricing strategies in terms of the QoS profiling for all the active users involved in the communication [16]. The work is organized as follows. Section 2 describes both the requirements and the operating targets of the QoS assessment in real-time multimedia communications. The signal-processing algorithm used for the QoS evaluation is depicted in Section 3, while simulation results are shown and discussed in Section 4. Finally, the conclusions of the work are briefly summarized in Section 5.

2   QoS Requirements in Real-Time Multimedia Communications

The problem of 'matching' the application to the network is often described as a 'quality of service' (QoS) problem [17]. The problem is twofold: there is the QoS required by the multimedia application, which relates to visual quality perceived by the user, and the QoS reachable by the transmission channel and the network, which depends on the physical link's capabilities [12]. The key parameters for QoS requirements about transmission are the mean bit rate and the variation of the bit rate [17]. The mean rate depends on the compression algorithm as well as the characteristics of the source video (such as frame size, number of bits per sample, frame rate, amount of motion, etc.). Many video coding techniques include a certain degree of compression tuning that bound of the mean rate after encoding. However, the achievable mean compressed bit rate is actually limited for a given source (with a particular frame size and frame rate). Real-time video is sensitive to delay by nature. The QoS requirements in terms of delay depend on whether video is transmitted one way (e.g. broadcast video, streaming video, playback from a storage device) or two ways (e.g. video conferencing) [17]. The original temporal relationships between frames have to be preserved in *simplex* (one-way) video transmission because any associated media that is 'linked' to the video frames should remain synchronized, but a certain degree of congruent delay is admitted. The most common example is accompanying audio, where a loss of synchronization of more than about 0.1 s can be obvious to the viewer [17]. *Duplex* (two-way) video transmission has the above constraints (constant delay in each direction, synchronization between related media) plus the obligation that the total delay from acquisition to display must be low enough. A 'rule of thumb' for video conferencing is to keep the total delay less than 0.4 s [17]. If the delay is longer than this, video-conversation appears like non-natural. Interactive applications also have a requirement of low delay. In such a case, an eventual too long delay between the user's action and the corresponding effect on the video source makes the application unresponsive. Moreover, in case of multimedia teleconference service, for example, users not only use real-time media services such as voice and video but also stored media services for their communication and media presentation simultaneously. Therefore, the system must process stored media and real-time media at the same time. In order to provide the stored multimedia information to users, the system must not only integrate both continuous media such as audio, video and discrete media such as still image, graphics and text, but also perform end-to-end QoS functions and different media processing functions inherit to each media data [11].

Therefore, real-time QoS management represents an important requirement during real-time interactive multimedia communications. The aim of QoS management is to adapt to changing conditions during runtime and to maintain committed QoS levels or to react upon changes in these levels. This contribution proposes a novel procedure to evaluate the QoS of a communication link. In practice, the receiving mobile station is able to verify whether the negotiated quality level has been actually maintained by the effective link or not. In such a case, the operator can lower the user's fare and bustle about in order to guaranty an improved service level by modifying the operating settings of the communication link. The blind operating system at the receiver of a communication link needs to decide in order to match the negotiated quality level and the effectively provided one. If the user had requested a secure (high-level) transmission, he has to be accordingly charged for the relative fare. Otherwise, the connection can be low-priced, even becoming free of charge. Moreover, like discussed in [13], it can be observed that in real-time interactive multimedia communication systems requiring error control, such as video telephony and streaming, a number of consequences follow a poor quality decision from the mobile station: first, the bit rate (and the video resolution) transmitted by the base station is lowered in a few seconds (by a higher spreading factor in the UMTS systems); second, if the mobile station declares a null quality, the link is broken down by the operator; third, frequent declarations of poor or null quality will be a reason for the operator's admission call manager to refuse the access for further calls of the same user for a given time period, i.e. until the mobile station moves into a region with higher signal-to-noise ratio (SNR). These should avoid fraudulent declarations of bad quality from the mobile station in order to transmit free of charge. In fact, the result of possibly false declarations on the QoS is to suddenly cut the communication in a few seconds, assuming the risk of no admission for further calls. Thus, by knowing the end-to-end QoS, it is possible to adopt optimal pricing strategies in terms of the QoS profiling for all the active users involved in the communication. In the next Section, the signal-processing algorithm for the QoS evaluation is described and we show how it is effectively possible to blindly estimate the quality of a transmission system (including the coder quality) without affecting the quality of the video-communications.

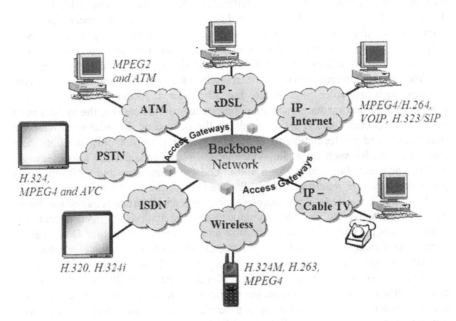

Fig. 1. Multimedia Communications: the desire to be connected *"any-time, any-where, and any-way"* has led to an increasing array of heterogeneous systems, multimedia applications, devices, and service providers.

3 Digital Watermarking for Real-Time QoS Assessment

The rational behind the approach is that the alterations suffered by the watermark are likely to be suffered by the data, since they follow the same communication link. Therefore, the watermark degradation can be used to evaluate the alterations endured by the data. At the receiving side, the watermark is extracted and compared with its original counterpart. Spatial spread-spectrum techniques perform the watermarking embedding. In practice, the watermark (narrow band low energy signal) is spread over the image (larger bandwidth signal) so that the watermark energy contribution for each host frequency bins is negligible, which makes the watermark near imperceptible. Following the same methodological approach of [13], a set of uncorrelated pseudo-random noise (PN) matrices (one per each frame and known to the receiver) is multiplied by the reference watermark (one for all the transmission session and known to the receiver): $w_i^{(s)}[k_1, k_2] = w[k_1, k_2] \cdot p_i[k_1, k_2]$, where $w[k_1, k_2]$ is the original watermark, $p_i[k_1, k_2]$ the PN matrices and $w_i^{(s)}[k_1, k_2]$ the spread version of the watermark to be embedded in the i-th frame. The embedding is performed in the DCT domain according to the following:

$$
F_i^{(w)}[k_1, k_2] = \begin{cases} F[k_1, k_2] + \alpha \cdot w_i^{(s)}[k_1, k_2], & (k_1, k_2) \in S \\ \\ F[k_1, k_2], & \text{otherwise} \end{cases}
\tag{1}
$$

where $F_i[k_1, k_2] = DCT\{f_i[k_1, k_2]\}$ is the DCT transform of the i-th frame; S is the region of middle-high frequencies of the image in the DCT domain, while α is a scaling factor that determines the watermark strength and $F_i^{(w)}[k_1, k_2]$ is the DCT of the i-th watermarked frame. In the following, the value $\alpha = 0.04$ has been always used. The i-th watermarked frame is then obtained by performing the IDCT transform of $F_i^{(w)}[k_1, k_2]$, finally the whole sequence is coded and then transmitted through a noisy channel, like shown in the principle scheme of figure 2.

The receiver implements video decoding as well as watermark detection. In fact, at the same time after decoding of the video stream, a matched filter extracts the (known) watermark from the DCT of each n-th received I-frame of the sequence. The estimated watermark is matched to the reference one (despread with the known PN matrix). The matched filter is tuned to the particular embedding procedure, so that it can be matched to the randomly spread watermark only. It is assumed that the receiver knows the initial spatial application point of the mark in the DCT domain. The Quality of Service is evaluated by comparing the extracted watermark with respect to the original one. In particular, its mean-square-error (MSE) is evaluated as an index of the effective degradation of the provided QoS. A possible index of the degradation is simply obtained by calculating the mean of the error energy:

$$
E_e = \frac{1}{M} \sum_{n=1}^{M} \left(\sum_{k_1=1}^{K_1} \sum_{k_2=1}^{K_2} (w_n[k_1, k_2] - \hat{w}_n[k_1, k_2])^2 \right)
\tag{2}
$$

where $w_n[k_1, k_2]$ and $\hat{w}_n[k_1, k_2]$, represent the original watermark, the extract watermark respectively, and $n = 1..M$ is the current frame index. Another video quality measure usually used for QoS assessment is the peak signal-to-noise ratio (PSNR) defined as the ratio between peak signal and root mean square (rms) noise observed between the original video and the watermarked video, where RMSE is the square root of the mean square error previously defined:

$$PSNR = 20\log_{10}\left(\frac{255}{RMSE}\right) \tag{3}$$

In particular, such QoS indexes can be usefully employed for a number of different purposes in wireless multimedia communication networks such as: control feedback to the sending user on the effective quality of the link; detailed information to the operator for billing purposes and diagnostic information to the operator about the communication link status [12].

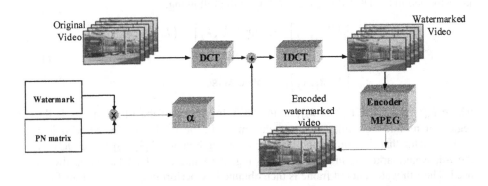

Fig. 2. Principle scheme of the watermark embedding process in the DCT domain.

4 Simulation Results

From the operating viewpoint, we are interested in the actual video over a given communication link determined by the (booked) maximum bit rate. In other words, the target quality is fixed by the negotiated channel capacity, while the actual quality depends also on the symbol errors introduced in the received data stream by the physical link, due to background noise as well as multi-path channel and interference effects. Such transmission errors are modeled as a random Poisson process, characterized by a parametric probability of symbol error (proportional to the bit error rate). Then, the actual quality needs to be detected by an in-service quality assessment measure, such as the tracing watermarking technique here discussed. For this purpose, a number of simulations have been carried out: the first simulations regard still images coded at different compression rates (i.e. at different qualities). In all the subsequent simulations the scaling factor regarding the watermark intensity, α, has been set equal to 0.04. Increasing this value, the mark becomes more evident and a visual degradation of the image (or video) occurs. On the contrary, by diminishing its value, the mark can be easily removed by the coder and/or channel's errors.

Therefore, in the application scenario of our simulation trials, the scaling factor α has been chosen in such a way to compromise between the two aforementioned requirements.

Figure 3 reports here the PSNR of the watermarked image "Lena" (fig. 3a) JPEG coded at different qualities ($Q = 100$ is the best quality) while fig. 3b shows the MSE of both the image and the watermark versus the bit error rate of a simulated link for several (target) bit-rates. Regarding the video-sequences, several simulations have been made to detect the sensitivity of digital watermark to two different degradation sources: the noisy channel affects the video quality and the co-decoder itself affects the perceived image quality. To evaluate the sensibility of the method to the degradations due only to the channel, tests have been performed with different values of BER using the same coder quality. In particular, figure 4a shows the mean square error of the decoded QCIF (176x144 pixels) "Akyio" video and the extracted watermark versus the bit error rate of the channel, using a MPEG-2 coder at 100 kbit/s. Figure 4b shows the MSE of the "News" video using a MPEG-4 coder at 100 kbit/s. The above trials confirm that the watermark error is roughly proportional to the coding quality, which conversely depends on the available channel capacity. The presented approach allows one to blindly estimate the QoS provided by a coder/channel system without affecting the quality of the video-communications.

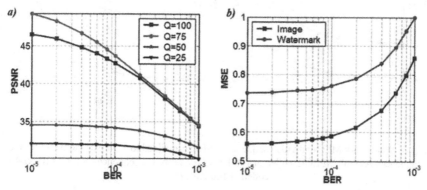

Fig. 3. PSNR of the watermarked "Lena" image JPEG-coded at different compression rates *(a)*; watermark and image MSE coded at Q=75 *(b)* versus the BER of a simulated communication link.

Moreover, the proposed technique can be usefully employed for a number of different purposes in real-time interactive multimedia communications such as: control feedback to the sending user on the effective quality of the link; detailed information to the operator for billing purposes and diagnostic information to the operator about the communication link status. In fact, in a real-world system, for example, the estimated quality of the received video can also be used by the service provider as a feedback information for billing purposes. This raises questions on the security of the system and it is worth pointing out that in order to prevent a fraudulent user to obtain any benefit from any false declaration about the QoS of the supplied service, a possible approach is briefly specified as follows. In real-time interactive video communications, let us consider the situation when the mobile station declares a received quality lower than the provided one.

This would imply that the channel is not suited for the current bit rate for the given BER, and therefore, the bit rate emitted by the base station is lowered in a few seconds. Whether the MS declares a null quality, the operator interrupts the call. Moreover, frequent declarations of poor or null quality is a valid reason for the admission call manager to refuse the access to further calls of the same user, at least until the MS has moved to a region with less noise or interference. As a consequence, the result of possibly false declarations on the QoS is to lower the bit rate or to cut the communication link in a few seconds, thus preventing any fraudulent action. In conclusion, The experimental results that have been presented confirm the hypothesis that the watermark alterations can be used to evaluate the video degradation amount and the transfer quality.

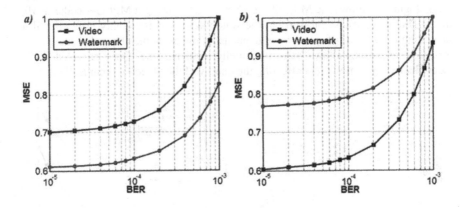

Fig. 3. Watermark MSE and video sequence MSE (normalized to 1) versus the BER: for the sequence "Akyio" MPEG-2 coded at 100 Kb/s *(a)* and for the sequence "News" MPEG-4 coded at 100 Kb/s *(b)*.

5 Conclusions

In this contribution, motivation and operating guidelines for quality of service assessment in new generation wireless video communications have been discussed and investigated. More specifically, real-time interactive multimedia communications, in which dynamic quality of service (QoS) reconfiguration constitutes an important constraint, have been analyzed. In these environments, the aim of QoS management and reconfiguration is to adapt to changing conditions during runtime and to maintain committed QoS levels or to react upon changes in these levels. This work focuses on the evaluation of a QoS index in order to realize a control feedback to the sending user on the effective quality of the link, to realize optimal strategies for QoS provisioning and control. In particular, we have considered the end-to-end service quality methodologies proposing a QoS evaluation technique by means of an unconventional use of fragile watermarking. We blindly estimate the quality of a transmission system (including the coder quality) without affecting the quality of the video-communications using a tracing watermarking approach for QoS assessment. Thus, by knowing the end-to-end QoS, it is possible to adopt optimal pricing strategies in terms of the QoS profiling for all the active users involved in the communication.

References

1. Gao, X., Wu, G., Miki, T., "End-to-end QoS provisioning in mobile heterogeneous networks", IEEE Wireless Comm., vol. 11, no. 3, pp.24-34, June 2004.
2. Lu, W.W., "Compact multidimensional broadband wireless: the convergence of wireless mobile and access", IEEE Comm. Mag., vol. 38, no. 11, pp. 119-123, Nov. 2000.
3. Jiang, H., Zhuang, W., "Quality-of-service provisioning to assured service in the wireless Internet", IEEE Global Telecomm. Conf., vol. 6, pp.3078-3082, 1-5 Dec. 2003.
4. Abedi, S., "Improved stability of QoS provisioning for 3G systems and beyond: optimum and automatic strategy selection for packet schedulers", IEEE Int. Conf. on Comm., vol. 4, pp. 1979-1985, June 2004.
5. Islam, M.M., Murshed, M., Dooley, L.S., "Enhanced cell visiting probability for QoS provisioning in mobile multimedia communications", Proc. of Int. Conf. on Information Technology: Coding and Computing, vol. 2, pp. 258-262, 5-7 April 2004.
6. Neto, C.S.S., Rodrigues, R.F., Soares, L.F.G., "Architectural description of QoS provisioning for multimedia application support", Proc. of Int. Conf. on Multimedia Modelling, pp.161-166, Jan. 2004.
7. Jha, S., Seneviratne, A., "Synchronisation skew: a QoS measurement study", Conf. on Local Computer Networks, pp.77-78, 18-20 Oct. 1999.
8. Chan, A.L., Law, K.L.E., "QoS negotiations and real-time renegotiations for multimedia communications", Proc. of Int. Conf. on Computer Comm. and Networks, pp. 522-525, Oct. 2002.
9. Pyun, J.-Y., Kim, Y., Jang, K.H., Park, J. Ae, Ko, S.-J., "Wireless measurement based resource allocation for QoS provisioning over IEEE 802.11 wireless LAN" IEEE Trans. on Consumer Electronics, vol. 49, no. 3, pp.614-620, Aug. 2003.
10. Bao, Y., Sethi, A.S., "OCP-A: an efficient QoS control scheme for real time multimedia communications", IEEE Global Telecomm. Conf., vol. 2, pp. 741-745, 3-8 Nov. 1997.
11. Hashimotot, K., Shibatat, Y., and Shiratorit, N., "Flexible Multimedia System Architecture with Adaptive QoS Guarantee Functions", Proc. of 7th Int. Conf. on Parallel and Distributed Systems, 2000, pp.119-126, July 2000.
12. Giunta, G., 'Quality of service assessment in new generation wireless video communications', in: Digital Image Sequence Processing, Compression, and Analysis, chap.5, pp.135-150, July 2004, CRC press.
13. Campisi, P., Carli, M., Giunta, G., Neri, A., "Blind Quality Assessment System for Multimedia Communication Using Tracing Watermarking", IEEE Trans. Signal Proc., vol. 51, no. 4, pp. 996-1002, 2003.
14. Campisi, P., Carli, M., Giunta, G., Neri, A., "Tracing watermarking for multimedia communication quality assessment", IEEE Int. Conf. on Comm., New York, 2002.

15. Campisi, P., Carli, M., Giunta, G., Neri, A., "Object based quality of service assessment for MPEG-4 videos using tracing watermarking", IEEE Int. Conf. on Image Proc., Rochester, NY, 2002.
16. Ileri, O., Mau, S.-C., and Mandayam, N. B., "Pricing for Enabling Forwarding in Self-Configuring Ad Hoc Networks", IEEE J. on Sel. Areas in Comm., vol. 23, no. 1, pp. 151-162, Jan. 2005.
17. Richardson, I.E.G., 'Transmission of Coded Video', in: Video Codec Design, chap. 11, 2002, J. Wiley & Sons.

Chapter II

Monitoring, Management And Configuration Of Networks And Networking Devices

A Critical View Of The Sensitivity Of Transit ASs To Internal Failures

Steve Uhlig* and Sébastien Tandel**

Computing Science and Engineering Department
Université catholique de Louvain, Belgium
E-mail:{suh,sta}@info.ucl.ac.be

(Invited Paper)

Abstract. Recent work on hot-potato routing [1] has uncovered that large transit ASs can be sensitive to hot-potato disruptions. Designing a robust network is felt as overly important by transit providers as paths crossed by the traffic have both to be optimal and reliable. However, equipment failures and maintenance make this robustness non-trivial to achieve. To help understanding the robustness of large networks to internal failures, [2] proposed metrics aimed at capturing the sensitivity of ASs to internal failures. In this paper, we discuss the strengths and weaknesses of this approach to understand the robustness of the control plane of large networks, having carried this analysis on a large tier-1 ISP and smaller transit ASs. We argue that this sensitivity model is mainly useful for intradomain topology design, not for the design the whole routing plane of an AS. We claim that additional effort is required to understand the propagation of BGP routes inside large ASs. Complex iBGP structures, in particular route-reflection hierarchies [3], affect route diversity and optimality but it an unclear way.

Keywords: network design, sensitivity analysis, control and data planes, BGP, IGP.

1 Introduction

Designing robust networks is a complex problem. Network design consists of multiple, sometimes contradictory objectives. This problem has been fairly discussed in the literature, in particular [4, 5]. Examples of desirable objectives during network design are minimizing the latency, dimensioning the links so as to accommodate the traffic demand without creating congestion, adding redundancy so that rerouting is possible in case of link or router failure and, finally, the network must be designed at the minimum cost. Recent papers have shown that large transit networks might be sensitive to internal failures. In [1], Teixeira et al. have shown that a large ISP network might be sensitive

---

  * Corresponding author. Steve Uhlig is "chargé de recherches" with the FNRS (Fonds National de la Recherche Scientifique, Belgium). This research was partly carried while Steve Uhlig was visiting Intel research Cambridge.
** Sébastien Tandel is funded by a grant from France Télécom.

to hot-potato disruptions. [6] extended the results of [1] by showing that a large tier-1 network can undergo significant traffic shifts due to changes in the routing. To measure the sensitivity of a network to hot-potato disruptions, [2] has proposed a set of metrics that capture the sensitivity of both the control and the data planes to internal failures inside a network.

To understand why internal failures are critical in a large transit AS, it is necessary to understand how routing in a large AS works. Routing in an Autonomous System (AS) today relies on two different routing protocols. Inside an AS, the intradomain routing protocol (OSPF [7] or ISIS [8]) computes the shortest-path between any pair of routers inside the AS. Between ASs, the interdomain routing protocol (BGP [9]) is used to exchange reachability information. Based on both the BGP routes advertised by neighboring ASs and the internal shortest paths available to reach an exit point inside the network, BGP computes for each destination prefix the "best route" to reach this prefix. For this, BGP relies on a "decision process" [10] to choose its a single route called the "best route among several available ones. The "best route" can change for two reasons. Either the set of BGP routes available has changed, or the reachability of the next-hop of the route has changed due to a change in the IGP. In the first case, it is either because some routes were withdrawn by BGP itself, or that some BGP peering with a neighbor was lost by the router. In the second case, any change in the internal topology (links, nodes, weights) might trigger a change in the shortest path to reach the next hop of a BGP route. In this paper we consider only the changes that consist of the failure of a single node or link inside the AS, not routing changes related to the reachability of BGP prefixes.

Our experience with the metrics proposed in [2] provided insight into the sensitivity of the network to internal failures. However, having used this sensitivity analysis on a large tier-1 network also revealed weaknesses of this model to understand the routing plane of transit ASs. We discuss further work required to help operators to understand the control plane of their network, particularly the behavior of iBGP.

Section 2 first presents the methodology to model an AS. Section 3 then introduces the sensitivity model to internal failures. Section 4 discusses the limitations of the model for realistic iBGP structures and Section 5 concludes and discusses further work in the area.

2 Methodology

In this section, we describe our approach to build snapshots of real ISP networks. The main point of our relatively heavy methodology is to make the model as easy as possible to match with the context of real transit ASs. We do not make assumptions on the internal graph of the iBGP sessions, even though in the case of GEANT there is a iBGP full-mesh between all border routers. Most large transit ASs do rely on route-reflection, hence putting assumptions on the sensitivity model restricts its applicability. The route solver on which we rely, C-BGP [11], has no restriction on the structure of the iBGP sessions inside an AS. C-BGP has been designed to help the evaluation of changes to the design of the BGP routing inside an AS [12]. Changes to the routing policies of an AS, or the internal configuration of its iBGP sessions is easy with C-BGP.

In this section we describe the methodology we use to model a transit AS. A more detailed discussion on how to model an AS with C-BGP can be found in [12]. We explain in this section how we build snapshots of the routing and traffic matrix of large ASs. The main steps of our methodology consist in producing snapshots of the routing inside a transit AS, consider that this view of the routing is valid for the whole bin between two time intervals, and based on the routing snapshot and the traffic statistics available to build a traffic demand for the considered time bin. Building more precise views of the routing and traffic matrices is possible by using small enough time intervals.

2.1 Related work

The most closely related works from the literature are [13] and [14]. The aim of [13] was to provide the networking industry with a software system to support traffic measurement and network modeling. This tool is able to model the intradomain routing and study the implications of local traffic changes, configuration and routing. [13] does not model the interdomain routing protocol though. [14] proposed a BGP emulator that computes the outcome of the BGP route selection process for each router in a single AS. This tool does not model the flow of the BGP routes inside the AS, hence it does not reproduce the route filtering process occurring within an AS. Finally, [12] discusses the problem of modeling the routing of an autonomous system and presents an opensource route solver aimed at allowing the study of networks with a large number of BGP routers.

2.2 Modeling the routing of an AS with C-BGP

To build snapshots of the routing inside a transit AS, we rely in this paper on C-BGP [11]. C-BGP is an open-source routing solver aimed at reproducing precisely BGP and making possible the modelling of networks containing a large number of BGP routers. The reason for choosing C-BGP instead of simply gathering BGP RIBs is that we aim at providing a tool that allows a network operator to build "what-if" scenarios, for instance by changing its topology or routing policies, and investigating their impact on the sensitivity of their network [12]. By relying on C-BGP, one can relatively simply change the internal network topology, the configuration of the routers, or even filter the set of BGP routes available inside the network.

To model a transit network with C-BGP, we use the following steps. These steps reflect the typical way we see how modeling the routing of an AS would be done, even though on particular network instances these steps might have to be slightly changed, depending on the available routing data. First, we infer the topology of the network based on its ISIS data. We create the internal POP-level topology and for each time bin, we read all the routing messages received during the time bin and inject them into C-BGP. For ISIS, we apply all link state packets (LSPs) so that both the reachability and the IGP weights at the end of the time bin are consistent with them. Note that replaying the events at their exact time is useless in C-BGP as the simulator is not even-driven, i.e. it only propagates the routing messages and lets BGP converge inside the BGP router. When only a single solution towards which the real BGP converges, C-BGP will find

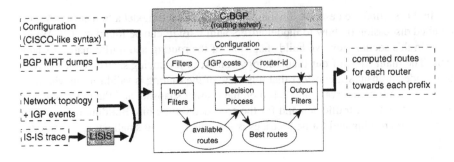

Fig. 1. C-BGP routing solver.

the same solution. However, when several distinct solutions exist in the real BGP [15], then it is unclear how C-BGP will behave. Sometimes is will converge towards of the the solutions, sometimes it will never converge. The model used by C-BGP provides scalability since it prevents from having to care with the computationally consuming state machines and timers of the real routing protocols.

We first inject the static topology of the transit network into C-BGP. This topology contains the routers, links, and IGP weights of the links. The shortest IGP paths inside each AS are computed at each router and BGP sessions are established between the routers. The ISIS data of the two networks up to the beginning of the actual time the simulation is supposed to start are then read and the LSPs converted into changes of the C-BGP topology. The IGP changes are injected into C-BGP and the shortest paths are recomputed at all routers.

For BGP, we restrict the set of prefixes to those largest ones, in a similar way to [16, 13, 17]. We infer by parsing the BGP RIBs where in the network the prefixes are announced by external peers. Depending on whether the transit AS relies on "next-hop-self" in its routers or not, the visibility of the eBGP routes might be different inside the iBGP sessions of the AS. It is important to understand that relying on BGP messages collected through iBGP sessions as often done in the literature implies that one does not know neither all the messages advertised by the peer routers of the network (eBGP), nor the full diversity of the BGP routes known by the routers. With the current data available today, one can only infer the state of the best routes of some router at some point in time given that it takes some time for messages to propagate inside iBGP. While this might be acceptable in some situations, not having the same set of routes as is available inside the BGP routers implies that simulating the failure of a router inside the network that is used as egress point by another router cannot be accurately simulated since the latter router will choose another best BGP route that one does not know based on the available BGP routing data.

3   Network sensitivity to internal failures

In this section and the following, we describe the main points of our version of the model proposed in [2]. A more detailed presentation of this model can be found in [18].

Let $G = (V, E, w)$ be a graph, V the set of its vertices, E the set of its edges, w the weights of its edges. A graph transformation δ is a function $\delta : (V, E, w) \rightarrow (V', E', w')$ that deletes/adds a vertex or an edge from G. In this paper we consider only the graph transformations δ that consist in removing a single vertex or edge from the graph. In practice, an AS may want to consider more complex failures corresponding to failures that are occur together for instance, or to consider only the failures that have actually occurred in the network for the last few weeks or months. For consistency with [2], we denote the set of graph transformations of some class (router or link failures) by ΔG. The new graph obtained after applying the graph transformation δ on the graph G is denoted by $\delta(G)$. In this paper, we restricted the set of graph transformations as well as the definition of a graph compared to [2], as we do not consider the impact of changes in the IGP cost. Changes to the IGP cost occur rarely in real networks.

To perform the sensitivity analysis to graph transformations, one must first find out for each router how graph transformations impact the egress point it uses towards some destination prefix p. The set of considered prefixes is denoted by P. The BGP decision process $dp(v, p)$ is a function that takes as input the BGP routes known by router v to reach prefix p, and returns the egress point corresponding to the best BGP route. The *region index set RIS* of a vertex v records this egress point of the best route for each ingress router v and destination prefix p, given the state of the graph G: $RIS(G, v, p) = dp(v, p)$.

We introduced the state of the graph G in the *region index set* to capture the fact that changing the graph might change the best routes of the routers. In AS that consist of a full-mesh of iBGP sessions, the impact of a node or router failure on the best route used by a particular router is straightforward to find out. Removing a node or link changes the shortest paths to reach the egress points. Simulating the decision process of the router is enough to predict the best route after the graph transformation. In more complex networks on the other hand, the failure of a link or node not only changes the shortest paths towards the egress points, but new routes might also be advertised in replacement of a previously known route. In such a case, the BGP convergence inside the AS must be replayed to know the exact outcome of the BGP decision process.

The next step towards a sensitivity model is to compute for each graph transformation δ (link or router deletion), whether a router v will change its egress point towards destination prefix p. For each graph transformation δ, we recompute the all pairs shortest path between all routers after having applied δ, and record for each router v whether it has changed the egress point for its best BGP route towards prefix p. We denote the new graph after the graph transformation δ as $\delta(G)$. As BGP advertisements are made on a per-prefix basis, the best route for each (v, p) pair has to be recomputed for each graph transformation. It is the purpose of the *region shift function H* to record the changes in the egress point corresponding to the best BGP route of any (v, p) pair, after a graph transformation δ:

$$H(G, v, p, \delta) = \begin{cases} 1, \ if \ RIS(G, v, p) \neq RIS(\delta(G), v, p) \\ 0, \ otherwise \end{cases}$$

The *region shift function H* is the building block for the metrics that will capture the sensitivity of the network to the graph transformations.

To summarize how sensitive a router might be to a set of graph transformations, the *node sensitivity* η computes the average *region shift function* over all graph transformations of a given class (link or node failures), for each individual prefix p:

$$\eta(G, \Delta G, v, p) = \sum_{\delta \in \Delta G} H(G, v, p, \delta) \cdot Pr(\delta)$$

where $Pr(\delta)$ denotes the probability of the graph transformation δ. Note that we assume that all graph transformations within a class (router or link failures) are equally likely, i.e. $Pr(\delta) = \frac{1}{|\Delta G|}, \forall \delta \in \Delta G$, which is reasonable unless one provides a model for link and node failures. Further summarization can be done by averaging the *vertex sensitivity* over all vertices of the graph, for each class of graph transformation. This gives the *average vertex sensitivity* $\hat{\eta}$:

$$\hat{\eta}(G, \Delta G, p) = \frac{1}{|V|} \sum_{v \in V} \eta(G, \Delta G, v, p)$$

The *node sensitivity* is a router-centric concept that performs an average over all possible graph transformations, measuring how much a router will change its egress point for its best routes after the graph transformations on average. Another viewpoint is to look at each individual graph transformation δ and measure how it impacts all routers of the graph. The *impact of a graph transformation* θ is computed as the average over vertices of the *region shift function*:

$$\theta(G, p, \delta) = \frac{1}{|V|} \sum_{v \in V} H(G, v, p, \delta)$$

The *average impact* of a graph transformation $\hat{\theta}$ summarizes the information provided by the *impact* by averaging it over all graph transformations of a given class:

$$\hat{\theta}(G, \Delta G, p) = \sum_{\delta \in \Delta G} \theta(G, p, \delta) \cdot Pr(\delta)$$

4   Discussion

Our use of the sensitivity model sketched in section 3 on different networks has shown the interest and limitations of this model. The first thing to be noted is that the model is very "hot-potato centric". For networks that do not have a complex iBGP structure (no route-reflection), the model reveals the critical links and routers [18]. That is, the model captures the sensitivity due to the concentration of many internal paths that will change after a graph transformation. However, using the model for complex transit networks using route-reflection is more tricky. Contrary to the case of an AS with an iBGP full-mesh, route-reflection introduces opacity into the selection of the best BGP route by a router. In the case of an iBGP full-mesh, each ingress router, i.e. a router that is not itself an exit point inside the AS for the considered destination prefix, chooses as its best route the one that is the "best" one among all the routes advertised by the external peers of the AS. If the routing policies applied inside the AS are consistent among all internal routers, then for each ingress router there exists only one best BGP route that will be chosen by this router, and the choice of this route will not depend on the best route choice of the other routers of the AS. In the case of route-reflection, the best route chosen by a router depends on the best route choice of the route reflectors on the BGP routes propagation path inside the AS.

When an iBGP full-mesh is used, it is relatively straightforward to predict the outcome of an internal change on the best route choice performed by BGP. Among several alternative routes having the same quality for BGP[1], the cost of the IGP path to reach the exit point inside the AS will be used to decide which BGP route will be considered as best. Hence in an iBGP full mesh, there is a direct relationship between the IGP shortest paths and the best route choice made by BGP. For instance, Figure 2 shows the internal topology of a simple transit AS relying on an iBGP full-mesh. Inside the iBGP, BGP routers only propagate to other iBGP peers routes they did not receive from an iBGP peer. We do not show all the iBGP sessions on Figure 2, but all routers have an iBGP session with every other router of the topology. We also consider a single destination prefix p. The transit AS has three ingress routers $i1$, $i2$, and $i3$, two egress routers $e1$ and $e2$, and a router located in the middle of the topology. All routers run both the IGP and BGP as in most ASs. The arrows on Figure 2 provide the choice of the best route by BGP towards p at each router of the AS. On Figure 2, two eBGP routes are learned to reach prefix p, one through egress router $e1$ and another through egress router $e2$. With an iBGP full-mesh, $i1$ and $i2$ will use the BGP route learned through $e1$, while $i3$ will use the BGP route learned through $e2$, both due to the IGP cost rule of the decision process. Each router hence relies for its best BGP route the smallest IGP cost path to exit the network.

Now suppose that our AS relies on route-reflection [3] as shown on Figure 3. The sole difference between Figure 2 and Figure 3 in terms of the routing configuration concerns the iBGP sessions. With route-reflection, all BGP routers are not directly connected anymore by an iBGP session, but all border routers are clients of the route-reflector RR. In this case, each routers knows the routes it learned from eBGP sessions, and a single route advertised by RR. The issue with route-reflection is that the best route chosen by RR depends on its own IGP cost to reach the exit point inside the AS, not the one of its clients. RR will choose as its best BGP route the one learned through $e1$ since its IGP cost is smaller than the one through $e2$. In that case, $i1$, $i2$ and $i3$ will all choose as their best route to reach p the route advertised by RR, and thus will use $e1$ as their exit point towards p. The IGP cost of the best route chosen by $i3$ is not optimal in terms of the IGP cost anymore compared to the iBGP full-mesh. Here we used a simple situation with a single route-reflector. In practice, large transit ASs can rely on a hierarchy of route-reflectors. Route-reflection trades-off the number of iBGP sessions inside the AS with a drastic reduction in the diversity of the routes known by the routers, and by a potentially suboptimal IGP cost of the best routes chosen by the routers.

A route reflector today chooses for all its clients routers a single best BGP route. The impact of this choice on the client routers is that the latter's will be sensitive to the graph transformations that affect the path of this best route chosen by their route reflector. Depending on how these client routers are located inside the AS, a particular

---

[1] Same value of local-pref, AS path length, MED, and route origin type.

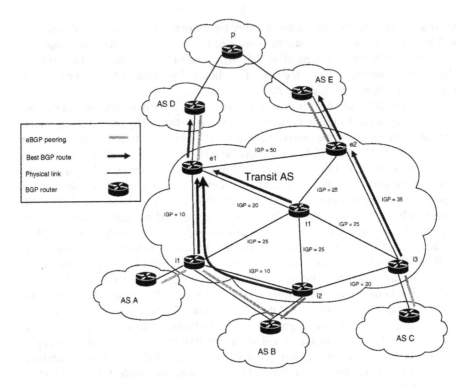

Fig. 2. Example topology with iBGP full-mesh.

subgraph of the AS will be used by the clients to reach the egress point of the BGP route advertised by the route reflector.

The implication for the network sensitivity is that the impact of the graph transformations and the node sensitivity will depend very much on how route-reflectors redistribute the routes they learn to their clients, and which routes they consider as best. If each route reflector was to compute for each of its client the latter would have selected in the case of an iBGP full-mesh and redistribute this route on a per-client basis as proposed in [19], then the sensitivity of the network in the two cases would be the same. However, if route reflectors are left to redistribute their best route without respect to how their clients would have chosen their best route in the iBGP full-mesh, then it is difficult to foresee how this will change the paths used to carry the traffic inside the AS since it depends both on the placement of the route reflectors, the interconnection structure of the iBGP sessions, the diversity of the BGP routes learned from peer ASs, and wherefrom the external routes are learned.

Our experience with a large tier-1 network have shown that the results of the sensitivity analysis between the actual network configuration and an iBGP full mesh are quantitatively much different but qualitatively very alike. However, understanding why differences in the sensitivity arise when the iBGP structure is changed is not answered by the sensitivity model of [2]. For that, one must first understand what filtering is

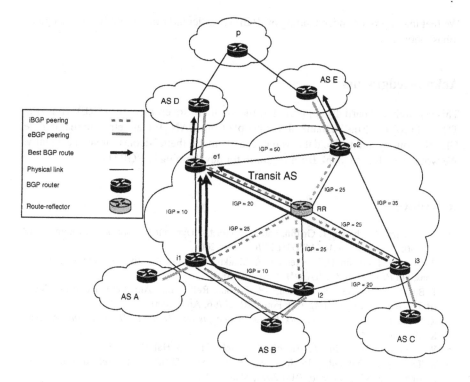

Fig. 3. Example topology with route reflection.

performed by a given iBGP structure compared to the iBGP full mesh. The reason to compare with an iBGP full mesh is that the best route visibility is achieved by the full mesh, while introducing route reflection reduces the diversity of the routes that can be chosen by the BGP routers inside the AS.

5 Conclusions and further work

In this paper we have presented our version of the sensitivity model to internal failures based on the work of [2]. We sketched our methodology to study the sensitivity of an AS based on this model and discussed the limitations of this model to provide insight into the behavior of the control plane of large transit ASs due to the presence of route reflectors.

The obvious further work we see is to study the actual impact of route reflection on the diversity and sensitivity of the routers inside a large transit AS. We are currently investigating this problem on a large tier-1 network. The lack of study in the literature concerning the actual impact of route reflection on the diversity and the filtering of the routes inside an AS indicates that our current understanding of iBGP is still very poor.

We feel that a proper understanding of iBGP is particularly important for the design of robust networks.

Acknowledgments

This research was partially supported by the Walloon Government (DGTRE) within the TOTEM project [20]. We thank all the people from DANTE that helped in making the GEANT routing and traffic data available, and among them Nicolas Simar specifically. We would also like to thank Bruno Quoitin for developping C-BGP [11].

References

1. R. Teixeira, A. Shaikh, T. Griffin, and J. Rexford, "Dynamics of hot-potato routing in IP networks," in *Proc. of ACM SIGMETRICS*, June 2004.
2. R. Teixeira, T. Griffin, G. Voelker, and A. Shaikh, "Network sensitivity to hot potato disruptions," in *Proc. of ACM SIGCOMM*, August 2004.
3. T. Bates, R. Chandra, and E. Chen, "BGP Route Reflection - An Alternative to Full Mesh IBGP," Internet Engineering Task Force, RFC2796, April 2000.
4. R. S. Cahn, *Wide Area Network Design: Concepts and Tools for Optimisation*, Morgan Kaufmann, 1998.
5. W. D. Grover, *Mesh-Based Survivable Networks*, Prentice Hall PTR, 2004.
6. R. Teixeira, N. Duffield, J. Rexford, and M. Roughan, "Traffic matrix reloaded: impact of routing changes," in *Proc. of PAM 2005*, March 2005.
7. J. Moy, *OSPF : anatomy of an Internet routing protocol*, Addison-Wesley, 1998.
8. D. Oran, "OSI IS-IS intra-domain routing protocol," Request for Comments 1142, Internet Engineering Task Force, Feb. 1990.
9. J. Stewart, *BGP4 : interdomain routing in the Internet*, Addison Wesley, 1999.
10. Cisco, "BGP best path selection algorithm," http://www.cisco.com/warp/public/459/25.shtml.
11. B. Quoitin, "C-BGP, an efficient BGP simulator," http://cbgp.info.ucl.ac.be/, September 2003.
12. B. Quoitin and S. Uhlig, "Modeling the routing of an Autonomous System with C-BGP," *IEEE Network Magazine*, November 2005.
13. Anja Feldmann, Albert Greenberg, Carsten Lund, Nick R eingold, and Jennifer Rexford, "NetScope: Traffic Engineering for IP Networks," *IEEE Network Magazine*, March 2000.
14. N. Feamster, J. Winick, and J. Rexford, "A model of BGP routing for network engineering," in *Proc. of ACM SIGMETRICS*, June 2004.
15. T. Griffin and G. Wilfong, "An analysis of BGP convergence properties," in *Proc. of ACM SIGCOMM*, September 1999.
16. A. Feldmann, A. Greenberg, C. Lund, N. Reingold, J. Rexford, and F. True, "Deriving traffic demands for operational IP networks: methodology and experience," in *Proc. of ACM SIGCOMM*, September 2000.
17. J. Rexford, J. Wang, Z. Xiao, and Y. Zhang, "BGP Routing Stability of Popular Destinations," in *Proc. of ACM SIGCOMM Internet Measurement Workshop*, November 2002.
18. S. Uhlig, "On the sensitivity of transit AShs to internal failures," in *Proc. of the fifth IEEE International Workshop on IP Operations and Management (IPOM2005)*, Barcelona, Spain, October 2005.

19. O. Bonaventure, S. Uhlig, and B. Quoitin, "The case for more versatile BGP Route Reflectors," Work in progress, draft-bonaventure-bgp-route-reflectors-00.txt, July 2004.
20. "TOTEM: a TOolbox for Traffic Engineering Methods," `http://totem.info.ucl.ac.be/`.

The CoMo Project: Towards A Community-Oriented Measurement Infrastructure

Gianluca Iannaccone and Christophe Diot

Intel Research, Cambridge, UK

(Invited Paper)

Abstract. CoMo (Continuous Monitoring) is a passive monitoring system. CoMo has been designed to be the basic building block of an open network monitoring infrastructure that would allow researchers and network operators to easily process and share network traffic statistics over multiple sites. This paper identifies the challenges that lie ahead in the deployment of such an open infrastructure. These main challenges are: (1) the system must allow any arbitrary traffic metric to be extracted from the packet streams, (2) it must provide privacy and security guarantees to the owner of the monitored link, the network users and the CoMo users, and (3) it must be robust to anomalous traffic patterns or traffic query loads. We describe the high-level architecture of CoMo and, in greater detail, the resource management, query processing and security aspects.

Keywords: Network measurements and monitoring.

1 Introduction

Despite the great interest of recent years in measurement based Internet research, the number and the variety of data sets and network monitoring viewpoints remains unacceptably small. Several players in the network measurement area, like CAIDA [2], NLANR [18], RouteViews [23], RIPE [21], Internet2 [13] and, recently, GEANT [9] have provided data sets with various degrees of accuracy and completeness. Some large commercial ISPs also deploy private infrastructures and share limited information with the rest of the research community [6, 8]. Other network operators (e.g., corporate networks, stub ISPs, Universities) instead tend to lack a measurement infrastructure or, in case they do have one, do not share any data or even report the existence of such infrastructure.

This situation constitutes a major obstacle for network researchers. It hampers the ability to validate the results over a wide and diverse range of datasets. It makes it difficult to generalize results and identify the presence of traffic characteristics that are invariant and common to all networks. For example, it is difficult to quantify the magnitude of a denial of service attack or a worm infection, to evaluate the relevance of network pathologies such as in-network packet duplication and reordering, or to simply identify the dominant applications in the Internet.

So far, several barriers have limited the ability of researchers to deploy and share large number of network datasets:

- The cost of the monitoring infrastructure that should be present on a large set of links with speeds varying from the few Mbps of small organizations up to the 10Gbps of ISPs' networks.
- The lack of software tools for managing a passive monitoring infrastructure. Today, monitoring systems use ad-hoc tools that are not appropriate for large infrastructures. For example, they tend to provide poor and inefficient query interfaces.
- The lack of a standard open interface to access the monitoring system. Although, several efforts exist to provide a single packet trace format (e.g., tcpdump/libpcap [22], IETF IPFIX working group [14]), no standard exists to access high speed network monitoring devices.
- The difficulty in controlling the access to the data avoiding to disclose private information about network users or organizations.

We have designed CoMo, an open software platform for passive monitoring of network links. CoMo has been designed to be flexible, scalable, and to run on commodity hardware. With CoMo, we intend to lower the barriers described above and encourage the deployment of a large scale monitoring infrastructure. We believe this effort constitutes a necessary first step towards a better understanding of network protocols and traffic. For example, the extensible nature of CoMo allows early deployment of novel methods for traffic analysis, anomaly diagnosis or network performance evaluation.

The rest of the paper describes the challenges posed by the design of the CoMo platform, its resulting architecture, and some specific aspects of CoMo such as the query engine, resource management and security.

2   Challenges

The design of a system such as CoMo is driven by the trade-off between openness and resilience. The system should be open enough to allow authorized users to compute allowed metrics on the observed traffic. At the same time, it should allow link owners to specify constraints on how the collected data can be used. Finally, the system should be robust to abuses by unauthorized users, and have a predictable and graceful performance degradation in periods when it cannot sustain the incoming packet rate, We can summarize the system requirements as follows:

Openness. The users should be able to customize the system and the software platform to their specific needs and deployment environment. For example, a user interested in intrusion detection or performance analysis may only need limited storage; on the other hand, an ISP's network operator interested in post-mortem analysis might require to store and retrieve a large and detailed packet-level or flow-level information.

The metrics also need to be dynamically configurable in order to address a large range of applications such as network trouble-shooting, anomaly and intrusion detection, SLA computation, etc. In addition, the system should allow users to easily interact and interface with it in order to start or stop some metric computation or to run a query on the collected datasets.

Resilience. This requirement is often orthogonal to the previous one. First and most important, the system should be able to monitor *and* analyze the traffic in any load condition, and in particular in the presence of unexpected traffic anomalies that may overload the system resources. The system should control its resources carefully. For example, the computation of one metric should not monopolize the use of resources, starving other crucial system tasks (e.g., packet trace collection). The owner of the system needs to be able to control the access to the system. Various type of users will want to access a monitoring system, including malicious ones. Because of its exporting capabilities, a system can impact the network it measures. Different request will be processed with different priorities. The system should also make sure it does not compute the same metric twice for two different users or applications.

2.1 Design Challenges

Given the requirements described above, we identify four main design challenges:

Ease of deployment. The success of the CoMo infrastructure will be a function of how simple it is for user to access the infrastructure, specify and implement traffic metrics and analysis methods, query the data from the system. As we will point out in Section 3, many of the design decisions are driven by the need to trade architecture simplicity for efficiency and performance.

Query interface. Designing a query interface with the constraints defined in the requirement section poses two problems: (i) how to express the query; (ii) how to run the query without explicit built-in support into the system. Expressing a query may be particularly hard given that the system is supposed to be used by a large range of users, defining new traffic metrics and analysis methods. It is unlikely that all metrics will be specified well enough to be translated in a standard query language; metrics may require new constructs that are not present in the original query language. However, the system should still allow custom-built queries to run. We will address this problem in Section 4.

Resource Management. Opening the system to a potentially large number of users requires a very careful resource management, i.e. CPU, memory, I/O bandwidth or storage space. Indeed, allowing users to compute any metric on the traffic stream may result in analysis that are particularly computing intensive, and in a large amount of data to be exported. Therefore, the system needs to define strict policies for analysis metrics and needs to be able to enforce them and possibly to adapt based on the load of the system. We will address the problem of resource management in Section 5.

Security issues. CoMo users will have different rights on the system depending also on the system environment. For example, a network operator may be allowed to inspect the entire packet payload in order to spot viruses or worms, while a generic user may only be able to access the packet header (probably anonymized). A second security aspect is related to the vulnerability of the monitoring system. CoMo systems contain confidential information. They can also export large amounts of traffic and impact the network where they are installed. An attacker may also target directly the CoMo system by preventing users to access the system or by corrupting the collected data. In Section 6 we will describe the threat model for CoMo and propose initial solutions.

3   Architecture

This section presents a high-level description of the architecture and an overview of the major design choices. We call *data* any measurement related information. Data include original packets captured on the monitored links, as well as statistics computed on the packets and other representations of the original packet trace such as flow records.

3.1   High level architecture

The system is made of two principal components. The *core processes* control the data path through the CoMo system, including packet capture, export, storage, query management and resource control. The *plug-in modules* are responsible for various transformations of the data.

The data flow across the CoMo system is illustrated in Figure 1. The white boxes indicate plug-in modules while gray boxes represent the core processes. On one side, CoMo collects packets (or subsets of packets) on the monitored link. These packets are processed by a succession of core processes and end stored onto hard disks. On the other side, data are retrieved from the hard disk on user request (by the way of queries addressed to a CoMo system). Before being exported to users, those data go through an additional processing step.

As explained earlier, the modules execute specific tasks on data. The core processes are responsible for the "management" operations, common to all modules (e.g., packet capture and filtering, data storage). The following tasks also fall under the responsibility of the core component: (*i*) resource management, i.e., deciding which plug-in modules are loaded and running, (*ii*) policy management to manage the access privileges of the modules, (*iii*) on-demand requests handling to schedule and respond to user queries, and finally (*iv*) exception handling to manage the situation of traffic anomalies and the possible graceful degradation of system performance.

The modules take data on one side and deliver user-defined traffic metrics or processed measurement data on the other side. One of the challenges identified

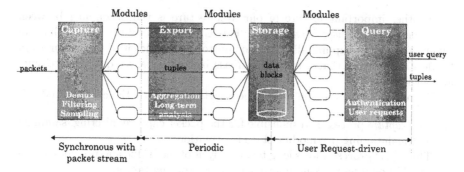

Fig. 1. Data flow in the CoMo system

in Section II is to keep modules very simple. All complex functions should be implemented within the core component. This strict division of labor allows us to optimize the core component[1], while the modules can run sub-optimally and can be implemented independently by any CoMo user.

3.2 The core processes

The core processes are in charge of data movement operations (i.e., from the packet capture card to memory and to the disk array). Moving data in a PC is the most expensive task given memory, bus and disk bandwidth limitations. Therefore, in order to guarantee an efficient use of the resources, it is better to maintain a centralized control of the data path. However, one of the goal of the architecture is to allow the deployment of CoMo as a cluster using dedicated hardware systems (such as network processors) for high performance monitoring nodes.

Communication between core processes is governed by a unidirectional message passing system to enable the partition of functionality over a cluster.

In a single system, a CoMo node uses shared memory and Unix sockets for the signaling channel. The use of processes instead of threads is justified by the need of high portability of the software over different operating systems.

Two basic guidelines have driven the assignment of the functionalities among the various processes. First, functionalities with stringent real-time requirements (e.g., packet capture or disk access) are confined within a single process (*capture* and *storage*, respectively). The resources assigned to these processes must be able to deal with worst case scenarios. Other processes instead operate in a best-effort manner (e.g., *query* and *supervisor*) or with less stringent time requirements (e.g., *export*). Second, each hardware device is assigned to a single process. For example, the *capture* process is in charge of the network sniffer, while *storage* controls the disk array.

---

[1] The CoMo code is open source and we aim to build an open community of developers in charge of the core components.

Another important feature of our architecture is the decoupling between real-time tasks and user driven tasks. This is visualized by the vertical lines in Figure 1. This decoupling allows us to control more efficiently the resources in CoMo and to avoid that a sudden increase in traffic starves query processing, and vice-versa.

We now describe the five main processes that compose the core of CoMo:

- The *capture* process is responsible for the packet capture, filtering, sampling and maintaining per-module state information;
- The *export* process allows long term analysis of the traffic and provides access to additional networking information (e.g., routing tables);
- The *storage* process schedules and manages the accesses to the disks;
- The *query* process receives user requests, applies the query on the traffic (or reads the pre-computed results) and returns the results;
- The *supervisor* process is responsible for handling exceptions (e.g., process failures) and to decide whether to load, start or stop plug-in modules depending on the available resources or on the current access policies.

The *capture* process receives packets from the network card (that could be a standard NIC card accessed via the Berkeley Packet Filter [16], or using dedicated hardware such an Endace DAG card [7]). The packets are passed through a filter that identifies which modules are interested in processing the packets. Then the *capture* process communicates with the modules to have them process the packets and update their own data structures. Note that those data structure may also be maintained by the *capture* process in order to keep the module simple.

Periodically, *capture* polls the data structures updated by the modules and sends its content to the *export* process. These data structures are then ready to be processed again by the modules. As explained earlier, this way, we decouple real-time requirements of the *capture* process that deals with incoming packets at line rate from storage and user oriented tasks. Flushing out the data structure periodically also allows *capture* to maintain limited state information and thus reduce the cost of insertion, update and deletion of the information stored by the modules.

The *export* process mimics the behavior of *capture* with the difference that *export* handles state information rather than incoming packets. Therefore, *export* communicates with the modules to decide how to handle the state information. A module can request *export* to store the information and/or to maintain additional, long term information. Indeed, as opposed to *capture*, *export* does not flush periodically its data. Instead, it needs to be instructed by the module to get rid of any data.

The *storage* process takes care of storing *export* data to the hard disk. The *storage* process is data agnostic and treats all data blocks equally[2]. It can thus

---

[2] A viable alternative is to allow *storage* to filter some of the data blocks as early as possible to reduce data movement and processing, following an approach similar to Diamond [12].

focus on an appropriate scheduling of disk accesses and on managing disk space. The *storage* process understands only two type of requests: "store" requests from *export* and "load" requests from *query*.

The *query* process manages users' requests, and if access is granted to the user, it gets the relevant data from disk via the *storage* process and returns the query results to the user. If the requested data is not available on disk, the *query* process can (*i*) perform the analysis on the packet trace stored on the disk (if the request refers to a period in the past) or (*ii*) request the initialization of a new module by the *supervisor* process to perform the query computation on the incoming packet stream.

Finally, the *supervisor* monitors the other processes and decides which modules can run based on the available resources, access policies and priorities. The *supervisor* communicates with all the other processes to share information about the overall state of the system.

3.3 Plug-in Modules

The traffic metrics and statistics are computed within a set of plug-in modules. The modules can be seen as a pair *filter:function*, where the filter specifies the packets on which the function should be performed. For example, if the traffic metric is "compute the number of packets destined to port 80", then the filter would be set to capture only packets with destination port number 80, while the function would just increment a counter per packet. Note that all modules do not necessarily compute statistics. Modules can simply transform the incoming traffic, like for example transform a set of packets in a flow record.

The core processes are responsible for running the packet filters and communicate with the modules using a set of callback functions. Actually there are several sets of callback functions, one for each of the core processes (represented by the three columns of white boxes in Figure 1). Going back to the previous example, the *capture* process will use a callback (`update()`) to let the module increment the counter. Then the *export* process will use a different callback (`store()`) to move the counter value to the disk. Also, *export* could use another callback to allow the module to apply, for example, a low pass filter on the counter values. The *Query* process will then use a different callback (`load()`) to retrieve the counter value from disk.

It is important to observer that the core processes are agnostic to the state that each module computes. Core processes just provide the packets to the module and take care of scheduling, policing and resource management tasks.

4 Querying Network Data

The query engine is the CoMo gateway to the rest of the world. The main function of queries is to request CoMo to export data. The range of data to be exported can vary significantly, from raw packet sequences to aggregated traffic statistics.

The processing of a query can be divided in three steps: (i) validate and authorize the query (as well as its origin), (ii) find and/or process the data and, finally, (iii) send the data back to the requester.

The amount of data stored on a CoMo system can be very large (in the order of 1TB on current prototypes). It is thus desirable to reduce the amount of processing needed to answer a query to a minimum. This is, indeed, the main purpose of the CoMo modules: pre-compute data to minimize the cost of processing incoming queries.

In CoMo we identify three types of queries:

- *Triggers* defined in the system configuration together with the relevant module. This kind of query will appear in the form `'send-to <IP address>:<port>'` and follow a push information model, i.e. as the module computes the metric on the traffic stream, data is sent to the specified IP address to trigger new computation or an alert.
- *On-demand queries* explicitly specify the relevant module. This could happen in two ways: indicating the name of the module in the query itself or sending directly the module source code. The query would then have to indicate the packet filter to be applied to the packet stream and the time window of interest. The response consist in the output of the module. On reception of this query, the CoMo system has to authenticate the module and the requester, and then figure out if the same module has already been installed. If the same module has been running during the time of interest, then this query revert to a static query. Otherwise, it requires the module to run on the stored packet-level trace[3] with an obvious impact on the query response time.
- *Ad-hoc queries* have no explicit module defined. These queries are written in a specific query language and code for the modules is generated on the fly. A similar approach is followed in systems like Aurora [3], TelegraphCQ [4], Gigascope [6] or IrisNet [10]. The caveat of this kind of approach is that it cannot always exploit existing modules that have been running on the packet stream. One solution is to have each module generate a common representation of the data it has computed so that the query process can compare the received query with all the modules to find a module that is computing a *super-set* of the response. One common representation could be a packet stream that all modules are designed to interpret correctly. Therefore, each module could include routines to "regenerate" a packet stream that resembles the one originally monitored. If the partial information contained in this regenerated stream is sufficient for the new query then it can run without any modification.

One of the big challenges of the system is to manage queries both in terms of computing resources and data transfers. Processing a query should not jeopardize

---

[3] Note that every CoMo system is supposed to keep a packet-level trace at all times. The duration of the packet trace will depend on the available storage space and the link speed.

the ability of the system to collect the packet-level trace without losses. Thus, it is important for the system to predict the usage of resources of a query before scheduling it to run. This can be easily done for trigger queries (the module is known in advance to the system), while it is more challenging for other type of queries.

Moreover, the system should take into account the priority of queries and that of all the other modules in the system that are pre-processing data for future queries. Indeed, some queries might be urgent (e.g., real-time security related information) and it might be appropriate not to delay the computation of the query (at the cost of other CoMo tasks).

Finally, it is most probable that multiple CoMo systems will be present in a network. Consequently, more than one system may be needed to answer a query or multiple systems may coordinate to identify the most appropriate subset of them over which to run the query. For example, a system could be more appropriate than others depending on the relevance of the monitored traffic for the query, (e.g., when tracking a denial of service attack) or the current system load. Moreover, some of these systems will have data to export, some will not. Therefore, an additional challenge is to design a query management system that minimizes the search and export cost in the context of distributed queries. For this specific challenge, we will also investigate innovative solutions for query management proposed in the context of sensor networks [1, 15, ?].

5   Resource Management

Managing system resources (i.e. CPU, hard disk, memory) in CoMo is challenging because of two conflicting system requirements. On the one hand, the system should be open to users to add plug-in modules for new traffic metrics and to query the system. On the other hand, the system needs to be always available, guarantee a minimum performance level and compute accurately the metrics that are of interest at a given time.

We divide the major causes of resource consumption in three categories: traffic characteristics, measurement modules, and queries. In the following we will address each of them separately.

Traffic characteristics. Resource utilization depends heavily on the incoming traffic, affecting both the core processes, which have per-packet processing costs, and the modules, whose consumption of storage, CPU cycles and (indirectly) disk I/O usage varies depeding on traffic characteristics and the type of metrics computed.

For example, a module could be idle for long periods of time and then have a burst of activity when packets hit its filter, possibly resulting in turn in large amounts of data to be stored to disk. A module computing flow statistic would become very memory-greedy in case of DoS attack. The amount of computations per packet will often depend on the packet content, e.g. in the case of IDS's.

Modules. Apart from the fixed per-packet processing costs, which only depend on the input traffic flows, most of the resource utilization costs depend on the activity of the modules. To control the resources used by modules, we impose a number of constraints on their operation:

- *Modules can be started and stopped at any time.* Modules are prioritized based on resource consumption and managed accordingly. We also rely on the fact that most module can be run off-line at a later time on the packet trace.
- *Modules have access to a limited set of system calls.* They cannot allocate memory dynamically and have no direct access to any I/O device. This allows us to maintain the control over the resource usage within the core processes and, at the same time, it keeps the module's source code simple.
- *Modules do not communicate with each other.* Modules are independent. They do not share any information with other modules. This constraint simplifies resource management, although it introduces redundancies. For example, one module cannot pass pre-processed information to another module. This way, different modules might perform the same computations on the same packets.

These three module requirements allow the CoMo system to regulate the amount of computing resources used at a given time. However, the decision on whether to start or stop a module depends on two parameters: (i) the measured resource usage of the module and (ii) the relevance of the metric computed by the module. The optimization of the sets of modules run at a given time is also a challenging issue. It is very difficult to estimate the resource usage (that depends on incoming traffic) and the relevance of a given metric can also vary significantly over time.

Queries. Queries can come at anytime and cause resource consumption for authentication, module insertion or removal, data processing. The query engine will have to manage the impact of the query processing on the system resources and active modules. The main issue with queries is that they should have higher priority than existing modules. A query indicate that a user exists and is waiting for results, while modules are just pre-computing queries based on the assumption that some users will be interested. On the other hand, in presence of a large number of queries, running them at high priority may force the CoMo system to be a simple packet collector, reducing the usefulness of the modules.

5.1 Resource control options

As described above, the prediction of resource utilization is almost impossible given the different factors that affect it. Therefore, we have no other option than accurately measuring resource utilization and timely react to overload situations. We need to identify what "knobs" can be turned to control resource utilization.

In CoMo, resource management is threshold-based. The resource usage of a CoMo system is monitored continuously in term of CPU cycles, memory usage and disk bandwidth. If the usage exceeds the pre-defined high-threshold, the following actions can be taken: (i) sample the incoming traffic for certain modules; (ii) stop some modules; (iii) block incoming queries.

The specific action to be taken should depend on the module priorities. As we said earlier, most modules can be delayed and run at a later time on the stored packet trace. The priority of a module should depend on the requested response time of the potential query that needs the data computed by the module. For example, a module computing a metric for anomaly detection should have high priority given that a query for anomalies requires in general a very short response time.

The module priority should then adapt to the query currently running on the system as well as on the historical queries that the system has received. At configuration time, the system administrator will define a static priority for each module that will then vary depending on how often the data processed by a module are actually read.

6 Security Issues

There is no doubt that the greatest challenge in providing an open monitoring service to users consists in enforcing access policies and safeguarding the privacy of network users.

One static policy applied to all systems is not enough given the large number of different uses that we envision for the monitoring infrastructure. For example, the network operator that owns the monitored link may have complete access to the traffic, including the payload. Other users should instead only be allowed to view the packet header or even just an anonymized versions of it. Finally, some queries could be limited to a subset of the users in order to avoid a constant overload of the system or to increase network traffic (e.g., all queries that require large data transfers). Hence, a rich and descriptive policy language is needed. The policy should define which modules can be plugged into the system, which users can plug modules in, and which users can post queries to the system together with the type of queries they can post.

6.1 Access policies

The only access points to the system for users is the plug-in interface where new modules are added and the query interface. However, note that a query always results in a module being plugged into the system. Therefore, in the rest of this section we will consider only the case of a user requesting to plug in a module.

The module is described by the two components *filter:function*. It is possible to assign *access request levels* to each component. These access requests levels indicate the privilege level at which the module has to run (and that has to match the user access privileges). For example, a filter that specifies "anonymized packet headers" will have the lowest access request level, while a filter that desires to have access to "not anonymized packet payloads" will be marked with the highest access request level.

Assigning access requests level for the *functions* of the modules is a much harder problem. It requires a deep understanding of what the module is computing and storing to disk. The solution that is often adopted is to allow the function to perform any computation but to restrict significantly the filter. NLANR, for example, provides only anonymized packet header traces but does not impose any condition on what user do with the traces [18]. Unfortunately, this approach is not appropriate in general. For example, worm signature detection requires full inspection of the packet stream although the state information it maintains (worm traffic) has little relevance and would certainly have a low access request level.

The approach followed by CoMo is to allow to load on the system only "signed" modules, for which the original developer can be authenticated. Then the module's function will inherit the developer privileges. User access privileges will initially depend on the CoMo system itself: (*i*) public access system will allow any user to plug-in modules and query the system; (*ii*) restricted system where only a subset of the users are allowed to plug-in new modules and the rest of the users can only perform queries on metrics for which a module already exists; (*iii*) private access, where only a subset of users can plug-in modules and query the system. In the future, we envision that each user will have individual privilege levels that will decide whether a *filter:function* pair is allowed to run on CoMo.

6.2 Infrastructure Attacks

So far, we have only addressed the security of the data. We now discuss the possible attacks on the monitoring infrastructure. We consider two types of attacks:

Denial of service attacks. Attacks in this class may come in the form of a module that uses a disproportionate amount of resources or that corrupts the data owned by other modules. The former type of attacks could be dealt directly with the resource manager and the use of "module black list" to forbid a module to run again in the system. Also, the use of a sandbox or of signed modules may help in avoiding this class of attacks and finding out a module's real "intentions". In fact, because modules process incoming traffic, on which we have little if any control, even a perfectly legal module could be driven into consuming large amount of resources in response to certain input patterns. In general this problem is dealt with by the generic resource control mechanisms discussed in Section 5.

The second class of attacks (data corruption) is harder to defend against. One first immediate solution is to provide memory isolation between modules. This can be achieved running modules as separate processes or moving some of the CoMo functionalities in the kernel. This introduces some overhead on the system but would guarantee that two modules will not interfere with each other.

Attack on the access policies. This class of attacks include attacks on the user privileges or on the access request level of a filter or function. For example, one can envision an attack on the packet anonymization scheme that would allow a user with low privileges to run a filter with high access level. An attack on the anonymization scheme could consist in sending carefully-crafted packets to the system and use a module that captures the anonymized version of those packets in an attempt to break the anonymization scheme [24].

7 Related Work

The list of software and techniques for active monitoring and network performance metrics computation is long. NIMI [20] is the pioneer in the deployment of active monitoring infrastructures, while CAIDA [2] has made available a large set of tools for active monitoring.

The area of passive monitoring is much less rich than active monitoring, mostly because of the deployment constraints of passive monitoring systems (i.e. active monitoring systems can be deployed at the edge of networks, when passive monitors must be deployed inside networks). The first generation of passive measurement equipment has been designed to collect packet headers at line speed on an on-demand basis. This generation of monitoring systems is best illustrated by the OC3MON [18], Sprint's IPMON [8] or NProbe [17] experience. Pandora [19] allows to specify monitoring components and this way provides greater flexibility in specifying the monitoring task. However, it differs from our approach in that it enforces a strict dependency among components and does not allow to dynamically load/unload some components.

Routers also embed monitoring software such as for example Cisco's Netflow [5]. Netflow collects flow level statistics on router line cards. Given the severe power and space constraints on routers, Netflow cannot store large amount of records but it exports all the information it capture to an external collector. This forces network operators to apply aggressive packet sampling (in the order of 0.1%) to reduce the data transfer rate from the routers.

Recently, the database community has approached the problem of Internet measurements. Several solutions have been proposed that deploy stream databases techniques. AT&T's Gigascope [6] is an example of a stream database that is dedicated to network monitoring. The system support a subset of SQL but it is proprietary and no measurement data is made publicly available. Other systems such as Telegraph [4] or PIER [11], IrisNet [10], Aurora [3] address the problem of continuous and distributed queries and as such are very relevant to CoMo.

8 Conclusion

We have presented the architecture of an open system for passive network monitoring. We have justified the design choices and indicated the three main open issues that are crucial for the success of the monitoring infrastructure: query engine, resource management and security of the system.

There is a number of other issues that have not been addressed in this paper but are currently under investigation: (*i*) coordination of multiple CoMo system to respond to a query or balance the computation load; (*ii*) optimal placement of CoMo system as well as modules to guarantee visibility on the traffic even in presence of network failures or re-routing events; (*iii*) use of sampling for reducing the load on the system in a controlled fashion; (*iv*) how to port the current architecture to other hardware systems, such as routers or network processors.

Acknowledgments

We thank Luigi Rizzo, Larry Huston, Pere Barlet, Euan Harris, Lukas Kencl, and Timothy Roscoe for their valuable feedback and comments on this work.

References

1. A. R. Bharambe, M. Agrawal, and S. Seshan. Mercury: Supporting scalable multi-attribute range queries. In *Proceedings of ACM Sigcomm*, Sept. 2004.
2. CAIDA: Cooperative Association for Internet Data Analysis. http://www.caida.org.
3. D. Carney et al. Monitoring streams - a new class of data management applications. In *Proceedings of VLDB*, 2002.
4. S. Chandrasekaran et al. TelegraphCQ: Continuous dataflow processing of an uncertain world. In *Proceedings of CIDR*, 2003.
5. Cisco Systems. NetFlow services and applications. White Paper, 2000.
6. C. Cranor, T. Johnson, O. Spataschek, and V. Shkapenyuk. Gigascope: A stream database for network applications. In *Proceedings of ACM Sigmod*, June 2003.
7. Endace. http://www.endace.com.
8. C. Fraleigh, S. Moon, B. Lyles, C. Cotton, M. Khan, R. Rockell, D. Moll, T. Seely, and C. Diot. Packet-level traffic measurements from the Sprint IP backbone. *IEEE Network*, 2003.
9. GEANT. http://www.dante.net.
10. P. B. Gibbons, B. Karp, Y. Ke, S. Nath, and S. Seshan. IrisNet: An architecture for a world-wide sensor web. *IEEE Pervasive Computing*, 2(4), Oct.
11. R. Huebsch, J. M. Hellerstein, N. Lanham, B. T. Loo, S. Shenker, and I. Stoica. Querying the Internet with PIER. In *Proceedings of VLDB*, 2003.
12. L. Huston, R. Sukthankar, R. Wickremesinghe, M. Stayanarayanan, G. Ganger, E. Riedel, and A. Ailamaki. Diamond: A storage architecture for early discard in interactive search. In *Usenix FAST*, Mar. 2004.
13. Internet2. http://www.internet2.org.
14. IP Flow Information eXport Working Group. Internet Engineering Task Force. http://www.ietf.org/html.charters/ipfix-charter.html.
15. X. Li, Y. J. Kim, R. Govindan, and W. Hong. Multi-dimensional range queries in sensor networks. In *Proceedings of the ACM Sensys*, Nov. 2003.
16. S. McCanne and V. Jacobson. The BSD Packet Filter: A new architecture for user-level packet capture. In *USENIX Winter*, Jan. 1993.

17. A. Moore, J. Hall, E. Harris, C. Kreibech, and I. Pratt. Architecture of a network monitor. In *Proceedings of Passive and Active Measurement Workshop*, Apr. 2003.
18. NLANR: National Laboratory for Applied Network Research. http://www.nlanr.net.
19. S. Patarin and M. Makpangou. Pandora: A flexible network monitoring platform. In *Usenix*, 2000.
20. V. Paxson, J. Mahdavi, A. Adams, and M. Mathis. An architecture for large-scale internet measurement. *IEEE Communications*, 36(8), 1998.
21. RIPE Rèseaux IP Europèens. http://www.ripe.net.
22. tcpdump/libpcap. http://www.tcpdump.org.
23. University of Oregon Route Views Project. http://www.routeviews.org.
24. J. Xu, J. Fan, M. Ammar, and S. Moon. Prefix-preserving IP address anonymization: Measurement-based security evaluation and a new cryptography-based scheme. Nov. 2002.

TCP Anomalies: Identification And Analysis

Marco Mellia, Michela Meo and Luca Muscariello *

Dipartimento di Elettronica, Politecnico di Torino, Torino, Italy
{mellia,meo,muscariello}@mail.tlc.polito.it

(Invited Paper)

Abstract. Passive measurements have recently received large attention from the scientific community as a mean, not only for traffic characterization, but also to infer critical protocol behaviors and network working conditions. In this paper we focus on passive measurements of TCP traffic, main component of nowadays traffic. In particular, we propose a heuristic technique for the classification of the anomalies that may occur during the lifetime of a connection. Since TCP is a closed-loop protocol that infers network conditions and reacts accordingly by means of losses, the possibility of carefully distinguishing the causes of anomalies in TCP traffic is very appealing and may be instrumental to the deep understanding of TCP behavior in real environments and the protocol engineering.

Keywords: Internet Traffic Measurements, TCP Protocol, Computer Networks.

1 Introduction

In the last ten years, the interest in data collection, measurement and analysis to characterize Internet traffic behavior increased steadily. Indeed, by acknowledging the failure of traditional modeling paradigms, the research community focused on the analysis of the traffic characteristics with the twofold objective of understanding the dynamics of traffic and its impact on the network elements, and of finding simple, yet satisfactory, models, like the Erlang teletraffic theory in telephone networks, for designing and planning packet-switched data networks.

The task of measuring Internet traffic is particularly difficult for a number of reasons. First, traffic analysis is made very hard by the strong correlations both in space and time due to the closed-loop behavior of TCP, the TCP/IP client-server communication paradigm, and the fact that the highly variable quality provided to the end user influences the user behavior. Second, the complexity of the involved protocols, and of TCP in particular, is such that a number of phenomena can be studied only if a deep knowledge of the protocol details is

* This work was founded by the European Community "euroNGI" Network of Excellence.

exploited. And, finally, some of the traffic dynamics can be understood only if the forward and backward directions of flows are jointly analyzed.

From what mentioned above, it is clear that, since TCP plays a central role in the generation of Internet traffic, measurement tools should be equipped with modules for the analysis of TCP traffic, which do not neglect the occurrence of all those anomalies that strongly influence TCP behavior. In this context, the objective of this paper is to propose a new heuristic classification technique of anomalies that may occur during the lifetime of a TCP connection. In [1] a simple but efficient classification algorithm for out-of-sequence TCP segments is presented. The classification proposed in this paper is a modification and extension of that classification which allows the identification of a number of phenomena which were not previously considered, such as unneeded retransmissions and flow control mechanisms.

The proposed classification technique is applied to a set of real traces collected at our institution. The results show that a number of interesting phenomena can be observed through the proposed classification, such as the impact of the use of TCP SACK on the occurrence of unnecessary retransmissions, the relative small impact of the daily variation of the load on the occurrence of anomalies, the quite large amount of network reordering.

2   Methodology

The methodology adopted in this paper is a modification and extension of the one proposed for the first time in [1], in which authors proposed a simple but efficient classification algorithm for out-of-sequence packets in TCP connections and presented measurement results within the Sprint IP backbone. Similarly to what proposed by them, we adopt a passive measurement technique rather than using active probe traffic. The classification in [1] identifies out-of-sequence events due to i) necessary or unnecessary segment retransmissions by the TCP sender, ii) network duplicates, or iii) network reordering. Building over the same idea, we complete the classification by distinguishing other possible causes of out-of-sequence or duplicate packets. In particular, we also focus on the *cause* that triggered the segment retransmission by the sender. We analyze packet traces which record packets in both directions of a TCP connection: both data segments and ACKs are recorded.

Figure 1 sketches the evolution of a TCP connection: connection setup, data transfer, and connection tear down phases are highlighted. The measurement point (sniffer) is located in some point in the path between the Client and the Server. Both the IP layer and TCP layer overhead are observed by the sniffer, so that the TCP sender status can be tracked. A TCP connection starts when the first SYN from the client is observed, and terminates after either the tear-down

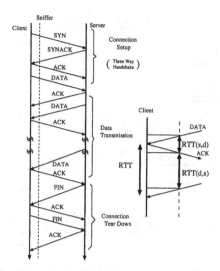

Fig. 1. Evolution of a TCP connection and the RTT estimation process.

sequence (the FIN/ACK or RST messages), or when no segment is observed for an amount of time larger then a given threshold [1].

By tracking the segment trace of a connection in both directions[2], the sniffer correlates the sequence number of TCP data segments to the ACK number of backward receiver acknowledgments and classifies the segments as,

- *In-sequence*: if the sender initial sequence number of the current segment corresponds to the expected one;
- *Duplicate*: if the data carried by the segment have already been observed before;
- *Out-of-sequence*: if the sender initial sequence number is not the expected one, and the data carried by the segment have never been observed before.

The last two classifications refer to an *anomaly* during the data transfer, that can have been caused by several reasons. We propose a fine heuristic classification of the anomalies. The classification has been implemented in [6, 7], and then used to analyze collected data. During the decision process that is followed to classify anomalies, several variables are used:

RTT_{min} Minimum RTT: is the estimated minimum RTT observed since the flow started;

---

[1] Given that a tear-down sequence of a flow under analysis is never observed, a timer is needed to avoid memory starvation; the timer is set to 15 minutes, which is sufficiently large according to the findings in [2].

[2] In case only one direction of traffic is observed by the sniffer, the heuristic will not be applicable.

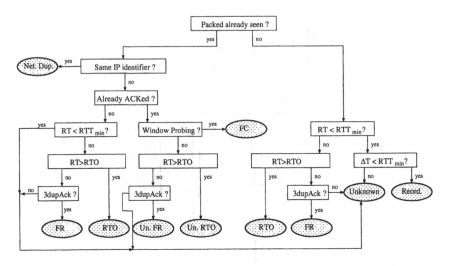

Fig. 2. Decision process of the classification of anomalous segments.

RT Recovery Time: is the difference between the time the current anomalous segment has been observed and the time the segment with the *largest* sequence number has been seen;

ΔT Inverted-packet gap: is the difference between the observation time of the current anomalous segment and the previously received one;

RTO Retransmission Timeout: is estimated sender retransmission timer value (in seconds) according to [3] as $RTO = max(1, E[RTT] + 4std(RTT))$;

DupAck Number of DupAcks: is the number of duplicate ACKs observed on the reverse path.

Following the flow diagram of Figure 2, we describe the classification heuristic. Given an anomalous segment, the process starts by checking if the segment has already been seen by comparing the TCP sequence number of the current segment with the one carried by segments observed so far. Thus, the segment can be classified as either duplicate or out-of-sequence. In the first case (left branch of the decision process), the IP identifier field of the current packet is compared with the same field of the original copy of the packet. If they are the same, then the packet is classified as **Network Duplicate** (Net. Dup.). Network duplicates may stem from malfunctioning apparatuses, routing loops, mis-configured networks (e.g., Ethernet LANs with diameter larger than the collision domain size), or, finally, by unnecessary retransmissions at the link layer (e.g., when a MAC layer ACK is lost in a Wireless LAN forcing the sender to retransmit the frame). Compared with the decision process adopted in [1], we classify as duplicate segments all those duplicate packets with the same IP identifier, regardless of the *ΔT* value. Indeed, there is no reason to exclude that a network duplicate may be observed at any time, and there is no relation between the *RTT* and the time a network can produce some duplicate packets.

When the IP identifiers are different, the TCP sender may have performed a retransmission. If all the bytes carried by the segment have already been acknowledged, then the receiver has already received the segment, and therefore this is an unneeded retransmission. The flow control mechanism adopted by TCP uses false unneeded retransmissions to perform *window probing*, i.e., to force the receiver to immediately send an ACK so as to probe if the receiver window $RWND$ (which was announced to be zero on a previous ACK) is now larger than zero. Therefore we classify as **Flow Control** (FC) retransmissions the retransmitted segments for which the following three conditions hold: i) the sequence number is equal to the expected sequence number decreased by one, ii) the segment size is of zero length, and iii) the last announced $RWND$ in the ACK flow was equal to zero. This is a new possible cause of unneeded retransmissions which was previously neglected in [1].

If the anomaly is not classified as flow control, then it must be an unnecessary retransmission, which could have been triggered because of either a Retransmission Timer (RTO) has fired, or the fast retransmit mechanism has been triggered, i.e., three or more duplicate ACKs have been received for the segment before the retransmitted one. We identify three situations: i) if the recovery time is larger than the retransmission timer ($RT > RTO$) the segment is classified as an **Unneeded Retransmission by RTO** (Un. RTO); ii) if 3 duplicate ACKs have been observed, the segment is classified as an **Unneeded Retransmission by Fast Retransmit** (Un. FR); iii) otherwise, if none of the previous conditions holds, we do not know how to classify this segment, and therefore we label it as **Unknown** (Unk.). Unneeded retransmissions may be due to a misbehaving sender, a wrong estimation of the RTO at the sender, or, finally, to ACKs lost on the reverse path. However, distinguishing among these causes is impossible by means of passive measurements.

Let us now consider the case of segments that have already been seen but have not been ACKed yet. This is possibly the case of a retransmission following a packet loss. Given the recovery mechanism adopted by TCP, a retransmission can occur only after at least a RTT, since duplicate ACKs have to traverse the reverse path and trigger the Fast Retransmit mechanism. Therefore, if the recovery time is smaller than RTT_{min}, the anomalous segment can only be classified as **Unknown**[3]. Otherwise, it can either be a **Retransmission by Fast Retransmit** (FR) or **Retransmission by RTO** (RTO); the classification being based on the same criteria adopted previously for unneeded retransmissions. Retransmissions of already observed segments may be due to i) data segments lost on the path from the measurement point to the receiver, and to ii) ACK segments delayed or lost before the measurement point.

Consider now the right branch of the decision process, which refers to out-of-sequence anomalous segments. In this case, the classification criterion is simpler. Indeed, out-of-sequence segments can be due to either the retransmission of

---

[3] In [1] the authors use the average RTT. However, being each RTT possibly different than the average RTT (and in particular smaller), we believe that using the average RTT forces a larger amount of misclassification.

lost segments, or to network reordering. Again, since retransmissions can only occur if the recovery time RT is larger than RTT_{min}, by double checking the number of observed duplicate ACKs and by comparing the recovery time with the estimated RTO, we can distinguish retransmissions triggered by RTO, by FR or we can classify the segment as an unknown anomaly. On the contrary, if RT is smaller than RTT_{min}, then a **Network Reordering** (Reord) is identified if the inverted-packet gap is smaller than RTT_{min}. Network reordering can be due to either load balancing on parallel paths, or to route changes, or to parallel switching architectures which do not ensure in-sequence delivery of packets [4].

2.1 Dealing with wrong estimates

The classification algorithm uses some thresholds whose values must be estimated from the packet trace itself, which may not be very accurate or even valid when classifying the anomalous event. Indeed, all the measurements related to the RTT estimation are particularly critical, since they are used to determine the RTT_{min} and the RTO estimation. RTT measurement is updated during the flow evolution according to the moving average estimator standardized in [3]. Given a new measurement m of the RTT, we update the estimate of the average RTT by mean of a low pass filter $E[RTT]_{new} = (1 - \alpha)E[RTT]_{old} + \alpha\, m$ where α $(0 < \alpha < 1)$ is equal to $1/8$.

Since the measurement point is not co-located at the transmitter, nor at the receiver, the measure of RTT is not available. Therefore, in order to get an estimate of the RTT values, we build over the original proposal of [1]. In particular, denote by $RTT(s, d)$ the *Half path RTT sample*, which represents the delay at the measurement point between an observed data segment flowing from the transmitter, or source, to the receiver, or destination, and the corresponding ACK on the reverse path, and denote by $RTT(d, s)$ the delay between the ACK and the following segment, as shown in Figure 1. From the measurement of $RTT(s, d)$ and $RTT(d, s)$ it is possible to derive an estimate of the total round trip time RTT,

$$RTT = RTT(s, d) + RTT(d, s)$$

The estimation of the average RTT is not biased, given the linearity of the expectation operators. Therefore, it is possible to estimate the *average RTT* by,

$$E[RTT] = E[RTT(s, d)] + E[RTT(d, s)]$$

Moreover, the standard deviation of the connection's RTT, $std(RTT)$ can be estimated following the same approach, given that $RTT(s, d)$ and $RTT(d, s)$ are independent measurements, which usually holds.

Finally, given $E[RTT]$ and $std(RTT)$, it is possible to estimate the sender retransmission timer as in [3]:

$$RTO = E[RTT] + 4std(RTT)$$

For what concerns the estimation of minimum RTT, RTT_{min}, we have

$$RTT_{min} = \min(RTT(s, d)) + \min(RTT(d, s))$$

In general, this estimator gives a conservative estimation of the real minimum RTT, as $RTT_{\min} \leq \min(RTT)$ holds. This leads to a conservative classification algorithm, which increases the number of anomalies classified as unknown, rather than risking some misclassifications.

2.2 Handling Particular Cases

No RTT Sample Classification. There are some cases in which the RTT measurement is not available, but an anomalous event is detected. This happens in particular at the startup of a TCP connection, as no valid RTT samples may be available at the very beginning of the connection. Since most of TCP flows are very short [7], these events are quite frequent and cannot be neglected. Moreover, the choice of the initial values of RTO and RTT_{min} results to be critical, and inappropriate estimations of these variables may lead to wrong classifications. We adopt the following approach:

- if no valid RTT samples have been collected, the heuristic uses $RTO = 3s$ and $RTT_{min} = 5ms$ as default values
- the RTO estimation is forced to assume values larger or equal to $1s$, according to [3]
- the RTT_{min} estimation is forced to be larger than $1ms$

Batch Classification. Given that TCP can transmit more than one segment per RTT, it may happen that more than one anomalous segments are detected back-to-back. This occurs, for example, when the TCP sender adopts the SACK extension and retransmits more than one segment per RTT, or, when packets belonging to the same window on a path in which packets are reordered, arrive with "strange" patterns difficult to be identified. In such cases, the measurement of RT and ΔT may be wrong and lead to incorrect classifications. We therefore implement a filter in the classification heuristic that correlates the classification of the current anomaly with the classification of the previous segment. In particular, if the current recovery time RT is smaller than $E[RTT]$ (suggesting that the segment is belonging to the same window as the previous one) and the previous segment was not classified as in sequence, we then classify the current anomalous segment as the previous one.

For example, consider a simultaneous SACK retransmission of two segments triggered by a Fast Retransmit. The first retransmitted segment is correctly classified given that three duplicate ACKs have been observed on the reverse path, and the RT is larger than RTT_{min}. However, the second retransmitted segment cannot be correctly classified, given that no duplicate ACK has ever been observed, and $RT < RTT$. By explicitly considering the classification of the first segment, it is possible to correctly identify this segment as a retransmission triggered by Fast Retransmit.

3 Measurement Results

Our measurements have been gathered from the external Internet edge link of our institution. Our campus network behaves like an Internet stub, because the access router is the sole gate to the external network. Our institution counts more than 7,000 hosts, whose great majority is constituted by clients. Some servers are regularly accessed from outside as well. We collected all packets flowing into the access link that connects the campus border router to the GARR network [5], the nation-wide ISP for research and educational centers. We measured for several months during several time periods and gathered the most interesting statistics related to the anomalous traffic. We present only a subset of results, and in particular:

- from the 6th to the 7th of February 2001. The bandwidth of the access link was 14 Mbit/s;
- from the 29th of April to the 5th of May 2004. The bandwidth of the access link was 28 Mbit/s.

3.1 Impact of RTT_{min} and RTO

We first lead a set of measurements to double-check the impact of the choices described in Sec 2.1. In particular, we are interested in the impact of the measurement of RTT_{min} and RTO. The first one is involved on the classification of the network reordering anomalies that may occur when identifying two out-of-sequence segments separated by a time gap smaller than RTT_{min}. Figure 3 (a) plots Cumulative Distribution Function (CDF) of the ratio between the inverted packet gap ΔT and the value of the RTT_{min} considering only TCP anomalies classified as network reordering. Measurements referring to 2004 are reported, and similar results are obtained considering the 2001 dataset. The CDF clearly shows that ΔT is much smaller than the RTT_{min}. This suggests that the initial choice of RTT_{min} is appropriate, and the conservative estimation of RTT_{min} does not affect the classification.

Figure 3 (b), reports part of the CDF of the ratio between the actual Recovery Time RT and the corresponding estimation of the RTO when considering anomalous events classified as retransmissions by RTO. Also in this case we report results referring to the 2004 dataset. The CDF shows that $RT > RTO$ holds, which is a clear indication that the estimation of the RTO is not critical. Moreover, it can be noted that about 50% of the cases have a recovery time which is more than 5 times larger than the estimated RTO. This apparently counterintuitive result is due to the RTO back-off mechanism implemented in TCP but not considered by the heuristic which doubles the RTO value at every retransmission of the same segment. Not considering the back-off mechanism during the classification lead to a robust and conservative approach.

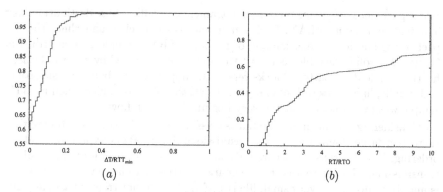

Fig. 3. (a) CDF of the ratio between the inverted packet gap ΔT and RTT_{min}. (b)CDF of the ratio between the recovery time RT and the actual estimation of the RTO.

3.2 Aggregate Results

In the following we report results obtained by running the classification heuristic over the two datasets we selected and by measuring the average number of occurred anomalies during the whole time period. The objective is twofold. First, we quantify the different causes that generated anomalous segment delivery; second, we double check the heuristic classification. Indeed, it is impossible to test the validity of the classification algorithm, given that the real causes of the anomaly are unknown. We therefore run the classification over real traces, and try to underline some expected and intuitive results that confirm the validity of the heuristic design.

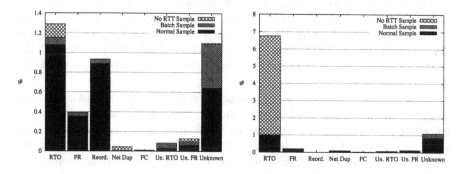

Fig. 4. Classification of anomalous events: incoming traffic on the left , outgoing traffic on the right.

Figure 4 reports the percentage of identified anomalous events during the 2004 period over the total amount of observed segments. Left plot refers to incoming traffic, i.e., traffic whose destination host is inside our campus LAN; right plot reports measurements on outgoing traffic, i.e., traffic whose destination host is outside our campus LAN. Each bar in the plot explicitly underlines the impact of the batch classification and of the lack of RTT samples, as described in Section 2.1: solid black blocks report the anomalies classified by a normal classification, while dark pattern blocks report the impact of the batch classification, and, finally, light pattern blocks report the classification obtained when no RTT sample was available, i.e., at the very beginning of each flow.

Considering the incoming anomalies classification (left plot), we observe that there is a large dominance of retransmissions due to RTO expiration and reordering. Fast Retransmit occurs only for a very small portion of the total retransmissions. This is related to the characteristics of today data traffic, which is mainly composed of very small file transfers that cannot trigger Fast Retransmit. This effect is further stressed by the fact that our campus LAN traffic is mainly made of web browsing applications, whose (short) HTTP requests travel on the outgoing directions.

A small percentage of unnecessary retransmissions is also present. A rather large percentage of anomalies that could not be classified but unknown is collected. Inspecting further, we observed that for most of them the recovery time is smaller than the estimated RTO (therefore missing the retransmission by RTO classification), but larger than the RTT_{min} (therefore missing the reordering classification) and the number of duplicate ACKs is smaller than 3 (therefore missing the retransmission by FR classification). We suspect that they may be due to either i) transmitters that trigger the Fast Retransmit with just 1 or 2 duplicate ACKs, or ii) servers with aggressive RTO estimation that triggers the retransmission earlier than the RTO estimation at the measurement point. Given the conservative approach that guided the classification heuristic, we prefer to classify them as unknown rather than misclassifying them.

For what concerns the impact of the batch classification and of the lack of a valid RTT sample, observe that the first is evenly distributed among all classification cases, while the latter one has a large impact on the identification of retransmissions by RTO. This is due to the lack of valid RTT samples at the very beginning of the TCP connection, when the sender can only detect packet losses by RTO.

When considering the outgoing anomalies (right plot), the heuristic correctly identifies the anomalies, and neither Network Reordering nor Duplicates are identified. Indeed, this is quite obvious given that our institution LAN is a switched Ethernet LAN. In this network, IP packets can be duplicated or reordered only in case of malfunctioning. This confirms the validity and robustness of the classification heuristic we developed.

Considering the average percentage of total anomalies identified,, we have that about 4% and 8% of incoming and outgoing traffic respectively is affected by an anomaly. We will see in the next section that the average values is not representative at all, given the non-stationaries of the anomalies.

Finally, in order to double check the validity of our heuristic, we split the flow into three different classes based on their segment length. *Short* flows (also

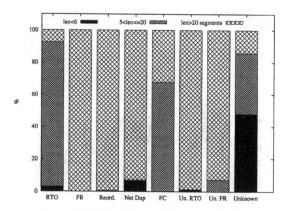

Fig. 5. Classification of anomalous events for different flow length; incoming traffic for 2004 measurements.

called mice in the literature) have payload size (in segments) no longer than 5 segments. *Long* flows (the so called elephants) have payload size (in segments) larger than 20 segments, and the *middle* length flows payload size is larger than 5 segments, but shorter or equal to 20 segments. Figure 5 reports the classification of the anomalous events split among the three different classes. Incoming segment classification is considered for the 2004 time period. Solid black refers to short flows, dark gray pattern refers to middle flows, and light gray pattern refers to long flows. For the sake of clarity, we omit the further batch or no rtt sample classification.

As expected, retransmissions due to RTO expiration are distributed among all flows, while retransmissions due to Fast Recovery are only triggered for long flows. This is intuitive, as already said, because of the limit in triggering Fast Retransmit by short flows. The majority of packet reordering affects long flows, for which the chance to suffer from a reordering is much larger. Neglecting the network duplicates, flow control and unnecessary retransmissions as they are very marginal, we observe that the anomalies classified as unknown are largely related to short flows. Indeed, this is an hint that the $E[RTT]$ estimation is affected by a larger error for short flows, while long flows have the chance to get a better estimation of the RTT and therefore to better classify the anomaly. This confirms the intuition that the unknown classification is related to possible different estimation of the RTO at the transmitter and at the measurement point.

Results relatively to outgoing flows and to the 2001 dataset are very similar and therefore not reported here.

3.3 Behavior in Time

We report in this section the results of the occurrence of anomalies in time. Again, we omit the sub-classification due to batch or no RTT samples for the

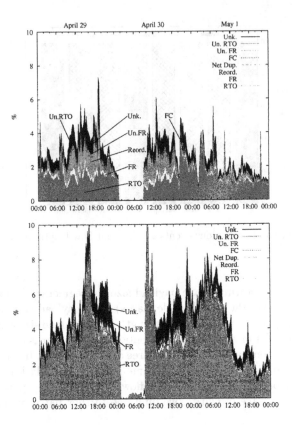

Fig. 6. Classification of anomalous events versus time: incoming traffic on the left, outgoing traffic on the right for 2004 measurements.

sake of clarity. Figures 6 and 7 depict the time evolution of the volume (in percentage, normalized on the total flowed traffic in each direction) of anomalous measured segments classified by the proposed heuristic. Measurements aggregate anomalies over an interval of time equal to 15 minutes. The detailed classification is outlined in colored slices whose size is proportional to the percentage of that particular event. Top plot refers to the incoming traffic; bottom plot reports measurements considering outgoing traffic. Figure 6 refers to the three days evolution of such traffic continuously monitored and classified during the 2004 period, while, similarly, Figure 7 refers to the two day long subset of measurements in 2001.

Apart from the network outage that is evident, the first unambiguous results is that TCP anomalies are highly non stationary over several time scales. There are some peaks of very significant magnitude that reach 10% during 2004 and 15% during 2001. Considering the incoming segments, Retransmissions by RTO, Network Reordering and Unknowns are the largest part of the anomalies, while TCP Flow control seldom kicks in, and negligible Unnecessary Retransmissions are identified. Surprisingly, the typical night and day effect, which is commonly present on the total traffic volume (and is valid also in the considered link), is not anymore visible when considering TCP anomalies. Notice also that the last two measurement days are weekend-days. In particular, the Labour Day is celebrated on Sunday the 1st of May in Italy. Therefore the link load during that weekend was particular low. Nonetheless, the RTO fraction is almost equal to the one observed during busy hours of weekdays. Only the Network Reordering seems to disappear. This hints to a weak correlation between link load and TCP anomalies.

Considering the outgoing traffic (bottom plots of Figure 6 and 7), observe that the heuristic correctly identifies the anomalies as retransmission by RTO. Given that hosts in our campus LAN are mainly clients of TCP connections, the outgoing flow size is very short, and therefore in case of a packet loss, the only way to recover is to fire the RTO.

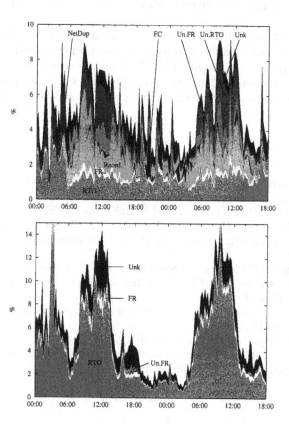

Fig. 7. Classification of anomalous events versus time: incoming traffic on the top, outgoing traffic on the bottom for 2001 measurements.

If we compare 2004 and 2001 incoming time plots, it can be noted that after three years the number of Unnecessary Retransmissions almost disappears. This fact is explained by the vast popularity of TCP SACK flows that were only the 21% of total flows during 2001 while it increased to 90% of total flows during 2004. At the same time, a reduction of the fraction of TCP anomalies is noticeable comparing 2001 and 2004 measurements. This could be related to the corresponding increase in the access link capacity, which doubled in 2004.

4 Conclusions

In this paper, we proposed a heuristic technique for the classification of TCP anomalies. The classification identifies seven possible causes of anomalies and extends previous techniques already proposed in the literature. We were also able to quantify the quality of our classification by studying the sensitivity of our approach in estimating the RTT which strongly influences the proposed methodology which gathers every flow state information from passive measurements and the quantitative analysis show its effectiveness in the identification procedure.

The technique was implemented and tested on the external Internet link of our institution and allowed the observation of a number of interesting phenomena such as the impact of the use of TCP SACK on the occurrence of unnecessary retransmissions, the relative small impact of the daily variation of the load on the occurrence of anomalies, the quite large amount of network reordering anomalies.

References

1. S. Jaiswal, G. Iannaccone, C. Diot, J. Kurose, D. Towsley, "Measurement and Classification of Out-of-Sequence Packets in a Tier-1 IP Backbone", IEEE Infocom, San Francisco, March 2003.
2. G.Iannaccone, C.Diot, I.Graham, and N.McKeown, "Monitoring very high speed links," *ACM Internet Measurement Workshop, San Francisco*, November 2001.
3. V. Paxson, M. Allman, "Computing TCP's Retransmission Timer", *RFC 2988*, November 2000.
4. J.C.R.Bennett,C.C.Partridge, N.Shectman, "Packet reordering is not pathological network behavior" em IEEE/ACM Transactions on Networking, Vol. 7, N. 6, pp.789-798, December 1999.
5. "GARR   Network"   http://www.noc.garr.it/mrtg/RT.TO1.garr.net /polito.garr.net.html
6. "Tstat Web Page" http://tstat.tlc.polito.it/
7. M. Mellia, R. Lo Cigno, F. Neri,"Measuring IP and TCP behavior on edge nodes with Tstat", Computer Networks 47(3): 1-21, Jan. 2005.

A High Performance IP Traffic Generation Tool Based On The Intel IXP2400 Network Processor

Raffaele Bolla, Roberto Bruschi, Marco Canini, and Matteo Repetto

Department of Communications, Computer and Systems Science,
DIST–University of Genova, Via Opera Pia 13, 16145 Genova, Italy
{raffaele.bolla, roberto.bruschi}@unige.it, marco@reti.dist.unige.it,
matteo.repetto@unige.it

Abstract. Traffic generation is essential in the processes of testing and developing new network elements, such as equipment, protocols and applications, regarding both the production and research area. Traditionally, two approaches have been followed for this purpose: the first is based on software applications that can be executed on inexpensive Personal Computers, while the second relies on dedicated hardware. Obviously, performance in the latter case (in terms of sustainable rates, precision in delays and jitters) outclasses that in the former, but also the costs usually grow of some order of magnitude. In this paper we describe a software IP traffic generator based on a hardware architecture specialized for packet processing (known as Network Processor), which we have developed and tested. Our approach is positioned between the two different philosophies listed above: it has a software (and then flexible) implementation running on a specific hardware only slightly more expensive than PCs.

Keywords: IP traffic generation, Network Processors, Network performance evaluation.

1 Introduction

The current Internet is characterized by a continuous and fast evolution, in terms of amount and kind of traffic, network equipment and protocols. Heterogeneous equipment, protocols and applications (also referred to as "network elements"), high transmission rates in most of Internet branches and the need of fast development (to quickly fulfil new services' requirements and to reduce time to market) are at present the most critical issues in Internet growth. In this context, every network element should be carefully tested before being used in real networks: an error in a device or a protocol could easily result in technical problems (e.g., packet losses, connection interruptions, degradation in performance) and it could lead to significative economic damage in production environments.

Many of the tests on a new network element require the ability to generate synthetic traffic off-line; such traffic should be sufficiently complex, fast (high-rates) and, in a word, realistic to cover a good deal of operating conditions. Thus, the possibility to create realistic network traffic streams is very useful today, and it is essential to reduce development times and bugs in new network elements. Moreover, the availability of good synthetic traffic sources helps researchers in understanding network dynamics and in designing suitable modifications and improvements in current protocols and equipment.

The generation of traffic streams is a quite "simple" task in simulated environments; on the contrary, the generation of real traffic on a testbed setup involves some critical issues related to precision, scalability and costs. Until now, two antithetical approaches have been traditionally followed for network traffic generation. The first one is based on software applications that do not require specific hardware, but can be executed on general purpose machines, such as inexpensive Personal Computers (PC). The second one is based on the development of specialized hardware.

The characteristics and functionalities of software tools can be very sophisticated but, at the same time, they can maintain a high flexibility level and (often) offer an open source approach: in other words, they are modifiable and adaptable to the specific requirements of the experiments to carry out. The main drawback of this approach is the architecture of the PC, which limits the precision and the maximum reachable performance; this is a heavy limitation today, with Internet traffic that increases continuously and network equipment that must manage a huge amount of traffic (from Gigabits to Terabits per second).

Several techniques can be used in PC-based network traffic generation, depending on the final goal. Quite often, it is useful to generate only a single stream of IP packets, characterized by the inter-arrival time between two consecutive packets; this stream can be used to test network equipment such as routers and switches with standard performance analyses, as those defined in RFC 2544 [1] and 2889 [2]. Examples of applications oriented to this type of generation are *Iperf* [3] and *Netperf* [4]. These two tools generate IP packets by starting from a fixed structure and by varying some fields (e.g., the source and destination addresses, the TOS, etc.); more advanced applications are able to keep into account also the behaviour of the transport protocols, by representing the traffic generation as a set of file transfers over TCP or UDP (see, for example, *Harpoon* [5]). All these tools do not care about packet contents; however, when the element under test is an application (such as a Web server or a DNS), this is not acceptable. Thus, other tools have been designed with this goal, for example *WebPolygraph* [6]. One common drawback of the above cited software tools is that each of them can generate only a specific type of traffic stream, which is not representative of the heterogeneous traffic flowing on a generic Internet link. Software such as *Tcpreplay* and *Flowreplay* [7] log the traffic on some links and then generate identical streams, by possibly changing some parameters in the IP packets. Unfortunately, in this case, the traffic streams are always the same and the registration of long sequences of traffic requires large amounts of fast memory.

The second approach, namely the custom hardware devices, is the most popular solution in the industrial environment. Suitable architectures have been designed (often by using also standard components) to minimize delays and jitters introduced in packet generation; here all the functionalities are implemented "near" the hardware, without the intermediation of a general purpose operating system (that may be present at a higher level to facilitate the configuration and the control of the instrument and the management of the statistical data). The development of such equipment is usually very expensive, based on proprietary solutions and carried out for professional and intensive use only. Thus, the final result is that the tools are configurable, but not modifiable or customizable, and very expensive. All these elements, together with the high final cost of each device, make this kind of tools not much attractive, especially for academic researchers or small labs. Examples of such products are Caldera Technologies *LANforge-FIRE* [8], CISCO IOS *NetFlow* [9], Anritsu *MD1231A IP/Ethernet Analyser* [10] and SmartBits *AX/400* [11].

Starting from these considerations, we have decided to build an open source traffic generator that could be situated in the middle between the existent approaches: it should be flexible and powerful enough to be useful in most of the practical situations, but less expensive than professional equipment. To realize all these objectives we need to work near the hardware, but we also need to implement all the functionalities in software, to reduce development costs and make customization effective. Among different technologies we have taken into account, we found our ideal solution in the Network Processors: these are devices conceived for fast packet processing, often with the high degree of flexibility needed for implementing the desired algorithms; moreover their cost is lower than custom hardware devices.

Several Network Processor architectures are available from different manufacturers [12]; we have found the Intel IXP2400 to be the best compromise between computation power and flexibility. Moreover, this chip is available on an evaluation board, the Radisys ENP-2611, which represents the cheapest way to access this kind of Network Processors. Until now, we know only another similar approach to the problem of traffic generation [13], but it implements a very simple structure, which is unable to generate realistic traffic. In this work we describe a structure for a high-speed IP/Ethernet traffic generator based on the Intel IXP2400 Network Processor that we have developed and tested. The requirements of this device are to be able to transmit more than 1 Gbps of traffic (to saturate all the Gigabit Ethernet lines of the ENP-2611) and to generate multiple traffic streams simultaneously, each of them with different statistical characteristics and header contents. Our aim is to model the more representative traffic classes in the Internet, such as real-time and Best Effort; moreover, we would like to have the chance to manage Quality of Service (by using the TOS field).

In the following, after briefly describing the IXP2400 architecture, we focus on two main issues. The first concerns the question if the Network Processor can be effectively utilized for traffic generation: the IXP2400 has been designed for packet processing, not explicitly for packet generation, and we were not sure that it can saturate the Gigabit interfaces, as we would like for a high-performance tool. The second issue regards the design of an application framework that exploits the Network Processor architecture to build a flexible tool for traffic generation. We have named such framework PktGen and, as will described in detail later, it consists of several applications running on different processors of the IXP2400. PktGen is very simple to use, as it provides a comfortable user graphical web interface; moreover, it can also be modified without much effort, as it is mostly written in C. Preliminary tests have been carried out to demonstrate the correctness and the performance of such tool.

The rest of the paper is organized as follows: Sect. 2 gives a brief overview of the Intel IXP2400, whereas the successive Sect. 3 explains how the Network Processor architecture is used to build a packet generator. Section 4 describes PktGen in details by analysing all its components, whereas in Sect. 5 we report some performance tests about PktGen and a comparison with UDPGen, a well-known software PC-based traffic generator. Finally, in Sect. 6 we give our conclusions and we report some ideas for future work.

2   The IXP2400 Architecture

Figure 1 shows the main components of the Intel IXP2400 architecture and the relationships among them. The IXP2400, as some other Network Processors, provides several processing units, different kinds of memory, a standard interface towards MAC components and some utility functions (e.g. hash and CRC calculation). The "intelligence" resides in the processing units, which can use the other peripherals though several internal buses.

For what concerns processing, the IXP is equipped with two kinds of microprocessors which play very different roles in the overall architecture: one XScale CPU and eight MicroEngines. The XScale is a generic RISC 32-bit processor, compatible with the ARM V5 architecture, but without the floating point unit. This processor is mainly used to control the overall system and to process network control packets (such as ICMP and routing messages). Due to its compatibility with the industry based standards (ARM), it is possible to use on it both the *Linux* and *VxWorks* [14] (derived from *Windows*) Operating Systems (OS); this is a great advantages as many existent applications and libraries can be compiled and used on this processor (greatly decreasing development time and increasing efficiency). For what concerns Linux, two different distributions are available at the moment: *Montavista* [15, 16], based on a 2.4.18 kernel, and *Fedora* [17], with a more recent 2.6.9 kernel.

Accordingly to our aim of building an open-source tool, we have opted for Linux OS; in particular we have chosen the Montavista distribution, which was already available at the beginning of the project (Fedora has been released only recently). This choice enables us to develop software for this processor in standard C or C++ languages.

The other processing units of the IXP are the Microengines (ME). They are minimal RISC processors with a reduced set of instructions (about 50), optimized for packet processing; they provide logical and arithmetic operations

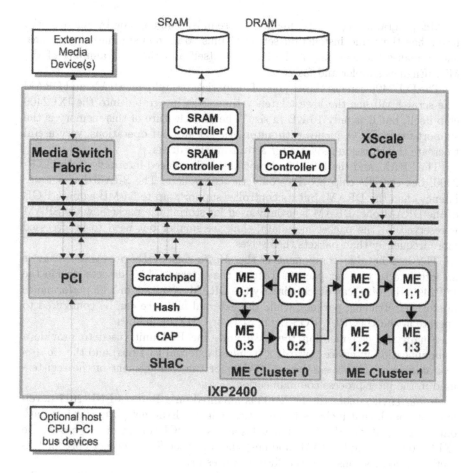

Fig. 1. The IXP architecture.

on *bits, bytes* and *dwords* but not the division nor the floating point functions. The MEs have access to all the shared units (SRAM, DRAM, MSF, etc.) and they are used along the *fast data path* for packet processing; they can be used in different ways (e.g., in parallel or sequentially) to create the framework that best fits the computational needs of network equipment.

No operating system is required on the ME, thus no software scheduler is available for multi-threaded programming. A simple round robin criterion is used to execute more threads (up to 8) on a ME; the programmer has the burden to write the code for each thread in a way that periodically releases the control of the ME to the other ones. Internally, each ME is equipped with specific registers for at most 8 hardware contexts (corresponding to the threads) and a shared low latency memory; moreover, some dedicated units are available for specific tasks (CRC computation and pseudo-random number generation).

ME programming can be made by assembler language or by microC; the latter has the same instruction set as C plus some non-standard extensions, and it makes the learning of the language itself and the programming of the MicroEngines simpler and faster.

The IXP2400 can use three types of memories: scratchpad, SRAM and DRAM. The scratchpad has the lowest latency time, it is integrated into the IXP2400 chip itself, but it is only 16 KB in size. The main feature of this memory is the support for 16 FIFO queues with automatic get and put operations, very useful to create transmission requests for a transmission driver.

The SRAM and the DRAM are not physically placed internally to the IXP-2400, but the NP chip provides only the controllers. The SRAM has a lower latency than the DRAM, but its controller supports up to 64 MB against 1 GB of the DRAM one. DRAM is usually used for storing the packets, while SRAM is reserved for the packet metadata, that are smaller but need to be accessed more frequently than packets themselves.

The last notable component is the *Media Switch Fabric* (MSF), that is an interface to transfer packets to/from the external MAC devices. Inside the IXP2400, the MSF can access directly the DRAM, resulting in high performance in storing/retrieving packets, while the external interface can be configured to operate in the UTOPIA, POS-PHY, SPI3 or CSIX modalities.

Finally, we can mention the hash unit, the PCI unit (used to provide a communication interface towards a standard external PCI bus) and the *Control and status register Access Proxy* (CAP), for the management of the registers used in the inter-process communication.

In our environment the IXP2400 is mounted on a Radisys ENP-2611 development board, that includes three optical Gigabit Ethernet ports (for fast-path data plane traffic), SRAM e DRAM sockets, a PCI connector (to access the IXP2400 SRAM and DRAM modules) and a further FastEthernet port for "direct" communications with the XScale processor.

A more detailed knowledge of the architectures of the IXP2400 and the Radisys development board is useless for the purpose of the paper. On the contrary, it is interesting to analyse how a standard packet processing phase takes place in the fast-path of the Network Processor.

Figure 2 shows a standard fast-path structure: packets are received from one Ethernet interface by a RX driver, passed to the Microengines (and eventually to the XScale) for the required processing and finally delivered to the TX driver for transmission over the physical line. The events that occur inside the Network Processor during this path crossing are the following:

1. A packet is received from the physical interface and delivered to the MSF.
2. The MSF buffers the packet and it notifies the RX driver, running on a MicroEngine.
3. The RX driver commands the MSF to transfer the buffered packet in the DRAM; then it creates a metadata structure for that packet, which is inserted in the SRAM. An identificator of the packet is put in a specific ring queue of the scratchpad, for successive processing.

Fig. 2. A standard IXP2400 fast-path.

4. A Microengine thread gets the packet identificators (queued from the RX driver) and starts the packet processing phase. The operations that occur on the packet are application-specific and they may range from a simple routing between ports to more advanced features, such as *firewalling*, *natting*, etc; these operations can be carried out by a single thread or by multiple threads (each of them should perform a simple task), and sometime by the XScale. The latter is usually involved very rarely (e.g., it processes the routing update packet exchanged between routers or the control information directed to the Network Processor), as the presence of an operating system enables more complex operations, but results in slower execution times. During this phase the data are stored in the SRAM and both the packet data and metadata could be modified; finally the packet identifier is inserted in a ring queue for transmission (again in the scratchpad memory).

5. The transmission driver continuously checks the transmission ring queue, looking for new identifiers of the packets to be transmitted; each time it finds a new identifier, it gets the metadata from the SRAM and it instructs the MSF for transmission.

6. The MSF gets the packet from the DRAM and transfers it to the physical interface for the transmission.

3 Traffic Generation

In order to generate network traffic, the very basic tasks that we need to realize are the creation and transmission of the packets. The first issue to solve is the choice of where to locate the functions for creating and then transmitting the packets; in particular the two alternatives are the Xscale processor, which offers developers a well known environment and programming language, or the Microengines, faster for this task but less easily programmable.

In this context we have carried out several tests, in which the same algorithm (which transmits the same 60[1] bytes sized packet for a given number of times) has been implemented as:

---

[1] The actual packet size is 64 bytes, but the last 4 bytes are the CRC computed and appended by the PM3386 MAC device located on the Radisys board.

Table 1. Comparative results for different generation methods.

Generation method	Average pps³ rate
From XScale - kernel mode	468 kpps
From XScale - user mode	658 kpps
From Microengine	842 kpps

1. a kernel space application running on XScale, which has been developed by using the Resource Manager[2] library;
2. a user space application running on XScale, which uses the *"mmaped"* memory to access hardware features;
3. a microcode application running on a Microengine.

In the first and second case the algorithm has been implemented as C code and compiled by using the available GCC compiler with all the optimizations enabled. The effort required to develop a kernel module is greater then that required to build a user space application. However, given the availability of the Resource Manager library, the overall readability of code results enhanced with respect to the direct mapped memory usage in the user space application. On the other hand the Resource Manager library is specifically engineered for control plane tasks, thus it is not optimized to handle the transmission of packets efficiently.

For the third approach we have chosen to implement the algorithm in microC language, which is certainly easier than microengine assembly, while it preserves the same potential strength. In this case we were faced with a new programming model and payed extra time to get a base skill for it. Indeed, the results (Table 1) were not fulfilling the expectations and the attempts to tune the code did not give effective performance gains.

The main difference between generation from XScale (in both ways) and from Microengines is that in the first case we write a packet in DRAM for each transmission, while in the second case we always use the packets already present in the DRAM: actually, the high latency times of this kind of memory prevent the XScale code to keep up with higher packet generation rates.

For our purpose, it is more important to send similar packets at high rates rather than sending potentially completely different packets at low rates. Thus, we choose to have a set of packets (in the simplest case only one) always present in memory and to use Microengines to transmit each time one of the available packets. With this approach we can reach the maximum physical rate for a single port: 1.488 kpps, which for a packet size of 64 bytes is equivalent to 1 Gbps.

---

[2] Resource Manager is part of Intel IXA Portability Framework; although the IXA is not supported by the Radisys ENP-2611, in this case we succeeded in using part of it for our code.

[3] Packets per second

Obviously, transmitting the same packet can be of limited interest (especially for what regards header field contents, such as the source/destination addresses and the TOS); moreover, to store a great number of packets differing only for a few fields (or some combination of them) is not a clever solution. Thus, we introduced the concept of packet template to indicate a common structure of packet with a few fields that can be assigned dynamically at each transmission (according to some rules or some predefined sets of values), while the others have a fixed value.

We can make the concept of packet template clearer with an example. Suppose we want to generate a stream of UDP packets. Now suppose that we assign a fixed value to the source IP address and leave the destination IP address unspecified, since we have a list of possible destination addresses. Without the template mechanism we would create a duplicate of the same packet for each destination IP address; instead, with the template mechanism, only one packet is buffered and the list of possible destination IP addresses has to be passed to the packet generator software. The same software, at run time, will put one of those addresses in the packet's destination IP address field. In this way, a lot of memory has been saved and can be used for other packet templates.

The great benefit of this model is that we can reach the highest transmission rate while preserving all the flexibility required for a network traffic generator. The disadvantage is that all packet templates need to be available in memory before the generation can start.

This implies that the available physical memory represents an upper bound to the number of usable packet templates, but this is not a great issue, since the size of DRAM memory is large enough to store an amount of packet templates adequate for the characterization of many different traffic streams (the actual number of packets depends on their size).

4 The PktGen Framework

The IXP2400 architecture is well suited to be used to create a network traffic generator framework, including both a packet generator engine and a flexible and intuitive configuration interface. We have realized a software tool including three main components (see Fig. 3):

- the packet generator, an application written in microC and running on the MicroEngines, whose main task is the generation of packets with predefined statistical characteristics; using multiple instances of the application it is possible to simultaneously generate a set of traffic streams with homogeneous or heterogeneous characteristics;
- the PktGen core controller, an application running on the XScale, with the goal to initialize and control the packet generators on the MEs, including the packet templates to be used;
- a graphical interface, HTML based and accessible by a standard web browser, used to easily configure the packet generator parameters.

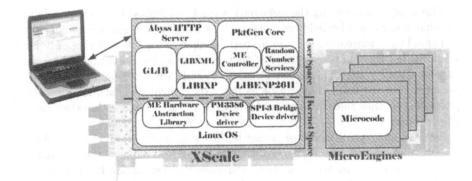

Fig. 3. Software architecture of PktGen.

At present PktGen is able to generate two kinds of traffic, which we consider very significant in testing IP-devices; they can be identified as *Constant Bit Rate* (CBR) and *Best Effort* (BE) or *bursty* streams.

For the BE streams we use a bursty model in which packets are generated in random-sized bursts with random inter-arrival times. The statistics of the two random variables can be modified quite simply, by passing the related probability density functions to PktGen; in our tests, we have used an exponential density function for both the inter-arrival times and the burst sizes. We think to introduce more kinds of traffic in the future; the structure of PktGen is flexible enough to perform this task in a simple and quick manner.

The Packet Generator

The packet generator is a microC code compiled to run on a single MicroEngine in 4-threads mode. This means that actually there are 4 instances of packet generators concurrently executing on the same MicroEngine. Multiple instances of packet generators can run on more MicroEngines without conflicting with one another; in our configuration 5 MicroEngines are reserved for the packet generation task, thus we are free to simultaneously use up to 20 packet stream generators. Indeed, up to 7 MicroEngines can be used for packet generators; one ME must be reserved to the TX driver.

This PktGen component has been engineered to efficiently handle the generation of packets for both CBR and BE traffic models. Independently of the traffic model, the packet generator provides a setup and an operational interface controlled by the core component. The references to packet templates are forwarded to the packet generator through the setup interface, while the operational interface is used to control the generator state (running, stopped).

The two main tasks of the packet generators are the insertion of the variable fields in packet templates and the generation of random variables for statistical traffic characterization (inter-arrival times and burst sizes). At present, we have chosen to work with 4 variable fields in the packet template: Source and Destination IP addresses, TOS value and Total Length. Indeed, there is a fifth dynamic field, the Header Checksum, but it is defined indirectly by the values of the other fields. Moreover, each of the four instances running on the same MicroEngine is able to store up to 160 packet template references: thus, we can obtain a great number of packet patterns that can differ, for example, for payload contents (protocol type and data).

To enable statistical traffic characterization, a hardware uniform pseudo-random number generator is used in conjunction with a set of off-line samples for an arbitrary probability distribution function (that are evaluated by the core controller at setup time, as explained in the following) to dynamically generate values for a given probability density function.

The Core Controller

The core controller component of PktGen running on the XScale processor is built in C language. As shown in Fig. 3, this component is located on top of the software stack that starts from the Linux kernel. The two major sub-components are the ME Controller and Random Number Services. The former allows the Pkt-Gen Core Controller to start, stop and setup packet generators, by encapsulating all the hardware details (features coming from kernel modules as ME Hardware Abstraction Library, PM3386 and SPI-3 device drivers) and by interfacing with the user space libraries provided with the IXP2400 (*libixp* and *libenp2611*, see Fig. 3). The latter is dedicated to handle all the mathematics involved in computing the samples for a given probability density function that are passed to packet generators and used at run time to give a statistical characterization to each traffic stream.

The flexibility of changing the probability density function is one of the main features of PktGen; however, the lack of a floating point unit and the scarceness of memory make the random number generation one of the most critical issues of the entire framework. For this reason a few more words on this topic are needed.

By using the graphical interface the user can specify the analytical formula of any desired density function; the core controller can compute the distribution function (by means of the *libmatheval* [18] and *gsl*[4] [19] libraries) used to find a set of values with the inversion method [20]. Such values are first converted into an integer representation (despite the lack of the floating point unit the XScale can work with this kind of numbers by means of software libraries, but the MEs cannot) and then passed to the packet generator that randomly picks up one of them with a uniform distribution.

The set of values representative of the desired probability density function must be stored in the SRAM memory, because the latency of the DRAM is too high. Unfortunately, the SRAM on the Radisys board is only 8 MB; thus we cannot store a great number of probability samples; the user can choose to use 256 or 64K samples (corresponding to one or two byte per sample).

---

[4] Gnu Scientific Library

Fig. 4. Testbed 1: Measuring the maximum performance of PktGen.

Clearly, the integer conversion and the use of a limited number of samples (up to 64K) introduce a precision loss in the final traffic statistical characterization; we are still investigating the effects on the traffic generation and looking for techniques to minimize the errors introduced.

The Graphical Interface

A simple, yet powerful, WEB-based interface is provided to facilitate the use of PktGen. This interface allows the user to create a different configuration for each desired traffic profile, to save the configuration for later reuse and to run simulations.

To ensure fast code development and maintenability we have used the Abyss HTTP server to provide the graphical interface, while to conform to standard file formats used by many existent tools we have adopted an XML coding of the configuration files.

5 Results

The main result of our activity is PktGen itself: it is a proof that our initial objectives were feasible and we succeeded in achieving them. As a matter of fact, we were able to fully saturate all the three Gigabit interfaces of the Radisys board.

Nonetheless, we are interested in a thorough evaluation of the performance of our tool, expecially for what concerns precision in generation (conformance of packet flows to the statistical description); most of these tests are planned for the immediate future, but some of them have been already carried out and can give an idea of the potentialities of the framework.

The main difficulties we have met concern our lack of a measurement powerful enough testing our generator. The PC architecture cannot sustain the packet rate from PktGen and thus no software tools can be used as meters; indeed, we would need hardware devices that are currently out of the project's budget.

Thus, we have carried on only a few simple measurements on PktGen performance, and to do this we have written a simple packet meter in microC to run it on one MicroEngine. This tool can measure the mean packet rate for any kind of traffic and the jitter for CBR traffic only.

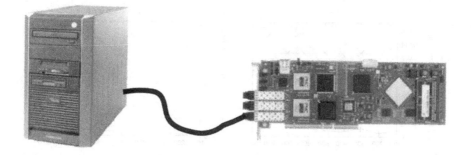

Fig. 5. Testbed 2: Comparison with generation from UDPGen.

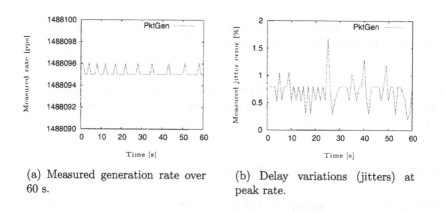

(a) Measured generation rate over 60 s.

(b) Delay variations (jitters) at peak rate.

Fig. 6. Maximum performance of PktGen.

Our measurement tests have been devoted to two main issues: the first was to evaluate the PktGen maximum performance (see Fig. 4), and the second was to compare PktGen with a software tool running on a high-end PC (Fig. 5).

The PktGen maximum performance has been evaluated by means of the testbed shown in Fig. 4, where PktGen runs on one Network Processor and the other is used as the meter. In the worst case of minimum packet size (64 bytes), as reported in Fig. 6, we were able to generate a maximum Constant Bit Rate traffic of 1488.096 kpps, corresponding to 1 Gbps, with a high level of precision: the maximum measured jitter is below 2% of packet delay interarrival time in the CBR stream.

In the second testbed, we compared PktGen with a well-known software tool, namely UDPGen. The latter is used to transmit UDP packets and thus is quite similar to our application, which, however, works at the IP layer. We can see from Fig. 7 how PktGen overcomes UDPGen in terms of precision for what concerns both mean rate and jitter. Again, tests have been carried out for CBR traffic in the worst case of 64 byte packet size. Finally, it is worth noting that

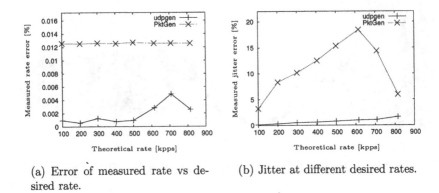

(a) Error of measured rate vs desired rate.

(b) Jitter at different desired rates.

Fig. 7. Comparison between PktGen (running on the Network Processor) and UDPGen (running on a high-end PC) with Constant Bit Rate traffic.

this analysis is limited to a maximum mean rate of 814 kpps, because this is the peak rate obtainable with UDPGen (whereas PktGen can reach about 1488 kpps).

6 Conclusions and Future Work

We have shown that high-performance traffic generation with a Network Processor (in particular with the Intel IXP2400) is possible.

The main result of our activity, namely PktGen, is an open source framework written mostly in C and is easily customizable by everyone; also the microcode used in the MicroEngines is quite intuitive and simple to learn. PktGen is also able to fully saturate the three Gigabit interfaces of the development board (the ENP-2611), reaching a maximum aggregate packet generation rate of about 4464 kpps (3 Gbps).

We can therefore conclude that our aim is completely achieved: we have demonstrated how it is possible to build an open source, customizable, high-speed and precise packet generator without specific hardware; the cost is less expensive than that of a professional device (the Radisys ENP-2611 costs about $ 5000).

We consider this only a first important step in this field: in fact, we think that more work has to be done in this direction.

First of all we have planned to extend the PktGen functionalities to include also a packet meter, able to collect detailed and precise information on network traffic streams (mean rate, jitter, packet classification, etc.); this is necessary to have a counterpart to the generator, with the same noteworthy characteristics (open source, low cost, high performance and precision).

The second important goal is to port PktGen to a more performing architecture, for example to the IXP 2800, in order to realize a more and more powerful traffic generator.

Other minor evolutions are foreseen, such as the introduction of additional traffic shapes in PktGen, the comparison with professional devices and its utilization in our research activities on Quality of Service, high-speed networks, and Open Router architecture [21].

7 Acknoledgements

We would like to thanks Intel for its support to our work. Intel gave us the two Radisys ENP-2611 development board with IXP2400 Network Processor that we have used for the development of PktGen and have funded our research in the last year.

References

1. Bradner, S., McQuaid, J.:   Benchmarking methodology for network interconnect devices.   RFC 2544, IETF (1999) Available online, URL: http://www.ietf.org/rfc/rfc2544.txt.
2. Mandeville, R., Perser, J.:   Benchmarking methodology for LAN switching devices.   RFC 2889, IETF (2000) Available online, URL: http://www.ietf.org/rfc/rfc2889.txt.
3. Tirumala, A., Qin, F., Dugan, J., Ferguson, J., Gibbs, K.: Iperf. Available online, URL: http://dast.nlanr.net/Projects/Iperf/ (2005)
4. Jones, R., Choy, K., et al.:   Netperf.   Available online, URL: http://www.netperf.org/ (2005)
5. Sommers, J., Barford, P.: Self-configuring network traffic generation. In: Proceedings of the 4th ACM SIGCOMM Conference on Internet Measurement 2004, Portland, OR - USA, IMT 04 (2004)
6. Rousskov, A., Wessels, D.:   Web Polygraph.   Available online, URL: http://www.web-polygraph.org/ (2004)
7. Turner, A.: Tcpreplay. Available online, URL: http://tcpreplay.sourceforge.net/ (2005)
8. Caldera Technologies, C.:   LANforge-FIRE.   Available online, URL: http://www.candelatech.com/ (2005)
9. Cisco Systems, C.:   IOS netflow feature.   Available online, URL: http://www.cisco.com/en/US/tech/tk812/tsd_technology_support_protocol_home.html (2005)
10. Anritsu, C.: IP/Ethernet analyser, Model: MD1231A. Available online, URL: http://www.eu.anritsu.com/products/default.php?p=97&model=MD1231A (2005)
11. Smartbits: AX/400. Available online, URL: http://www.netcomsystems.com/ (2005)
12. Comer, D.E.: Network Systems Design using Network Processors - Agere version. Pearson Prentice Hall, Upper Saddle River, New Jersey - USA (2005)
13. University of Kentucky, L.f.A.N.: IXPKTGEN project. Available online. URL: http://protocols.netlab.uky.edu/ esp/pktgen/ (2004)

14. Wind River, C.: Wind River Operating Systems. Available online, URL: http://www.windriver.com/products/device_technologies/os/ (2005)
15. Montavista: Montavista Linux Preview kit. Available online, URL: http://www.mvista.com/previewkit/index.html (2004)
16. Montavista: Montavista Linux Professional Edition. Available online, URL: http://www.mvista.com/products/pro/ (2004)
17. Buytenhek, L.: Port of Fedora for XScale processor". Available online, URL: http://skrybele.wantstofly.org/ (2005)
18. GNU: Libmatheval library. Available online, URL: http://www.gnu.org/software/libmatheval/ (2005)
19. GNU: Gsl library. Available online, URL: http://www.gnu.org/software/gsl/ (2005)
20. L'Ecuyer, P. In: Random Number Generation. Handbook of Computational Statistics. Springer-Verlag (2004) pp. 35-70
21. Bolla, R., Bruschi, R.: A high-end linux based open router for IP QoS networks: tuning and performance analysis with internal (profiling) and external measurement tools of the packet forwarding capabilities. In: Proc. of the 3rd International Workshop on Internet Performance, Simulation, Monitoring and Measurements (IPS MoMe 2005), Warsaw - Poland, Institute of Telecommunications, Warsaw University of Technology (2005)

IP Forwarding Performance Analysis In The Presence Of Control Plane Functionalities In A PC-Based Open Router

Raffaele Bolla, Roberto Bruschi

Department of Communications, Computer and Systems Science (DIST)
University of Genoa
Via Opera Pia 13, I-16145 Genova, Italy
Email: raffaele.bolla, roberto.bruschi@dist.unige.it

Abstract. Nowadays, networking equipment is realized by using decentralized architectures that often include special-purpose hardware elements. The latter considerably improve the performance on one hand, while on the other they limit the level of flexibility. Indeed, it is very difficult both to have access to details about internal operations and to perform any kind of interventions more complex than a configuration of parameters. Sometimes, the "experimental" nature of the Internet and its diffusion in many contexts suggest a different approach. This type of need is more evident inside the scientific community, which often encounters many difficulties in realizing experiments. Recent technological advances give a good chance to do something really effective in the field of open Internet equipment, also called Open Routers (ORs). The main target approached in this paper is to extend the evaluation of the OR forwarding performance proposed in [1], by analyzing the influence of the control plane functionalities.

Keywords. Open Router, Linux kernel, IP forwarding performance

1 Introduction

During the past 20 years, Internet equipment has radically changed many times, to meet the increasing quantity and the complexity of new functionalities [2]. Nowadays, current high-end IP nodes belong to the 3rd generation, and are designed to reach very high performance levels, by effectively supporting high forwarding speeds (e.g., 10 Gbps). With this purpose, these devices are usually characterized by decentralized architectures that often include custom hardware components, like ASIC or FPGA circuits. On one hand, the dedicated hardware elements improve the performance considerably, while, on the other hand, they limit the level of flexibility. Moreover, with respects to this commercial equipment, it is very difficult both to have access to internal details and to perform any kind of interventions that would require more complex operations than those involved by a configuration of parameters: in this case, the "closure" to external modifications is a clear attempt to protect the industrial investment.

In many contexts the "experimental" nature of the Internet and its diffusion suggest a different approach. This type of need is more evident within the scientific community, which often finds many difficulties in realizing experiments, test-beds and trials for the evaluation of new functionalities, protocols and control mechanisms. But also the market frequently asks for a more open and flexible approach, like that suggested by the Open Source philosophy for software. This is especially true in those situations where the network functions must be inserted in products, whose main aim is not limited to realizing basic network operations.

As outlined in [1] and in [3] (that report some of the most interesting results obtained in the EURO project [4]), the recent technology advances give a good chance to do something really effective in the field of open Internet devices, sometimes called Open Routers (ORs) or Software Routers. This possibility comes, for what concerns the software, from the Open Source Operating Systems (OSs), like Linux and FreeBSD (which have sophisticated and complete networking capabilities), and for what concerns the hardware from the COTS/PC components (whose performance is always increasing, while their costs are decreasing). The attractiveness of the OR solution can be summarized in multi-vendor availability, low-cost and continuous update/evolution of the basic parts, as assumed by Moore's law.

The main target approached in this paper is to extend the evaluation of the OR forwarding performance proposed in [1], by analyzing the influence of the control plane functionalities. In fact, we have to outline that in a PC-based OR, unlike in much industrial dedicated equipment, the data and the control plane have to share the computational resources of the CPU(s) in the system. Thus, our main objective is to study and to analyze, both with external (throughput) and internal (profiling) measurement tools, if and how the presence of heavy control operations may affect the performance level of the IP forwarding process.

The paper is organized as in the following. The next Section describes the architecture of the open node for what concerns both hardware and software structure. Section III reports the parameter tuning operations realized on the OR to obtain the maximum performance. Section IV describes the external and internal performance evaluation tools used in the tests, while Sections V reports the numerical results of all the most relevant experiments. The conclusions and future activities are in Section VI.

2   Open Router Architecture

To define the OR reference architecture, we have established some main criteria and we have used them to a priori select a set of basic elements. The objective has been to obtain a high–end node base structure, able to support top performance with respect to IP packet forwarding and control plane elaborations. The decision process, the criteria and the final selection results are described in some detail in the following, separately for hardware and software elements.

2.a   Hardware Architecture
The PC architecture is a general-purpose one and it is not specifically optimized for network operations. This means that, in principle, it cannot reach the same

performance level of custom high-end network equipment, which generally uses dedicated HW elements to handle and to parallelize the most critical operations. This characteristic has more impact on the data plane performance, where custom devices usually utilize dedicated ASIC, FPGA, Network Processor and specific internal bus, to provide a high level of parallelism in the packet processing and exchange. On the other hand, COTS hardware can guarantee two very important features as the cheapness and the fast and continuous evolution of many of its components. Moreover, the performance gap might be not so large and anyway more than justified by the cost difference.

The PC internal data path uses a centralized I/O structure composed by: the I/O bus, the memory channel (both used by DMA to transfer data from network interfaces to RAM and vice versa) and the Front Side Bus (FSB) (used by the CPU with the memory channel to access to the RAM during the packet elaboration). It is evident that the bandwidth of these busses and the PC computational capacity are the two most critical hardware elements involved in the determination of the maximum performance in terms of both peak passing bandwidth (in Mbps) and maximum number of forwarded packets per second.

For example, a full duplex Gigabit Ethernet Interface makes immediately inadequate the traditional PCI bus (32 bits with a frequency of 33 MHz). With these high-speed interfaces we need at least one of the two newest evolutions of this standard, namely PCI-X and PCI-Express, which can assure the needed overall bandwidth to handle the traffic of several Gigabit Ethernet interfaces. So, the selection criteria have been very fast internal busses and a dual CPU system with high integer computational power.

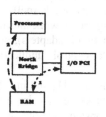

Fig. 1. Scheme of the packet path in a PC hardware architecture.

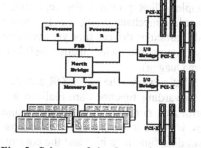

Fig. 2. Scheme of the Supermicro X5DL8-GG main board.

With this goal, we have chosen the Supermicro X5DL8-GG main board, mounting a ServerWorks GC-LE chipset, whose structure is shown in Fig. 2. This chipset can support a dual-Xeon system with a dual memory channel and a PCI-X bus at 133 MHz, with 64 parallel bits. The Xeon processor has the Pentium 4 core that allows multiprocessor configurations and Hyper-Threading (HT) capability enabled. It derives from the Intel NetBurst architecture and it is one of the most suitable 32 bit processors for high-end server architectures, thanks to its large L2/L3 cache size. In our tests we used a Xeon version with 2.4 GHz clock and 512 KB sized cache. The memory bandwidth, supported by this chipset, matches the system bus speed, but the most important point is that the memory is 2-way interleaved, which assures high performance and low average access latencies. The bus that connects the North Bridge to the PCI bridges, namely IMB, has more bandwidth (more than 25 Gbps) than the maximum combined bandwidth of the two PCI-X busses (16 Gbps) on each I/O bridge.

Network interfaces are another critical element in the system, as they can heavily condition the PC Router performance. As reported in [5], the network adapters on the market have different levels of maximum performance and a different configurability. In this respect, we have selected the Intel PRO 1000 XT Server network interfaces, which are equipped with a PCI-X controller, supporting the 133 MHz frequency and also a wide configuration range for many parameters, like, for example, transmission and receive buffer lengths, maximum interrupt rates and other important features [6].

2.b   Software Architecture

The software architecture of an OR has to provide many different functionalities: from the ones directly involved in the packet forwarding process to the ones needed for control functionalities, dynamic configuration and monitoring. In particular, we have chosen to study and to analyze a Linux based OR framework, as it is one of the open source OSs that have a large and sophisticated kernel-integrated network support, it is equipped with numerous GNU software applications. and it has been selected in the last years as framework for a large part of networking research projects.

Moreover, we have also decided not to take into account some software architectures, extremely customized to the network usage (e.g., Click [7], [8]): even if they often provide higher performance with respect to the "standard" frameworks, they often result compatible with few hardware components and standard software tools. For example, Click provides (by replacing the standard network kernel) a very fast forwarding mechanism [9], which is realized by polling the network interfaces, but it needs particular drivers (not yet available for many network adapters), and it is not fully compatible with many well-known network control applications (e.g., Zebra/Quagga).

For what concerns the Linux OR architecture, as outlined in [1] and in [3], while all the forwarding functions are realized inside the Linux kernel, the large part of the control and monitoring operations is running as daemons/applications in user mode. Thus, we have to outline that, unlike most of the commercial network equipment, the forwarding functionalities and the control ones have to share the CPUs in the system. In fact, especially the high-end commercial network equipment provides separated computational resources for these processes: the forwarding is managed by a switching fabric, which is a switching matrix, often realized with ad hoc hardware elements (ASICs, FPGAs and Network Processors), while all the control functionalities are executed by one ore more separated processors.

Let us now go into the details of the Linux OR architecture: as shown in Fig. 3 and previously sketched, the control plane functionalities run in the user space, while the forwarding process is entirely realized inside the kernel.

In particular, the networking code of the kernel is composed by a chain of three main modules, namely receiving API (RxAPI), IP Processing, transmission API (TxAPI), which manage respectively the reception of packets from the network interfaces, their IP layer elaboration and their transmission to the network devices. Moreover, the first of these modules, the RxAPI, has experienced many structural and refining developments, as it is a very critical element for performance optimization.

The NAPI (New API) [10] is the latest version of RxAPI architecture, it is available from the 2.4.22 kernel, and it has been explicitly created to increase the system scalability, as it can handle network interface requests with an interrupt moderation mechanism that allows to adaptively switch from a classical interrupt management of the network interfaces to a polling one.

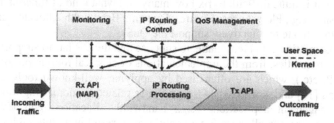

Fig. 3. Software architecture of a Linux-based OR.

Another important element of the kernel integrated forwarding architecture is the memory management. All the kernel code is structured with a zero-copy statement [10]: to avoid unnecessary and onerous memory transfer operations of packets, they are left to reside in the memory locations used by the DMA-engines of ingress network interfaces, and each operation on the packets is performed by using a sort of pointer to the packet and to its key fields, called descriptor or *sk_buff*. These descriptors are effectively composed by pointers to the different fields of the headers contained in the associated packets.

Moreover, the *sk_buff* data structure allows to realize all the buffers used by the networking layers, as it provides the general buffering and flow control facilities needed by network protocols. A descriptor is immediately allocated and associated to each received packet in the sub-IP layer, it is used in every subsequent networking operation, and finally de-allocated when the network interface signals the successful transmission.

The network code of the newest versions of the Linux kernel allows managing the forwarding process with Symmetric MultiProcessing (SMP) configurations. In fact, starting from version 2.0 on, the networking architecture of the Linux kernel has been developed by using a thread-based structure. Moreover, the kernel (>2.4.26) allows statically associating the elaboration of all the packets, which have to be transmitted or which have been received by a network interface, to a single CPU. This is a feasible solution to reduce concurrency effects, like cache contentions/invalidates due to locks, counters and ring buffer access, which can heavily decrease the average useful CPU cycles. By the way, as shown in [1], when the forwarding process involves two network interfaces associated to different CPUs, the packet descriptors, after that the routing decision has been taken and the egress interface selected, have to change the associated CPU, and the concurrency effects decrease the useful CPU cycles and give raise to a heavy performance reduction. Thus, in some particular cases (e.g., when it is quite likely that packets cross interfaces associated to different CPUs), increasing the number of CPUs may even decrease the overall forwarding performance. For these reasons, to maximize the forwarding performance, it is reasonable to bind the network interfaces that exchange high amounts of packets among them to a single CPU.

Moreover, for what concerns the control plane, it is very difficult to populate a list of control functionalities that have to be probably activated in the OR. In fact, except the classical IP routing protocols, it is reasonable to suppose that the control plane functionalities to activate on a router mostly depend on the whole network scenario. For example, it is quite difficult to fix how many and what kind of functionalities and of mechanisms (e.g., Flow Access Control, Dynamic Bandwidth Allocator, and so on) an OR has to activate to effectively support the QoS.

As to the control plane, it is reasonable to suppose that the OR has to support different types of functionalities: from the implementations of classical routing protocols (RIP, OSPF, BGP, etc.), which are often used by applying well-known tools like Zebra [11], Quagga [12] and Xorp [13][14], to the ones related to monitoring modules and to QoS management applications (e.g., bandwidth allocations and flow access control). The support to all these functionalities may result in a quite high number of processes, every so often scheduled and activated, which as a whole may represent a non-negligible computational load for the OR.

Note that our main objective is not to evaluate the performance of the control plane, whose functionalities and performance analysis require both a heavy effort and the usage of sophisticated tools, to emulate large and complex network environments, but to analyze how its presence may influence the forwarding performance. For these reasons, in our tests we have decided to not take explicitly into account the real OR control plane applications, but to use a set of dummy processes to emulate them by performing a reasonable set of operations (i.e., integer and floating point calculations, memory allocations, disk writing, etc.). Thus, we have decided to use a very "CPU hungry" set of always active dummy control processes that can give us a reasonable idea of the OR performance in a worst case.

3   Performance tuning

As shown in Section V and reported in [1], the computational capacity power appears to be the only real bottleneck in the packet forwarding process of our OR architecture. In this context, "computational power" means both the software architecture efficiency and the characteristics of some hardware components, like the CPUs, the North Bridge chipset and the RAM banks. In fact, while for medium-large sized datagrams the OR achieves the full Gigabit/s capacity, the maximum throughput is limited to only about 400K packets per second in the presence of an ingress flow composed by 64 byte-sized datagrams, with a standard 2.5 version of the Linux kernel.

To maximize the forwarding performance, a reduction of the computational complexity of the OR software architecture is clearly needed. Thus, in this respect, we have performed the tuning of some driver and kernel parameters and introduced some interesting architectural refinements.

For what concerns the driver parameter tuning, we have to take into account that many recent network adapters [6] allow to change the ring buffer dimension and the maximum interrupt rates. Both these parameters have a great influence on the NAPI performance. In this respect, [10] shows that the interrupt rate of network adapters should not be limited in NAPI kernels and that ring buffers should be large to prevent the packet drops in the presence of bursty traffic. Note that, as a too large ring buffer may causes high packet latencies, we have chosen to set the ring buffer sizes to the minimum value that permits to achieve a good performance level.

For what concerns the kernel parameters, we have decided to over-dimension all the internal buffers (e.g., IP layer and egress buffers), to avoid useless internal drops of already processed packets. Another important parameter to tune is the *quota* value that fixes, in the NAPI mechanism, the number of packets that each device can elaborate at every polling cycle.

It is also possible to act on some specific 2.5 kernel parameters, by adapting them to the specific networking usage: for example, the profiling results in [1] show that the kernel scheduler operations employ about 4-5% of the overall quantity of computational resources uselessly. To avoid this CPU time waste, the OS scheduler clock frequency should be decreased: by reducing its value to 100 Hz, the forwarding rate improves of about 20K packets per second.

The rationalization of memory management is another important aspect: as highlighted in the profiling results of Section V, a considerable part of the available resources is used in the allocation and de-allocation of packet descriptors (memory management functions). Reference [16] proposes a patch that allows to recycle the descriptors of the successfully sent packets: the basic idea is to save CPU resources during the receive NAPI operations, by reusing the packet descriptors inside the completion queue. The use of this patch can again improve the performance.

Summarizing, our optimized NAPI 2.5.75 kernel image includes the descriptor recycling patch and the 5.1.13-k1 version of e1000 driver, and it has been configured with the following optimized values:

I) the Rx and Tx ring buffers have been set to 512 descriptors;
II) the Rx interrupt generation has not been limited;
III) the egress buffer size for all the adapters has been dimensioned equal to 20,000 descriptors;
IV) the NAPI *quota* parameter has been set to 23 descriptors;
V) the scheduler clock frequency has been fixed to 100 Hz.

4   Testbed and measurement tools

The OR performance can be analyzed by using both internal and external measurement methods. The external measures can be performed by using both hardware and software tools, which are outside the OR, and which usually provide global performance indexes, such as, for the forwarding process, the throughput or the maximum delay. Internal measures are obtained by using specific software tools (called profilers) placed inside the OR, which are able to trace the percentage of CPU utilization for each software modules running on the node.

Internal measurements are very useful for the identification of the architecture bottlenecks. The problem is that many of these tools require a relevant computational effort that perturbs the system performance, and that makes the results not meaningful. In this respect, their correct selection is a strategic point. We have verified with many different tests that one of the best tools is Oprofile [17], an open source code that realizes a continuous monitoring of system dynamics with a frequent and quite regular sampling of CPU hardware registers. Oprofile allows evaluating, in a very effective and deep way, the CPU utilization of each software application and each single kernel/application function running in the system with a very low computational overhead.

For what concerns the external measurements, to test the forwarding performance we have chosen an open source software-benchmarking code running on PC. More in particular, since all the well-known and commonly used traffic generation applications (like NetPerf [18] or Rude&Crude [19]) do not achieve the scalability level needed to generate and to measure high packet rates, we have used a kernel level tool. It allows a high performance level, much higher than the classical kind of software generators and measurers. The latter are simpler and they work nearer to the hardware, but obviously they have a lower flexibility level. Most of the kernel level benchmarking codes are based on the same idea: the creation in the reserved kernel space of one or more packet patterns, and their recycle for multiple transmissions. Among the different ones available, (we can cite Kernel Generator [20], UDPGen [21]) we have chosen to use the Click environment with some generation and measurement modules [10].

Thus, by using the previously cited tools, we have defined a testbed composed by the OR itself and some other PCs dedicated to traffic generation and measurement. As shown in Fig. 4, we have also used a 3Com Office Connect Gigabit Switch to aggregate the traffic generated by 3 PCs on one or more router interfaces and 1 PC to measure the output traffic.

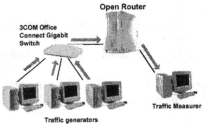

Fig. 4. Scheme of the testbed used for the OR performance evaluation.

5   Numerical Results

In this section, some of the numerical results that were obtained are shown. In particular, we have decided to report the performance evaluation results in terms of external and internal measurements for both a single processor and an SMP OR software architecture. We have performed several tests for each selected OR configuration, with and without the activation of the control plane dummy processes, by increasing the traffic offered load. Since, as outlined in [1], the computational capacity is the main bottleneck of the OR architecture, we have chosen to perform all the tests by using a traffic flow with 64 Byte sized datagrams.

Concerning the results reported in this Section, we show two types of indexes: external throughput (i.e., the throughput crossing the OR), and profiling information, represented by the CPU percentage used by the different kernel parts. In particular, about this last type of results, all the kernel functions have been grouped in few homogeneous sets: CPU idle, scheduler, memory management, IP processing, NAPI, Tx API, IRQ routines, Ethernet processing, Oprofile and Control Plane.

Thus, in the first test session we have used a single processor running a 2.5.75 kernel with different optimization levels: a standard version, a standard version with the tuning of driver parameters, and a version that includes the parameter tuning and the descriptor recycling patch.

As shown in Fig. 5, when the control plane processes are not running, the applied optimizations allow to nearly double the maximum throughput and to achieve a forwarding rate of about to 720K packets per second (that corresponds to about half of the Gigabit capacity). The presence of active control plane processes reduces the maximum throughput obtainable with all the three adopted versions in a considerable way. In particular, Fig. 6 shows that, in such environment, the three kernel versions achieve a maximum throughput value nearly equal to 315K, 415K and 500K packets per second.

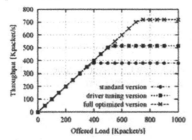 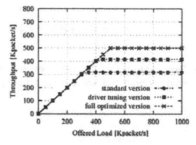

Fig. 5. Maximum throughput versus traffic load for different versions of single CPU kernel.

Fig.6. Maximum throughput versus traffic load for different versions of single CPU kernel in the presence of active control plane processes.

Figs 7, 8, and 9 report the profiling results of the tests without the control plane, while in Figs 10, 11 and 12 the ones including it are shown. In all these tests, the CPU resources used by the forwarding process (IP Processing, NAPI, Tx API, Ethernet Processing and Memory) increase their computational weight proportionally to the forwarding rate, while the IRQ management operations show a decreasing behaviour, which depends on the characteristic of NAPI kernels; when the traffic load offered to the ingress interfaces increases, the NAPI passes adaptively from an interrupt mechanism to a polling one, by reducing the interrupt rate.

Fig. 8 outlines that the driver parameter tuning causes an increase of IRQ management contribution. This effect is a clear indication that the driver parameter tuning produces a more aggressive behaviour of the network interfaces, which try to activate the kernel network stack more frequently.

Moreover, the descriptor-recycling patch used in the fully optimized kernel version allows to reach the best forwarding performance, by reducing to zero the memory management computational weight (that reaches, in the other kernel versions, nearly 55% of the available CPU resources).

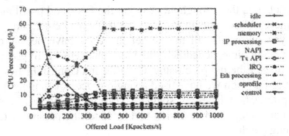

Fig. 7. Profiling results of the single processor standard kernel version without any processes active in the control plane. The forwarding rate is shown in Fig. 5.

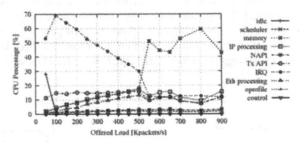

Fig. 8. Profiling results of the single processor kernel version with the driver parameter tuning and without any processes active in the control plane. The forwarding rate is shown in Fig. 5.

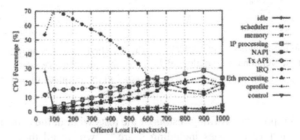

Fig. 9. Profiling results of the single processor fully optimized kernel version (that includes the driver parameter tuning and the descriptor-recycling patch) without any processes active in the control plane. The forwarding rate is shown in Fig. 5.

Figs 10, 11 and 12 show that, when there are active processes in the control plane, the forwarding operations appear to have a very similar behaviour, with a slightly reduced computational weight with respect to those in Figs. 7, 8 and 9. For what concerns the control plane, the active processes tend to use few computational resources in the presence of a high IRQ rate, while, when the OR gets saturated, they become stable at a computational weight of about 30-35% of the available resources.

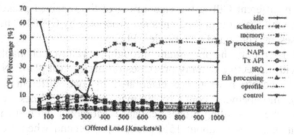

Fig. 10. Profiling results of the single processor standard kernel version without any processes active in the control plane. The forwarding rate is shown in Fig. 6.

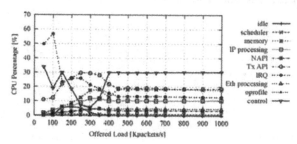

Fig. 11. Profiling results of the single processor kernel version with the driver parameter tuning and without any processes active in the control plane. The forwarding rate is shown in Fig. 6.

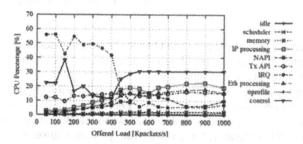

Fig. 12. Profiling results of the single processor fully optimized kernel version (that includes the driver parameter tuning and the descriptor-recycling patch) without any processes active in the control plane. The forwarding rate is shown in Fig. 6.

In the second test session we have used a SMP version of the optimized 2.5.75 Linux kernel (that includes the driver parameter tuning and the descriptor-recycling patch). In particular, we have chosen to use the following three different kernel configurations:

- no CPU assignments: the elaboration of both the packets received or transmitted by the different network interfaces and the control plane processes is not bound to a specific CPU. Note that, when there are no CPU-NIC bindings, the Linux kernel uses the same CPU to process all the packets.

- NICs on a single CPU: the elaboration of the received or transmitted packets is bound to CPU 0, while the control plane processes run on CPU 1.
- NICs on different CPUs: the elaboration of the received packets is bound to CPU 0, while the transmitted ones are processed by CPU 1. The control plane processes are not bound on a single CPU.

Figs. 13 and 14 report the maximum throughput values obtainable respectively without and with the active dummy processes on the control plane. Note that, when the forwarded packets have to be elaborated by two different CPUs (i.e., 3rd configuration), the memory concurrency management becomes critical and the performance of the forwarding process collapse. Moreover, with such configuration the throughput increases of about 15K packets per second when the control plane processes are active: in fact, the presence of other running processes in the system seems to reduce the concurrence of the two CPUs for the kernel reserved memory, where all the packet descriptors are located.

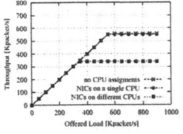
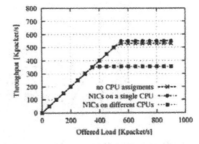

Fig. 13. Maximum throughput versus traffic load for different configurations of the SMP Linux kernel.

Fig.14. Maximum throughput versus traffic load for different configurations of the SMP Linux kernel in the presence of active control plane processes.

Figs. 15, 16 and 17 report the profiling results obtained without control plane, while Figs. 18, 19 and 20 report the profiling results of the tests while such processes are active with the three kernel configurations.

By comparing the results obtained with the first configuration (Figs. 15 and 18), we can note how the operation sets seem to have a behaviour almost similar to the single processor case: for example, also in this case, when the IRQ management functions increase their CPU occupancy, the control plane processes lower their performance. This effect is avoided by using the 2nd kernel configuration: the results in Figs. 19 do not outline any decay of the performance of control processes. Moreover, this SMP configuration seems to be the only feasible way to make the performance of the forwarding process uncorrelated with the one of the control plane.

By observing the profiling results of the 3rd configuration (Figs. 17 and 20), we can note that the CPU utilization of memory management function rises with respect to the other SMP case: as previously sketched, this particular effect is caused by a more critical memory management in the forwarding process, due to the concurrency between the CPUs to access the kernel reserved memory. Moreover, also the control plane processes seem to suffer low CPU resources.

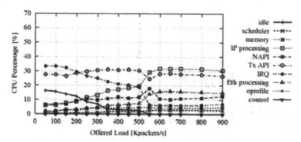

Fig. 15. Profiling results of the SMP fully optimized kernel version (that includes the driver parameter tuning and the descriptor recycling patch) with the 1st configuration and without any processes active in the control plane. The forwarding rate is shown in Fig. 13.

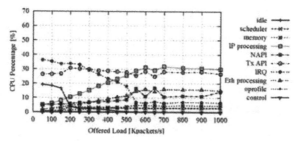

Fig. 16. Profiling results of the SMP fully optimized kernel version (that includes the driver parameter tuning and the descriptor recycling patch) with the 2nd configuration and without any processes active in the control plane. The forwarding rate is shown in Fig. 13.

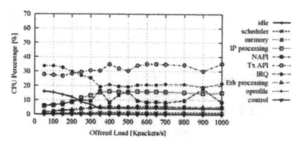

Fig. 17. Profiling results of the SMP fully optimized kernel version (that includes the driver parameter tuning and the descriptor recycling patch) with the 3rd configuration and without any processes active in the control plane. The forwarding rate is shown in Fig. 13.

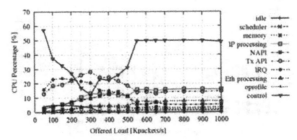

Fig. 18. Profiling results of the SMP fully optimized kernel version (that includes the driver parameter tuning and the descriptor recycling patch) with the 1st configuration and with the processes active in the control plane. The forwarding rate is shown in Fig. 14.

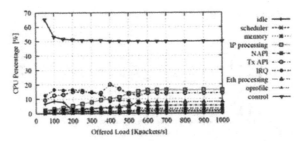

Fig. 19. Profiling results of the SMP fully optimized kernel version (that includes the driver parameter tuning and the descriptor recycling patch) with the 2nd configuration and with the processes active in the control plane. The forwarding rate is shown in Fig. 14.

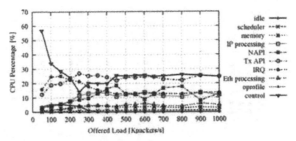

Fig. 20. Profiling results of the SMP fully optimized kernel version (that includes the driver parameter tuning and the descriptor recycling patch) with the 3rd configuration and with the processes active in the control plane. The forwarding rate is shown in Fig. 14.

6   Conclusion

The main contribution of this work has been reporting the results of a deep activity of optimization and testing realized on a PC Open Router architecture based on Linux software, and, more in particular, based on Linux kernel 2.5. The main objective has been the performance evaluation (with respect to packet forwarding) of an optimized OR, both with external (throughput) and internal (profiling) measurements.

The obtained results show that a single processor OR can achieve interesting performance for what concerns the forwarding process, but, in the presence of active processes in the control plane this performance can notably decrease. The results obtained with the SMP forwarding kernel show that, by binding the forwarding process and the control plane functionalities to different CPUs, the OR can reach a maximum throughput nearly equal to 550K packets per second, without any interference (in terms of performance) with the control plane processes.

Reference

1. Bolla, R., Bruschi, R.: A high-end Linux based Open Router for IP QoS networks: tuning and performance analysis with internal (profiling) and external measurement tools of the packet forwarding capabilities. Proc. of the 3rd International Workshop on Internet Performance, Simulation, Monitoring and Measurements (IPS MoMe 2005), Warsaw, Poland (2005) pp. 203-214
2. Comer, D.: Network Processors: Programmable Technology for Building Network Systems. The Internet Protocol Journal Vol. 7(4) (2004) pp. 2-12.
3. Bianco A., Finochietto, J. M., Galante, G., Mellia, M., Neri, F.: Open-Source PC-Based Software Routers: a Viable Approach to High-Performance Packet Switching. Proc. of the 3rd International Workshop on QoS in Multiservice IP Networks (QoS-IP 2005), Catania, Italy (2005) pp. 353-366
4. EURO: University Experiment of an Open Router, http://www.diit.unict.it/euro
5. Gray, P., Betz, A.: Performance Evaluation of Copper-Based Gigabit Ethernet Interfaces. Proc. of the 27th Annual IEEE Conference on Local Computer Networks (LCN'02), Tampa, Florida (2002) pp. 679-690
6. The Intel PRO 1000 XT Server Adapter,   http://www.intel.com/network/connectivity/products/pro1000xt.htm
7. The Click! Modular Router, http://www.pdos.lcs.mit.edu/ click/
8. Kohler, E., Morris, R., Chen, B., Jannotti, J., Kaashoek, M. F.: The Click modular router. ACM Transactions on Computer Systems Vol. 18(3) (2000) pp. 263-297
9. Bianco, A., Birke, R., Bolognesi, D., Finochietto, J. M., Galante, G., Mellia, M.: Click vs. Linux: Two Efficient Open-Source IP Network Stacks for Software Routers. Proc. of the IEEE Workshop on High Performance Switching and Routing (HPSR 2005), Hong Kong (2005)
10. Salim, J. H., Olsson, R., Kuznetsov, A.: Beyond Softnet. Proc. of the 5th annual Linux Showcase & Conference, Oakland CA (2001)
11. Zebra, http://www.zebra.org/
12. Quagga Software Routing Suite, http://www.quagga.net
13. Xorp Open Source IP Router, http://www.xorp.org/
14. Handley, M., Hodson, O., Kohler, E.: XORP: an open platform for network research. ACM SIGCOMM Computer Communication Review Vol. 33(1) (2003) pp. 53-57
15. Cox, A.: Network Buffers and Memory Management. Linux Journal (1996), http://www2.linuxjournal.com/ lj−issues/issue30/1312.html
16. The descriptor recycling patch,   ftp://robur.slu.se/pub/Linux/net-development/skb_recycling/
17. Oprofile, http://oprofile.sourceforge.net/news/
18. NetPerf, http://www.netperf.org/netperf/NetperfPage.html

19. Rude&Crude, http://rude.sourceforge.net/
20. UDPGen,   http://www.fokus.gmd.de/research/cc/berlios/employees/sebastian.zander/
    private/udpgen/
21. The Linux Documentation Project,   http://www.tldp.org/tldp-redirect.php?url=/HOWTO/
    Kernel-HOWTO.html

Chapter III

Data Acquisition And Aggregation
In Sensor Networks

On Data Acquisition And Field Reconstruction In Wireless Sensor Networks*

Carla-Fabiana Chiasserini, Alessandro Nordio, and Emanuele Viterbo

Dipartimento di Elettronica, Politecnico di Torino
C. Duca degli Abruzzi 24, I-10129 Torino (Italy)
e-mail: <name>@polito.it

(Invited Paper)

Abstract. Wireless sensor networks are often used for environmental monitoring applications. Sampling and reconstruction of a physical field is therefore one of the most important problems to solve. We focus on band-limited fields and investigate the relationship between the random topology of a sensor network and the quality of the reconstructed field. By reviewing irregular sampling theory, we derive some guidelines on how sensors should be deployed over a spatial area for efficient data acquisition and reconstruction. We analyze the problem using random matrix theory and show that even a very irregular spatial distribution of sensors may lead to a successful signal reconstruction, provided that the number of collected samples is large enough with respect to the field bandwidth.

Keywords: sensor networks, field reconstruction, irregular sampling.

1 Introduction

One of the most popular applications of wireless sensor networks is environmental monitoring. In general, a physical phenomenon (hereinafter also called sensor or physical field) may vary over both space and time. In this work, we address the problem of sampling and reconstruction of a one-dimensional, spatial field at a particular time instant. We focus on a band-limited field (e.g., pressure and temperature), and assume that sensors are randomly deployed over the area of interest. Also, nodes can represent each sample with a sufficient number of bits, so that the quantization error is negligible.

Data are transferred from the sensors to a common data-collecting unit, the so-called sink node. In this work, however, we are concerned only with acquisition and reconstruction of the sensor field, and we do not address issues related to information transport. Thus, although studying the effect of errors and losses due to data transfer is of great interest, we assume that all data is correctly received at the sink. Furthermore, we assume that the sensors position are known. This

* This work was supported through the PATTERN project

implies that nodes are either located at pre-defined positions, or, if randomly deployed, their location can be acquired.

Our objective is to investigate the relation between the network topology and the quality of the reconstructed field. More specifically, we aim at identifying the topology characteristics and, hence, a sample distribution that allows the sink node to reconstruct the signal of interest with the desired precision. The main contributions of this paper are the following:

(i) by reviewing irregular sampling theory, we derive some guidelines on the number of sensors to be deployed and on how they should be spatially spaced so as to successfully reconstruct the measured field;

(ii) by analyzing the problem using random matrix theory, we show that even a very irregular spatial distribution of sensors may lead to a successful signal reconstruction, provided that the number of collected samples is large enough with respect to the field bandwidth;

(iii) we identify the theoretical basis to estimate the required number of active sensors, given the field bandwidth.

2   Related work on data acquisition in sensor networks

To the best of our knowledge, few works have addressed the problem of sampling and reconstruction in sensor networks. Efficient techniques for spatial sampling in sensor networks are proposed in [4, 5]. In particular [4] presents an algorithm to determine which sensor subsets should be selected to acquire data from an area of interest and which nodes should remain inactive to save energy. The algorithm chooses sensors in such a way that the node positions can be mapped into a blue noise binary pattern. In [5], an adaptive sampling is described, which allows the central data-collector to vary the number of active sensors, i.e., samples, according to the desired resolution level. The problem of data acquisition is also addressed in [3], where the authors consider a one-dimension field, uniformly sampled at the Nyquist frequency by low precision sensors. The authors show that the number of sensors (i.e., samples) can be traded-off with the precision of sensors. Finally, the work in [6] proposes to use synthetic data generation techniques to generate irregular data topology from some available experimental data. The objective there is to obtain a field model to evaluate sensor network algorithms.

3   Irregular sampling of band-limited signals

Let us consider the one-dimensional model where r sensors, randomly located in the interval $[0, 1)$, measure the value of a band-limited signal $p(t)$. Let $t_j \in [0, 1)$ for $j = 1 \ldots, r$ be the locations of the sampling points ordered increasingly and $p(t_j)$ the corresponding samples.

A strictly band-limited signal over the interval $[0, 1)$ can be written as the weighted sum of M' harmonics in terms of Fourier series:

$$p(t) = \sum_{k=-M'}^{M'} a_k \exp(2\pi i k t) \tag{1}$$

Note that for real valued signals the Fourier coefficients satisfy the relation $a_k^* = a_{-k}$ so that

$$p(t) = \sum_{k=-M'}^{M'} \rho_k \cos(2\pi k t + \phi_k)$$

where $a_k = \rho_k \exp(i\phi_k)$.

The reconstruction problem can be formulated as follows: *given r pairs $[t_j, p(t_j)]$ for $j = 1, \ldots, r$ find the band-limited signal in (1) uniquely specified by the sequence of its Fourier coefficients a_k.*

Let the reconstructed signal be

$$\hat{p}(t) = \sum_{k=-M}^{M} \hat{a}_k \exp(2\pi i k t) \tag{2}$$

where the $\hat{a}_k$ are the corresponding Fourier coefficients up to the M-th harmonic. In general, the reconstruction procedure will minimize $\|p(t) - \hat{p}(t)\|^2$ if $M < M'$ and gives $p(t) = \hat{p}(t)$ if $M = M'$.

Consider the $(2M + 1) \times r$ matrix $\mathbf{F}$ whose (k, q)-th element is defined by

$$(\mathbf{F})_{k,q} = \frac{1}{\sqrt{r}} \exp(2\pi i k t_q) \quad \begin{matrix} k = -M, \ldots, M \\ q = 1, \ldots, r \end{matrix}$$

the vector $\hat{\mathbf{a}} = [a_{-M}, \ldots, a_0, \ldots, a_M]^T$ of size $2M + 1$ and the vector $\mathbf{p} = [p(t_1), \ldots, p(t_r)]^T$. We have the following linear system

$$\mathbf{F}\mathbf{F}^\dagger \hat{\mathbf{a}} = \mathbf{F}\mathbf{p} \tag{3}$$

where $(\cdot)^\dagger$ is the conjugate transpose operator. In the following we will denote $\mathbf{T} = \mathbf{F}\mathbf{F}^\dagger$ and $\mathbf{b} = \mathbf{F}\mathbf{p}$.

When the samples are equally spaced in the interval $[0, 1)$, i.e., $t_q = (q-1)/r$, we observe that the matrix $\mathbf{F}$ is a unitary matrix ($\mathbf{F}\mathbf{F}^\dagger = \mathbf{T} = \mathbf{I}_{2M+1}$) [1] and its rows are orthonormal row vectors of an inverse DFT matrix. In this case (3) gives the first M Fourier coefficients of sample sequence $\mathbf{p}$.

When the samples t_q are not equally spaced, the matrix $\mathbf{F}$ is no longer unitary and the matrix $\mathbf{T}$ becomes a $(2M + 1) \times (2M + 1)$ Hermitian Toeplitz matrix

$$\mathbf{T} = \mathbf{T}^\dagger = \begin{pmatrix} r_0 & r_1 & \cdots & r_{2M} \\ r_{-1} & r_0 & \cdots & r_{2M-1} \\ & & \ddots & \\ r_{-2M} & & \cdots & r_0 \end{pmatrix}$$

---

[1] The symbol $\mathbf{I}_n$ represents the n by n identity matrix

where

$$(\mathbf{T})_{k,m} = r_{k-m} = \frac{1}{r}\sum_{q=1}^{r}\exp(2\pi i(k-m)t_q) \qquad k,m = -M\ldots,M \qquad (4)$$

It follows that the Toeplitz matrix $\mathbf{T}$ is uniquely defined by the $4M + 1$ variables

$$r_\ell = \frac{1}{r}\sum_{q=1}^{r}\exp(2\pi i\ell t_q) \qquad \ell = -2M,\ldots 2M$$

The solution of (3), which involves the inversion of $\mathbf{T}$, requires some attention if the condition number of $\mathbf{T}$ (or equivalently of $\mathbf{F}$) becomes large. We recall that the condition number of $\mathbf{T}$ is defined as

$$\kappa = \sqrt{\frac{\lambda_{\max}}{\lambda_{\min}}}$$

where $\lambda_{\max}$ and $\lambda_{\min}$ are the largest and the smallest eigenvalues of $\mathbf{T}$, respectively. In practice, matrix inversion is usually performed by algorithms which are very sensitive to small eigenvalues, especially when smaller than the machine precision. For this reason in [1] a preconditioning technique is used to guarantee a bounded condition number when the maximum separation between consecutive sampling points is not too large. More precisely, by defining $w_q = (t_{q+1} - t_{q-1})/2$ for $q = 1\ldots,r$, where $t_0 = t_r - 1$ and $t_{r+1} = 1 + t_1$, and by letting $\mathbf{W} = \text{diag}(w_1,\ldots,w_r)$, the preconditioned system becomes

$$\mathbf{T}_w\hat{\mathbf{a}} = \mathbf{b}_w$$

where $\mathbf{T}_w = \mathbf{FWF}^\dagger$ and $\mathbf{b}_w = \mathbf{FWp}$. By defining the maximum gap between consecutive sampling points as

$$\delta = \max(t_q - t_{q-1}),$$

when $\delta < 1/2M$, it is shown in [1] that

$$\kappa(\mathbf{T}_w) \le \left(\frac{1 + 2\delta M}{1 - 2\delta M}\right)^2$$

This result generalizes the Nyquist sampling theorem to the case of irregular sampling, but only gives a sufficient condition for perfect reconstruction when the condition number is compatible with the machine precision.

In Figure 1 and 2, we give an example of the reconstruction from irregular samples of a band-limited signal, using (3). In Figure 1, we chose $M = 10$ and $r = 26$ and the samples have been randomly selected over the interval $[0, 0.8]$. The signal has been perfectly reconstructed even if large gaps are present ($\delta > 0.2$). In Figure 2, $r = 21$ samples of the same signal of Fig. 1 have been taken randomly over the entire window. Due to the bad conditioning of the matrix $\mathbf{T}$ (i.e., very low eigenvalues) the algorithm failed to reconstruct the signal due to machine precision underflow.

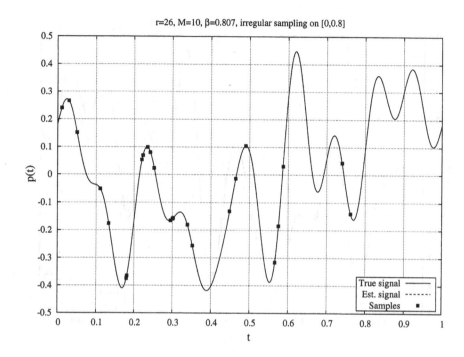

r=26, M=10, β=0.807, irregular sampling on [0,0.8]

Fig. 1. Example of a reconstructed signal from irregular sampling

4 The random matrix approach

We recall that the asymptotic eigenvalue distribution of a $\mathbf{HH}^\dagger$ matrix where $\mathbf{H}$ is a $K \times N$ matrix with independent zero-mean complex random variables with variance $1/N$ and fourth moments of order $O(1/N^2)$ was given by Marčenko and Pastur [2]. In particular, as $K, N \to \infty$ and $K/N \to \beta$ the empirical distribution of $\mathbf{HH}^\dagger$ converges almost surely to a nonrandom limiting distribution with density

$$f_\beta(x) = \left(1 - \frac{1}{\beta}\right)^+ \delta(x) + \frac{\sqrt{(x-a)^+(b-x)^+}}{2\pi\beta x}$$

where $a = (1 - \sqrt{\beta})^2$, $b = (1 + \sqrt{\beta})^2$ and $(x)^+ = \max(0, x)$. Unfortunately, this result does not apply to $\mathbf{T} = \mathbf{FF}^\dagger$ due to the dependence among the elements of $\mathbf{F}$, nevertheless it is useful for comparison.

As an example, we plot in Figures 3–5 the experimental eigenvalue distribution of $\mathbf{T}$, with $r = 600$ and $M = 100, 150, 180$ obtained by Monte Carlo simulation. We compare it with the Marčenko–Pastur asymptotic eigenvalue distribution. We observe that the two distributions have significantly different shape as β increases. The bin width is set to 0.1 and prevents from seeing the behavior of the distribution around zero. However, in all cases, the experiments

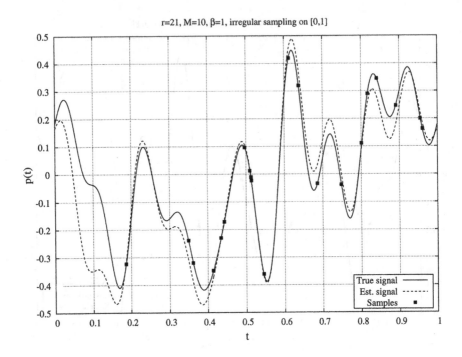

Fig. 2. Example of a badly reconstructed signal due to numerical instability

showed that the minimum eigenvalue is not bounded away from zero as in the Marčenko–Pastur distribution. This is critical for the condition number of $\mathbf{T}$, therefore we are interested in evaluating the probability that the minimum eigenvalue is greater than the machine precision, i.e., the probability of correct field reconstruction.

The dotted curves in Figure 6 show $\log_{10} F_\lambda(x)$, for β ranging from 0.1 to 0.8. The curves show an approximately linear behavior for $\log_{10} F_\lambda(x) < -1$, for all values of β. Notice that the machine precision is around 10^{-16}. Due to this limitation the dotted curves loose their linear behavior while approaching $x = 10^{-16}$ (see the case $\beta = 0.8$). The solid lines in Figure 6 are the tangent of the c.d.f computed in $\log_{10} F_\lambda(x) = -2$. By numerical interpolation we have obtained the following approximation (dashed lines)

$$\log_{10} F_\lambda(x) = \tan(\theta) \log_{10} x - 0.5 \qquad (5)$$

where $\theta = (1.14 - 0.5514\beta)^5$ which yields

$$F_\lambda(x) = \frac{1}{\sqrt{10}} x^{\tan(\theta)}$$

for $x \to 0$.

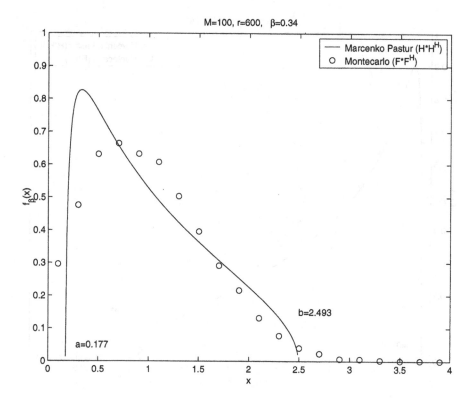

Fig. 3. Comparison between the Marčenko–Pastur distribution and the Monte Carlo experimental distribution for $\beta = 0.34$

The c.d.f computed according to the above expression is denoted in the figure by the dashed lines. The approximation accuracy is evident especially for β in the range from 0.2 to 0.7.

By considering the moment generating function of the eigenvalue distribution of $\mathbf{T}$

$$\Lambda(s) = \mathbb{E}\left[\mathrm{tr}\left(e^{s\mathbf{T}}\right)\right]$$

we were able to compute the following moments in closed form

$$\mathbb{E}[\lambda] = 1$$

$$\mathbb{E}[\lambda^2] = 2\frac{M}{r} + 1$$

$$\mathbb{E}[\lambda^3] = 4\frac{M^2}{r^2} - 2\frac{M}{r^2} + 6\frac{M}{r} + 1$$

$$\mathbb{E}[\lambda^4] = 8\frac{M^3}{r^3} + \frac{80}{3}\frac{M^2}{r^2} - \frac{44}{3}\frac{M^2}{r^3} + \frac{10}{3}\frac{M}{r^3} + 12\frac{M}{r} - \frac{28}{3}\frac{M}{r^2} + 1$$

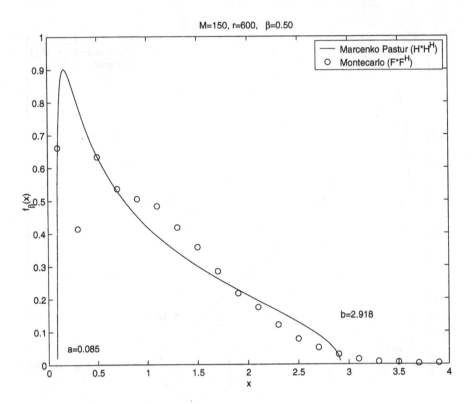

Fig. 4. Comparison between the Marčenko–Pastur distribution and the Monte Carlo experimental distribution for $\beta = 0.50$

$$\mathbb{E}[\lambda^5] = 16\,\frac{M^4}{r^4} + \frac{320}{3}\,\frac{M^3}{r^3} - \frac{224}{3}\,\frac{M^3}{r^4} + \frac{152}{3}\,\frac{M^2}{r^4} - 120\,\frac{M^2}{r^3} + \frac{280}{3}\,\frac{M^2}{r^2}$$
$$-\frac{26}{3}\,\frac{M}{r^4} + 20\,\frac{M}{r} - \frac{80}{3}\,\frac{M}{r^2} + \frac{70}{3}\,\frac{M}{r^3} + 1$$

$$\mathbb{E}[\lambda^6] = 32\,\frac{M^5}{r^5} - \frac{1644}{5}\,\frac{M^4}{r^5} + \frac{2044}{5}\,\frac{M^4}{r^4} + \frac{2392}{5}\,\frac{M^3}{r^5} + 608\,\frac{M^3}{r^3} - \frac{5032}{5}\,\frac{M^3}{r^4}$$
$$-528\,\frac{M^2}{r^3} + \frac{2691}{5}\,\frac{M^2}{r^4} - \frac{1051}{5}\,\frac{M^2}{r^5} + 240\,\frac{M^2}{r^2} - \frac{433}{5}\,\frac{M}{r^4} - 60\,\frac{M}{r^2}$$
$$+\frac{163}{5}\,\frac{M}{r^5} + 94\,\frac{M}{r^3} + 30\,\frac{M}{r} + 1$$

In the limit for $M \to \infty$ and $r \to \infty$ with constant $\beta = 2M/r$ we get the following central moments

$$\mathbb{E}[(\lambda - 1)] = 0$$

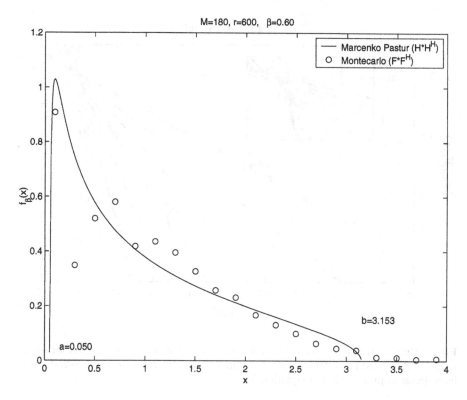

Fig. 5. Comparison between the Marčenko–Pastur distribution and the Monte Carlo experimental distribution for $\beta = 0.60$

$$\mathbb{E}[(\lambda - 1)^2] = 2\beta$$

$$\mathbb{E}[(\lambda - 1)^3] = \beta^2$$

$$\mathbb{E}[(\lambda - 1)^4] = \beta^3 + \frac{8}{3}\beta^2$$

$$\mathbb{E}[(\lambda - 1)^5] = \beta^4 + \frac{25}{3}\beta^3$$

$$\mathbb{E}[(\lambda - 1)^6] = \beta^5 + \frac{391}{20}\beta^4 + 11\beta^2$$

We observe how the eigenvalues are concentrated around 1 when β is small. Recall that M is the field bandwidth and r is the number of samples, i.e., the number of active sensors over the observation area. We are interested in deriving the eigenvalue distribution as $M \to \infty$ and $r \to \infty$ with constant $\beta = 2M/r < 1$. Indeed, given the eigenvalue distribution we could determine the condition

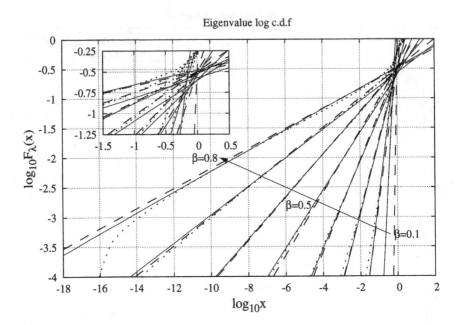

Fig. 6. Cumulative density function of λ obtained by Montecarlo simulation for some values of β. Dotted lines: Montecarlo; solid lines: tangents at $\log_{10} F_\lambda(x) = -2$; dashed lines: linear approximation in equation (5).

number of the matrix $\mathbf{T}$, hence the probability of correct field reconstruction. This will be the next step in our work.

5   Conclusions and Future Work

We considered a large-scale wireless sensor network sampling a physical field, and we investigated the relationship between the random network topology and the quality of the reconstructed field. We employed random matrix theory, and gave the basis to derive the ratio of the field bandwidth to the number of samples necessary for a successful reconstruction.

In our future research, we need to address several issues. First of all, we would like to obtain the asymptotic eigenvalue distribution of the matrix employed for the field reconstruction, so that the condition number, i.e., the probability of correct field reconstruction can be determined. A two-dimensional field should be analyzed, and several aspects should be taken into account. For instance, the fact that sensors can represent the detected information with a limited number of bits, data (i.e., sample) losses can occur during the information transfer to the sink, and the sensor locations can be unknown, should all be included in the analysis.

References

1. H.G. Feichtinger, K. Gröchenig, T. Strohmer, "Efficient numerical methods in non-uniform sampling theory," *Numerische Mathematik*, vol. 69, pp. 423–440, 1995.
2. V.A. Marčenko and L.A. Pastur, "Distributions of eigenvalues for some sets of random matrices," *Math. USSR-Sbornik*, vol. 1, pp. 457–483, 1967.
3. P. Ishwar, A. Kumar, K. Ramchandran, "Distributed sampling for dense sensor networks: a bit-conservation principle," *3rd International Symposium on Information Processing in Sensor Networks (IPSN 2003)*, Apr. 2003.
4. M. Perillo, Z. Ignjatovic, W. Heinzelman, "An energy conservation method for wireless sensor networks employing a blue noise spatial sampling technique," *3rd International Symposium on Information Processing in Sensor Networks (IPSN 2004)*, Apr. 2004.
5. R. Willett, A. Martin, R. Nowak, "Backcasting: adaptive sampling for sensor networks," *3rd International Symposium on Information Processing in Sensor Networks (IPSN 2004)*, Apr. 2004.
6. Y. Yu, D. Ganesan, L. Girod, D. Estrin, R. Govindan, "Synthetic data generation to support irregular sampling in sensor networks," *Geo Sensor Networks 2003*, Portland, Maine, Oct. 2003.

Cost Efficient Localized Geographical Forwarding Strategies For Wireless Sensor Networks

Michele Rossi[1] and Michele Zorzi[2]

[1] Department of Engineering, University of Ferrara,
via Saragat 1,
44100 Ferrara, Italy
mrossi@ing.unife.it
[2] Department of Information Engineering, University of Padova
via Gradenigo 6B,
35131 Padova, Italy
zorzi@ing.unife.it

(Invited Paper)

Abstract. In this paper we investigate the relationship between local next hop selection strategies and their efficiency in terms of both link related metrics, such as the mean packet delivery fraction, and network related metrics, such as the energy status of the node elected as relay. In standard geographical forwarding algorithms the relay selection is usually carried out by means of advancements toward the destination. However, channel attenuation phenomena often make pure geographical strategies ineffective as the quality of a transmission link is not necessarily deterministically related to the node coordinates. In order to achieve effective and cost efficient routing solutions, it is therefore crucial to couple advancements toward the destination with link quality aspects as well as network related metrics (e.g., node energies). This study is a preliminary step toward the design of local relay selection rules which jointly account for these aspects and whose aim is to cut the desired trade-off between delay and cost efficiency.

Keywords: Wireless sensor networks, routing, MAC techniques, cross-layer design, performance evaluation.

1 Introduction

Geographical routing is a key concept which is very often considered for data forwarding in multi-hop wireless sensor networks (WSNs) [1] and Ad Hoc networks [2]. Many routing solutions, in fact, exploit the concept of maximum advancements toward the destination [3–6] to effectively route packets in a best effort (greedy) manner. However, recent empirical measurements [7–9] have proved that the unit disk connectivity model [10], on which these solutions are based, often fails in real settings. In particular, channel attenuation phenomena such as e.g. multi-path fading [11], invalidate the unit disk connectivity assumption, thereby heavily affecting the good results obtained so far for pure geographical routing schemes. In this work, we remove the unit disk model assumption, by going in the direction of recent research [12] and studying the impact of a more accurate connectivity model on the metric to be used to implement geographical forwarding. The impact of fading on geographical random forwarding has been studied in [13]. Our aim here is to account for real fading statistics and derive exact formulae to properly weigh the nodes geographical advancements toward the destination in a faded channel. Subsequently, we use such statistics to drive the relay node (next hop) election by accounting for the "expected advancements", that we define here as the product of the actual geographical advancements and the related packet success rates [14]. In addition, we also account for the so called *network costs*, that we use in the present contribution to model node specific quantities such as residual energies and/or congestion states. In our framework, link specific costs are accounted for by the above mentioned expected advancements, whereas node specific costs, such as residual energies, are taken into account by the network costs.

In the present paper, we propose a novel relay contention scheme, where all nodes with a good expected advancement metric are first collected; our analytically derived curves on the optimal expected advancements are used to this end. Subsequently, these nodes are involved in the relay election phase, which is performed by means of a probabilistic back-off scheme and whose aim is to promote the node with the lowest network cost. The original aspects of our contribution consist of both the greater accuracy of our analytical derivations with respect to previous results [12] as well as of the novelty of the proposed channel contention procedure for the election of the relay node. Our derivations for the optimal advancement metric are in line with [14]; however, we do not consider the interference due to out of range nodes, and we also derive the statistics with a different perspective, i.e., conditioned on the actual advancement of a given node in the forwarding region. In fact, sensor networks are expected to deal with low traffic communications and therefore in these scenarios this type of interference is less important. We instead still focus on channel fading and its consequences to the achievable advancements within a given local relay election phase. Our work is also very much in line with [15], where the authors also stress the importance of keeping the packet error rate into account in geographical forwarding. The main differences of our approach with respect to [15] consist in the novel MAC contention procedure that we propose in Section 4, as well as the new probabilistic filtering procedure that we propose to pick the nodes with the highest expected advancement within range. Furthermore, we remark that in our study we explicitly consider the correlation among nodes costs by showing the impact of this metric on the relay election procedure. To the best of our knowledge, the cost correlation has never been considered before in the design of contention algorithms for WSNs. However, this is a crucial metric that has to be taken into account when the objective is to elect "good" relay nodes, where the node goodness might be related to residual energies, congestion levels as well as data aggregation aspects. In certain settings, in fact, in order to optimally exploit the network resources it might be beneficial to elect a next hop which has data to aggregate, in spite of other requirements.

The remainder of the paper is organized as follows: in Section 2 we present the system model that will be subsequently considered to carry out our analytical derivations. In Section 3 we analytically derive the relationship between geographical and expected advancements. In Section 4 we propose a novel channel contention phase for the local election of relay nodes while in Section 5 we discuss the impact of the cost correlation on the relay selection procedure. In Section 6 we report some results by comparing our scheme to previously proposed solutions that exploit the pure geographical advancement metric and finally, in Section 7, we report the conclusions of our work.

2   System Model

Throughout the paper we will make the following assumptions:

1. *Topology*: We model the network as a weighted graph $\mathcal{G} = (\mathcal{N}, \mathcal{L})$, consisting of a set $\mathcal{N}$ of nodes and a set $\mathcal{L}$ of arcs, where we refer to $l_{ij} \in \mathcal{L}, i, j \in \mathcal{N}$, as the link connecting node i with node j. We consider bi-directional links and we say that a link between the two nodes i and j exists with probability $P_s(d_{ij})$, where $P_s(\cdot)$ corresponds to the probability of successfully transmitting a data packet from node i to node j and is calculated as a function of the distance d_{ij} separating the two nodes. The characterization of $P_s(\cdot)$ is detailed below. For the topology, we assume that nodes are distributed according to a planar Poisson process with intensity ρ users per unit area [16]. That is, the probability of having $n \in \mathbb{N}$ devices within an area $\mathcal{A} \in \mathcal{R}^+$ is given by $\mathcal{P}(n, \rho, \mathcal{A}) = ((\rho\mathcal{A})^n/n!) \exp(-\rho\mathcal{A})$.

2. *Channel model*: for the channel model, we consider both path loss attenuation and fast fading, which is modeled here by means of the Rayleigh fading statistics [11]. For the sake of illustration, let us refer to the communication between node F (forwarder) and node N (next hop) in Fig. 1. If the distance between the two nodes is r, then the probability that node N will receive the packet transmitted by F is calculated as:

$$P_s(r) = \text{Prob}\{\alpha r^{-\eta} \geq b\} \tag{1}$$

where η is the path loss propagation exponent, usually within the range $\eta \in [2, 4]$, α is the fading value for a given packet transmission[3] and b is a technology dependent threshold used to model the probability that the received signal envelope is successfully decoded. We further define R as the transmission range value for which $P_s(R) = \zeta$, where ζ is a small probability value. We refer to R as the maximum transmission range, by probabilistically modeling the fact that for communication distances longer than R the transmitted data is likely to be corrupted. In practice, we use R to model the minimum acceptable level of QoS (quality of service). The following analytical framework will rely on this assumption, i.e., the derived results will be conditioned on assuming R as the maximum transmission range: nodes placed at longer distances are not considered as possible relay nodes.

3. *Radio activities*: We allow nodes to periodically switch between awake and sleeping modes, where they can switch off the radio activity for energy saving purposes. If we express the duty cycle t_{on} as the fraction of time in which nodes are in the active state, then at every data forwarding stage the only nodes that can be considered for data routing are the ones actually awake within the forwarding range R. By considering independent on/off radio cycles at every node, this fact is modeled through an equivalent Poisson process of density $\rho_{on} = \rho t_{on}$, which gives the average number of awaken nodes per unit area at a given instant.

---

[3] We reasonably assume that the attenuation due to fading remains constant during a packet transmission, but is uncorrelated among subsequent transmission events (block fading model).

4. *Nodes advancements*: Consider the node advancement diagram illustrated in Fig. 1, where we represent a snapshot of the routing process for a given data packet. In particular, node F has to select a next hop N to act as a relay for the current packet. In our setting, data forwarding is achieved on the fly, by only exploiting local knowledge about network topology and nodes costs. For what concerns the topology aspect, F should select the node leading to the maximum expected advancement toward the *sink* (destination). For illustration purpose, suppose there are M neighbors lying in the forwarding area (half circle with radius R, toward the sink) and that their distances from F are $(r_1, r_2, \ldots, r_M)$. Let $(z_1, z_2, \ldots, z_M)$ be the vector of projected distances toward the sink. A locally optimal geographical forwarding is therefore achieved by selecting node i^* such that:

$$i^* = \mathrm{argmax}_{j \in \{1,2,\ldots,M\}} \left\{ z_j P_s(r_j) \right\} \tag{2}$$

In fact, in our setting the correct way of dealing with geographical advancements is to account for expected advancements, which are achieved as $z_j P_s(r_j)$. Observe that this leads to a substantially different analysis from the unit disk [10] propagation model, where transmissions to the nodes placed within the transmission range are always successful and the only cause of error is packet collision. In the following derivations, we refer to the forwarding area, the half circle with radius R toward the sink in Fig. 1, as $\mathcal{F}$. It is important to stress that expected advancements will only be used in the initial phase of the protocol that we propose in the present paper and with the aim of picking in a distributed fashion the most suitable relay candidate. Furthermore, in a subsequent contention phase, these nodes will further contend for the relay election by means of a properly designed backoff algorithm, where the choice of the relay will be driven by the so called network costs, i.e., by jointly accounting for nodes geographical advancements and nodes residual energies, as addressed in the following point. In practice, the choice of the relay is a two-step process where we first discriminate among nodes with a good expected advancement metric and we subsequently refine our choice of the relay node by also accounting for network related aspects such as residual energies. This second phase is driven by the node costs presented below.

5. *Node cost*: These are the costs considered in the second contention phase of the joint MAC/routing protocol that will be presented in Section 4. We define nodes costs so as to encode several aspects of the communication. First of all, they must reflect geographical advancements, as our objective is to route packets toward the destination using node coordinates. (Note that in this case we use again the advancement, which is then considered in both contention phases.) However, it shall be observed that advancements are not the only quantities to be accounted for. In fact, one may also think of optimizing other factors such as residual energies and congestion levels. These metrics are indeed important to discriminate among nodes with the same advancement metric and therefore implement a "network" efficient choice of the relay node. This "network consciousness" refers to the fact of jointly accounting for possibly heterogeneous factors so as to pick the nodes with good advancements (current communication perspective) but also with other desirable properties (e.g. residual energies) and this is done with the aim of optimizing the network utilization (network perspective). In order to implement the above requirements, here we

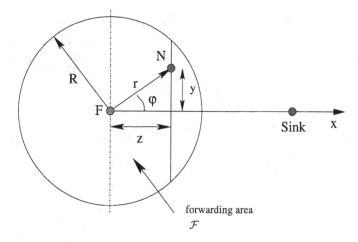

Fig. 1. Diagram for the considered nodes geographical advancement model.

associate a normalized finite cost c_i to every node $i \in \mathcal{N}$, where without loss of generality $c_i \in [0, 1]$. Observe that these costs are independent of the link quality and are node specific. For these costs, we introduce a flexible definition which accounts for both advancement and residual energy aspects. Accordingly and without loss of generality, in this contribution we will express the costs for the generic node $i \in \mathcal{N}$ as follows:

$$c_i = \xi(1 - E_i/E_{init}) + (1 - \xi)(1 - z_i/R) \qquad (3)$$

where $\xi \in [0, 1]$ is a factor used to weigh the relative importance of the two terms, E_i is the residual energy reserve at node i, E_{init} is the initial energy reserve and z_i is the advancement toward the destination associated with node i. It shall be observed that in the costs one might also encode further factors such as congestion levels; we refer here to energy levels only as an example. Further investigation in this direction is the object of our current and future work.

3   Characterization of Optimal Advancements

Let us refer to Fig. 1. If r is the distance between the sender (F) and a given receiver in the forwarding area $\mathcal{F}$, its pdf[4] is derived as $f(r) = 2r/R^2$. Moreover, if Z is the r.v. governing the projected advancement toward the sink, its pdf conditioned on r is given by [17]:

$$f_Z(z|r) = \begin{cases} 0 & r < z \\ \dfrac{2}{\pi r \sqrt{1 - z^2/r^2}} & 0 \leq z \leq r \leq R \end{cases} \qquad (4)$$

Now, we further define $\varXi$ as the r.v. of the actual advancement $\xi_i = z_i P_s(r_i)$ for the generic node i in the forwarding region $\mathcal{F}$. It follows that the cdf associated with $\varXi$,

---

[4] Conditioned on the maximum range R defined as above.

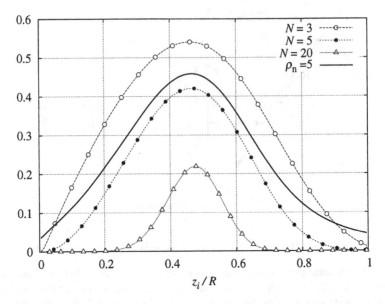

Fig. 2. $\Gamma(z_i, N)$ by varying $N \in \{3, 5, 20\}$ and $\Gamma(z_i)$ for $\rho_n = 5$. $\zeta = 0.01$, $\eta = 4$.

Prob$\{\Xi \leq x\}$, is given by:

$$F_\Xi(x) = \int_0^R f(r) \int_0^{x/P_s(r)} f_Z(z|r) \, dz \, dr = \tag{5}$$

$$= \frac{4}{\pi R^2} \int_0^R r \arcsin\left\{\frac{\min[r, x/P_s(r)]}{r}\right\} dr$$

Moreover, referring to y as the distance between node N and the line connecting F to the sink (Fig. 1) and applying the uniformity property of the Poisson process we have that the pdf $f_Y(y|z)$ conditioned on a given advancement z is:

$$f_Y(y|z) = \begin{cases} \dfrac{1}{2\sqrt{R^2 - z^2}} & y_{min}(z) \leq y \leq y_{max}(z) \\ 0 & \text{elsewhere} \end{cases} \tag{6}$$

where $y_{min}(z) = -\sqrt{R^2 - z^2}$ and $y_{max}(z) = \sqrt{R^2 - z^2}$. Now, if we consider a number N of users in $\mathcal{F}$, the probability $\Gamma(z_i, N)$ that a given device $i \in \{1, 2, \ldots, N\}$ with given geographical advancement z_i is the one leading to the highest expected advancement ξ_{i*} (see Eq. (2)) is obtained as follows:

$$\Gamma(z_i, N) = \begin{cases} 1 & N = 1 \\ \displaystyle\int_{y_{min}(z_i)}^{y_{max}(z_i)} f_Z(y|z_i) \left[F_\Xi\left(z_i P_s\left(\sqrt{z_i^2 + y^2}\right)\right)\right]^{N-1} dy & N > 1 \end{cases}$$

Observe that the above probability is conditioned on z_i and N. Moreover, the awake nodes in $\mathcal{F}$ can be modeled through a Poisson distribution with intensity ρ_{on}. Hence,

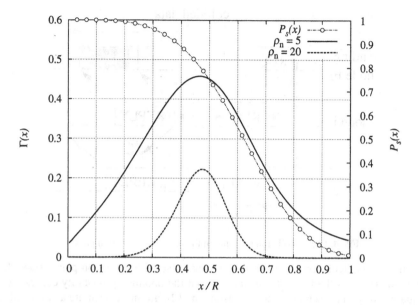

Fig. 3. $\Gamma(z_i)$ for $\rho_n \in \{5, 20\}$. $P_s(r)$ is plotted for comparison. $\zeta = 0.01$, $\eta = 4$.

we can use $\mathcal{P}(n, \rho_{on}, \pi R^2/2)$ to average $\Gamma(z_i, N)$ over the number of awake nodes N in $\mathcal{F}$.[5] Finally, we obtain $\Gamma(z_i) = \mathbb{E}_N[\Gamma(z_i, N)]$ which corresponds to the expected probability for a given node with advancement z_i to be the "best" node in $\mathcal{F}$ when the active nodes in $\mathcal{F}$ are Poisson distributed with density ρ_{on}.

For illustration, in Fig. 2 we report $\Gamma(z_i, N)$ for different values of N, where we normalize the advancement z_i to R. In the figure, we refer to the normalized density ρ_n which is defined as the average number of awake nodes in $\mathcal{F}$, i.e., $\rho_n = \rho_{on}(\pi R^2/2)$. As expected, with a fading channel the nodes close to the limit of the coverage range R are not good candidates to be selected as relays for the packet transmission, as they will likely lead to small success probabilities ($P_s(r)$ decreases as $r \to R$, see Fig. 3). On the other hand, if we pick a node i with a small advancement z_i, we have that $P_s(r)$ is close to one but again the node is not a good relay candidate as $\xi_i(z_i P_s(r)) \to 0$ as $z_i \to 0$. Instead, for intermediate values of z_i, we have a so called *transitional region* [8] where nodes lead to good expected advancements toward the sink. This is indeed the most reasonable region to consider for the selection of relay nodes in geographical routing. In the following sections, we will discuss a possible way to exploit such a probability curve to implement effective relay selection schemes. In Fig. 3, we report both the success probability curve $P_s(x)$ and $\Gamma(x)$ as a function of the normalized distance x/R. Clearly, $P_s(x)$ drops as $x \to R$ and this is the reason for which the often considered [6, 18] *maximum advancement within radius* metrics does not represent the optimal relay node selection criterion when fading is taken into account. It shall be observed that our derivation is an extension of previous results. In fact, differently from the deterministic and one-dimensional topology considered in [12] we carry out

---

[5] The average has to be carried out for $N \geq 1$, as at least node i must exist in $\mathcal{F}$.

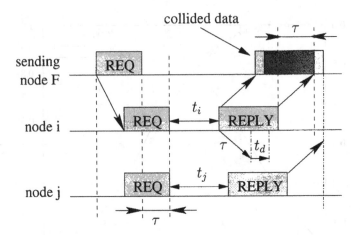

Fig. 4. Packet collision example in the relay node election phase.

the analytical calculation for a two dimensional case with a Poisson planar node distribution. This has the important advantage that the obtained probability curves, even if qualitatively in agreement with the results in [12], are more accurate as they reflect the true two-dimensional stochastic nature of a real forwarding environment and can therefore be directly used within practical forwarding schemes.

4 Proposal for a Coupled MAC/Routing Geographical Scheme

In this section, we present an integrated MAC/routing solution that exploits the probability curves derived above to implement an efficient relay node election procedure. We assume that the sending device has an estimate for the density of awake nodes in its coverage area, ρ_{on}. Moreover, we assume that every node knows its own geographical position as well as the geographical coordinates of the sink. The problem to be solved is to carry out the selection of the relay node by jointly meeting the following requirements: 1) the relay node should have a good expected advancement metric, according to what discussed in the previous section, 2) the selected node should also have a good network cost metric, 3) the relay selection should be implemented such as to limit, as much as possible, the number of collisions [6] associated with the relay election phase. In order to meet the first requirement, we advocate to use the function $\Gamma(\cdot)$ for implementing a probabilistic filtering of the number of nodes that will participate in the relay node selection phase. In particular, each node $j \in \mathcal{F}$ uses an estimate of the network density ρ_n to properly select a $\Gamma(\cdot)$ curve and subsequently calculates the probability of being a good candidate to act as the relay for the current packet transmission; this probability is derived as $\Gamma(z_j)$. Hence, the awake nodes in $\mathcal{F}$ decide to participate in the following contention phase according to the probability $\Gamma(z_j)$; we refer to the set of these nodes as $\mathcal{S}$. Hence, we use such a probabilistic filtering to exclude from the channel contention those nodes that will likely lead to poor expected advancements. After

---

[6] Reducing the number of collisions corresponds to reducing the delay as well as the energy wastage in the contention procedure.

this, we proceed with the actual election of the relay node. Within this second phase, each node in S calculates its own network cost, i.e., a mixture of residual energy and advancement as expressed by Eq. (3), and exploits this cost to derive the back-off value to be used in its subsequent access to the channel. In particular, we consider that a node $j \in S$ transmits a message back to the sending node (F in Fig. 1) with a time delay t_j which is computed as follows:

$$t_j = c_j T_1 + r_j T_2 \qquad (7)$$

where $c_j \in [0, 1]$ is the node cost as defined by Eq. (3), whereas r_j is a random number in $\mathcal{U}[0, 1]$, where with $\mathcal{U}[a, b]$ we indicate the uniform distribution in the interval $[a, b], a < b$. The parameters T_1 and T_2 can be set to adjust the performance of the channel contention by cutting the desired tradeoff between collision probability (duration of the contention) and quality of the solution found (cost of the node elected as relay). The setting of these parameters as well as their dependence on the cost statistics are addressed in greater detail in the following Section 5.

As a first step for the study of the contention scheme, we further consider the following two assumptions: 1) first of all, we do not account for the capture effect, i.e., we declare a collision whenever the reception of different packets overlaps at the receiver; 2) the second assumption relates to the carrier sensing, for which we assume that a node i in the forwarding region $\mathcal{F}$ can always sense the ongoing transmission of another node j in $\mathcal{F}$. Observe that this assumption is reasonable as the sensing range is in general higher than the transmission range. (If needed, the proposed protocol could be slightly modified to accommodate the uncommon situation in which assumption 2 is not verified.) Note that the effect of these two assumptions are a performance decrease (for point 1) and a performance increase (for point 2). Accurate evaluations by simulation have shown that the net effect is limited. A more detailed study is left for future research.

To track packet collisions, we refer to the channel propagation delay and to the minimum time required by the radio circuitry to detect an ongoing transmission as τ and t_d, respectively. Moreover, we express $t_d = n_d/B_r$, where n_d is the number of subsequent bits to be received in order to detect an ongoing transmission, whereas B_r is the communication bit-rate. If a collision occurs, i.e., the replies of two or more nodes in S partially overlap (see Fig. 4), then the collision is detected by the sending node that re-triggers a new contention round. In the new round, the sender also properly modifies T_1 and T_2 to decrease the collision probability, as will be discussed below. On the other hand, if node i is the one selecting the smallest backoff t_i and every other awake device in the forwarding area picks a back-off time t_j such that $t_j \geq t_i + \tau + t_d$, $\forall j \in S, j \neq i$, then the packet sent by node i is received by the sending node with a probability $P_s(d)$, where d is the distance between the sender and node i, and all the scheduled transmissions from any other node $j \in S, j \neq i$ are canceled.[7] The above procedure is repeated until a relay node is elected. After this, the sender forwards the current data packet to the selected relay. As an example, in Fig. 4 we plot the diagram for a collision event where the set S is composed by the two nodes i and j. First of all, the sending node F starts the channel contention by sending a REQ message. This REQ

---

[7] Here, we exploit assumption 2, as every other node in $\mathcal{F}$ is able to sense an ongoing transmission.

$i = 0$;
$\Delta p = 0$;

1:
Send REQT1$(\rho_{on}, T_1^i, T_2^i, \Delta p)$;

2:
if (*no nodes reply in* $(T_1^i + T_2^i)$ *seconds*) **then**
   | $\Delta p = \Delta p + \delta p$;
   |_ Go to **1**;

3:
if (*collision*) **then**
   | $i \leftarrow i + 1$;
   | Send REQT2(T_1^i, T_2^i);
   | Go to **2**;
else
   | Decode REPLY;
   |_ Send data packet;

Algorithm 1: Algorithm executed by the sending node.

triggers every node in $\mathcal{S}$ which independently computes its back-off time as explained above. Then, as the back-off expires, each node in $\mathcal{S}$ sends a REPLY back to F. In the figure, $t_j - t_i < \tau + t_d$ and therefore node j does not have a sufficient time to detect the ongoing communication; its transmission after t_j seconds from the reception of the REQ will therefore result in a collision at the sending node.

In Algorithms 1,2 and 3, we detail the relay selection procedure discussed above. Algorithm 1 describes the procedure executed by the sending node (F in Fig. 1). Node F starts the contention procedure by sending a REQ message of type 1 (REQT1), inclusive of the estimated node density ρ_n, of the two parameters T_1 and T_2 and of a constant Δp whose meaning will be soon clarified. Each node in $\mathcal{S}$, after receiving a REQT1 packet (Algorithm 2) selects a $\Gamma(\cdot)$ curve depending on the value of ρ_n contained in the request and decides to participate to the following contention phase with probability $\Gamma(z_i) + \Delta p$, where z_i is the nodes own advancement. If all nodes in $\mathcal{F}$ decide not to participate in the subsequent channel contention, then F will receive no REPLY. This situation should be unlikely as it means that all nodes lie either in a region very close to the forwarder (node F) or close to the maximum transmission range R. In either case, in fact, the expected advancements $\xi = z P_s(r)$ are small and therefore lead to small access probabilities that, in turn, may cause such an "empty" transmission round. If an empty transmission round is detected, i.e., no REPLYs are received within a time interval of $T_1^0 + T_2^0$ seconds, node F re-sends a further REQT1 message by inflating Δp by the fixed quantity $\delta p \in [0, 1]$. After this, every node in $\mathcal{F}$, upon receiving this second request, adds Δp to $\Gamma(z_i)$, thereby increasing its probability of participating to the contention. This, on the long run, will force every node in the forwarding area to take part in the channel contention. We observe that $\Gamma(\cdot)$ is used here to shape the participation probabilities as a function of the expected advancements toward the sink. In such a way, we probabilistically advantage those nodes in the transitional region, by extending the possibility to take part in the channel access to less desirable nodes only

---

$(\rho_n, z_j) \rightsquigarrow \Gamma(z_j);$

if (*random*() $< \min(\Gamma(z_j) + \Delta p, 1)$) **then**
 | access = TRUE;
else
 | access = FALSE;

if (*access* == *TRUE*) **then**
 | $t_j = c_j T_1^i + random() \, T_2^i;$
 |
 | **if** !(*ongoing TX is detected*) **then**
 | | Send a REPLY after t_j seconds;

Algorithm 2: On receiving a REQT1$(\rho_n, T_1^i, T_2^i, \Delta p)$ message at node $j \in \mathcal{S}$. *random*() generates a random number in $\mathcal{U}[0, 1]$.

---

if (*access* == *TRUE*) **then**
 | $t_j = c_j T_1^i + random() \, T_2^i;$
 |
 | **if** !(*ongoing TX is detected*) **then**
 | | Send a REPLY after t_j seconds;

Algorithm 3: On receiving a REQT2(T_1^i, T_2^i) message at node $j \in \mathcal{S}$.

if needed, i.e., if no candidates are found with a good expected advancement. After having decided to take part in the contention, a node $j \in \mathcal{S}$ initializes a back-off timer to t_j according to Eq. (7) (see Algorithm 2) and, if no ongoing transmission from any other node is detected, transmits a REPLY back to F as its back-off expires. In the case of collision (Algorithm 1), F re-resends a REQ message of type 2 (REQT2), where it specifies new values for T_1 and T_2. As will be clarified by the results discussed in the following section, the adaptation of these two parameters is necessary to decrease the collision probability when nodes costs are correlated.

5   Some Considerations on the Impact of the Statistical Properties of the Node Costs

As the aim of this section is to understand the impact of the above introduced parameters T_1 and T_2 on the performance of the contention algorithm, we focus here on a simplified analytical cost model. This is done to derive a meaningful analysis that will drive us in the choice of these parameters and that will reveal the importance of the *degree of correlation* between the costs of the nodes participating in the contention. The insight gained from this simplified analysis can then be used as heuristics in more general cases. The more realistic cost model in Eq. (3) will be considered again in the performance evaluation section.

Assume to have $K \leq N$ nodes in the set $\mathcal{S}$, where N is the number of devices in $\mathcal{F}$, and let us refer to c_j as the cost associated with node $j \in \mathcal{S}$. Moreover, in order to model the cost correlation among nodes, we assume that the r.v. C_j governing the cost

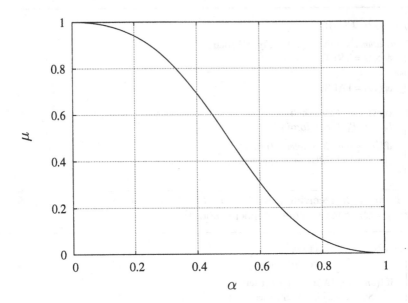

Fig. 5. Relationship between the cost correlation μ and α.

of node j (c_j) is achieved by summing two r.v.s $\overline{C}$ and Ω_j as follows, $C_j = \overline{C} + \Omega_j$, where $\overline{C} \in \mathcal{U}[0,1]$ and $\Omega_j \in \mathcal{U}[-\alpha\overline{c}, \alpha(1-\overline{c})]$, $\alpha \in [0,1]$ and $\overline{c}$ is the actual value of the r.v. $\overline{C}$. Therefore, the cost of a generic node $j \in \mathcal{S}$ is given by a common part $\overline{C}$, which is equal for all nodes in $\mathcal{S}$, and an additive random displacement (or disturbance) $\Omega_j \in [-\alpha\overline{c}, \alpha(1-\overline{c})]$, which is independently picked for every node in the set but that depends (is conditioned) on the actual value of the r.v. $\overline{C}$. $\overline{c}$ in our model is used to represent the common cost component of nodes in $\mathcal{S}$. Clearly, the limiting cases $\alpha = 0$ and $\alpha = 1$ correspond to the fully correlated case, i.e., where all nodes in $\mathcal{S}_N$ have the same cost $\overline{c}$, and to the independent case, i.e., where all costs are uncorrelated, respectively. This is a simple model that we introduce to mathematically derive a precise relationship between the cost correlation μ and the collision probability P_{coll}. We observe that the model is in general not accurate for every network condition. However, it allows to find the quantities of interest in analytical form as well as to derive useful insights on the impact of the cost correlation on the relay selection procedure. We define the correlation coefficient between any two nodes $r, s \in \mathcal{S}$ as

$$\mu_{r,s} = \frac{\mathbb{E}[C_r C_s] - \mathbb{E}[C_r]\mathbb{E}[C_s]}{\sigma_r \sigma_s} \tag{8}$$

where $\sigma_s^2 = \mathbb{E}[(C_s - \mathbb{E}[C_s])^2]$. By standard calculations $\mu_{r,s}$ can be derived as (see Fig. 5)

$$\mu_{r,s} = \frac{(1-\alpha)^2}{(1-\alpha)^2 + \alpha^2} \tag{9}$$

Now, for a given device $j \in \mathcal{S}$ let us refer to $\mathcal{T}_1$ and $\mathcal{T}_2$ as the r.v.s associated with the two terms composing the back-offs $c_j \mathcal{T}_1$ and $r_j \mathcal{T}_2$ (see Eq. 7), respectively. Their pdfs are given by:

$$f_{T_1}(x) = \begin{cases} \dfrac{1}{\alpha T_1} & x \in \mathcal{I} \\ 0 & \text{elsewhere} \end{cases} \tag{10}$$

where $\mathcal{I} = [T_1(1-\alpha)(1-\bar{c}), T_1(1-\bar{c}(1-\alpha))]$ and

$$f_{T_2}(x) = \begin{cases} \dfrac{1}{T_2} & x \in [0, T_2] \\ 0 & \text{elsewhere} \end{cases} \tag{11}$$

Moreover, if we refer to $\mathcal{T}$ as the r.v. $\mathcal{T}_1 + \mathcal{T}_2$, then we have that its pdf is given by the following Eq. (12) which is the result of the convolution of the two pdfs above:

$$f_{\mathcal{T}}(x) = \begin{cases} \dfrac{[\min(T_1(1-\bar{c}(1-\alpha)), x) - \max(T_1(1-\alpha)(1-\bar{c}), x - T_2)]}{\alpha T_1 T_2} & x \in \mathcal{I} \\ 0 & \text{elsewhere} \end{cases} \tag{12}$$

At this point, we are in the position of deriving the collision probability, P_{coll}. In particular, for a given number K of nodes in $\mathcal{S}$, for a given correlation value μ and for a given pair of parameters (T_1, T_2), P_{coll} in the worst case[8] is derived as

$$P_{coll} = 1 - K \int_0^{T_1+T_2} f_{\mathcal{T}}(x) \left[1 - F_{\mathcal{T}}(x + \tau + t_d)\right]^{K-1} dx \tag{13}$$

where $F_{\mathcal{T}}(x)$ is the cdf associated with the r.v. $\mathcal{T}$, whereas τ and t_d are the propagation delay and the time needed to detect an ongoing communication, respectively. By following the same rationale, one can easily derive the joint probability $P\{\text{success \& min}\}$ of having a successful contention, i.e., that a single node will access to the channel, and that this node is the one with the smallest cost in $\mathcal{S}$. Based on the above analytical model, in the sequel we present several important results and considerations on the impact of the cost correlation on both the collision and the success probability.

5.1 Impact of the Cost Correlation on the Relay Election Phase

In the following discussion, we refer to $\rho(\mathcal{S})$ as the average number of nodes in the set $\mathcal{S}$ and we average $P\{\text{success \& min}\}$ and P_{coll} over K, the number of nodes in $\mathcal{S}$,[9] and $\bar{c} \in \mathcal{U}[0, 1]$. As a first result, in Fig. 6 we report the metric $P\{\text{success \& min}\}$ as a function of the contention parameter T_2 for $T_2 + T_1 = 0.2$ seconds. As can be observed from the figure, the cost correlation heavily impacts on the system performance. In fact, for a given (T_1, T_2) pair, $P\{\text{success \& min}\}$ is initially decreasing as a function of μ, whereas when $\mu \to 1$ it starts increasing. It is also to be stressed that the importance of selecting the minimum cost node decreases with an increasing correlation as, by

---

[8] The worst case performance comes from the fact that in the following equation we assume to have a collision with probability one whenever more than one user sends a REPLY, i.e., we do not account for the capture and fading effects.

[9] This is achieved by considering the nodes in $\mathcal{S}$ to be Poisson distributed with the intensity $\rho(\mathcal{S})$.

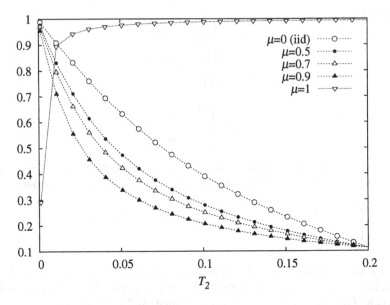

Fig. 6. $P\{$success & min$\}$ as a function of T_2 by varying the cost correlation μ for $\rho(\mathcal{S}) = 10$ and $T_1 + T_2 = 0.2$ seconds.

definition, in such case all nodes in $\mathcal{S}$ tend to be equivalent (node costs in the limiting case $\mu = 1$ become equal). In Fig. 7, we plot P_{coll} as a function of T_2 for the same settings. It is interesting to note that for this metric a good choice is given by $T_2 = 0.2$ s ($T_1 = 0$). These plots reflect the impact of the balancing between the two terms in Eq (7). On the one hand, when costs are independent it is beneficial to emphasize the first term ($c_j T_1$) so as to give priority to the lowest cost nodes. On the other hand, as costs become correlated it is worth to put more weigh on $r_j T_2$ so as to decrease the collision probability, that in this case in naturally increased due to the inherent degree of similarity among the costs (term $c_j T_1$). In other words, the correct balance between $c_j T_1$ and $r_j T_2$ depends on the desired trade-off between *probability of picking the lowest cost node* and *collision probability* which, in turn, depends on the underlying cost correlation structure.

The calculations in Section 5 may therefore be used to derive these metrics and select the appropriate values of T_1 and T_2 depending on our requirements (minimizing the cost associated with the relay or minimizing the collision probability). Note also that, in the most general case T_1 and T_2 might be varied between subsequent rounds of a single relay election procedure (see T_1^i and T_2^i in Algorithm 1). How these values can be effectively modified as a function of the round number is left for future research.

6 Performance Evaluation

In this section, we report some preliminary performance results by comparing our new approach with the GeRaf scheme proposed in [6]. The GeRaf framework consists of an integrated practical MAC/routing scheme based on pure advancements: its capacity of

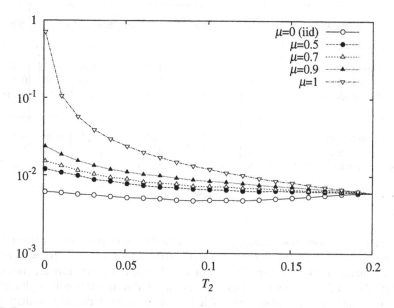

Fig. 7. P_{coll} as a function of T_2 by varying the cost correlation μ for $\rho(\mathcal{S}) = 10$ and $T_1 + T_2 = 0.2$ seconds.

approaching the maximum possible advancements toward the destination makes GeRaf a good candidate for our investigation. Results are obtained by means of accurate Monte Carlo simulation where all packet transmissions (requests sent by the forwarder, replies sent by the candidate relay nodes as well as the final packet transmission to the relay) are affected by fading. That is, for each packet we calculate the probability of correct reception according to Eq. (1). A study of GeRaF in the presence of fading has been presented in [13], where it was shown that the protocol is very robust to *slow* fading. Here, we focus on a different case, where we consider the fading channel to be completely uncorrelated between subsequent transmission/reception events. This is motivated by the following facts: 1) it represents the worst case scenario for the performance, as we can not pick a relay node with a good SNR (signal to noise ratio) metric and be sure that the good link quality will persist up to and including the actual packet transmission (forwarder ⤳ relay), 2) with the latest sensor devices produced so far, due to hardware limitations as well as to the channel contention algorithms, the minimum lapse of time between subsequent transmissions/receptions is likely in the order of 200 ms, 3) forward (forwarder ⤳ candidate relay nodes) and backward channels (relay nodes ⤳ forwarder) are likely uncorrelated due to both channel phenomena as well as hardware asymmetries [19]. Hence, the success probability associated with the transmission of a REQ by the sending entity to a node in its forwarding region likely differs from the success probability of the subsequent reply. In this case, we expect the GeRaF protocol to suffer since the successive signalling messages in a handshake are independently affected by propagation effects.

SCHEME	$\mathbb{E}[z\|\text{suc}]$	$\mathbb{E}[n_{tx}\|\text{suc}]$	$\mathbb{E}[n_{rounds}\|\text{suc}]$	$\mathbb{E}[n_{cont}]$	$P_{failure}$	$\mathbb{E}[z]$
GeRaf ($N_r = 2$)	0.463	8.665	3.799	3.835	0.272	0.337
GeRaf ($N_r = 4$)	0.517	7.989	3.754	2.660	0.382	0.319
GeRaf ($N_r = 8$)	0.574	8.461	4.118	1.966	0.550	0.258
New Scheme ($\delta p = 0.05$)	0.301	6.806	3.069	1.827	0.003	0.300
New Scheme ($\delta p = 0.1$)	0.284	6.157	2.639	2.181	0.0005	0.283
New Scheme ($\delta p = 0.2$)	0.268	5.684	2.328	2.814	0.0003	0.267

Table 1.

For the performance evaluation, we consider the following parameter settings: $T_1 = 100$ ms, $T_2 = 100$ ms, $\xi = 0.5$, $\bar{c} \in \mathcal{U}[0, 1]$, $\mu = 0.01$, $n_d = 128$ bits[10], $B_r = 64$ Kbps, $\rho_n = 20$, that is, on average 20 nodes are Poisson distributed over the forwarding region $\mathcal{F}$. With these values we verified that, besides the good results that will be illustrated in the following, our algorithm is also able to promote relay nodes with a small cost. In fact, the difference between the minimum cost among the nodes in S and the cost of the node elected as the relay is on average 0.08. Further results on this issue are one of the main objectives of our future research. For the pure geographical routing scheme, we consider the version of the GeRaf protocol proposed in [6], by subdividing $\mathcal{F}$ into a given number N_r of priority regions, according to the advancement toward the destination provided by the nodes therein. For the relay election, we consider the probabilistic contention as in [6], where the nodes in the non-empty region with the highest priority are the ones contending to act as relay.[11] For what concerns the performance metrics, we consider the normalized advancement (z_{relay}) provided by the relay node, the number of contention rounds (n_{rounds}) needed to elect a relay as well as the total number of packets transmitted (n_{tx}) within the entire relay election procedure, including the transmission of REQ/REPLY messages, collided packets and the final packet transmission from the forwarder to the relay node. Observe that this last metric is a good indication of the energy expenditure associated with the transmission of a single packet. Moreover, in each channel contention we account for a maximum of $N_{max} = 10$ rounds, i.e., after N_{max} failed requests (REQs) the relay election procedure is suspended and a failure is declared. $P_{failure}$ is used here to represent the failure event probability. $P_{failure}$ for the GeRaf scheme is defined similarly, i.e., as the number of packets sent by the forwarder up to the successful reply from a single node in $\mathcal{F}$ (the winner of the contention). Finally, we also track the number of devices taking part in a single con-

---

[10] Note that the number of bits for carrier sensing depends on the hardware characteristics. Since we do not make any specific assumption here, we purposely take a conservative value.

[11] Note that in our case, as the channel is faded, a region is found to be non empty by the forwarding node if its REQ is correctly decoded by at least one node in the region and if the subsequent REPLY is correctly received at the forwarder.

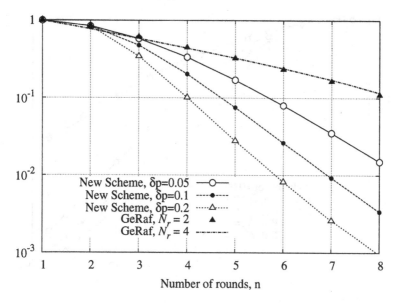

Fig. 8. Complementary cumulative distribution (ccdf) of the number of rounds (n_{rounds}) needed to elect a relay node, i.e., Prob$\{n_{rounds} \geq n\}$.

tention round, n_{cont}. In a good channel contention algorithm, n_{cont} should be limited, as much as possible, in order to keep the collision probability low.

In Table 1, we report the average values for the above performance metrics, where with $\mathbb{E}[\cdot|\text{suc}]$ we indicate the average of the considered metric conditioned on having a successful relay election (probability $1 - P_{failure}$), i.e., that the relay election is successfully accomplished in a number of rounds lower than or equal to $N_{max} = 10$. As can be seen from the table, the GeRaf protocol is the one showing the maximum advancement metric ($\mathbb{E}[z|\text{suc}]$). However, it must be observed that this metric is calculated by considering the cases where the contention is successful. In fact, the expected advancement $\mathbb{E}[z]$ is given by $\mathbb{E}[z|\text{suc}] \times (1 - P_{failure})$. As expected, for the GeRaf protocol an increasing N_r leads to the following consequences: 1) the average number of devices participating in the relay election (n_{cont}) decreases as the size of the priority region is also decreased, 2) the failure probability increases as the forwarding node tries to elect a relay among the nodes placed close to the limit of the transmission range, 3) conditioned on a successful contention, the advancement $\mathbb{E}[z|\text{suc}]$ also increases for the reasons illustrated in the previous point. However, as highlighted by the results shown here, when the channel is faded the maximization of the pure advancement metric has to be avoided, as the resulting success probability may become very low. In general, in the present scheme we trade pure geographical advancements for more reliability as well as a smaller number of packet transmissions (lower energy consumption) for each packet forwarding. In Fig. 8, we report the complementary cumulative distribution (ccdf) of the number of rounds needed to elect a relay node, whereas in Fig. 9 we plot the ccdf of the total number of transmissions (n_{tx}) involved in a single channel contention. These statistics are conditioned on having a successful contention. As can be observed from

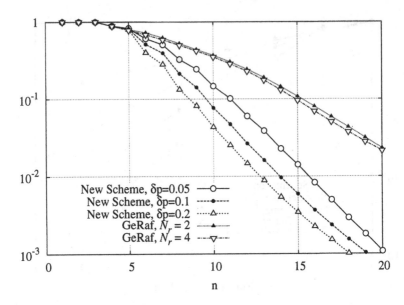

Fig. 9. Complementary cumulative distribution (ccdf) of the total number of packet transmitted (n_{tx}) in a single relay election phase (including collided packets), i.e., Prob$\{n_{tx} \geq n\}$.

Table 1 and Figs. 8 and 9, the available parameters (e.g., δp) can be varied in order to cut the desired tradeoff between advancements, reliability and energy consumption (number of transmitted packets).

We observe that the ξ parameter, which govern the cost-based contention, can also be set to further improve the advancement metric, this of course will be achieved at the expense of the residual energy (see Eq. (3)). The study of the effect of this parameter as well as a the investigation of 1) multi-hop performance of the scheme and 2) impact of ξ on the residual energy of the node elected as relay are the objective of future work. Finally, we can conclude that the obtained results indicate that our probabilistic filtering of the active nodes in $\mathcal{F}$ is effective in selecting the nodes with a good expected advancement metric and that the subsequent channel contention is also able to elect the relay node very quickly and considerably limiting the number of competitors accessing the channel. While here we highlighted the feasibility as well as the effectiveness of our approach, we also stress that further results on the setting of various parameters as well as a deeper investigation of the impact of the cost correlation are needed. These aspects will be addressed in our future research.

7 Conclusions

In this paper we discussed a novel integrated MAC/routing solution for geographical routing in wireless sensor networks. Differently from most previous contributions, we explicitly considered the fading channel statistics and we subsequently proposed a new method to deal with geographical advancements when the channel is faded. Our framework is based upon a probabilistic filtering of the awake nodes in the forwarding region.

That is, based on analytically derived curves, we rule out from the contention phase the nodes that will likely lead to either unsatisfactory advancements or poor link qualities. In addition, we couple this first filtering mechanism with a novel channel contention method where back-off timers are set depending on node costs, so as to control the trade-off between the cost (e.g., residual energy) of the elected relay and the collision probability, i.e., the delay associated with the channel contention. Finally, we compare our solution with a recent scheme based on pure geographical advancements showing that, by taking the fading statistics into account in the relay election, good improvements can indeed be achieved. Moreover, our results confirm that pure geographical advancement toward the destination is not a good policy to be used in the presence of independent multi-path fading.

References

1. I. F. Akyildiz, W. Su, Y. Sankarasubramaniam, and E. Cayirci, "A Survey on Sensor Networks," *IEEE Commun. Mag.*, vol. 40, no. 8, pp. 102–116, Aug. 2002.
2. M. Ilyas, *The Handbook of Ad hoc Wireless Networks.* CRC Press, 2002.
3. M. Mauve, J. Widmer, and H. Hartenstein, "A Survey on Position-based Routing in Mobile Ad Hoc," *IEEE Network Mag.*, vol. 15, no. 6, pp. 30–39, Nov. 2001.
4. B. Karp and H. T. Kung, "GPSR: Greedy Perimeter Stateless Routing for Wireless Networks," in *Proceedings of IEEE/ACM MobiCom 2000*, Boston, MA, Aug. 2000, pp. 243–254.
5. F. Kuhn, R. Wattenhofer, Y. Zhang, and A. Zollinger, "Geometric Ad–Hoc Routing: of Theory and Practice," in *Annual ACM Symposium on Principles of Distributed Computing*, Zurich, Switzerland, 2003, pp. 63–72.
6. M. Zorzi and R. R. Rao, "Geographic Random Forwarding (GeRaF) for Ad Hoc and Sensor Networks: Multihop Performance," *IEEE Trans. on Mobile Computing*, vol. 2, pp. 337–348, Oct-Dec 2003.
7. G. Zhou, T. He, S. Krishnamurthy, and J. Stankovic, "Impact of Radio Irregularity on Wireless Sensor Networks," in *ACM MobiSys 2004*, Boston, Massachusetts, US, June 2004.
8. Marco Zuniga and Bhaskar Krishnamachari, "Analyzing the Transitional Region in Low Power Wireless Links," in *IEEE SECON 2004*, Santa Clara, CA, Oct. 2004.
9. A. Cerpa, J. L. Wong, L. Kuang, M. Potkonjak, and D. Estrin, "Statistical Models of Lossy Links in Wireless Sensor Networks," in *ACM IPSN 2005*, Los Angeles, CA, US, Apr. 2005.
10. B. N. Clark, C. J. Colbourn, and D. S. Johnson, "Unit Disk Graphs," *Discrete Mathematics*, vol. 86, no. 1-3, pp. 165–177, Aug. 1991.
11. Gordon Stüber, *Principles of Wireless Communications*, 2nd ed. Kluwer Academic, 1896.
12. K. Seada, M. Zuniga, A. Helmy, and B. Krishnamachari, "Energy-Efficient Forwarding Strategies for Geographic Routing in Lossy Wireless Sensor Networks," in *IEEE SECON 2004*, Santa Clara, CA, Oct. 2004.
13. M. Zorzi and R. R. Rao, "Energy efficient forwarding for ad hoc and sensor networks in the presence of fading," in *IEEE International Conference on Communications*, Paris, France, June 2004.
14. M. Zorzi and A. Armaroli, "Advancement optimization in multihop wireless networks," in *IEEE VTC 2003*, Orlando, Florida, US, Oct. 2003.
15. S. Lee, B. Bhattacharjee, and S. Banerjee, "Efficient Geographic Routing in Multihop Wireless Networks," in *ACM MobiHoc 2005*, Urbana-Champaign, Illinois, US, May 2005.
16. D. Stoyan, W. S. Kendall, and J. Mecke, *Stochastic Geometry and its Applications*, 2nd ed. John Wiley & Sons, 1995.

17. R. Nelson and L. Kleinrock, "The Spatial Capacity of a Slotted ALOHA Multihop Packet Radio Network wth Capture," *IEEE Trans. Commun.*, vol. 32, no. 6, pp. 684–694, June 1984.
18. L. Kleinrock and J. Silvester, "Optimum Transmission Radii for Packet Radio Newtorks or why Six is a Magic Number," in *Proceedings of IEEE National Telecommunications Conference*, Birminghan, Alabama.
19. D. Kotz, C. Newport, and C. Elliot, "The mistaken axioms of wireless-network research," *Technical Report TR2003-467*, July 2003, Dept. of Computer Science, Dartmouth College.

Semi-Probabilistic Routing
For Highly Dynamic Networks

Paolo Costa and Gian Pietro Picco

Dipartimento di Elettronica e Informazione, Politecnico di Milano, Italy
{costa,picco}@elet.polimi.it

(Invited Paper)

Abstract. In this paper we describe a semi-probabilistic routing approach designed to enable content-based publish-subscribe on highly dynamic networks, e.g., mobile, peer-to-peer, or wireless sensor networks. We present the rationale and high level strategy of our approach, and then show its application in a link-based graph overlay as well as in a broadcast-based sensor network. Simulation results confirm that, in both scenarios, our semi-probabilistic approach strikes a balance between entirely deterministic and entirely probabilistic solutions, achieving high reliability with low overhead.

Keywords: Epidemic Algortihms, Publish/Subscribe, Peer-to-peer, Sensor Networks

1 Introduction

Modern distributed applications exhibit increasing degrees of dynamicity, as evidenced by the emergence of mobile computing, peer-to-peer networks, and wireless sensor networks. Programming distributed applications becomes therefore increasingly complex. In this context, publish-subscribe middleware is advocated by many as a viable solution thanks to its simple programming interface, and to its inherently decoupled interaction paradigm.

Publish-subscribe middleware is organized as a collection of *client* components, which interact by *publishing* messages and by *subscribing* to the classes of messages they are interested in. The core component of the middleware, the *dispatcher*, is responsible for collecting subscriptions and forwarding messages from publishers to subscribers. In the *content-based* incarnation of this publish-subscribe model, the filtering of relevant events is specified by the subscriber using predicates on the event content (e.g., using regular expressions and logic operators), therefore providing additional expressiveness and flexibility.

However, the potential of the publish-subscribe *model* can be fully unleashed in dynamic scenarios only if the underlying *system* is compatible with their requirements. Unfortunately, mainstream systems are typically geared to large-scale settings, and therefore focus on a distributed implementation of the event dispatcher—typically organized as a tree-shaped overlay network for improved scalability—but provide few or no mechanisms for dealing with topological reconfiguration. Moreover, these systems typically base their routing decisions on information that is disseminated deterministically—a strategy that, in a highly dynamic environment, is likely to frequently lead to stale routes.

The alternative approach described in this paper departs from the mainstream along two dimensions. First of all, we do not rely on the existence of a tree overlay, but only on the ability of a dispatcher to communicate with its neighbors. Moreover, our routing strategy exploits deterministic information only in the immediate vicinity of the subscriber, resorting to probabilistic forwarding otherwise. As our simulation results show, this semi-probabilistic approach strikes a balance between a fully deterministic approach (efficient, but not very resilient to reconfigurations) and a fully probabilistic one (very resilient to reconfiguration but characterized by a higher overhead).

The paper is structured as follows. Section 2 discusses the motivation and rationale behind our approach. Section 3 first illustrates the high-level idea underlying our semi-probabilistic approach, and then shows in detail how to exploit it in two different network scenarios: the first characterized by link-based communication on a graph-shaped overlay network, and the other by broadcast-based communication in a wireless sensor network. Section 4 reports on the evaluation through simulation of the aforementioned protocols, in their respective scenarios. Section 5 places our work in the context of related efforts. Finally, Section 6 ends the paper with brief concluding remarks.

2   Rationale and Motivation

In the application scenarios we target, the connectivity among hosts, and therefore dispatchers, can change freely and frequently. This characteristic is typical of mobile ad hoc networks, peer-to-peer networks, and wireless sensor networks. Earlier work on topological reconfiguration of publish-subscribe from our research group successfully tackled the problems posed by the topological reconfigurations occurring in this scenarios [13], e.g., showing that it is possible to reconcile routing information [22] and recover events lost during reconfiguration [9] efficiently. However, these efforts still assumed the availability of an underlying tree-shaped overlay network, as the vast majority of available content-based publish-subscribe systems relies on this assumption to provide high scalability. Nevertheless, this assumption is likely to be challenged when dynamicity is high. In fact, not only the tree maintenance protocols are likely to cause considerable overhead, but the very structure of the tree, providing exactly one route among any two dispatchers, is ill-suited to provide reliability.

In the work described here, instead, we abandon the tree-shaped overlay network and simply assume that the dispatchers are able to communicate with their neighbors. The notion of neighbor clearly depends on the network characterizing the application scenario at hand, and in this paper we consider two very common scenarios, representative of the kind of dynamicity we address. The first one assumes the existence of a graph-shaped overlay network, like those often characterizing peer-to-peer networks. In this scenario, the neighbors of a dispatcher are defined by its links on the overlay. Instead, the second scenario relies solely on the existence of wireless broadcast communication, and therefore defines the neighbors of a dispatcher in terms of its communication range. While this assumption encompasses several scenarios, in this paper we focus preminently on wireless sensor networks.

Besides the assumptions about the network, however, the defining feature of our approach is its peculiar approach to routing. Conventional content-based publish-subscribe systems adopt a deterministic routing strategy, where events are routed according to the information disseminated at subscription time. An example is the widely adopted *subscription forwarding* strategy [7], where a subscription is sent to all the dispatchers along the tree, and events follows the reverse path from the publisher to the subscriber. A direct application of this strategy on a graph overlay would frequently create loops, and is therefore impractical. Moreover, we contend that virtually *any* fully deterministic strategy is going to experience severe drawbacks in the highly dynamic scenario we target, where routing information quickly becomes stale. At the other extreme, probabilistic approaches like *epidemic* (or *gossip*) algorithms [4, 14] are known to satisfy many of the aforementioned requirements in the context of multicast communication. Inspired by the spreading of diseases, these algorithms forward information at random towards a small subset of available nodes, and rely on the availability of multiple routes to ensure that the "infection" carrying the information extends to a sufficient percentage of the receivers. Epidemic algorithms essentially trade the absolute guarantees provided by deterministic approaches for probabilistic ones, yielding in turn increased scalability and resilience to change, as well as reduced complexity. Unfortunately, these algorithms are well-versed for group communication or broadcast, where a message must be sent to *all* the members of a predetermined set of intended recipients. Instead, in our scenario subscribers can be a small fraction of the overall dispatcher network; moreover, each subscriber may be subscribed to a different set of subscriptions. In this case, a purely epidemic approach generates unnecessary overhead, since it proceeds by "blindly" infecting all the network.

In the rest of this paper we describe a semi-probabilistic routing strategy that borrows from both the aforementioned approaches to provide reliable and efficient routing in the context of content-based publish-subscribe. On one hand, we still maintain deterministic information about subscriptions but only in the vicinity of a dispatcher, therefore reducing the likelihood of loops and yet providing accurate—albeit limited—information for routing events. On the other hand, in the portion of the network where this localized information is unavailable we complement it with probabilistic routing decisions, by forwarding events at random towards neighbors. Essentially, we use an epidemic approach complemented by deterministic information. The former addresses reconfiguration, while the latter reduces indiscriminate propagation by "steering" events towards the subscribers. As we demonstrate in Section 4, our mix of deterministic and probabilistic routing enables high reliability and low overhead in both the scenarios considered in this paper.

3   Semi-probabilistic Routing

In our semi-probabilistic approach, routing is governed by two parameters. The *subscription horizon* ϕ represents the number of hops a subscription is propagated away from the subscriber. Therefore, in the area of radium ϕ around a subscriber, all dispatchers are aware of its interests. Instead, the parameter τ represents the *event propagation threshold*, which determines to what extent events are disseminated in the network. The higher the value of τ, the higher the number of copies of an event that are forwarded by a dispatcher.

In a nutshell, our approach works as follows. When an event gets routed through the network, the local subscription table of the dispatcher is examined at each hop. If it contains some subscription coming from subscribers nearby and $\phi \neq 0$, the event is forwarded towards them. Otherwise, the decision about whether to forward the event and to what extent is taken at random, based on the value of τ. If $\phi = 0$ no subscription is ever transmitted by the subscriber node, and therefore our approach degenerates in an entirely probabilistic one.

Clearly, a real protocol is slightly more complex. In the rest of this section we describe how the high level strategy we just described is instantiated into real protocols for the two network scenarios we mentioned in the previous section. The presentation is kept concise due to space constraints: more details are available in [10, 11].

3.1   Link-based Communication: Graph Overlay

In this section we describe how our approach can be exploited in the context of link-based communication. In particular, we assume here that communication takes place along the links of an undirected graph-shaped overlay network. We further assume that the overlay network layer is able to inform our protocol when the connectivity changes, i.e., when a new link appears or an old one vanishes.

Base Routing Scheme

Subscription Propagation   Propagation occurs similarly to subscription forwarding. When a subscription request is issued by a dispatcher, the corresponding message is forwarded to all of its neighbors, which update their subscription tables accordingly. If $\phi = 1$, no further action is taken. Otherwise, each dispatcher forwards the subscription message to all of its neighbors, except the one who sent the message. A subscription is never forwarded twice along the same link, unless an unsubscription occurs in between the two. Differently from subscription forwarding, however, the subscription tables maintain information not only about which subscription was received on which link, but also about the distance of the subscriber, ranging between zero (for a local subscription) and ϕ. Figure 1 shows the layout of subscriptions for a case where $\phi = 1$.

Topological reconfigurations, i.e., the appearance of a new link or the vanishing of an existing one, must induce a proper reconfiguration of subscription information. However, this is easily accomplished by relying on (un)subscription operations, as discussed in [22]. When a dispatcher detects the presence of a new link, it simply sends a subscription message along that link. Similarly, when a link vanishes, the dispatcher behaves as if it received an unsubscription message for all patterns associated to that link. These (un)subscriptions are then propagated based to the extent determined by ϕ.

Event Propagation Event propagation is where probabilistic decisions may come into play. Upon receiving an event, in principle[1] the following processing occurs. First, the subscription table is inspected for subscriptions matching the event. If a match is found, the event is routed along the link associated to the subscription. Subscriptions selection is prioritized according to ϕ: an event is forwarded based on a subscription at distance d only if there is no matching subscription at distance $d - 1$. As we verified through simulation, this strategy reduces the likelihood of forwarding the event along a stale route.

This step is iterated until the number of links used for propagation is greater than f. If the number of matching subscriptions is not sufficient, the propagation threshold is met by forwarding the event along as many links as needed to reach f, randomly selected among those that have not been used in the current forwarding step. The only exception is constituted by links associated to subscriptions at distance $d = 1$; a matching event is forwarded along *all* of these links, regardless of the propagation threshold. The rationale is the fact that subscriptions at $d = 1$ represent the most accurate routing information, and the most direct route towards the corresponding subscribers. Finally, it is important to note that, during the overall process, an event is never forwarded twice along the same link.

Figure 1 shows an example. Let us assume that $\tau = 0.5$ and that an event matching both subscriptions is published by dispatcher 0. This dispatcher has only one link and no subscription information: therefore, the event gets forwarded to dispatcher 1 as this is the only alternative. At dispatcher 1,

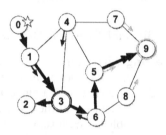

Fig. 1. Semi-probabilistic routing with $\phi = 1$ and $\tau = 0.5$. Numbered circles represent dispatchers. A colored circle around a dispatcher denotes it as a subscriber. The short colored arrows represent subscription information, and indicate the forwarding path for matching events. Dispatcher 0 publishes an event that gets forwarded either deterministically (double-headed thick arrows) or probabilistically (single-headed thick arrows).

---

[1] This sequence of steps serves only for illustration purposes: a number of optimizations are possible in reality.

two links are available. Nevertheless, the link towards 3 is associated with subscription information: it is therefore selected for forwarding and no further action is taken since the threshold is met. At dispatcher 3, the event is delivered locally. Moreover, the links towards 2, 4 and 6 are all viable routing options, and the event must be forwarded along $f = 2$ links. No deterministic information is available, therefore the decision is done entirely at random. The figure shows the case where the event is forwarded towards 2, where it stops propagating, and 6. There, the same situation occurs, with the links towards 5 and 8 as viable options. The figure shows the case where 5 is selected. At this dispatcher, the presence of deterministic information "captures" the events and steers it towards 9, where it gets locally delivered.

It is interesting to note that, at each hop, an event may be routed according to different criteria. As we already mentioned, "holes" in the dissemination of subscription information are bypassed by relying on random selection of links. However, the very nature of content-based systems is an asset for our routing approach, because an event matching multiple subscriptions may leverage of a bigger set of subscription information during its travel. Again, this is exemplified in Figure 1, where the event not only is routed by a mixture of deterministic and non-deterministic decisions, but deterministic ones (i.e., the hops from 1 to 3, and from 5 to 9) are generated by different subscriptions.

Additional Protocol Details

Dealing with loops   With reference to Figure 1, a choice of $\phi = 2$ would have created a routing loop among the nodes 4, 5, 7, and 9. Loops can be detected easily by relying on a unique identifier for every event, trivially implemented using the identifier of the event publisher and the value of a counter incremented at the publisher each time it publishes an event. Therefore, an event received is actually propagated by a dispatcher only if it has never been received before.

More sophisticated loop avoidance and detection algorithms are available in the literature. However, on one hand they are likely to be impractical in the highly dynamic scenario we target, while on the other hand they would introduce a lot of complexity in our algorithm, which instead we want to keep as lightweight as possible.

Avoiding unnecessary propagation   In Figure 1, we note how event forwarding does not really stop at dispatcher 9, since there is no way to know that no other subscriber exists in the system. Without a way to stop forwarding, events would be forwarded indefinitely—more precisely, until a loop is detected. Indefinite propagation is dealt with by attaching a time-to-live (TTL) field to each event message, and by decrementing its value at every hop. When an event is duplicated at a dispatcher along multiple routes, all the copies retain the same TTL. Therefore, an event is propagated only if its TTL is greater than zero.

A more refined mechanism consists of associating different TTLs to the two form of routing we exploit, therefore defining a *deterministic TTL* (TTL_d) and a *probabilistic TTL* (TTL_p), each limiting only the corresponding routing component. With this scheme, propagation ceases when both TTL values reach zero. The advantage of this scheme is that it provides a direct way to control both aspects of propagation, therefore enabling a more accurate tuning of the performance of our approach.

3.2 Broadcast-based Communication: Wireless Sensor Networks

In this section we show how to adapt our approach for a wireless sensor network scenario. In the following we assume wireless broadcast is the only communication media used, and also assume that each (active) sensor takes part in routing, regardless of whether it is currently interested in publishing or subscribing.

Base Routing Scheme

Subscription Propagation   When the application running on a node issues a subscription, our protocol broadcasts the corresponding filter. This information is rebroadcast by the subscriber neighbors to an extent defined by the subscription horizon ϕ. In the link-based approach, ϕ was measured as the number of hops travelled by a subscription message along the links of the graph overlay. Here, instead, ϕ represents the number of times the subscription message is (re)broadcast. Moreover, in Section 3.1, we exploited the standard technique of dealing with (un)subscriptions explicitly, by using control messages propagated whenever a node decides to (un)subscribe. The same technique is used to deal with appearing or vanishing links, by treating the disappearing endpoint as if it were, respectively, subscribing or unsubscribing. Here, we use a different strategy that associates *leases* to subscriptions, and requires the subscriber to refresh subscriptions by re-propagating the corresponding message[2]. If no message is received before a lease expires, the corresponding subscription is deleted.

Clearly, there are tradeoffs involved. Without a leased approach the (un)subscription traffic is likely to be significant, due to the need to reconcile routing information whenever a link appears or disappears. The leased approach remarkably reduces the communication overhead, by removing this need. On the other hand, if subscriptions are stable, bandwidth is unnecessarily wasted for refreshing leases. However, in sensor networks the former case is much more likely to happen than the latter, since nodes typically alternate work and sleep periods to save energy. Moreover, the combination of leased subscriptions and broadcast communication remarkably simplifies the management of the subscription table, and drastically reduces the associated computational and memory overhead. In the previous section, to properly reconcile subscription information upon connectivity changes, we kept a different table for each value of ϕ, where each row contained the subscription filter and the link the subscription referred to. Here, instead, all we need is to store the subscription filter together with a timestamp

---

[2] Optimizations are possible, e.g., to broadcast the subscription hash, and transmit the entire one only if missing on the receiving node.

used for managing leases. Differentiating according to ϕ is no longer needed, since subscriptions simply expire, and broadcast removes the need for maintaining information about links.

Event Propagation   In the link-based approach, the event propagation threshold τ controls the effectiveness of event routing by specifying a fraction of the links available at a given dispatcher. Nevertheless, here we assume broadcast communication, therefore this parameter assumes a different meaning. When an event is received[3] for which a matching filter exists in the subscription table, the event is simply rebroadcast. On the other hand, if no matching subscription is found, the event is rebroadcast with a probability τ. The parameter τ, therefore, still limits the extent of propagation, but more indirectly than in Section 3.1, as it comes into play only when no deterministic information is available.

The effectiveness of our approach is clearly proportional to the number of forwarders F, i.e., the neighbors receiving and retransmitting an event. In absence of deterministic information, in our approach $F = \tau \cdot \eta$ holds, being η the number of neighbors. As a consequence, a small value of η (e.g., in sparse networks) must be compensated by increased values of τ.

Moreover, using a link abstraction the event always got routed along the fraction of links mandated by τ, here instead we have a non-zero probability that none of the neighbors will rebroadcast the event. More precisely, in absence of deterministic information, if η is the average number of neighbors, the probability of stopping the propagation of the event is $(1 - \tau)^\eta$. If no subscriber is in the immediate vicinity of the event publisher and τ is small, there is a significant possibility that event propagation immediately stops. To ensure that a reasonable amount of event messages are injected into the network, we mark event messages with a flag stating whether they have been just published or instead they already travelled through the network. In the first case, the receiver behaves as if $\tau = 1$ and rebroadcasts the event in any case. This mechanism guarantees that at least η copies of the event message are injected in the network and propagate independently.

Additional Protocol Details

Dealing with Collisions   Wireless broadcast is subject to packet collisions, which occur when two or more nodes in the same area send data at the same time. Since in our approach the propagation of subscriptions and events both rely on wireless broadcast, it becomes crucial to reduce the impact of collisions and avoid wasting precious energy on useless retransmissions.

TinyOS [16] adopts a very simple scheme to recover from collisions where, after a broadcast message has been sent, the sender waits for an acknowledgment from at least one of its neighbors. If none is received before the associated timeout expires, the message is resent. The evident weakness of this solution is

---

[3] Clearly, events that have already been processed and that are received again because of routing loops are easily discarded based on their identifier.

that it does not take into account the actual number of neighbors. If only one neighbor received and acknowledged successfully the message, the transmission is assumed successful, regardless of the possibly many nodes that did not receive the message. Moreover, it does not try to limit in any way the number of collisions. More sophisticated MAC protocols has been proposed in literature [21] but none is currently supported by the Crossbow MICA2 [1], our target platform.

Therefore, we conceived a simple yet effective solution that decreases significantly the number of collisions, without requiring any synchronization among nodes. The idea can be regarded as a sort of simplified TDMA protocol where each node, upon startup, sets a timer whose value is a global configuration parameter. Sending messages (i.e., subscriptions and events) takes place only upon timer expiration, while receiving is in principle always enabled. Since each node in the network bootstraps at a different time, it is highly unlikely that two nodes in range of each other end up with synchronized timers. The simulations in Section 4 show that this trivial idea goes a long way in drastically reducing the amount of collisions.

Avoiding Unnecessary Propagation In Section 3.1 we limited the propagation of events using a TTL. However, our simulations showed that this solution is much less effective with broadcast propagation. In fact, even when an event travels for a small number of hops, the number of nodes it reaches is great, and therefore the impact of TTL is limited.

To address this issue, we modified slightly the retransmission strategy we just described. Let us assume a node A waiting to broadcast an event e hears one of its neighbors, say B, transmitting e before A's timer expires. If the set of A's neighbors partially overlaps with B's neighbors, it is likely that most of A's neighbors receive the event from B's transmission, therefore making A's broadcast largely useless. Some of A's neighbors may not hear about e from B but, given the epidemic nature of our algorithm, they are very likely to get it through other routes. Based on this observation, in our approach (which we called *delay-drop*) we would simply let A safely remove e from its transmission queue. In doing this, not only we limit propagation—our initial rationale for this modification—but also reduce communication and therefore save battery power. A downside of this approach is a potentially higher latency, as the event may go through longer routes before reaching its recipients. Nevertheless, in principle this delay-drop mechanism could be only one of many alternatives specified at the application or middleware layer, therefore enabling to tradeoff latency for overhead as needed.

4 Evaluation

In this section we report about the evaluation through simulation of our approaches in a wired, link-based scenario as well as a wireless, broadcast-based one. The original and complete evaluations can be found in [10,11].

The metrics we analyze are event delivery rate and overhead. The former is defined as the ratio between the number of subscribers that should receive a given event and those who actually get it. The overhead is constituted by subscription messages and by event messages that are either duplicated, never received by a subscriber, or routed along unnecessarily long routes. These contributions are difficult to separate, and in any case do not provide significant insights. Therefore, we analyze overhead by simply plotting the overall number of messages flying in the system.

In our simulations, an event is represented as a randomly-generated sequence of integers, determined using a uniform distribution. An event pattern associated to a subscription is represented by a single number. An event matches a subscription if it contains the number specified by the event pattern in the subscription. Each dispatcher is subscribed to two event patterns, drawn randomly from the overall number of patterns available in the system. For each event, the percentage of receivers is about 10% of the overall number N of dispatchers in the system as this is a commonly accepted "rule of thumb" for content-based systems (see e.g., [8]). Finally, simulations are run with dispatchers continuously publishing events on a network with stable subscription information, i.e., where no (un)subscriptions are being issued.

4.1 Graph Overlay

In this section we evaluate the protocol described in Section 3.1 over a graph overlay network. The graph is built with a constant degree $l = 5$, to eliminate as much as possible the bias induced by a random shape. To accommodate dynamicity, however, the actual number of links of a dispatcher is allowed to vary between $l - 1$ and $l + 1$. A topological reconfiguration consists of a link breakage, followed by the appearance of a new link. Graph repair is performed after a time interval (that we set to $0.1s$) modeling the delay necessary to the underlying layers to find the replacement link. When a link breakage occurs, our simulator looks for two nodes with a degree lesser than or equal to l, to maintain the average degree as steady as possible. Likewise, a link is selected for removal only if its endpoints' degree is greater than or equal to l. The time between two reconfigurations is $\rho = 0.03s$ and since a link is always replaced after $0.1s$ the system is undergoing continuous and frequent reconfiguration[4]. Also, we initially set TTL=∞ to first analyze the behavior of the system without limiting event propagation. Each simulation was run 10 times with different seeds and the values averaged. Simulations were run with the OMNeT++ discrete event simulator [26].

---

[4] Each reconfiguration involves two dispatchers, the link endpoints. At 300 reconfigurations per second, with a network size of $N = 300$ each dispatcher changes two neighbors per second. Since our simulations are run for over two seconds, at least 4 neighbors out of 5 get replaced.

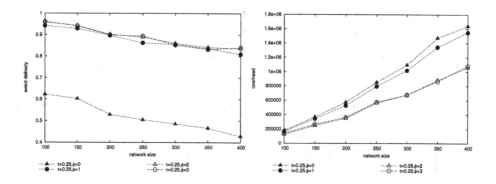

Fig. 2. Event delivery and overhead in a network with reconfigurations every $\rho = 0.03s$.

Network Size We first analyze the performance of our approach when the size of the system grows in a setting where the network topology undergoes reconfiguration. The left chart in Figure 2 shows the event delivery rate for a configuration with an event propagation threshold $\tau = 0.25$ and a subscription horizon varying between $\phi = 0$ (purely probabilistic) and $\phi = 3$. The chart evidences that indeed deterministic information boosts delivery, which almost doubles when moving from a purely probabilistic routing to one with a 1-hop subscription information, while there is no appreciable difference against $\phi > 1$. The reason for the overlapping of these latter curves is that, unlike $\phi = 1$, they are subject to the limitation on propagation set by τ.

It is worth noting that, as discussed in [11], the performance of our semi-probabilistic routing on a static topology is similar to the one we just described. Indeed, the perturbation induced by dynamicity is easily absorbed by the probabilistic component of routing and by the redundancy of the graph. Moreover, as known from probabilistic algorithms, the continuous restructuring of connectivity among dispatchers effectively helps spreading information, by allowing nodes to suddenly become in contact with a different set of neighbors, and spread messages from another point.

The overhead is shown in the right chart of Figure 2. The upper bound is provided by flooding ($\tau = 1$), which is not plotted because it generates an extremely high number of messages (e.g., 11,882,777 at $N = 400$). Therefore, Figure 2 shows that our overhead is extremely far from the upper bound. Indeed, more deterministic information ($\phi > 1$) enables savings up to 35% w.r.t. pure probabilistic and $\phi = 1$. We also verified that higher values of τ quickly bring the system to 100% delivery. This is clearly true for flooding. Also, $\tau = 0.5$ already brings all the curves to full delivery except for $\phi = 0$, which remains at about 96%. Nevertheless, in this latter case the overhead is five times higher w.r.t. to the case with $\tau = 0.25$ and $\phi = 1$ (around 5.5 million messages instead

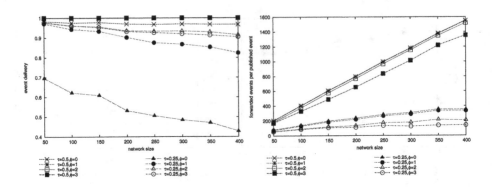

Fig. 3. Event delivery and overhead in a network with reconfigurations every $\rho = 0.03s$ and a fixed number of receivers.

of 1.6 at $N = 400$). The relative performance among the curves with $\tau = 0.5$ is unchanged, with $\phi > 1$ providing the smaller overhead.

Looking at Figure 2, one could notice how delivery drops as the scale increases. Nevertheless, it is worth noting that in the charts above we assume that each dispatcher added to the system is also a publisher emitting 5 events per second, and that the fraction of receivers for each event is always 10% of the dispatchers in the system. Instead, both the dispatcher's degree l and the event propagation threshold τ remain constant: the fanout f (i.e., the number of links along which events are forwarded) therefore remains constant as well. As a consequence, while the number of dispatchers and receivers increases the ability of the system to spread messages decreases. In a real deployment setting, the increase in scale should be compensated by increasing f, i.e., by intervening either on τ or l. In [11] we showed how increasing the degree to $l = 9$ boosts delivery which becomes close to (and for $\phi = 1$ exactly) 100%, due to the ability to spread messages over more links. On the other hand, the overhead is also largely increased and reaches the same order of magnitude of the configuration with $l = 5$ and $\tau = 0.5$. This is not surprising, since the product $\tau \cdot l$, which defines the number f of links available for routing and therefore ultimately constrains the effectiveness of routing, is roughly the same in both scenarios.

Density of Receivers vs. System Scale   To evaluate how the density of receivers in the network affects our strategy we studied a scenario where the number of dispatchers (still all publishing at 5 event/s) is increased while the number of receivers per event (10 in our case) remains constant. The density of receivers therefore decreases linearly, from 20% for $N = 50$ down to 2.5% for $N = 400$. This scenario elicits new issues w.r.t. the one we examined previously. With a constant density of receivers and a growing scale, we need to increment the number of forwarded events to reach a larger set of receivers, and therefore increasing the fanout is a viable solution. Instead, here the number of receivers does not change: therefore, what we are assessing is how "selective" is our routing towards the receivers.

Results are in Figure 3, with the same simulation parameters as in Figure 2. The chart shows both the event delivery and the number of forwarded events divided by the number of published events, where the latter characterizes the effort, in terms of forwarded events, required to deliver a single event to a fixed set of receivers in a growing network[5].

Figure 3 shows that a high fanout ($\tau = 0.5$) always achieves high delivery but basically saturates the network by reaching almost every dispatcher, thus increasing the traffic linearly with scale. However, even in this case deterministic information ($\phi = 3$) achieves some savings, as it enables a more selective routing. Instead, a lower fanout ($\tau = 0.25$) yields a very different behavior. Event delivery is a lower than with $\tau = 0.5$, since the probability to reach a receiver depends on the product of the probabilities to select the right neighbor at each hop— which in turn depends on fanout, $f = 1$ in this case—and therefore decreases as the routes connecting publishers to receivers become longer. However, while this negative effect is evident for pure probabilistic routing, it is almost entirely compensated in terms of delivery by the deterministic information available when $\phi > 0$, which reduces the reliance on random forwarding by "steering" forwarded events more efficiently through the network. This increased efficiency is mirrored in the overhead chart, where the number of forwarded events per published event still increases, due to the longer routes towards receivers, but this time remains well below a linear trend.

Limiting Propagation As we discussed in Section 3.1, introducing a TTL enables considerable savings in overhead. Clearly, low TTL values may constrain propagation too much, and negatively impact event delivery. In our simulation scenario we found out that with TTL=14 the event delivery with $\phi > 0$ remains about the same, while the purely probabilistic routing gets about 20% worse. This is not surprising, given that the deterministic component leads to significantly shorter routes, and therefore is not affected significantly by the TTL. Moreover, the overhead drops considerably, as we expected. For $\phi > 1$, overhead is reduced of about 8%, while for the others reduction is around 30%.

The use of two different values TTL_d and TTL_p we mentioned in Section 3.1 enables further optimizations, as shown in Figure 4 for $TTL_d = 10$ and $TTL_p = 8$. Differently from the chart we showed thus far, this one represents a single run plotted against (simulated) time, and $N = 300$. The values for TTL_d and TTL_p enable only minor—albeit positive—variations over the single TTL=14. Instead, the tradeoff in terms of overhead is profoundly different. The overhead of $\phi > 1$ is slightly increased, but justified by the small increase in the delivery rate. On the other hand, the overhead of $\phi = 1$ is greatly improved, dropping from about 820,000 messages to about 640,000 (more than 20% less), while $\phi = 2$ and $\phi = 3$

---

[5] We do not consider the traffic generated by (un)subscriptions since in our heavy publishing scenario it is negligible w.r.t. the number of events.

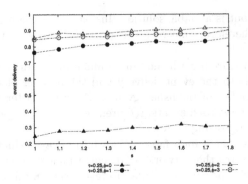

Fig. 4. Event delivery with $\text{TTL}_d = 10$ and $\text{TTL}_p = 8$.

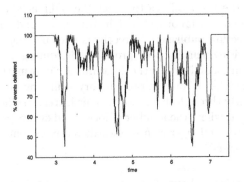

Fig. 5. Event delivery on a dynamic tree topology (from [22]).

are at about 680,000 and 730,000, respectively. This configuration makes routing with $\phi = 1$ more appealing than in earlier scenarios, making it a valid alternative to $\phi = 2$. A different choice for TTL_d and TTL_p, e.g., further increasing the gap between the two, would favor routing with $\phi > 1$.

Stability of Event Delivery   Figure 4 enables us to evidence also another interesting phenomenon whose generality goes beyond the use of TTL, that is, the event delivery is quite stable over time. This is particularly relevant especially if compared against similar results obtained by approaches that rely on a tree as in Figure 5. Although the comparison is entirely qualitative, since the simulators and scenarios differ, it is worth observing how event delivery has wide and very frequent changes, ranging from 100% down to 40%. The reason for the remarkable improvement of our approach can be attributed to the use of a graph and the ability to exploit alternative routes thanks to its probabilistic component.

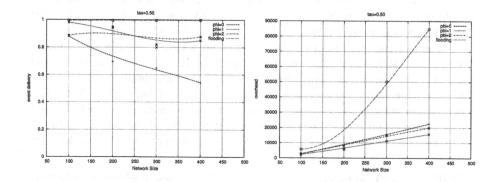

Fig. 6. Event delivery and overhead using $\tau = 0.5$.

4.2 Wireless Sensor Networks

In this Section, we report about the performance of the approach described in
Section 3.2 using TOSSIM [18], the simulation tool provided with TINYOS [16].
TOSSIM emulates all the operating systems layers and therefore works by reusing
directly the code deployed on the sensor nodes—Crossbow's MICA2 motes [1]
in our case. Here, we report only results for $\tau = 0.5$, as it yields the best tradeoff
in our simulations. Each simulation run lasted 60 simulated seconds, with an
extra second devoted to "booting" the network, as performed automatically
by TOSSIM. Transmission occurs by using our simple delay technique to avoid
collisions. The impact of this technique, as well as of its delay-drop variant, is
analyzed later in this section. Simulations are performed on a stable network,
except for the analysis of the behavior of the sensors duty cycle. Finally, we
assume each node has $\eta = 5$ neighbors.

Network Size The first parameter we analyze is the size of the network,
which we ranged from 100 to 400. To maintain a steady publishing load and
receiver density, we increased them proportionally by ranging the former from 1
to 4 published events per second, and keeping the latter at 10% (yielding from
10 to 40 receivers).

The results in Figure 6(a) show[6] that event delivery depends only marginally
on network size. This is not surprising, since the probabilistic component of our
approach tends to distribute the load equally on each node and, therefore, the
more the network grows (and the more receivers need to be reached), the more
nodes participate in delivering the events. Notably, in some cases event delivery
is even increased as more routes become available. On the other hand, as shown
in Figure 6(b) the overhead increases too, since the number of receivers and the
publishing load augments linearly, i.e., there are more events to deliver to more

---

[6] We use Bezier interpolation to better evidence the trends.

recipients. Nevertheless, the two increments share the same trends, that is, no additional overhead is introduced by the size. This, again, stems from the fact the the effort imposed on each node by our algorithm is constant.

It is interesting to see that $\phi = 1$ and $\phi = 2$ exhibit a different behavior. When $N = 100$, $\phi = 2$ performs worse than $\phi = 1$, most likely due to the fact that the smaller size increases the likelihood of creating loops. As N increases, however, the additional deterministic information provided by $\phi = 2$ becomes precious in steering events towards the receivers in a sparser network. Finally, event delivery with $\phi = 1$ and $\phi = 2$ is about the 90% of the delivery obtained through flooding, but the overhead is only 25% of the one introduced by flooding.

Collisions and Rebroadcast   In Section 3.2 we described two simple techniques for, respectively, reducing collisions and avoiding useless rebroadcasts.

The effect of these techniques on the system is shown in Figure 7 for $\tau = 0.5$ and $\eta = 10$. Figure 7(a) shows that the delivery is largely unaffected, with a small decrease in the case of delay-drop. On the other hand, Figure 7(c) shows that our simple mechanism for avoiding collisions is very effective, since it more than halves the number of collisions. The delay-drop mechanism does not improve much in terms of collisions. Instead, by avoiding useless rebroadcasts, this latter technique drastically reduces overhead, as shown in Figure 7(b). Although we do not have simulations linking directly these results to the power consumption, it is evident how the combination of these two simple techniques not only improves the performance of our approach, but also yields remarkable savings in communication, therefore enabling a longer life of the overall sensor network.

Duty Cycle   A prominent feature of our approach is the resilience to changes in the underlying topology and connectivity. Most approaches for content dissemination and group communication for sensor networks rely on exact routes that must be recalculated each time the topology is modified. This is an important limitation, since sensors are often supposed to regularly switch from active to sleeping, to preserve battery and extend the system lifetime. Therefore, unless some kind of synchronization is in place, routes become invalid and must be recomputed, with consequent overhead. Conversely, our approach does not make any assumption on the underlying topology, as it "explores" it semi-probabilistically. Therefore, it can tolerate sleeping (or even crashed, or moving) nodes, without any particular measure.

In the simulations in Figure 8, we used a simple model where each node is active for a period T_a, followed by a sleeping period T_s. All nodes are initially active: after a random time (which temporally scatters them) they are regularly switched off and reactivated after T_s. To obtain meaningful results, sleeping nodes are not considered in the event delivery, which is then computed by taking into account only the active subscribers. Also, since the temporal scattering among nodes is completely random, it may happen that under certain combination of T_a, T_s and η, the network becomes not connected. Then, a delivery of 100% is not meaningful because, if no path exists among two nodes, there is no

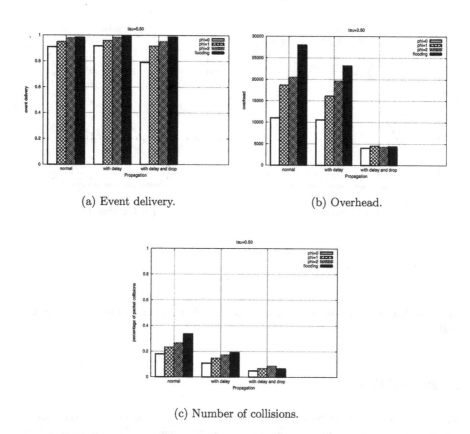

(a) Event delivery.

(b) Overhead.

(c) Number of collisions.

Fig. 7. Collisions and delay-drop ($\tau = 0.5, \eta = 10$).

way to correctly deliver the event. Consequently, our upper bound is represented by the delivery of the flooding approach.

In most scenarios found in literature, sensor nodes sleep for most time and switch on only for a short amount of time. However, in our scenario, sensor nodes are essential not only to acquire data from the environment but also to participate in their propagation. Hence it seems reasonable that the ratio $\frac{T_a}{T_s}$ is greater than (or at least equal to) 1. Clearly, if too many nodes are sleeping at the same time, delivery falls abruptly since the number of forwarders is too low. However, the delivery of flooding also falls abruptly, and some of our solutions remain comparable to it.

These results are not surprising, since what we stated earlier about density holds here as well. Indeed, the effect of sleeping nodes is to reduce the density, expressed in terms of the number η of neighbors. Therefore, since our algorithm

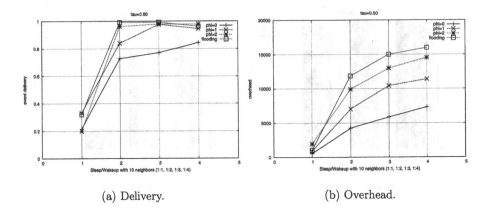

(a) Delivery.   (b) Overhead.

Fig. 8. Performance with sleeping nodes.

tolerates low densities up to a given extent, it is resilient to sleeping nodes as well.

5   Related Work

The majority of content-based publish-subscribe systems are built upon a tree-shaped overlay, with some of them addressing the easier problem of supporting client mobility. Recent work by the authors' research group [9,12,22] deals instead with reconfigurations affecting the dispatchers in the tree. Other recent approaches [8,23] exploit a graph-based topology upon which a set of dispatching trees are superimposed. However, these papers do not provide any detail about if and how dynamicity is taken into account, and at which cost.

In the related area of MANET routing, none of the approaches is directly reusable because of the peculiar challenges posed by content-based routing, but some rely on similar ideas. In the Zone Routing Protocol (ZRP) [15] for unicast, a node proactively maintains routing information about its neighborhood, and reactively requests information about destinations outside of it. In our approach, long distance propagation is instead achieved in a probabilistic way. On the other hand, route driven gossip [19] exploits epidemic algorithms to maintain and disseminate a localized view of the system, enhanced with routing information.

In the context of sensor networks, researchers mostly focused on efficiently delivering the sensed data from the sensors to a fixed base station or, alternatively, on enabling communication from the base station towards all the sensors (e.g to perform a query or force a network re-programming). Our work, instead, is directly applicable to more general scenarios where multiple data sinks (e.g., multiple base stations, but also actuators as in *wireless sensor and actor networks* [2]) are present.

Traditional approaches (e.g., [20,24]) rely on a tree-based structure to deliver messages. This approach minimizes data traffic, but tree maintenance and updates require many control messages and, more importantly, a stable network. Alternative approaches (e.g., [5,17]) spread the nodes' interests across the whole network to create a reverse path from a publisher to receivers. However, again, no details are provided about how to deal with a dynamic network, as in the case of mobile or sleeping sensors, and failures.

A recent work [3] exploits probabilistic forwarding combined with knowledge of the network topology to route messages from sensors to a special node acting as collector. The forwarding probability depends on various parameters, in particular the current distance from the collector. The probabilistic component allows to tolerate stale information on the global topology. Despite the different aim of the work, targetting at single sink application, this approach differs from ours in that we require a much smaller knowledge of the network, namely, only the subscribers ϕ hops away.

The possibility of temporarily switching off nodes is particularly amenable in sensor networks as the battery is not easily replaceable. At the same time, however, the network must maintain its functionality through a connected subnetwork, i.e., it should be able to correctly deliver events despite the lack of some nodes. Some works (e.g., [6,25]) address this issue by introducing synchronization of the sleeping patterns to minimize the energy spent without affecting network connectivity. The weakness of this solution, however, is that other kinds of topological reconfiguration (e.g., mobility or failures) are not tolerated. In these cases, the (expensive) synchronization procedure must be restarted, with added overhead. Conversely, our approach does not require any synchronization protocol and yet tolerates arbitrary reconfigurations.

6 Conclusions

Modern distributed applications exhibit increasing degrees of dynamicity, due to topological reconfigurations occurring at the physical or logical level. Supporting application development through middleware in dynamic scenarios in many cases demands new approaches to routing the applicative information managed by the middleware. Current approaches exploit either a deterministic approach, relying on the dissemination of routing information, or a probabilistic one, inspired by the diffusion of epidemics. However, both have drawbacks.

In this paper we described a semi-probabilistic routing approach targeted to publish-subscribe middleware for highly dynamic networks. Our routing strategy strikes a balance between the two aforementioned approaches, by combining the scalability and resilience to change of probabilistic approaches with the ability to quickly steer events towards the intended receivers typical of deterministic approaches. Our current experience with simulating this routing strategy in both link-based and broadcast-based network scenarios confirms that semi-probabilistic routing indeed achieves high delivery rates with low overhead in presence of frequent topological reconfigurations.

References

1. Crossbow Technology Inc. http://www.xbow.com.
2. I. F. Akyildiz and I. H. Kasimoglu. Wireless sensor and actor networks: Research challenges. *Ad Hoc Networks Journal (Elsevier)*, 2(4):351–367, October 2004.
3. Christopher L. Barrett, Stephan J. Eidenbenz, Lukas Kroc, Madhav Marathe, and James P. Smith. Parametric probabilistic sensor network routing. In *Proceedings of the 2nd ACM international conference on Wireless sensor networks and applications (WSNA)*, 2003.
4. K. P. Birman et al. Bimodal multicast. *ACM Trans. on Computer Systems*, 17(2):41–88, 1999.
5. D. Braginsky and D. Estrin. Rumor routing algorithm for sensor networks. In *Proc. of the 1st Int. Wkshp. on Wireless Sensor Networks and Applications*, pages 22–31, 2002.
6. J. Carle and D. Simplot. Energy-efficient area monitoring for sensor networks. *IEEE Computer*, 37(2):40–46, 2004.
7. A. Carzaniga, D.S. Rosenblum, and A.L. Wolf. Design and Evaluation of a Wide-Area Event Notification Service. *ACM Trans. on Computer Systems*, 19(3):332–383, August 2001.
8. A. Carzaniga, M.J. Rutherford, and A.L. Wolf. A routing scheme for content-based networking. In *Proc. of INFOCOM*, March 2004.
9. P. Costa, M. Migliavacca, G.P. Picco, and G. Cugola. Epidemic Algorithms for Reliable Content-Based Publish-Subscribe: An Evaluation. In *Proc. of the 24th Int. Conf. on Distributed Computing Systems (ICDCS04)*, pages 552–561, 2004.
10. P. Costa, G. P. Picco, and S. Rossetto. Publish-subscribe on sensor networks: A semi-probabilistic approach. In *Proc. of the 2nd IEEE Int. Conf. on Mobile Ad-Hoc and Sensor Systems (MASS05)*, 2005. To appear.
11. P. Costa and G.P. Picco. Semi-probabilistic content-based publish-subscribe. In *Proc. of the 25th IEEE Int. Conf. on Distributed Computing Systems (ICDCS05)*, Columbus (Ohio, USA), June 2005.
12. G. Cugola, D. Frey, A.L. Murphy, and G.P. Picco. Minimizing the Reconfiguration Overhead in Content-Based Publish-Subscribe. In *Proc. of the 19th ACM Symp. on Applied Computing (SAC04)*, pages 1134–1140, March 2004.
13. G. Cugola, A.L. Murphy, and G.P. Picco. Content-Based Publish-Subscribe in a Mobile Environment. In P. Bellavista and A. Corradi, editors, *Mobile Middleware*. CRC Press, 2005. To appear.
14. A. Demers et al. Epidemic algorithms for replicated database maintenance. *Operating Systems Review*, 22(1):8–32, 1988.
15. Z. Haas and M. Pearlman. The Zone Routing Protocol (ZRP) for Ad Hoc Networks. IETF draft, June 1999.
16. J. Hill et al. System architecture directions for networked sensors. In *Proc. of the 9th Int. Conf. on Architectural Support for Programming Languages and Operating Systems (ASPLOS-IX)*, pages 93–104. ACM Press, 2000.
17. C. Intanagonwiwat, R. Govindan, and D. Estrin. Directed diffusion: a scalable and robust communication paradigm for sensor networks. In *Proc. of MobiCom*, 2000.

18. P. Levis, N. Lee, M. Welsh, and D. Culler. TOSSIM: accurate and scalable simulation of entire TinyOS applications. In *Proc. of the 1st Int. Conf. on Embedded Networked Sensor Systems (SenSys'03)*, pages 126–137. ACM Press, 2003.

19. J. Luo, Patrick Eugster, and J.P. Hubaux. Route Driven Gossip: Probabilistic Reliable Multicast in Ad Hoc Networks. In *Proc. of INFOCOM'03*, April 2003.

20. S.R. Madden, M.J. Franklin, and W. Hong. The design of an acquisitional query processor for sensor networks. In *Proc. of SIGMOD 2003*, 2003.

21. P. Naik and K.M. Sivalingam. A survey of MAC protocols for sensor networks. In *Wireless Sensor Networks*. Kluwer Academic Publishers, 2004.

22. G. P. Picco, G. Cugola, and A. L. Murphy. Efficient Content-Based Event Dispatching in the Presence of Topological Reconfigurations. In *Proc. of the 23rd Int. Conf. on Distributed Computing Systems (ICDCS03)*, pages 234–243, 2003.

23. P. Pietzuch and J. Bacon. Hermes: A Distributed Event-Based Middleware Architecture. In *Proc. of the 1st Wkshp on Distributed Event-Based Systems*, 2002.

24. C. Srisathapornphat, C. Jaikaeo, and C.-C. Shen. Sensor information networking architecture. In *Proc. of the Int. Workshop on Parallel Processing*, 2000.

25. D. Tian and N.D. Georganas. A coverage-preserving node scheduling scheme for large wireless sensor networks. In *Proc. of the First ACM Int. Workshop on Wireless Sensor Networks and Applications*, pages 32–41, 2002.

26. A. Varga. OMNeT++ Web page. www.omnetpp.org.

S-WiNeTest: Remote Access To A Sensor Testbed

L. Galluccio, A. Leonardi, G. Morabito, A. Panto', and F. Scoto

{name.surname}@diit.unict.it
Wireless Networking Laboratory (Wi-Ne Lab)
Dipartimento di Ingegneria Informatica e delle Telecomunicazioni
University of Catania, Italy

Abstract. Sensor networks are composed of tiny, cheap, low power devices equipped with small batteries. Depending on the application scenarios, sensors collect some parameters and send them to the sink. This device processes and filters the information sending them to the remote network controller. In this paper we discuss the implementation and the results of the activity carried out and currently on work at the WiNe-Lab of the University of Catania. This activity consists in developing a sensor testbed, using MICAz motes, which can be remotely accessed via a web server by users. The latter can interact with the applications already running on the network and modify the operation mode of a single or all the sensor devices. Users can also run-time monitor the activity carried out by the sensors through a graphical interface and some plots of the sensed parameters. This activity aims either to allow sensor resources sharing among remote users or research groups that do not have the sensors available or to guarantee a high level of interaction between users and sensor devices accordingly this testbed introduces innovative features with respect to other existing and remotely accessible sensor network testbeds.

Keywords: Sensor Networks, Testbed, MICA motes, Remote Control

1 Introduction

Wireless sensor networks are composed of low-cost, low-processing tiny devices which perform sensing, data processing, data fusion and routing [1–5].

Data are monitored by sensor devices which periodically send the collected information to a more powerful device called *sink*. The sink may reside very far from the monitored area and may communicate the collected information to the remote research laboratory using a GPRS or a satellite network.

Novel design challenges arise when dealing with sensor network protocols. More specifically,

- Reduction in power consumption: sensor network's lifetime is strongly dependent on battery since a wireless sensor node is typically equipped with a limited power source to reduce size and weight. When exhausting batteries, sensor devices are abandoned, because their batteries cannot be replaced.

So, power saving and power management are of primary importance because, when sensors disappear, this influences network lifetime and thus the
communication support capability.

- Dynamic network topology management: due to possible mis-functioning and
  low available energy related to batteries' exhaustion, reliable protocols which
  re-route packets and re-organize the network in case of failures should be designed and implemented.
- Fault tolerance and Reliability: protocols must be implemented with a high
  fault tolerance capability not only because of the problem of batteries exhaustion which leads to dynamic network topology but also because the environment where sensor nodes are deployed can be characterized by a high
  level of interference or be hostile.
- Scalability: node density depends on the particular application. Protocols
  designed for these networks should scale well also with an increase in the
  number of the considered devices.

Due to the increasing interest on sensor networks, in the last years numerous
research projects involving sensor networks have emerged [6–8]. However, only
some of these projects have lead to the development of a sensor network testbed
which can be remotely controlled, [7, 8], while other projects have been mainly
concerned with the miniaturization of the components which constitute the sensor device, [9–11]. Also the standardization institutions have carried out a lot of
work aiming at the definition of a standard suitable for sensor network communications. On the one hand the ZigBee Alliance proposed a protocol stack for low
data rate networks ($\in [20, 250]$ Kbps) aimed at supporting very low power consumption and targeted towards automation and remote control applications. On
the other hand, almost in the same period of time, the IEEE 802.15.4 working
group began to work on a low data rate standard focused on the specification of
the first two layers of the network architecture. Accordingly, the ZigBee Alliance
and the IEEE decided to join forces leading to the introduction of the ZigBee
technology [12].

ZigBee technology is aimed at providing connectivity for equipments that
need battery to last for several months to several years, but do not require high
data rate to be supported.

In this paper we discuss the results of an ongoing activity carried out at the
WiNe-Lab [13] which has been dealt with the implementation of a sensor testbed
using MICAz motes [14], mainly focusing on networking functionalities, called
S-WiNeTest.

The main focus of this activity has been to develop a sensor testbed remotely
available to users connected via the web server, which allows to either run-time
monitor and plot the values of light and temperature parameters measured by
sensors in our lab, or to vary the operation mode of a single or all the sensor
devices. Moreover, users can choose to change the routing protocol used by sensor
nodes to send data to the sink, according to the two protocol currently supported
by our testbed.

The main differences with other available remotely accessible testbeds [7, 8] reside in the possibility given to the user to remotely control the applications run by our testbed via the web server, modify the sensor operation of a single or all the sensors and test different available routing protocols for forwarding information from the nodes to the sink. More specifically, in [7], no possibility to inject packets in the network for controlling sensor activities is given since devices can only be re-programmed in the persp ective of sharing mote devices with other research groups. In [8] remote users can monitor light and temperature measured by MICA2 motes disseminated in the various departments of the EPFL and see the related graphs. However,no possibility to switch on and off a single node is given and the routing protocol cannot be changed.

Other features which render our work different from the previous ones are the use of MICAz motes which are the new generation of MICA Crossbow products working in the 2.4 Ghz ISM band and commercialized in August 2004 allowing also to test the interference caused by the coexistence with other technologies working on the same bandwidth such as IEEE 802.11 and Bluetooth. Moreover, through our portal, users can also interact and modify the preconfigured sensor applications currently running in the network. In addition, as in [8], users can see the light and temperature monitored time by time by either all the sensors or each individual sensor device and download the log file in order to possibly reuse data in the future. As a future work we plan to enlarge the set of functionalities which can be remotely accessed by users allowing also to test new applications.

The rest of this paper is organized as follows.

In Section 2 some of the most interesting activities aimed at developing a sensor network testbed are presented, either in the view of letting it be accessed by a remote user or not.

In Section 3 the features and components of the S-WiNeTest testbed are described and discussed in detail. In Section 4 some consideration on sensed parameters which can be measured via our testbed and accessed by remote users through the web interface are given. Finally in Section 5 some concluding remarks on the activity carried out and on the future work are drawn.

2   Related work

In this section, some of the most significant testbeds for sensor networks will be analyzed.

More in detail, in Section 2.1 the Sensonet testbed will be described in detail, while in Section 2.2 the Motelab project will be introduced. Finally, in Section 2.3, the basic features of the Sensorscope testbed will be discussed.

2.1   Sensonet

The target of the Sensonet [6] research project, developed at the Georgia Institute of Technology, is to introduce various protocols, working at different layers of the network architecture, which are aimed at:

1. Adaptation in the transmission power and dynamic scaling of voltage at the physical layer.
2. Localization based on the cooperation between nodes instead that on the availability of signals, such as GPS, that could be expensive and not available in certain environments.
3. Time-synchronization which is needed in order to allow data fusion and reordering of audio and/or video information collected by different nodes.
4. Medium Access Control (MAC) aimed at reducing the overall network energy consumption through the introduction of some wake-up round mechanisms and exploitation of spatial correlation for reducing the amount of correlated data sent at the sink and collected by nodes monitoring in the same area.
5. Routing, aimed at maintaining data topology update, which is a challenge since devices have scarce capabilities and thus signaling should be kept as low as possible.
6. Transport since new light transport paradigms and semantics suitable for reliable event-to-sink detection are needed.

The Sensonet Project, conceived as a research activity, has also lead to the development of a sensor testbed as shown in Figure 1 which however cannot be remotely accessed by users.

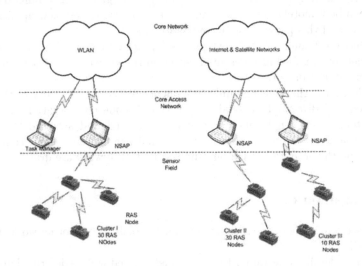

Fig. 1. Sensonet Testbed.

It consists of three parts: the *core network*, the *core access network* and the *sensor field*.

The core network is the backbone of the overall sensor network. It consists of the Wireless Local Area Network (WLAN), the Internet and the Satellite network.

The core access network is made up of NSAPs which are sinks for the sensor network and are emulated by laptops.

The sensor field is the area, where the RAS (route, access, and sense) nodes are deployed. Each RAS node can be either mobile or static but includes the following components:

- MPR300CA MICA [14] process/radio board shown in Figure 2 using the TinyOS distributed operating system. TinyOS is the event-driven operating system used by MICA motes employing a component-based architecture.
- Atmel Atmega 103L processor (operating at 4 MHz with 128 KB of internal flash), 4 KB external SRAM, 4 KB external EEPROM, RFM TR1000 916 MHz radio transceiver at maximum 50 Kb/s, and 51-pin expansion connector for MICA sensors.
- MTS310CA MICA light, temperature, acoustic actuator, magnometer, and accelerometer sensors.
- Video and audio capturing devices.
- Differential Global Positioning System (GPS).
- Remote control cars or brownian motion cars.

Fig. 2. MICA motes devices.

2.2 MoteLab

MoteLab is a wireless sensor network testbed developed at the Maxwell Dworkin Laboratory at Harvard University. MoteLab consists of a sensor testbed which allows reprogramming of devices and testing of applications developed by remote users via a web interface. The MoteLab activity, differently from the one developed within the Sensonet Project is considered as an implementation work allowing remote users to exploit the resources available at the Harward laboratory. Users can upload files and execute them on the MoteLab testbed, so testing and developing new applications, provided they re-program the Cross-bow motes using the TinyOS. MICA mote devices are programmed in the NesC programming language which is a component-variant of the C language.

The applications could be developed using the TinyOS mote simulator, called TOSSIM [15], which has been introduced at the Berkeley University. TOSSIM is scalable and compiles directly from TinyOS code; TOSSIM simulates the TinyOS network stack at the bit level, allowing experimentation with low-level protocols in addition to top-level application systems.

However, no tools for interaction with running applications are provided.

The hardware employed in this project consists of

- 40 Mica2 motes with an Atmel ATMEGA128L processor running at 7.3MHz, 128KB of read-only program memory, 4KB of RAM, and a Chipcon CC1000 radio operating at 433 MHz with an indoor range of approximately 100 meters.
- A sensor board attached to each sensor allowing to sense many variables and to collect environment images
- An Ethernet connection available at each sensor as well as a wall plug

2.3 SensorScope

SensorScope is a project developed at the LCAV (Audiovisual Communications Laboratory) at EPFL

The wireless sensor network testbed consists of about 20 MICA2/DOT motes equipped with a variety of sensors which monitor light, temperature, acoustic vibrations, magnetic field and acceleration, mounted on MTS310 and MTS510 sensorboards.

The target of the SensorScope project is to build a stable, long-running system with all the constraints of a real deployment i.e. motes run on batteries and no wired back-channel is available. The system employs an end-to-end system going from the motes to the web. The initial version uses mostly standard TinyOS components for data routing and command broadcast.

Through the web site of the SensorScope project [8] the user is allowed to access the current sensor readings. The graphs are updated every 5 minutes and represent the actual state of the sensor network and its measurements.

The routing tree is also shown through a graphical interface. An example taken from the SensorScope site is shown in Figure 3.

The Commands web page allows the remote user to get/set RF power, to set sampling rate of all the sensor devices, to put all the mote in sleep/wakeup mode but not the single device, to start/stop sensing of a single or all the devices and to gather routing statistics using the pre-defined motes routing protocol.

Differently from the S-WiNeTest system, the Sensorscope testbed does not allow users to change sleep/wakeup status of a single device as well as to change its sampling rate because the sensor network is conceived as a whole. Differently from [8], in the S-WiNeTest system, each device can be addressed through its ID and its operation mode changed indipendently of the other motes. Moreover,

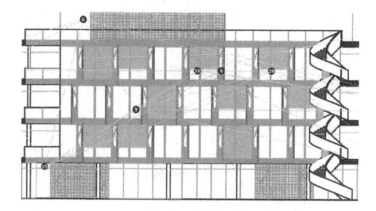

Fig. 3. A sector of the SensorScope routing tree.

the routing protocol being used can be run-time changed and chosed between the Surge and the GeoSense protocol.

3   Scenario characterization

The target of our work has been to develop a working testbed set in our laboratory that users can remotely access and control via the interactive web site. More in detail, users can either use the applications already running on the system, monitor temperature or light or perform packet injection and test different routing protocols.

The testbed has been set in the laboratory as shown in Figure 4 using MICAz sensor devices placed in the positions shown in the boxes.

MICAz motes represent the new generation of MICA2 devices. The latter work in the 315, 433 or 900 MHz bandwidth and employ multi-channel transceivers. Moreover, a Direct Sequence Spread Spectrum physical layer technique is used to reduce RF interference and increase security. The data rate which can be supported is up to 250 kbps.

Differently from the above mentioned MICA2 motes, the MICAz motes offer an IEEE 802.15.4/ZigBee compliant RF transceiver and work in the 2.4 ISM band similarly to IEEE 802.11b and Bluetooth devices.

Numerous obstacles and time-varying interference sources are considered in the testbed: people working or going around the lab, wardrobes and walls due to the irregular form of the area as well as IEEE 802.11b devices and Bluetooth equipments disseminated in the lab. The hardware devices being used constitute the MICA MOTE KIT 2400 which includes:

- 8 MICAz Processor/Radio Boards
- 4 MTS310 sensor boards which monitor acceleration, magnetic field, light, temperature, acoustic vibrations and sound.

Fig. 4. Map of the laboratory where the S-WiNeTest has been set.

- 3 MTS300 sensor boards which monitor light, temperature, acoustic vibrations and sound.
- 1 MDA300 Data Acquisition Board
- 1 MIB600 Ethernet Interface Board used to connect the board directly to the Ethernet
- 1 MIB510 Programming and Serial Interface Board with DB-9 (Male-Female) RS-232 Cable

In order to allow remote users to interact with the sensor networking system two components have been developed as shown in Figure 5: the Proxy System and the Sensor Environment.

- The Proxy System consists of the following elements:
  - Web server which includes the PHP module
  - Parsing and Monitoring Module
  - Serial Forwarder
  - Packet Injection Module
- The Sensor Environment instead consists of the following elements:
  - an Injector Node

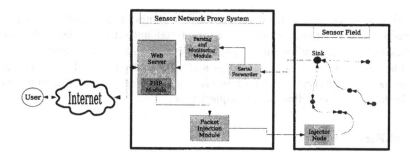

Fig. 5. System Architecture (data flow direction is represented by arrows).

- 7 MICAz motes (including the sink) implementing different Routing Modules

The Proxy System is obtained through a PC running the GNU/Linux operating system; the Linux distribution is the Red Hat 9 with 2.4.20-8 kernel version. The TinyOS version used to develop the testbed is the 1.1.9. All the above cited components will be dealt with in detail in the following sections.

3.1 Proxy System

Web Server

The user remote access to the sensor network is obtained through a web interface, using standard HTML pages. We have used the Apache web server version 1.3.33 with standard settings employing the PHP module 4.3.10. We are currently developing a solution using a Java applet to render the remote access more flexible and user friendly.

The user can choose two different access modes to the Proxy System:

- Can send a monitoring request, that will be processed by the Parsing and Monitoring Module;
- Can request to change one or more of the sensor network parameters. The command is handled by the PHP module and then the action request is sent to the Packet Injection Module.

The PHP module handles the form filled in the HTML page and provides the interface to the Packet Injection Module.

Parsing and Monitoring Module

The data are can be accessed through the web server interface in both text and graphic format. After the user request, considering that the data monitored by sensor devices are collected in an unconventional format which results very hard to read, information need to be parsed and filtered, according to the requested output format.

We have developed a stand alone tool that monitors the wireless sensor networks' log files written in a raw format and filters output data, classifying them in different files depending on the payload type or on the source node.

The tool is written in C language and runs in background when the web server is active. Everytime a user sends a monitoring request the web server retrieves the file according to the request type.

Serial Forwarder

The TinyOS developing environment provides some simple Java programs to help the development of more complex applications and simplify the interaction between the sensor network and the device to which it is connected.

An important Java tool we used is the *SerialForwarder*. The Serial Forwarder application is used to allow multiple clients to communicate with a mote device operating as a sink. The Serial Forwarder works as a proxy on the host connected to the mote device acting as a sink via a serial port, thus providing a bi-directional packet stream to the client applications. Accordingly, the Serial Forwarder provides a socket based abstraction for connecting to the motes.

Basically, it works as a relay point between any data source device and the remote source device. It's intended use is to provide TCP/IP connectivity to a locally connected TinyOS device. The SerialForwarder provides the capability of multiplexing many network connections into a single one.

Packet Injection Module

In order to allow remote users to inject packets in our sensor network testbed, a Java tool, denoted as *Packet Injection Module*, has been developed. The program builds TinyOS packets and injects them into the network using the Serial Forwarder. Accordingly, the injector generates packets that will cause the receiving mote to start/stop its sensing activity, to adjust its sampling rate, to tune the sleep/wake up state, to turn on/off its LEDs etc. Furthermore our motes have been programmed to use one out of two different routing schemes which will be described in the next section so that the remote user is able to inject a packet and to change the routing protocol for forwarding this packet into the network.

The Figure 6 shows the local GUI of the Packet Injection Module. The Commands page in the web site provides a remote interface, in HTML format, having the same aspect of the local one.

Fig. 6. Local interface of the Packet Injection Module.

3.2 Sensor Environment

Injector Node

Once Packet Injection Module has built a TinyOS packet containing the user command, the packet is sent, via serial port or Ethernet connection, to a mote that has to be programmed to act as a packet injector. Currently, the Injector Node in our system executes the *TosBase* TinyOS program. This reprogrammed node receives the command message, checks the header fields and puts it in its sending queue, managed using a FIFO policy.

At this time such mechanism does not use acknowledgments but we are developing a technique which ensures higher reliability in the packet injection scheme, so that the remote user will be able to know if his command request was correctly delivered or not.

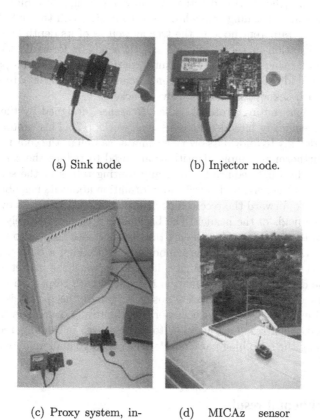

(a) Sink node (b) Injector node.

(c) Proxy system, injector node and sink. (d) MICAz sensor node.

Fig. 7. Testbed components.

Routing Modules

In our testbed, two different routing protocols are available to the users which remotely interact with the network. For what concerns the data sent from the injector node to the rest of the network nodes, a broadcast technique which employs flooding is used. Instead, when sensors send the data they monitor to the sink, two routing protocols are available in the S-WiNeTest system:

- the tree based routing protocol implemented in the *Surge Reliable* [16] which is an open-source data gathering application developed by Crossbow that extends the features of TinyOS Surge application. Both Surge and Surge Reliable use the library components included in TinyOS-1.1.0 release and later, that provide ad-hoc multi-hop routing for sensor network applications. The implementation uses a shortest-path-first algorithm with a single destination node (the sink) and active two-way link estimation. Surge motes are organized in a spanning tree whose root coincides with the sink. Each mote is aware of its parent node in the tree as well as of its depth and advertises this information in each radio message it sends or forwards. Furthermore each mote, using link quality measures, periodically selects its parent node and it periodically, i.e. once a second, sends its sensed data to the parent node which acknowledges the received packets.
- a geographic routing protocol, called *Geosense*, designed by the WiNeLab Team, specifically intended for sensor networks [17]. This protocol requires each node only to known its own coordinates as well as the coordinates of the sink. Furthermore simple algorithms are used to reduce the computational load and does not require to store any routing tables at the sensor nodes. Moreover the generic node needs no information about its neighbors in order to decide to forward the received packets or not, evaluating its own position and some fields of the header. The basic idea is to progressively forwarding the data towards the sink reducing the distance with respect to the previous forwarder. Accordingly, a sensor node upon receiving a data packet which requires to be forwarded, evaluates if it can forward the packet guaranteing that the distance to the destination is reduced with respect to the previous node. In case this check gives a positive answer, the packet is forwarded, otherwise it is not. The protocol works without any control packets so that nodes save their resources avoiding to waste energy in sending, receiving and processing the signalling packets. Moreover a very small amount of memory is needed.

4 Experimental results

Users remotely accessing to our testbed can exploit all the information monitored by the motes devices. To this purpose, the web interface offers the possibility to view real-time graphs reporting the light and temperature measured inside the laboratory where the S-WiNeTest is set. There are several visualization methods; the user can choose to view only the graphs plotting the values of some parameters monitored by a specific sensor or by all the sensors at the same time. Moreover, users can browse the raw log file and retrieve the information for later use.

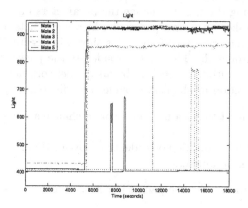

Fig. 8. Light values measured by some MICAz motes in the S-WiNeTest (start time 7:00 am - end time 12:00 am).

Figure 8 shows the light values measured by five MICAz motes during a time interval of five hours starting from 7 a.m. to 12 a.m. The two motes measuring low values of light have been put into two closets. The peaks correspond to the opening of the closet's shutters. In the other three curves, the steps are in correspondence of the switching on of the light in the laboratory.

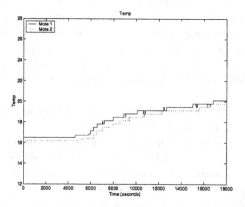

Fig. 9. Temperature values measured by some MICAz motes in the S-WiNeTest (start time 5:00 am - end time 10:00 am).

Figure 9 shows the temperature values measured by two MICAz motes during a time interval of five hours starting from 5 a.m to 10 a.m. Here the curves have increasing trends according to the switching on of the air conditioning system.

The above figures represent only a monitoring activity but our testbed allows also a deep interaction with the sensors parameters and functionalities. This can be obtained allowing the user to change the parameters filling an HTML form. The web server wakes up a java program which constructs a packet and send it, by UART communication, to a mote (as shown in Figure 7(b)) programmed to act as an injector node. This mote broadcasts the packet which, through a flooding routing algorithm, is received by all the network nodes.

The packet injection mechanism provides two different capabilities:

– broadcast a packet to set a parameter or to change a functionality in all the sensor nodes
– send a packet to a specific node identified by its ID in order to change one of its operation parameters

The packets provided by the packet injection mechanism allow the following actions:

– to toggle sleep/wake up mode
– to start/stop the sensing activity
– to change the sensing and packet sending rate.
– to change the routing protocol.
– to replace one source (when the Geosense runs).

The remote user is able, like in [8], to send a command to toggle the sleep/wake up state of all the motes in the network. S-WiNeTest allows also the remote user to sleep-wake up a single mote. When a mote is in sleep state it stops all its main activities (including the forwarding task). Instead when the start/stop command is sent it will involve only the sensing activity. Another command we provide allows the remote user to change the frequency with which sensors are polled, as in [8]. In S-WiNeTest, however, the sensing frequency of a single mote is indipendent from the sensing frequency of the other motes so that the remote user can change the sensing frequency of a single device by specifying its ID.

An innovative characteristic of our testbed resides in the possibility to change the routing algorithm used by the sensor network, without reprogramming the motes.

Figure 10 shows the web page that allows the remote user to choose the routing protocol or to change the sampling rate.

The user can also view a graphical representation of the routing tree changing run-time, after the routing algorithm switch; this can be supported because every packet carries information about the path from the source node to the sink.

Furthermore when the network uses the Surge routing, all the motes poll the sensors in their sensor board and send packets to the sink. Instead, when the Geosense routing protocol is enabled the network starts only with one mote acting as a packet source; while the Geosense runs the system allows the remote

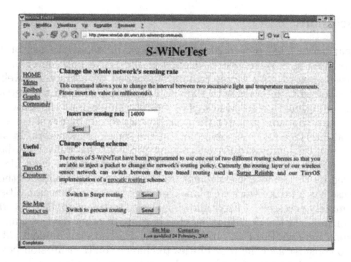

Fig. 10. S-WiNeTest Commands Page: commands for changing the sampling rate and the routing scheme.

user to send commands to replace the source node with another one, to increase or decrease the number of motes that poll their sensors and to specify, for each device, the sampling frequency. The web interface that allows the remote user to send these commands is represented in Figure 11.

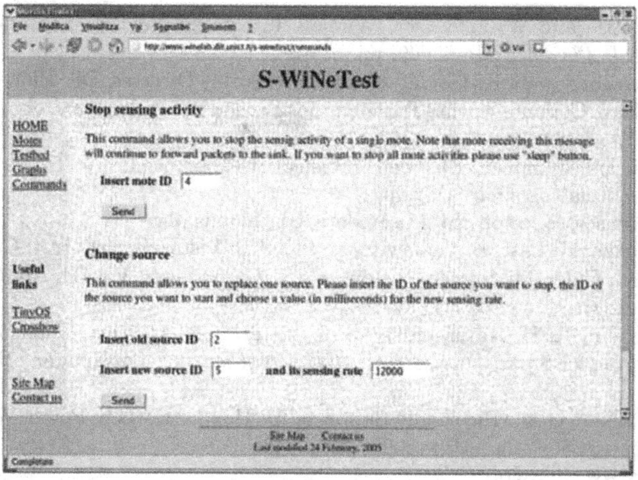

Fig. 11. S-WiNeTest Commands Page: commands for starting/stopping the sensing activity and changing the packet source when routing data with the Geosense protocol.

5 Conclusions

In this paper we have presented and discussed in detail the sensor network testbed denoted as S-WiNeTest developed at the WiNe-Lab center in Catania. This testbed uses MICAz motes remotely accessed by users via a web server. Users are allowed to interact with the applications running on the network and modify the sampling rate or the activity mode of a single or all the sensor devices. Also, users can run-time monitor the activity carried out by the sensors through a graphical interface and see some plots of the sensed parameters. In addition, a remote user can employ different routing protocols for delivering data from the sensor devices to the sink and compare their performance. Log files can be downloaded by the remote users to ensure also the reproducibility of the results for later use. As a future work we plan to enlarge the set of functionalities which can be remotely accessed by users allowing also to test new applications designed by users or monitor the degree of interference due to other devices transmitting on the same bandwidth or to obstacles in the testbed environment. Moreover, we plan to consider also the possibilities to change the transmission power level of the sensor devices, investigating on the performance which can be achieved.

References

1. I. F. Akyildiz, W. Su, Y. Sankarasubramaniam, and E. Cayirci. A Survey on Sensor Networks. *IEEE Communications Magazine*. Vol. 40 , No. 8. August 2002.
2. C. Shen, C. Srisathapornphat, and C. Jaikaeo. Sensor Information Networking Architecture and Applications *IEEE Pers. Commun.* Aug. 2001, pp. 52-59.
3. G. Hoblos, M. Staroswiecki, and A. Aitouche. Optimal Design of Fault Tolerant Sensor Networks. *IEEE Int'l. Conf. Cont. Apps., Anchorage, AK* Sept. 2000, pp. 467-72.
4. G. J. Pottie and W. J. Kaiser. Wireless Integrated Network Sensors. *Commun. ACM, vol. 43, no. 5* May 2000, pp. 551-58.
5. C. Intanagonwiwat, R. Govindan, and D. Estrin, Directed Diffusion: A Scalable and Robust Communication Paradigm for Sensor Networks. *Proc. ACM MobiCom '00, Boston, MA* 2000, pp. 56-67.
6. http://users.ece.gatech.edu/ weilian/Sensor/testbed.html
7. http://motelab.eecs.harvard.edu/
8. http://sensorscope.epfl.ch./sensor_data/continuous.php
9. B. Warneke, M. Last, B. Leibowitz, and K. S. J. Pister. Smart Dust: Communicating with a Cubic-Millimeter Computer. *IEEE Computer.* Vol. 34, No. 1. pp. 43-51. January 2001.
10. J. M. Kahn, R. H. Katz, and K. S. J. Pister. Next Century Challenges: Mobile Networking for Smart Dust. *Proc. ACM MobiCom '99, Washington, DC* 1999, pp. 271-78.
11. J. M. Rabaey et al., PicoRadio Supports Ad Hoc Ultra- Low Power Wireless Networking. *IEEE Comp. Mag.* 2000, pp. 42-48.
12. http://www.zigbee.org/
13. http://winelab.diit.unict.it/s-winetest
14. http://www.xbow.com
15. P. Levis, N. Lee, M. Welsh, and D. Culler. TOSSIM: Accurate and Scalable Simulation of Entire TinyOS Application. *Proc. of the First ACM Conference on Embedded Networked Sensor Systems* (SenSys 2003)

16. D. Gay, P. Levis, R. von Behren, M. Welsh, E. Brewer, and D. Culler. The nesC Language: A Holistic Approach to Networked Embedded Systems. *Proc. of Programming Language Design and Implementation (PLDI)*. June 2003.
17. D. Ferrara, L. Galluccio, A. Leonardi, G. Morabito, and S. Palazzo. MACRO: An Integrated MAC/Routing Protocol for Geographical Forwarding in Wireless Sensor Networks. *Proc. of IEEE INFOCOM 2005*. March 2005.

Decentralized Detection In Sensor Networks With Noisy Communication Links*

Gianluigi Ferrari and Roberto Pagliari

University of Parma, Parco Area delle Scienze 181A - 43100 Parma, Italy
gianluigi.ferrari@unipr.it,pagliari@tlc.unipr.it
WWW home page: http://www.tlc.unipr.it/ferrari

Abstract. This paper presents a general approach to distributed detection with multiple sensors in network scenarios with noisy communication links between the sensors and the fusion center (or access point, AP). The sensors are independent and observe a common phenomenon. While in most of the literature the performance metrics usually considered are missed detection and false alarm probabilities, in this paper we follow a Bayesian approach for the evaluation of the probability of *decision error* at the AP. We first derive an optimized fusion rule at the AP in a scenario with ideal communication links. Then, we consider the presence of noisy links and model them as binary symmetric channels (BSCs). This assumption leads to a simple, yet meaningful, performance analysis. Under this assumption, we show, both analytically and through simulations, that if the noise intensity is above a critical level (i.e., the cross-over probability of the BSC is above a critical value), the lowest probability of decision error at the AP is obtained if the AP selectively discards the information transmitted by the sensors with noisy links.

Keywords: Decentralized detection, sensor networks, noisy communication links, multiple observations, cross-layer design.

1 Introduction

Distributed detection has been an active research field for a long time [19]. In particular, several approaches have been proposed to study this problem, in the realms of information theory [10], target recognition [16,17], and several other areas. The increasing interest, over the last decade, for sensor networks, has spurred a significant research activity on distributed detection techniques in this context [21,4,7,9].

In recent years, wireless sensor networks are becoming more common, as, for example, in terrain monitoring applications [18]. In a wireless communication scenario, links between sensors and the access point (AP) are likely to be

* This work was supported in part by Ministero dell'Istruzione, Università e Ricerca (MIUR), Italy, under the PRIN project "CRIMSON" (Cooperative Remote Interconnected Measurement Systems Over Networks).

faded [15, 6]. In this case, most of the results proposed in the literature are not immediately applicable, since they are based on the assumption that communication links between sensors and AP are "ideal," i.e., the information transmitted by sensors is received correctly by the AP. The characteristics (in terms of capacity) of the radio multiple access channel in wireless sensor networks are taken into account in [5], where optimal configurations for decentralized detection are analyzed. Study of decentralized detection taking into account realistic communication constraints is also considered in [1, 12].

In this paper, we first revisit the basic principles of distributed detection with binary decisions at the sensors. In order to model a scenario where some of the links between sensors and AP are non-ideal, we assume that a link can be modeled as a binary symmetric channel (BSC) [14]. We show that selective elimination of noisy links may lead to a performance improvement when the *cross-over* (or *bit-flipping*) probability of the BSC increases. In particular, for each value of the common signal-to-noise ratio (SNR) at the sensors we determine a critical bit-flipping probability which discriminates between two network operating regimes: for values of the bit-flipping probability above the critical value, the best performance is obtained when the AP excludes the sensors with noisy links. In particular, selective exclusion of sensors with noisy links could be obtained, for instance, by using a clever medium access control (MAC) protocol at the AP. Therefore, our results suggest that the use of a *cross-layer approach* to the design of sensor networks with unreliable communication links (e.g., wireless sensor networks) is the best choice.

This paper is structured as follows. In Section 2, we provide the reader with preliminaries on distributed detection principles, referring to a classical distributed detection scheme with *parallel* schedule. In Section 3, the presence of noisy links, modeled as BSCs, is considered, and the corresponding sensor network performance is analyzed. Conclusions and future research directions are presented in Section 4.

2 Preliminaries on Distributed Detection

We consider a classical sensor network scenario where all sensors are connected to a single AP [21]. Two main approaches for combining the information gathered by multiple sensors have been proposed.

- The first approach is referred to as *centralized*: all sensors observations are transmitted to a central processor that performs a global decision.
- The second approach is referred to as *decentralized*: each sensor makes a local decision and a fusion processor, i.e., the AP, makes the final decision, by applying a suitable *fusion rule*.

In this paper, all sensors make an observation of a common binary phenomenon. In other words, we consider the binary hypothesis signal detection problem [13], with statistically independent observations from sensor to sensor.

We will refer to the two hypotheses as H_1 and H_0, respectively. The true hypothesis will be simply denoted as H. We will assume that the two hypotheses are equally likely. The extension of this work to the case of correlated sensors [3] is currently under investigation.

Suppose that there are N sensors and that they observe the same phenomenon at a given point in time (for notational simplicity, we do not explicitly consider the time instant of the observation). The discrete-time observation at the i-th sensor can be expressed as

$$r_i = y_i + n_i \qquad (1)$$

where

$$y_i \triangleq \begin{cases} 0 & \text{if} \quad H_0 \\ s & \text{if} \quad H_1 \end{cases}$$

with $i = 1, 2, \ldots, N$. Assuming that the noise samples $\{n_i\}$ are independent and identically distributed with the same Gaussian distribution $\mathcal{N}(0, \sigma^2)$, the *common* signal-to-noise ratio (SNR) at each sensor can be defined as follows:

$$\text{SNR}_{\text{sensor}} \triangleq \frac{[\text{E}\{y_i|H_1\} - \text{E}\{y_i|H_0\}]^2}{\sigma^2} = \frac{s^2}{\sigma^2}. \qquad (2)$$

For the sake of notational simplicity, we assume that $\sigma = 1$, so that $\text{SNR}_{\text{sensor}} = s^2$. We also assume that the SNR is the same at all sensors, i.e., the sensors are equivalent.

In a classical distributed detection scheme with parallel schedule, each sensor makes an observation of the common phenomenon, decides for one of the two hypotheses, and then sends its binary decision, denoted as u_i, to the AP. In general, the decision rule at each sensor (common for all sensors) can be written as $u_i = \gamma(r_i)$, where $\gamma(\cdot)$ is a suitable *decision function*. Usually, the communication link between each sensor and the AP is *ideal*, i.e., the AP receives correctly the bit transmitted by each sensor. In order to make a decision, the i-th sensor compares the observation r_i with a threshold value τ and computes its binary decision as follows:

$$u_i = \gamma(r_i) = \begin{cases} 1 & \text{if} \quad r_i < \tau \\ 0 & \text{if} \quad r_i > \tau. \end{cases} \qquad (3)$$

Equivalently, one can write $\gamma(r_i) = U(r_i - \tau)$, where $U(\cdot)$ is the unitary step function. It is possible to show that this decision rule is equivalent to a local likelihood ratio test [11]. In [20], it is shown that selecting the same value of τ for all sensors is an asymptotically (for large values of N) optimal choice for minimizing the probability of incorrect decision. Moreover, in [20] the author shows also that selecting the same value of τ also for a relatively small number N of sensors leads a negligible performance loss with respect to an optimal threshold selection among the sensors. Motivated by this observation, in the remainder of this paper we will assume that the threshold value τ for local decision is the same for all sensors.

Once all sensors have made their local decisions $\{u_i\}$, the AP receives an array of N binary values, and makes a final decision u_0 according to a *fusion rule*

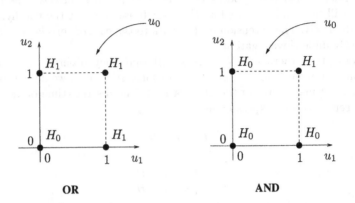

OR   **AND**

Fig. 1. Decision regions for majority-like fusion rules in the case with $N = 2$ sensors: OR (left) and AND (right) rules. In the axes there are the local decisions (denoted as "0" and "1") at the two sensors, while within the diagram there is the final decision at the AP.

$u_0 = \Gamma(u_1, \ldots, u_N)$. As shown in the literature, the fusion rule must be based on a binary monotonic increasing function of the decisions array of length N [21]. Given N, even if there are 2^{2^N} possible fusion rules, one can limit herself/himself at investigating only binary monotonic increasing functions [16, 21]. Under the assumption that the SNR is the same at all sensors, these fusion rules can be given the following general *majority-like* expression:

$$\Gamma(u_1, \ldots, u_N) = \begin{cases} 1 & \text{if } \sum_{i=1}^{N} u_i \geq k \\ 0 & \text{if } \sum_{i=1}^{N} u_i < k \end{cases} \tag{4}$$

where $k = 1, \ldots, N$. In general, if $k = 1$ the OR fusion rule is obtained, while if $k = N$ the AND fusion rule is obtained. In a network with $N = 2$ sensors, only the OR and AND fusion rules are possible and a pictorial description of these rules is shown in Fig. 1.

Provided that the fusion rule is in the form given by (4), the key problem consists in determining the value of k that minimizes the probability of error under a Bayesian criterion, defined as

$$P_e \triangleq P\{u_0 \neq H\}.$$

Based on our assumption of equally likely hypotheses ($P(H_0) = P(H_1) = 1/2$), the probability of error can be written as

$$P_e = \frac{1}{2}P(u_0 = H_0|H_1) + \frac{1}{2}P(u_0 = H_1|H_0). \tag{5}$$

In the general case with $N \geq 2$ sensors, the two terms at the right side of (5) can be evaluated as follows:

$$P(u_0 = H_0|H_1) = P\{\text{less than } k \text{ sensors decide for } H_1|H_1\}$$

$$= \sum_{i=0}^{k-1} \binom{N}{i} P(u_i = H_1|H_1)^i P(u_i = H_0|H_1)^{N-i}$$

$$= \sum_{i=0}^{k-1} \binom{N}{i} [1 - \Phi(\tau - s)]^i \Phi^{N-i}(\tau - s) \qquad (6)$$

$$P(u_0 = H_1|H_0) = P\{\text{at least } k \text{ sensors decide for} H_1|H_0\}$$

$$= \sum_{i=k}^{N} \binom{N}{i} P(u_i = H_1|H_0)^i P(u_i = H_0|H_0)^{N-i}$$

$$= \sum_{i=k}^{N} \binom{N}{i} [1 - \Phi(\tau)]^i \Phi^{(N-i)}(\tau) \qquad (7)$$

where $\Phi(x) \triangleq \frac{1}{\sqrt{2\pi}} \int_{-\infty}^{x} e^{-\frac{y^2}{2}} dy$. Therefore, using (6) and (7) into (5), one obtains

$$P_{\mathrm{e}} = \frac{1}{2} P(u_0 = H_0|H_1) + \frac{1}{2} P(u_0 = H_1|H_0)$$

$$= \frac{1}{2} \sum_{i=0}^{k-1} \binom{N}{i} [1 - \Phi(\tau - s)]^i \Phi^{N-i}(\tau - s) + \frac{1}{2} \sum_{i=k}^{N} \binom{N}{i} [1 - \Phi(\tau)]^i \Phi^{N-i}(\tau).$$

$$(8)$$

The behavior of the probability of error, as a function of the threshold value τ, is shown in Fig. 2, in the case with $\mathrm{SNR}_{\mathrm{sensor}} = s^2 = 0\,\mathrm{dB}$. As one can observe from Fig. 2, for each decision rule the probability of error is a quasi-convex function of τ and has an absolute minimum. Numerically, one can characterize the absolute minimum depending on the value of N.

- N odd: the optimal value of τ is $s/2$ and the best fusion rule is the *majority rule*, i.e., $k = \lfloor N/2 \rfloor + 1$.
- N even: between the optimal value for the threshold τ and $s/2$ there is an offset that, in general, depends on (i) the number of sensors N, (ii) the sensor SNR s^2, and (iii) the fusion rule. In particular, the best fusion rules are obtained selecting $k = N/2 + 1$ (i.e., adopting a majority rule) or $k = N/2$. For both fusion rules, by properly selecting the threshold value τ the probability of decision error is the same.

As intuitively expected, increasing the number N of sensors and choosing the corresponding optimal fusion rule, the performance (in terms of P_{e}) improves dramatically.

3 Sensor Networks with Noisy Communication Links

While all previous results apply to a sensor network scenario where the communication links between sensors and AP are ideal, in a realistic scenario (e.g., a

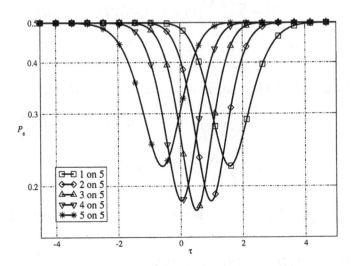

Fig. 2. Probability of error, as a function of the threshold value τ, in a scenario with $N = 5$ sensors and $\text{SNR}_{\text{sensor}} = 0$ dB. Various values of k, corresponding to different fusion rules, are shown.

wireless sensor network) it might happen that these links are noisy (e.g., they are *faded* [6]). Studying such a scenario is difficult, since the presence of fading might also create correlations among the sensors [15]. The analysis and optimization of wireless sensor networks is, therefore, a complicated problem. In [6], the authors propose fusion algorithms that take into account channel fading statistics. In order to derive significant insights into the problem of decentralized detection in sensor networks with realistic communication links, we now consider a simplified model for a noisy communication link. More precisely, a noisy link between a sensor and the AP is modeled as a BSC with parameter p, corresponding to the channel cross-over probability[1] [8]. In other words, the bit transmitted by the sensor has a probability p of being "flipped." The parameter p will depend on the specific characteristics of the sensors-AP communication links (e.g., modulation format, presence of channel coding, presence of fading, detection strategy at the AP, etc.). Assuming binary hard decision at each sensor, if u_i is the decision sent by the i-th sensor, the AP will receive the following information:

$$u_i^{\text{received}} = \begin{cases} u_i & \text{with probability } 1 - p \\ 1 - u_i & \text{with probability } p. \end{cases}$$

---

[1] We remark that the sensor SNR, i.e., $\text{SNR}_{\text{sensor}} = \sqrt{s}$, is the SNR at each sensor relative to the local detection of the common phenomenon (or state of nature). A realistic communication link between a sensor and the AP could be characterized by an SNR at the AP. In this paper, however, we do not explicitly consider the communication link SNR, since we concisely describe the communication link as a BSC, which is completely characterized by the single parameter p.

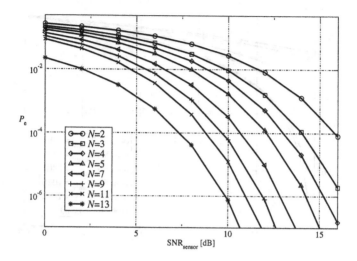

Fig. 3. Probability of error for various values of the number of sensors N. In each case, the optimal fusion rule is considered.

We extend the derivation of the probability of error proposed in Section 2 in order to encompass the presence of noisy links. More precisely, we want to evaluate the final probability of error (5) in a sensor network with noisy links. We consider a majority-like fusion rule at the AP, as described in Section 2, with optimized values of k and τ. We first consider a scenario where all N links are noisy. Then, we generalize the obtained results to the case where $d \leq N$ links are noisy. For example, this scenario could correspond to a wireless sensor network where some of the sensors do not have, temporarily, a "clear" communication path to the AP. Note that the proposed approach could be extended to a scenario where the noise intensity is not the same in all noisy links.

3.1 Sensor Networks with All Noisy Communication Links

After proper algebraic manipulations, it is possible to show that the first conditional probability in (5) can be written as

$$P(u_0 = H_0|H_1) = P\{i < k \text{ sensors for } H_1|H_1\} = \sum_{i=0}^{k-1} \binom{N}{i} P_{c1}^i P_{e1}^{N-i} \quad (9)$$

where

$$P_{c1} = (1-p)P(s+n > \tau) + pP(s+n < \tau) = (1-p)[1 - \Phi(\tau - s)] + p\Phi(\tau - s)$$

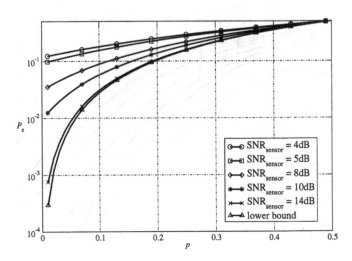

Fig. 4. Probability of error, as a function of the cross-over probability p, for different values of the sensor SNR. The number of sensors is $N = 3$. The curve labeled "lower bound" corresponds to the theoretical limit with $\text{SNR}_{\text{sensor}} = \infty$.

and $P_{e1} = 1 - P_{c1}$. Similarly, the second conditional probability in (5) can be written as

$$P(u_0 = H_1|H_0) = P\{i \geq k \text{ sensors for } H_1|H_0\} = \sum_{i=k}^{N} \binom{N}{i} P_{e2}^i P_{c2}^{N-i} \quad (10)$$

where

$$P_{e2} = (1-p)P(n > \tau) + pP(n < \tau) = (1-p)[1 - \Phi(\tau)] + p\Phi(\tau)$$

and $P_{c2} = 1 - P_{e2}$.

The probability of error (5) can then be evaluated numerically, by using the derived expressions (9) and (10). In particular, the probability of error depends on (i) the decision threshold value τ at the sensors, (ii) the sensor SNR s^2, and (iii) the cross-over probability p.

In Fig. 4, the probability of error is shown as a function of the cross-over probability p, for various values of $\text{SNR}_{\text{sensor}}$, in a scenario with $N = 3$ sensors. As one can observe, regardless of the sensor SNR, for increasing values of p the probability of error becomes unacceptable. The *lower bound* corresponds to a theoretical case where the sensor SNR is infinite. This lower bound, denoted as $P_{e-\text{lb}}(p)$ (to underline its dependence on the cross-over probability p), can be given the following analytical expression:

$$P_{e-\text{lb}}(p) = \lim_{s \to \infty} P_e = \frac{1}{2} \left[\sum_{i=0}^{k-1} \binom{N}{i} (1-p)^i p^{N-i} + \sum_{i=k}^{N} \binom{N}{i} p^i (1-p)^{N-i} \right].$$

Table 1. Analytic expressions of $P(u_0 = H_1|H_0)$ in the following cases: (a) $d \geq k$, $N - d \geq k$, (b) $d \geq k$, $N - d < k$, (c) $d < k$, $N - d < k$ and (d) $d < k$, $N - d \geq k$.

Case	$P(u_0 = H_1|H_0)$
(a)	$\sum_{d_e=0}^{k} \left[\binom{d}{d_e} P_{e2}^{d_e} P_{c2}^{d-d_e} \sum_{i_e=k-d_e}^{N-d} \binom{N-d}{i_e} P_{eH_0}^{i_e} P_{cH_0}^{N-d-i_e} \right]$ $+ \sum_{d_e=k+1}^{d} \binom{d}{d_e} P_{e2}^{d_e} P_{c2}^{d-d_e} \cdot U(d - k - 1)$
(b)	$\sum_{d_e=k+d-N}^{k} \left[\binom{d}{d_e} P_{e2}^{d_e} P_{c2}^{d-d_e} \sum_{i_e=k-d_e}^{N-d} \binom{N-d}{i_e} P_{eH_0}^{i_e} P_{cH_0}^{N-d-i_e} \right]$ $+ \sum_{d_e=k+1}^{d} \binom{d}{d_e} P_{e2}^{d_e} P_{c2}^{d-d_e} \cdot U(d - k - 1)$
(c)	$\sum_{d_e=k+d-N}^{k} \left[\binom{d}{d_e} P_{e2}^{d_e} P_{c2}^{d-d_e} \sum_{i_e=k-d_e}^{N-d} \binom{N-d}{i_e} P_{eH_0}^{i_e} P_{cH_0}^{N-d-i_e} \right]$
(d)	$\sum_{d_e=0}^{k} \left[\binom{d}{d_e} P_{e2}^{d_e} P_{c2}^{d-d_e} \sum_{i_e=k-d_e}^{N-d} \binom{N-d}{i_e} P_{eH_0}^{i_e} P_{cH_0}^{N-d-i_e} \right]$

From the results shown in Fig. 4, one can conclude that, for any value of p, increasing the sensor SNR beyond a critical threshold does not lead to any significant performance improvement. This might have practical implications on the design of sensors, in terms of their detection accuracy. In fact, one should not increase the sensor sensitivity without limit, but, rather, should find the critical sensitivity at which the ultimate theoretical performance is practically obtained.

3.2 Sensor Networks with a Generic Number of Noisy Links

We now extend the previous analysis to encompass the case with a generic number $d \leq N$ of noisy links—and, consequently, $N - d$ ideal links. The fusion rule is the majority-like rule given in (4), with an optimized value of k.

In order to evaluate the probability of error, we first compute the conditional probability $P(u_0 = H_1|H_0)$ at the right-hand side of (5). Let us denote by $d_e \leq d$ the number of noisy links associated to sensors in error, i.e., sensors which decide for H_1 when H_0 has happened, and by $i_e \leq N - d$ the number of ideal links associated to sensors in error, i.e., sensors which decide for H_1 when H_0 has happened. With these definitions, the AP *might make*[2] a final erroneous decision if $d_e + i_e \geq k$, with $d_e \in \{0, \ldots, d\}$ and $i_e \in \{0, \ldots, N - d\}$. Depending on the relations between the integers N, k and d, one can distinguish the following four cases, respectively: (a) $d \geq k$, $N - d \geq k$, (b) $d \geq k$, $N - d < k$, (c) $d < k$, $N - d < k$ and (d) $d < k$, $N - d \geq k$. After tedious manipulations, the final expressions for $P(u_0 = H_1|H_0)$, in the four considered cases, are shown in Table 1, where

$$P_{eH_0} \triangleq P(u_0 = 1|H_0, p = 0) = 1 - \Phi(\tau)$$

and $P_{cH_0} = 1 - P_{eH_0}$.

We now consider the second conditional probability at the right-hand side of (5), i.e., $P(u_0 = H_0|H_1)$. In this case, the AP makes a final decision error

---

[2] The reader should observe that if a sensor is in error and the bit transmitted to the AP is flipped, then the bit actually received by the AP is correct.

Table 2. Analytic expressions of $P(u_0 = H_0|H_1)$ in the four cases corresponding to (a) $d \leq k-1$, $N-d \leq k-1$, (b) $d \leq k-1$, $N-d > k-1$, (c) $d > k-1$, $N-d > k-1$ and (d) $d > k-1$, $N-d \leq k-1$.

Case	$P(u_0 = H_0|H_1)$
(a)	$\sum_{d_c=k+d-N}^{d} \left[\binom{d}{d_e} P_{c_1}^{d_c} P_{e_1}^{d-d_c} \sum_{i_c=0}^{k-1-d_c} \binom{N-d}{i_c} P_{cH_1}^{i_c} P_{eH_1}^{N-d-i_c} \right]$ $+ \sum_{d_c=0}^{k-1+d-N} \binom{d}{d_c} P_{c_1}^{d_c} P_{e_1}^{d-d_c}$
(b)	$\sum_{d_c=0}^{d} \left[\binom{d}{d_e} P_{c_1}^{d_c} P_{e_1}^{d-d_c} \sum_{i_c=0}^{k-1-d_c} \binom{N-d}{i_c} P_{cH_1}^{i_c} P_{eH_1}^{N-d-i_c} \right]$
(c)	$\sum_{d_c=0}^{k-1} \left[\binom{d}{d_e} P_{c_1}^{d_c} P_{e_1}^{d-d_c} \sum_{i_c=0}^{k-1-d_c} \binom{N-d}{i_c} P_{cH_1}^{i_c} P_{eH_1}^{N-d-i_c} \right]$
(d)	$\sum_{d_c=k+d-N}^{k-1} \left[\binom{d}{d_e} P_{c_1}^{d_c} P_{e_1}^{d-d_c} \sum_{i_c=0}^{k-1-d_c} \binom{N-d}{i_c} P_{cH_1}^{i_c} P_{eH_1}^{N-d-i_c} \right] \cdot U(N-d-1)$ $+ \sum_{d_c=0}^{k-1+d-N} \binom{d}{d_c} P_{c_1}^{d_c} P_{e_1}^{d-d_c}$

when $n \leq k-1$ sensors decide for H_1. Let us denote by d_c and i_c the number of sensors in errors (i.e., they decide for H_0 even if H_1 has happened) connected with noisy and ideal links to the AP, respectively. A final decision error *might happen* if $d_c + i_c \leq k-1$, with $d_c \in \{0,\ldots,d\}$ and $i_c \in \{0,\ldots,N-d\}$. As for the computation of $P(u_0 = H_1|H_0)$, four possible cases can be distinguished, depending on the values of N, k and d: (a) $d \leq k-1$, $N-d \leq k-1$, (b) $d \leq k-1$, $N-d > k-1$, (c) $d > k-1$, $N-d > k-1$ and (d) $d > k-1$, $N-d \leq k-1$, respectively. Reasoning as before, one obtains the expressions for $P(u_0 = H_0|H_1)$ shown in Table 2, where

$$P_{eH_1} \triangleq P(u_0 = H_0|H_1) = \Phi(\tau - s)$$

and $P_{cH_1} = 1 - P_{eH_1}$.

In Fig. 5, the probability of decision error is shown for $N = 3$ sensors. All possible values (from 0 to N) of the number d of noisy links are considered. Obviously, for increasing number of noisy links the performance degrades, and this degradation is more pronounced (relatively) for high values of the sensor SNR. This means that when sensors are very reliable, i.e., the sensor SNR is high, the impact of noisy communication links is (proportionally) higher.

3.3 Selective Exclusion of Sensors with Noisy Links

As we have observed from the results in Fig. 5, for increasing number of noisy links, the sensor network performance (in terms of probability of decision error at the AP) degrades rapidly. At this point, one might consider an "intelligent" AP, which neglects the decisions of a sensor if the link is noisy. For example, in a wireless sensor network, each sensor could send a pilot symbol to the AP, which, consequently, could determine the status of the corresponding link. Obviously, if some sensors are excluded, there is a loss of information. Therefore, selective elimination of the noisy links will lead to a performance improvement depending on the value of p, i.e., on the noise intensity in noisy links.

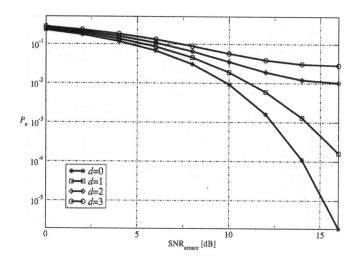

Fig. 5. Probability of error, as a function of the sensor SNR, in a scenario with $N = 3$ sensors.

In order to understand *when* exclusion of noisy links leads to a performance improvement, we evaluate the probability of decision error, as a function of the bit-flipping probability p, for a given value of the sensor SNR. In Fig. 6, the probability of decision error is shown in a scenario with $N = 5$ sensors and $\text{SNR}_{\text{sensor}} = 12$ dB, for different values of the number of noisy links $d \geq 1$. In a scenario with $N = 3$ sensors and no noisy link ($d = 0$), from the results in Fig. 5 one concludes that the probability of error is 1.6×10^{-3}. Considering, in Fig. 6, the curve relative to the case with $d = 2$ noisy links out of $N = 5$, it follows that $P_e = 1.6 \times 10^{-3}$ corresponds to a value $p = 0.12$. Therefore, one can distinguish the following two network operating regions (depending on the value of p).

– If $p < 0.12$, the probability of decision error in a sensor network with $N = 5$ sensors and $d = 2$ noisy links is *lower* than that of a sensor network with $N = 3$ sensors and ideal links. Therefore, using the local decisions of all sensors (even if $d = 2$ communication links are noisy) is the best strategy.
– If $p > 0.12$, the probability of decision error in a sensor network with $N = 5$ sensors and $d = 2$ noisy links is *higher* than that of a sensor network with $N = 3$ sensors and ideal links. In this case, the AP should neglect the information originated by the sensors corresponding to noisy links, and use only the bits coming from the sensors with ideal links.

In a scenario with $d = 3$ noisy links, the critical value of the bit-flipping probability which discriminates between use of all sensors or selection of the subset of sensors with ideal links is (obviously) lower than the critical value in a scenario with $d = 2$ links.

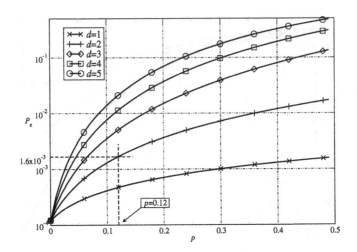

Fig. 6. Probability of error, as a function of cross-over probability p, in a sensor network with $N = 5$ sensors and $\text{SNR}_{\text{sensor}} = 12$ dB.

In general, given a particular sensor network structure (N sensors and d noisy links), for each value of the sensor SNR it is possible to determine the critical bit-flipping probability which discriminates between (i) using all sensors or (ii) using only the subset of sensors with ideal links. In the example previously considered with $N = 5$ sensors and $d = 2$ noisy links, the critical bit flipping probability is shown, as a function of the sensor SNR, in Fig. 7. The diagram has to be interpreted as follows. Given a particular sensor network scenario with a particular sensor SNR and a cross-over probability p (which will depend on the characteristics of the channel between the sensor and the AP), one can determine the ($\text{SNR}_{\text{sensor}}, p$) network operating point: if this point falls above the critical curve, then the AP should neglect the sensors with noisy links; otherwise, if this point falls below the critical curve, then the AP should use all sensors. For ease of understanding, we have also indicated the critical ($\text{SNR}_{\text{sensor}}, p$) operating points corresponding to the probabilities of error between 10^{-2} and 10^{-6}. For example, consider the sensor SNR corresponding to $P_{\text{e}} = 10^{-3}$: if $p < 0.16$, then using all sensors will lead to a probability of error lower than 10^{-3}; for $p \geq 0.16$, the lowest possible probability of error (equal to 10^{-3}) is obtained by using only the sensors with ideal links.

Finally, we remark that the results in Fig. 6 show that the critical bit-flipping probability decreases for increasing values of the sensor SNR. In other words, whenever sensors are very sensitive (i.e., the sensor SNR is high), then the presence of even a limited link noise has a significant impact on the network performance—in fact, the best operating regime is the one corresponding to selective exclusion of the sensors with noisy links. On the constructive side, sensors

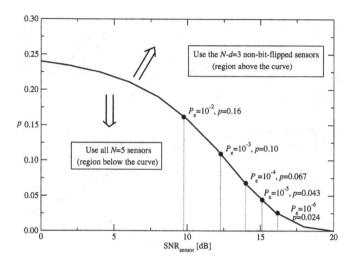

Fig. 7. Critical cross-over probability p as a function of the sensor SNR, relative to a sensor network with $N = 5$ sensors and $d = 2$ noisy links. The curve divides two regions: in the upper region the best performance is obtained by selecting only the $N - d = 3$ sensors with ideal links, whereas in the lower region the best performance is obtained using all $N = 5$ sensors.

which are selectively excluded could be temporarily turned off (e.g., by properly estimating the fade duration in a wireless communication scenario), prolonging the sensor network lifetime. The analysis of this network performance metric, i.e., the network lifetime, is the subject of on-going research activity.

3.4   Multiple Observations at the Sensors

In [2], it has been shown that the use of multiple consecutive and independent observations of the same phenomenon at each sensor has a beneficial effect on the performance, i.e., it reduces the probability of decision error at the AP. While in [2] multiple observations have been considered for sensor networks with *ideal* communication links, we now evaluate the effect of multiple observations in sensor networks with *noisy* communication links. After tedious manipulations (not reported for lack of space), it is possible to extend the previous analysis (carried out in a scenario with single observations at the sensors) and derive analytical expressions for the probability of decision error at the AP. More precisely, in a sensor network scenario with N sensors and d noisy communication links, given a number M of multiple observations, it is possible to evaluate the critical bit flipping probability which discriminates between (a) using all sensors and (b) discarding the sensors with noisy communication links.

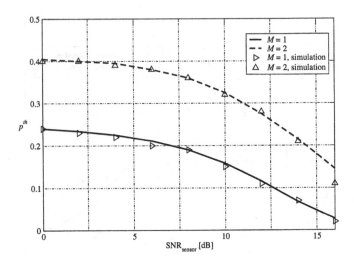

Fig. 8. Critical bit-flipping probability p as a function of the sensor SNR, relative to a sensor network with $N = 5$ sensors and $d = 2$ noisy links with $M = 1$ and $M = 2$ observations per sensor, respectively.

In a scenario with $M = 2$ observations at each sensor, the critical bit flipping probability curve is shown in Fig. 8. In the same figure, for the sake of comparison, the critical bit flipping probability curve of Fig. 7 (relative to a scenario with $M = 1$ observation per sensor) is also shown. It is immediate to observe that the critical bit flipping probability increases when $M = 2$ observations per sensor are used (roughly speaking, it doubles). In the same figure, both analysis and simulation results are shown: as one can see, there is excellent agreement.

In order to consider a higher number of observations per sensor, for numerical reasons we have reduced the number of sensors to $N = 3$. The obtained analytical results, for various numbers of observations, are shown in Fig. 9. As expected, increasing the number of observations has a beneficial effect on the performance in terms of probability of decision error. In other words, our results suggest that use of multiple observations (which comes at the cost of (i) increased delay in the final decision and (ii) increased energy consumption at the sensors and AP) makes the sensor network more robust against impairments in the sensor-AP communication links.

From the results in Fig. 9, it is also interesting to observe that the improvement brought by the use of 2 observations instead of 1 is higher than the improvement obtained by considering 4 observations instead of 3. More precisely, since the optimal fusion rule is a majority-like rule, (i) significant performance improvements are obtained for larger *odd* values of M and (ii), considering the next even value (i.e., $M + 1$), the relative improvement becomes negligible for

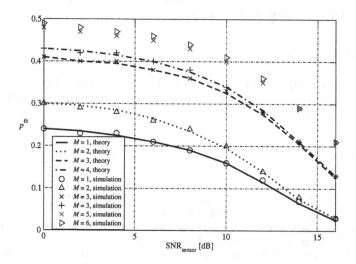

Fig. 9. Critical bit-flipping probability p as a function of the sensor SNR, relative to a sensor network with $N = 3$ sensors and $d = 2$ noisy links with multiple observations, compared to the performance of one sensor with ideal link.

increasing values of M. In other words, the number of observations, if sufficiently high, should be *odd*.

4 Conclusions and Future Work

In this paper, we have considered the problem of distributed detection in sensor networks where some of the communication links between the sensors and the AP may be noisy. First, we have revisited basic principles of distributed detection with binary decisions at the sensors, discussing optimal fusion rules at the AP. Then, we have introduced a simple BSC model for noisy communication links between sensors and AP, and we have analyzed the corresponding network performance, in terms of probability of decision error at the AP. For each value of the sensor SNR, we have shown the existence of a *critical bit-flipping probability*: for values of p *higher* than this critical value, network performance is optimized by discarding the decisions coming from sensors with noisy links; for values of p *lower* than this critical value, network performance is optimized by using the decisions from all sensors. Our results show that the critical bit-flipping probability is a monotonically decreasing function of the sensor SNR.

The different sensor network operating regimes, depending on the number of noisy links and the noise intensity over such links, could be forced by the use of a suitable MAC protocol (with channel sensing) at the AP and we are currently working on its design. We are also extending our approach to encompass the presence of quantization at the sensors.

Acknowledgments

The authors would like to thank Prof. L. A. Rusch (University of Laval, Canada), on sabbatical at the University of Parma, Italy, during Spring semester 2005, for useful discussions.

References

1. S. Aldosari and J. Moura. Fusion in sensor networks with communication constraints. In *Proc. IEEE/ACM Symposium on Information Processing in Sensor Networks* (IPSN'04), Berkley, CA, USA, October 2004.
2. S. Althakeem and P.K. Varshney. Decentralized bayesian detection with feedback. *IEEE Trans. on Systems*, pages 503–513, July 1996.
3. R. S. Blum and S. A. Kassam. Optimum distributed detection of weak signals in dependent sensors. *IEEE Trans. Inform. Theory*, 38(3):1066–1079, May 1992.
4. R. S. Blum, S. A. Kassam, and H. V. Poor. Distributed detection with multiple sensors–Part II: Advanced topics. *Proc. IEEE*, 85(1):64–79, January 1997.
5. J.-F. Chamberland and V. V. Veeravalli. Decentralized detection in sensor networks. *IEEE Trans. Signal Processing*, 51(2):407–416, February 2003.
6. B. Chen, R. Jiang, T. Kasetkasem, and P. K. Varshney. Channel aware decision fusion in wireless sensor networks. *IEEE Trans. Signal Processing*, 52(12):3454–3458, December 2004.
7. C.-Y. Chong and S. P. Kumar. Sensor networks: evolution, opportunities, and challenges. *Proc. IEEE*, 91(8):1247–1256, August 2003.
8. T. M. Cover and J. A. Thomas. *Elements of Information Theory*. John Wiley & Sons, Inc., New York, 1991.
9. H. Gharavi and K. Ban. Multihop sensor network design for wide-band communications. *Proc. IEEE*, 91(8):1221–1234, August 2003.
10. I. Y. Hoballah and P. K. Varshney. An information theoretic approach to the distributed detection problem. *IEEE Trans. Inform. Theory*, 35(5):988–994, September 1989.
11. W.W. Irving and J.N. Tsitsiklis. Some properties of optimal thresholds in decentralized detection. *IEEE Trans. Automat. Contr.*, pages 835–838, 1994.
12. K. Liu and A. M. Sayeed. Optimal distributed detection strategies for wireless sensor networks. Monticello, IL, USA, October 2004.
13. H. V. Poor. *An Introduction to Signal Detection and Estimation*. Springer-Verlag, New York, NY, USA, 1994.
14. J. G. Proakis. *Digital Communications*, 4th Edition. McGraw-Hill, New York, 2001.
15. T. S. Rappaport. *Wireless Communications. Principles & Practice*, 2nd Edition. Prentice-Hall, Upper Saddle River, NJ, USA, 2002.
16. A.R. Reibman and L.W. Nolte. Detection with distributed sensors. *IEEE Trans. Aerosp. Electron. Syst.*, pages 501–510, December 1981.
17. A.R. Reibman and L.W. Nolte. Design and performance comparison of distributed detection networks. *IEEE Trans. Aerosp. Electron. Syst.*, pages 2474–2478, December 1987.

18. S. N. Simic and S. Sastry. Distributed environmental monitorning using random sensor networks. In *Proc. 2-nd Int. Work. on Inform. Processing in Sensor Networks*, pages 582–592, Palo Alto, CA, USA, April 2003.

19. J. N. Tsitsiklis. *Adv. Statist. Signal Process.*, volume 2, chapter Decentralized detection, pages 297–344. 1993. Eds.: H. V. Poor and J. B. Thomas.

20. J.N. Tsitsiklis. Decentralized detection by a large number of sensor. *Mathematics of Control, Signals and Systems*, pages 167–182, 1988.

21. R. Viswanathan and P. K. Varshney. Distributed detection with multiple sensors– Part I: Fundamentals. *Proc. IEEE*, 85(1):54–63, January 1997.

Chapter IV

GRID Structures For Distributed Cooperative Laboratories

Real Time Streaming Data Grid Applications

Geoffrey C. Fox, Mehmet S. Aktas, Galip Aydin, Hasan Bulut, Shrideep
Pallickara, Marlon Pierce, Ahmet Sayar, Wenjun Wu, and Gang Zhai

Community Grids Lab, Indiana University
501 North Morton Street, Suite 224
Bloomington, IN 47404
{gcf, maktas, gaydin, hbulut, spallick, mpierce, asayar,
wewu, gzhai}@cs.indiana.edu

(Invited Paper)

Abstract. We review several aspects of building real-time streaming data Grid
applications. Building on general purpose messaging system software
(NaradaBrokering) and generalized collaboration services (GlobalMMCS), we
are developing a diverse set of interoperable capabilities. These include
dynamic information systems for managing short-lived collaborative service
collections ("gaggles"), stream filters to support the integration of Geographical
Information Systems services with data analysis applications, streaming video
to support collaborative geospatial maps with time-dependent data, and video
stream playback and annotation services to enable scientific collaboration.

Keywords. Collaboration, geographical information systems, grid information
systems, message-oriented middleware, service-oriented architecture.

1   Introduction

This paper describes research work of the Community Grids Laboratory on Grids
built around streaming data sources. This work builds upon general purpose
messaging middleware (NaradaBrokering [1, 2, 3]) and incorporates a diverse set of
services that include audio/video conferencing (GlobalMMCS [4]) and Geographical
Information System services [5]. The architecture and core services are summarized
in a recent companion publication [6]. Here, we examine more closely applications
and additional functionality that are being integrated into the overall system.

A critical idea in our approach is to view both services and messages (and streams
as ordered set of messages) as "first class" entities. The law of the millisecond [7]
suggests one should use this type of message oriented middleware (software overlay
network) when one can afford latencies of a millisecond or more. This is
characteristic of all systems with significant geographic distribution and non-
specialized interconnect. As discussed in [6-8], NaradaBrokering is capable of
supporting millisecond messaging in a diverse range of applications, ranging from
binary data streams to XML-based Web Service messages.

For scientific applications, the data deluge [8] suggests the growing importance of real time data assimilation with the integration of sensors, databases and simulation codes. Similarly, the geographically distributed nature of much current research requires collaboration tools. Our research thus focuses on the reuse of concepts and software as well as integration of data-driven and collaboration-driven real-time problems. As discussed in [6], systematic use of Grid and web services gives us interoperability and access to commodity (industry) capabilities. In our system, we generalize the well-known "system of systems" concept to a "Grid of Grids" [9] and show how one can build Grid applications by using appropriate services drawn from Grid service families.

In this paper, we concentrate on Grid service families that are applicable to real-time data Grid applications. These include general purpose metadata and discovery services, Geographical Information System services suitable for streaming data and information, and services to support playback and shared annotation of video streams. These diverse services may be integrated into Grid of Grids applications through the use of management, orchestration, and workflow services, such as described in [10].

We have identified the importance of supporting both the worldwide Grid and smaller sessions or "gaggles of grid services" that support local dynamic action. We discuss in Sec. 2 the meta-data services optimized for these different requirements. This work also shows the need for three distinct types of XML data/metadata services. UDDI exemplifies a scalable repository; WS-Context a general dynamic store and Web Feature Service, a domain specific repository. Our implementations of these do not use XML databases but for efficiency convert the XML to SQL and store in a conventional MySQL database. This illustrates the important difference between semantics and representation. We preserve the XML Infoset (semantic meaning) but for efficiency do not use a conventional XML representation.

This paper surveys several aspects of real time streaming data grids. We first discuss the metadata management requirements of these systems (Sec. 2). Collaborative streaming systems involve both large, mostly static information systems as well as much smaller, highly dynamic information systems. We refer to these latter collections as "gaggles." We next review the integration of streaming data and Geographical Information systems. This may involve both streaming data (suitable for data mining and other applications), described in Sec3, as well as streaming map imagery (suitable for user interfaces), described in Sec 4. Map servers with streaming video capabilities can also be integrated with GlobalMMCS's general purpose collaborative infrastructure. We may thus inherit many additional features such as replay and collaborative annotation and whiteboard systems, as described in Sec. 5. We summarize this paper and future research activities in Sec. 6.

2   Gaggle-Like Metadata Support in GIS and Sensor Grid

Geographical Information Systems and Sensor Grids present an environment where many geo-resources and geo-processing applications are packaged as services and put together for a particular functionality such as forecasting earthquakes [10]. Here, Grid Information Services [11] maintains metadata about these geo-services and provides standardized methods for publishing and discovery. An Information Service can be thought of as a solution to general problem of managing information about Grid/Web Services, yet it should also support domain-specific information requirements such as geospatial domain information requirements.

GIS/Sensor Grid information may be classified as either a) session metadata or b) static, interaction-independent metadata [12]. Session metadata is the dynamically generated information as result of interactions of Grid/Web Services. Static and interaction-independent metadata is the information describing Grid/Web Service characteristics. Information Services should support handling and discovery of both session-based and static metadata associated to services.

The GIS/Sensor Grids may be thought of as an actively interacting (collaborating) set of managed services where services are put together for particular functionality [12]. Here each collection of services maintains most dynamic information which is the session related metadata. Handling and discovery of such dynamic information requires high performance, fault tolerant and distributed systems.

We use and extend UDDI [13] and WS-Context [14] specifications to build Information Services supporting extensive metadata requirements of GIS Grids [15], [16]. We also design distributed metadata management architecture to support dynamically assembled GIS/Sensor Grid applications where metadata is widely-scattered and dynamically generated [17]. Our Information Services implementation has been focused in two main branches. First, we designed and implemented an advanced UDDI Information Model to support metadata-oriented service registries. Here, we base this on an implementation of Web Service interfaces for publishing and discovering metadata associated to Grid/Web services. We extend basic UDDI functionality by implementing advanced functionalities in order to process service metadata and keep the Registry entries up-to-date. Some examples of these functionalities are a) monitoring capabilities such as leasing, b) Geographical Information System-specific taxonomies to support geo-spatial services, and c) XPATH query capabilities to support domain-specific Information Services.

For dynamic session data, we have designed and implemented a Context Service that deals with handling and discovery of dynamic, session-related metadata. Here, session related metadata is short-lived and dependent on the client. We base this on the WS-Context specification from the standards organization OASIS. We extended WS-Context Specifications to provide advanced capabilities to manage session metadata between multiple participants in Web Service interactions. To decentralize the Context Service, we currently implement a distributed and dynamic metadata management architecture which is described in [16] in greater detail.

3 High Performance Web Service Grid Architecture to Support Real-Time Sensor Streams

Recent technological developments have allowed sensors to be deployed in a variety of application domains. Environmental monitoring, air pollution and water quality measurements, detection of the seismic events and understanding the motions of the Earth's crust are only a few areas where extent of the deployment of sensor networks can easily be seen. Extensive use of sensing devices and deployment networks of sensors that can communicate with each other to achieve a larger sensing task will fundamentally change information gathering and processing [17]. However, the rapid proliferation of sensors presents unique challenges different than the traditional computer network problems.

Several studies have discussed the technological aspects of the challenges with the sensor devices, such as power consumption, wireless communication problems, autonomous operation, adaptability to the environmental conditions, etc [18] [19]. Here we describe architecture to support real-time information gathering and processing from Global Positioning System (GPS) sensors by leveraging principles of service oriented architectures and open GIS standards.

3.1   GPS Networks

The Global Positioning System has been used in geodesy to identify long-term tectonic deformation and static displacements while Continuous GPS has proven very effective for measurement of the interseismic, coseismic and postseismic deformation [20]. Today networks of individual GPS Stations (monuments) are deployed along the active fault lines, and data from these are continuously being collected by several organizations. One of the first organizations to use GPS in detection of the seismic events and for scientific simulations is Southern California Integrated GPS Network (SCIGN) [21]. One of the collaborators in SCIGN is Scripps Orbit and Permanent Array Center (SOPAC) [22] which maintains several GPS networks and archives high-precision GPS data, particularly for the study of earthquake hazards, tectonic plate motion, crustal deformation, and meteorology. Real time sub-networks maintained by SOPAC include Orange County, Riverside County (Metropolitan Water District), San Diego County, and Parkfield. These networks provide real-time position data (less than 1 sec latency) and operate at high rate (1 – 2 Hz).

3.2   Real-Time Streaming Access to GPS Position Messages

As shown in Fig. 1, raw data from the SOPAC GPS stations are continuously collected by a Common Link proxy (RTD server) and archived in RINEX files. The RTD server outputs the position of the stations in real-time in a binary format called RYO. To receive the position messages, clients are expected to open a socket connection to the RTD server. An obvious downside of this approach is the extensive load this might introduce to the server when multiple clients are connected. These data streams are collections from entire networks rather than individual stations, and for many applications will require per station separation and reformatting. As described below, we are developing general purpose solution to these filtering problems.

To make the position information available to the clients in a real-time streaming fashion we are using the NaradaBrokering messaging system. Additionally we developed filters to serve position messages in ASCII and Geography Markup Language [23] formats.

3.3 Chains of Filters

Since the data provided by RTD server is in a binary format we developed several filters to decode and present it in different formats. Once we receive the original binary data we immediately publish this to a NaradaBrokering topic (null filter), another filter that converts the binary message to ASCII subscribes to this topic and publishes the output message to another topic. We have developed a GML schema to describe the GPS position messages. Another filter application subscribes to ASCII message topic and publishes GML representation of the position messages to a different topic. This approach allows us to keep the original data intact and different formats of the messages accessible by multiple clients in a streaming fashion.

Our GML Schema is based on RichObservation type which is an extended version of GML 3 Observation model [24]. This model supports Observation Array and Observation Collection types which are useful in describing SOPAC Position messages since they are collections of multiple individual station positions. We follow strong naming conventions for naming the elements to make the Schema more understandable to the clients.

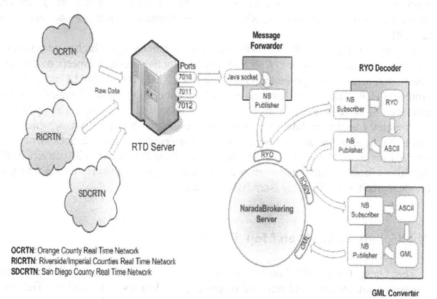

OCRTN: Orange County Real Time Network
RICRTN: Riverside/Imperial Counties Real Time Network
SDCRTN: San Diego County Real Time Network

Fig. 1 – SOPAC GPS Services

Currently the system is being tested for the following networks: San Diego Counties, Riverside/Imperial Counties and Orange County GPS networks. For more information, see [5]

3.4   SensorGrid Research Work

We are developing a Web Service-based Grid called SensorGrid for supporting archived and real-time access to sensor data. GPS services we mentioned above will be part of this system. SensorGrid will help us integrate sensor data with scientific applications such as simulation, visualization or data mining software by leveraging GIS standards and Web Services methodologies. The system will use NaradaBrokering as the messaging substrate and this will allow high performance data transfer between data sources and the client applications. The Standard GIS interfaces and encodings like GML will allow data products to be available to the larger GIS community. We will extend OGC's [25] Sensor Collection Service to support streaming data. The Sensor Collection Service provides sensor metadata encoded in SensorML and the actual data products. We use a Web Service version of OGC WFS to distribute archived geospatial data.

3.5   Negotiation Protocol for High Performance Data Transport

Several studies [26, 27, 28, 29] have shown that transport of XML and SOAP messages encoded in conventional "angle-bracket" representation is too slow for applications that demand high performance. At the same time several groups are developing ways of representing XML in binary formats for fast message exchange [28, 29].

We are developing a Web Service negotiation language for higher performance Web Services, a protocol for Web Services to negotiate several aspects of the data transportation such as representation scheme (Fast XML, Binary XML) and transport protocol (TCP, UDP). Initial negotiation will be done using standard angle-bracketed messages to determine the supported representation and transport capabilities. We will employ handlers to take care of the conversion and transport issues, which will make the negotiation and transport process transparent to the services. Once the services agree on the conditions of the data exchange, handlers will convert XML data into an appropriate binary format and stream it over a high performance transport protocol (such as UDP) using NaradaBrokering.

4   Collaborative Web Map Services

The Open Geospatial Consortium defines several related standards for the representation, storage, and retrieval of geographic data and information. The Web Map Service (WMS) [30] produce maps from the geographic data. Geographic data are kept in Web Coverage Server (WCS) or Web Feature Service (WFS). WCS stores raster data in image tiles and WFS stores feature vector data in GML formats. WMS produces maps from these raw geographic data upon requests from the WMS clients. These maps are the static representations of geospatial data. Representations are in pictorial formats such as PNG, SVG, JPEG, GIF, etc. [31]

4.1 Streaming Video and Map Servers

Standard map servers produce static images, but many types of geographic data are time dependent. In order to understand geographic phenomena and characteristics of temporal data it is necessary to examine how these patterns change over time for these types of data. We are therefore investigating the problems of creating streaming video map servers based upon appropriate standard collaboration technologies [4].

In our approach, visualizing changes over time is achieved by integrating temporal information on a map. Usually the result is a series of static maps showing certain themes at different moments. In addition to creating static maps, WMS also has the ability to combine the static maps correspond to a specific time interval data and combine them in an animated movie. Movies created by WMS are composed of a certain number of frames. Each frame represents a static map that corresponds to a time frame defined in request.

WMS is not able to create movie for all of its supported layers listed in its capabilities file. If WMS supports movie functionality for a layer it adds some attributes under this layer. Before making the request, the client first makes a "get capabilities" request and after getting information about the WMS, it makes requests according to capabilities of the WMS. If a client makes a request to get a movie for a specific layer, to succeed, this layer should have a time dimension defined under this layer element in the capabilities file. Clients should make the "get map" standard request to WMS to get the movie for a specific layer. WMS does not provide any other request types for the movie creation functionalities. Clients set the "format" variable to string "movie/<movietype>" and "time" variable to a value in an appropriate format to make a request to get movie from the WMS.

We initially examined two approaches to creating movies: one is client oriented and the other is server oriented. For the sake of performance issues, we have chosen the second one for the collaborative map movies. Since our aim is to create collaborative map movies, we do not need to get the movie back to the client. As soon as the movie frames are created at WMS side for each time slice for the same data layer, WMS publishes them as streams to specific Real Time Protocol (RTP) [32] sessions. RTP sessions are represented as <IP Address, Port Number> pairs. A video stream published to a RTP session can be visualized by any video client connecting to the same RTP session. RTP session can be configured at properties file. Map images are dynamically generated from raw geographic data and those images are transcoded into video streams. The supported video stream formats are H.261 and H.263, which are mostly used formats in AccessGrid sessions. Map video stream can be played in collaborative environments such as AccessGrid and GlobalMMCS sessions.

If the client oriented approaches is used, the client waits the movie to return and once it returns client either starts to play the movie or archive or both. Archived movies can be replayed without making another request to WMS. As it is shown in this scenario there are some tradeoffs between these two approaches.

To solve these tradeoffs we have developed a third approach. This is the best way for the performance issues that we will be using in the near future for the movie mapping. By this approach, creating and publishing movies to collaboration sessions will be achieved at the client side. Client will not make request for the movie. In other words, client will never set "format" variable to movie/<movietype>. It will make threaded "get map" requests for the defined periodicity between starting date and ending date. All the returned static maps in images will be stored locally based on their time intervals and periodicities. When client needs to replay some or all parts of stored images then, he does not need to make a "get map" request for these layers. If there is no match for some time intervals then clients make the "get map" requests just for these unmatched frames.

The map video stream has several parameters that can be adjusted. These parameters affect the quality of the produced map video stream. Among these configurable parameters are frame rate and video format of the stream, update rate of the map images in the video stream. In our experiments we updated map images for every 0.5 seconds while we kept the video frame rate at 10 frames per second (fps). This provides a high quality of the video stream at the receiving side. This is necessary because some clients might not be capable of visualizing video streams with low frame rate or can visualize them with very low quality.

Map video streams produced are published to RTP sessions, whether unicast or multicast session. AccessGrid clients use multicast sessions to send/receive video. Once a video is published to a multicast session, it can be received by any client listening that multicast session as long as the underlying network lets client receive multicast packets. GlobalMMCS can also provide this map video stream to its clients as unicast video stream, as discussed in Sec. 5.

4.2   Map Service Research Issues

We will continue to integrate the streaming map server with GlobalMMCS's archiving capabilities and will examine higher performance encodings. This will enable useful functionality, such as allowing users to select a movie from the archive. To make the WMS available for the collaborative conferences or online education, administrative users will be able to update map streams on the fly while they are playing.

We will be creating movies at the client side even in case of collaborative movie creating environment. There will be significant performance gain when we use this approach. Clients can archive both previously created frames and movies. If a client needs the same type of frames for the same matching time intervals then it does not need to go back to WMS and spend time getting the movie frames.

5   e-Annotation Collaborative Architecture

In this section we describe video archiving and annotation services that we are adding to the GlobalMMCS collaboration system. GlobalMMCS (Global Multimedia Collaboration System) is described in more detail in Ref. [4]. In brief summary, it is a scalable, robust, service-oriented collaboration system. GlobalMMCS integrates various services including videoconferencing, instant messaging and streaming, and is interoperable with multiple videoconferencing technologies. This collaboration system is developed based on the eXtensible Generic Session Protocol (XGSP) collaboration framework and NaradaBrokering messaging middleware. Such a service-oriented collaboration environment greatly improves the scalability of traditional videoconferencing system, benefits users with diverse multimedia terminals through different network connections, and simplifies the further extension and interoperability.

We use a component based design on top of NaradaBrokering shared event collaboration model to make e-Annotation system scalable and extensible for new functional plug-ins. e-Annotation is designed to run in grid computing environment, which spans different organizations across different countries. The architecture is shown in Fig. 2.

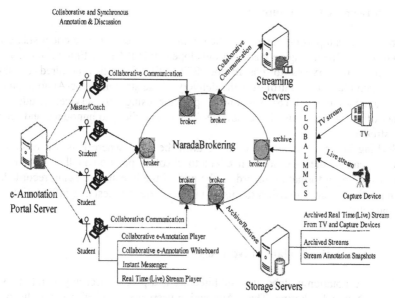

Fig. 2 e-Annotation Architecture

5.1 Streaming Server

The Real Time Streaming Protocol (RTSP) [33] is a protocol specified in the Internet Engineering Task Force's RFC 2326 for control over the delivery of live or stored data with real-time properties. RTSP is similar in syntax and operation to HTTP/1.1, which allows it to be extended by HTTP, but as opposed to HTTP, RTSP maintains state by default. States involved in RTSP are init, ready, playing/recording. Basic RTSP control functionalities are play, pause, seeking (absolute positioning to a specific point in the stream) etc.

Our collaboration system provides a Web Services-based streaming server which provides RTSP semantics with control functionalities mentioned above. In addition, the streaming service also provides seeking of live streams, which is not supported by the current RTSP implementations such as RealNetworks Helix Server [34] and QuickTime Streaming Server [35].

The streaming service uses NaradaBrokering middleware to transport data across network. Streaming clients (e-Annotation client) receives and sends data through NaradaBrokering nodes. Data sent across the network is wrapped inside native NaradaBrokering events. Any type of data, audio, video, still images or text messages can be wrapped inside events and sent across the broker network. Based on the stream metadata information, client processes accordingly whether it is video, audio or any other data type.

5.2   Streaming Client Support

In order to support e-Annotation client with RTSP semantics, a client side service called RTSP Client Engine has been developed. RTSP Client Engine interacts with the streaming services to ensure the RTSP functionalities required by the e-Annotation client for control over the delivery of the stream. The e-Annotation client can initiate as many RTSP sessions as possible using this service. In addition to replay of the stream, this also enables e-Annotation client to announce and record a new stream.

Seeking in live streams can be done until the point where it is last recorded. In order to enable the e-Annotation client to absolute position in the live stream, it requires timing information regarding the part of the live stream being recorded. The e-Annotation client subscribes to this service to retrieve timestamp information of the last data stored.

5.3   e-Annotation Collaborative Tools

The e-Annotation collaborative architecture is a peer-to-peer collaboration system based on NaradaBrokering. The e-Annotation users are identical peers in the system, they communicate with each other to collaboratively annotate a real time or archived video stream. Each peer shares the same collaborative applications which include e-Annotation Player and e-Annotation Whiteboard.

5.4   e-Annotation Player

As shown in Fig. 3, the e-Annotation player is composed of four components:
1. Stream list panel
2. Real time live video panel
3. Streaming player panel
4. Video annotation snapshot player panel.

The stream list panel contains the real time live video stream list, archived video stream list and the composite annotation stream list. All of the stream lists information is gotten from streaming server by subscribing the streaming control info topic from a broker. Real time video play panel plays the real time video that is selected by user in the real time video list. It subscribes the real time video stream data topic to get real time video data published by GlobalMMCS [globalmmcs]. Streaming player panel plays the archived video stream. Streaming player supports pausing, forwarding, rewinding a video stream with dynamic length (live video) or fixed length (archived one). It can also take snapshots on a video stream whiling playing. When taking a snapshot, the timestamp is associated with that snapshot. These snapshots are loaded to whiteboard to be annotated collaboratively. When playing back the composite annotation stream, the original video stream is played in the streaming player panel, at the same time, video annotation snapshot player play the annotation snapshots synchronized with the original video stream by the timestamps.

The e-Annotation Player is collaborative: each participant will view the same content in his/her e-Annotation Player, the only difference is that only the participant with the 'coach' role can control the play of a video stream(to pause, rewind, take snapshot).

5.5   e-Annotation Whiteboard

The whiteboard works collaboratively in peer to peer mode as the e-Annotation Player does, each peer has the same view of the current whiteboard content. One user's draw on the whiteboard can be seen immediately in all other users' whiteboard. Each action in one peer's whiteboard will generate a NaradaBrokering event that will be broadcasted using whiteboard communication topic through NaradaBrokering to all other peers in this session, so all the peers get a consistent and synchronized view of the shared whiteboard. The user with the 'leader' role can control the save and erasure of whiteboard content, other 'student' users can only add comments (text, any shape, pictures etc) to the whiteboard, the whiteboard is where all the users do the annotation on a snapshot of a video stream taken from e-Annotation player, annotated snapshots will be saved as JPEG images with timestamps to generate a new composite annotation stream, this composite annotation stream is composed of the video stream plus the annotation snapshots, they are synchronized using the timestamps.

The e-Annotation Player and Whiteboard collaborative tools user interfaces are shown in Fig. 3.

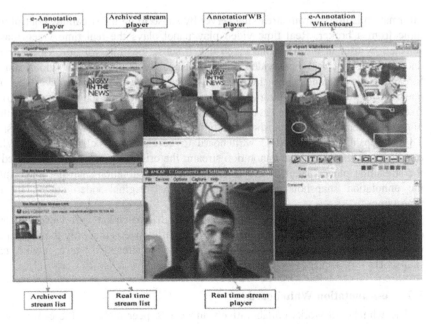

Fig. 3 e-Annotation Player and e-Annotation Whiteboard User Interface

5.7   e-Annotation Status and Future Work

e-Annotation system supports synchronous collaborative video annotation. It supports annotation about a live real time stream from capturing devices or TVs. The annotation of video stream can be played back synchronously with the original stream by creating a new composite stream. The e-Annotation system provides a framework for integrating different multimedia collaborative tools for e-Coaching and e-Education.

The future work includes design and employs quality-of-service control mechanism to improve the performance and develop a new metadata framework for collaborative multimedia annotation system using XML as the metadata exchange and description

6   Summary and Future Work

We have reviewed in this paper several applications that build upon the Community Grid Laboratory's general Web Services messaging infrastructure (NaradaBrokering) and general collaboration services (GlobalMMCS). Our generalized metadata and information management research concentrates on extending static metadata services (UDDI) to support richer descriptions and discovery capabilities. We also distinguish between these session-independent services and session-dependent metadata services needed to support dynamic groups of services and users. We next reviewed our work on streaming data for Geographical Information Systems, which are described by the information services. GIS applications include both streaming data (GPS data streams and filters) for data analysis codes as well as video streams for human users. In the latter case, we build upon video streaming standards, which allow us to integrate collaborative mapping tools into more general audio/video systems. This in turn enables us to combine streaming map tools with general purpose video stream playback and annotation tools.

We have identified many areas of ongoing research at the end of each section. Of particular general interest to us is the investigation of high performance Web Services that take advantage of efficient message representations and higher performance transport protocols. This research will impact all of the services that we have described.

References

1. Shrideep Pallickara and Geoffrey Fox. NaradaBrokering: A Middleware Framework and Architecture for Enabling Durable Peer-to-Peer Grids. Proceedings of ACM/IFIP/USENIX International Middleware Conference Middleware-2003.
2. Shrideep Pallickara and Geoffrey Fox. A Scheme for Reliable Delivery of Events in Distributed Middleware Systems. Proceeding sof the IEEE International Conference on Autonomic Computing. 2004.
3. Shrideep Pallickara, Geoffrey Fox and Harshawardhan Gadgil. On the Creation & Discovery of Topics in Distributed Publish/Subscribe systems. (To appear) Proceedings of the IEEE/ACM GRID 2005. Seattle, WA.
4. Wenjun Wu, Geoffrey Fox, Hasan Bulut, Ahmet Uyar, Harun Altay "Design and Implementation of A Collaboration Web-services system", Journal of Neural, Parallel & Scientific Computations (NPSC), Volume 12, 2004. See also http://www.globalmmcs.org.
5. GIS Research at Community Grids Lab, web site: www.crisisgrid.org
6. Geoffrey Fox, Galip Aydin, Harshawardhan Gadgil, Shrideep Pallickara, Marlon Pierce, and Wenjun Wu. Management of Real-Time Streaming Data Grid Services. Invited talk at Fourth International Conference on Grid and Cooperative Computing (GCC2005), held in Beijing, China, during Nov 30-Dec 3, 2005.
7. Geoffrey Fox. The Rule of the Millisecond in CISE Magazine. Vol 6, No 2, pp 93-96 (2004).
8. Berman, F., Fox, G., and Hey, T., (eds.). Grid Computing: Making the Global Infrastructure a Reality, John Wiley & Sons, Chichester, England, ISBN 0-470-85319-0 (2003). http://www.grid2002.org.
9. Geoffrey Fox, Shrideep Pallickara, and Marlon Pierce. Building a Grid of Grids: Messaging Substrates and Information Management. To appear as chapter in book "Grid Computational Methods" Edited by M.P. Bekakos, G.A. Gravvanis and H.R. Arabnia.
10. Galip Aydin, Mehmet S. Aktas, Geoffrey C. Fox, Harshawardhan Gadgil, Marlon Pierce, Ahmet Sayar SERVOGrid Complexity Computational Environments (CCE) Integrated Performance Analysis Technical report June 2005 accepted as poster and short paper in Grid2005 Workshop, 2005.

11. B. Plale, P. Dinda, and G. Von Laszewski., Key Concepts and Services of a Grid Information Service. In Proceedings of the 15th International Conference on Parallel and Distributed Computing Systems (PDCS 2002), 2002.
12. Mehmet S. Aktas, Geoffrey Fox, Marlon Pierce Managing Dynamic Metadata as Context Istanbul International Computational Science and Engineering Conference (ICCSE2005) June 2005 also please see: http://www.opengrids.org/fthpis.
13. Bellwood, T., Clement, L., and von Riegen, C. (eds) (2003), UDDI Version 3.0.1: UDDI Spec Technical Committee Specification. Available from http://uddi.org/pubs/uddi-v3.0.1-20031014.htm.
14. Bunting, B., Chapman, M., Hurlery, O., Little M., Mischinkinky, J., Newcomer, E., Webber J., and Swenson, K., Web Services Context (WS-Context), available from http://www.arjuna.com/library/specs/ws_caf_1-0/WS-CTX.pdf.
15. Mehmet S. Aktas, Galip Aydin, Geoffrey C. Fox, Harshawardhan Gadgil, Marlon Pierce, Ahmet Sayar, Information Services for Grid/Web Service Oriented Architecture (SOA) Based Geospatial Applications, Technical Report, June, 2005.
16. Mehmet S. Aktas, Geoffrey C. Fox, Marlon Pierce, "An Architecture for Supporting Information in Dynamically Assembled Semantic Grids", Technical report August 2005. Available at http://grids.ucs.indiana.edu/ptliupages/publications/SKG2005_Aktas.pdf.
17. D. Estrin, R. Govindan, J. Heidemann and S. Kumar, "Next Century Challenges: Scalable Coordination in Sensor Networks," In Proceedings of the Fifth Annual International Conference on Mobile Computing and Networks (MobiCOM '99), August 1999, Seattle, Washington.
18. Akyildiz, I. F., Su, W., Sankarasubramaniam, Y., and Cayirci, E., "A Survey on Sensor Networks" IEEE Communications Magazine, August 2002.
19. Archana Bharathidasan, Vijay Anand Sai Ponduru, "Sensor Networks: An Overview".
20. Bock, Y., Prawirodirdjo, L, Melbourne, T I. : "Detection of arbitrarily large dynamic ground motions with a dense high-rate GPS network" GEOPHYSICAL RESEARCH LETTERS, VOL. 31, 2004.
21. Southern California Integrated GPS Network web site: http://www.scign.org/.
22. Scripps Orbit and Permanent Array Center web site: http://sopac.ucsd.edu/.
23. The Geography Markup Language described in an Encoding Specification from Open Geospatial Consortium available from http://portal.opengeospatial.org/files/?artifact_id=4700.
24. Open Geospatial Consortium Discussion Paper, Editor Simon Cox: "Observations and Measurements". OGC Document Number: OGC 03-022r3.
25. The Open Geospatial Consortium, Inc. web site: http://www.opengeospatial.org/.
26. Chiu, K., Govindaraju, M., and Bramley, R.: Investigating the Limits of SOAP Performance for Scientific Computing, Proc. of 11th IEEE International Symposium on High Performance Distributed Computing HPDC-11 (2002) 256.
27. Oh, S., Bulut, H., Uyar, A., Wu, W., Fox G., Optimized Communication using the SOAP Infoset For Mobile Multimedia Collaboration Applications. In proceedings of the International Symposium on Collaborative Technologies and Systems CTS05 (2005).
28. R. Berjon, chair, "XML Binary Characterization Working Group Public Page." Available from http://www.w3c.org/XML/Binary/
29. Paul Sandoz, Alessando Triglia, and Santiago Pericas-Geertsen, "Fast Infoset." Avaialble from http://java.sun.com/developer/technicalArticles/xml/fastinfoset/.
30. Jeff De La Beaujardiere, OpenGIS Consortium Web Mapping Server Implementation Specification 1.3, OGC Document #04-024, August 2002.
31. Ahmet Sayar, Marlon Pierce, Geoffrey Fox OGC Compatible Geographical Information Services Technical Report (Mar 2005), Indiana Computer Science Report TR610.

32. H. Schulzrinne, S. Casner, R. Frederick, and V. Jacobson, RTP: A Transport Protocol for Real-Time Applications. Internet Engineering Task Force Request for Comments 3550 (2003). http://www.ietf.org/rfc/rfc3550.txt

33. H. Schulzrinne, A. Rao, and R. Lanphier, Real Time Streaming Protocol (RTSP), Internet Engineering Task Force Request For Comments 2326 (1998). http://www.ietf.org/rfc/rfc2326.txt

34. RealNetworks Helix Server http://www.realnetworks.com/

35. QuickTime Streaming Server http://www.apple.com/quicktime/streamingserver/

The GRIDCC Project

Istituto Nazionale di Fisica Nucleare, Legnaro, Italy
Institute of Accelerating System and Application, Athens, Greece
Brunel University, Uxbridge, UK
Consorzio Nazionale Interuniversitario per le Telecomunicazioni, Italy
Sincrotrone Trieste S.C.P.A.,Trieste, Italy
IBM,Haifa, Israel
Imperial College London,London, UK
Istituto di Metodologie per l'Analisi ambientale, Consiglio nazionale delle ricerche, Italy
Università degli studi di Udine, Udine, Italy
Greek Research and Technology Network S.A. Athens, Greece

Contact for this paper: **Francesco Lelli, Gaetano Maron**

INFN- Laboratori Nazionali di Legnaro, Italy:
Francesco.Lelli@lnl.infn.it, Gaetano.Maron@lnl.infn.it

(Invited Paper)

Abstract: The GRIDCC [1] project will extend the use of Grid computing to include access to and control of distributed instrumentation. Access to the instruments will be via an interface to a Virtual Instrument Grid Service (VIGS). VIGS is a new concept and its design and implementation, together with middleware that can provide the appropriate quality of service, is a key part of the GRIDCC development plan. An overall architecture for GRIDCC has been defined and some of the application areas, which include distributed power systems, remote control of an accelerator and the remote monitoring of a large particle physics experiment, are briefly discussed.

Keywords. Grid, Web Services, Instrument Element, Systems Control, Remote Control.

1 Introduction

Grid computing refers to the coordinated and secured sharing of computing resources among dynamic collections of individuals, institutions and resources. It involves the distribution of computing resources among geographically separated sites (creating a "grid" of resources), all of which are configured with specialized software for routing jobs, authenticating users, monitoring resources, and so on. Shared, site-based computing resources may include computing and/or storage nodes, software, data, a variety of scientific instruments, and so on. The aim to provide reliable and secure access to widely scattered resources for authorized users located virtually anywhere in the world.

In this scenario the goal of GridCC [1] is to build a widely distributed system that is able to remotely control and monitor complex instrumentation that ranges from a set of sensors used by geophysical stations monitoring the state of the earth to a network of small power generators supplying the European power grid.

There are several areas in which the GridCC goes beyond work being carried out in other Grid projects. These include:

- Control of instrumentation within a Grid environment;
- Quality of service requirements for all components of the GridCC Grid;
- Automated problem solving within a Grid environment;
- Human interaction with Grids through the Virtual Control Room;
- Enactment of complex workflows.

While it covers areas that are not part of other Grid projects, GridCC does not exist in isolation and must inter-operate with other Grids – most notably that of the EGEE project [5].

This has forced a pragmatic approach when considering areas of the project that overlap with similar areas in other projects. It has also resulted in an architecture that is modular with interactions between modules being via web services. These applications introduce requirements for real-time and highly interactive operation of computing Grid resources.

The rest of the paper is outlined as follows: Section 2 will present an overview of the GridCC project. In session 3 we will show in detail the Instrument Element (IE) component. In session 4 we will see an overview on the GridCC pilot application and, finally, in session 5 we will show our conclusions.

2   Project Overview

The GRIDCC project is a 3 year project funded by the European Union which started in September 2004. There are 10 project partners from Greece, Italy, Israel and the United Kingdom. The first complete release of the software will be during the second year of the project. The main goals are:

- To develop generic Grid middleware, based on existing building blocks (Grid Services) which will allow the remote control and monitoring of instruments such as distributed systems.
- To incorporate the new middleware into a few significant applications that will allow the software to be validated both in terms of functionality and quality of service aspects: European Power Grid, Meteorology, Analysis of Neurophysiological data, Remote Operation of an Accelerator Facility, High Energy Physics Experiment.
- To verify the feasibility of a Grid-based remote control of systems requiring real-time response with real applications running on existing Grid test beds over both national and international network infrastructures.
- To disseminate widely the new software technology, the results of the evaluation on the test-beds and to encourage a wide range of enterprises to evaluate and adopt our Grid oriented approach to real-time control and monitoring of remote instrumentation.

The project is currently in its initial stages and work is primarily concentrated in the areas of workflow and architecture definition, refining the use-cases for the applications and the testing and evaluation of existing Grid middleware to determine whether it meets our requirements.

2.1 Architecture Overview

The overall architecture of GRIDCC has now been defined. The Virtual Control Room (VCR) is the GRIDCC user interface. It allows users to build complex workflows which are then submitted to the Execution Service; it can connect to resources directly and be used to control instruments in real time. The VCR also extends human interactions with Grid resources through its Multipurpose Collaborative Environment. The following figure shows the main components of the GridCC Project:

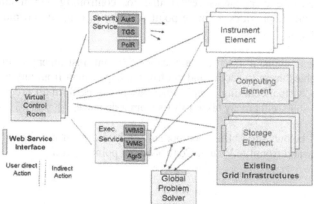

Fig. 1: GridCC architecture

The **Virtual Control Room** (VCR) is the GridCC user interface. It allows users to build complex workflows which are then submitted to the Execution Service (ES). it can connect to resources directly and be used to control instruments in real time. The VCR also extents human interaction with Grid resources through its Multipurpose Collaborative Environment (MCE).

The **Instrument Element** consists of a coherent collection of services which, between them, provide all the functionality to configure partition and control the physical instrument behind the IE.

The **Compute and Storage Elements** (CE & SE) within GridCC are similar to those with in other Grid projects. However these are extended to encompass the quality of service requirements that are central to GridCC.

The **Execution Services** (ES) are at the heart of GridCC. The Execution Services include: the Workflow Management System (WfMS) for workflow management, the Workload Management System (WMS) for logging and bookkeeping and service discovery, and the Agreement Service (AS) for reservation management.

The **Problem Solver** consists in a system that provides automated problem detection/solving in a Grid environment. Two levels of problem solver can exist within the GridCC Grid: the first local to a given IE, the Local Problem Solver (LPS).The second global to the entire system, the Global Problem Solver (GPS).

In the next paragraph we will see in detail the Instrument Element architecture showing how it can allow remote control of a generic instrument or a set of them.

3   The Instrument Element

The Instrument Element (IE) is a concept unique to the GridCC project. It is the collection of software components responsible for controlling and monitoring a generic instrument. It provides a single point of entry to operate and monitor the controlled resources.

The instrument element interface are called Virtual Instrument Grid Service (VIGS) and enables users to access and control the experiment from any part in the world using the "virtual counting room", where users can perform all programmable actions on the system, effectively taking shifts from a distance.

The figures below show this component architecture from the instrument information flow and from the Virtual Control Room point of view:

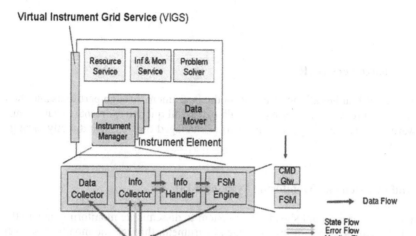

Figure 2: Instrument Element Information Flow

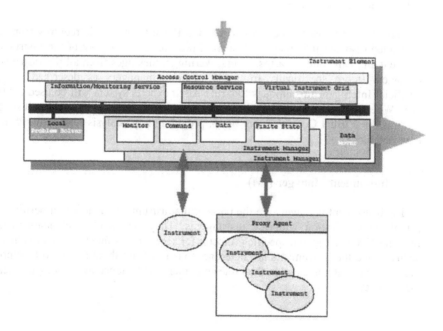

Figure 3: Instrument Element from the VCR point of view

3.1 Resource Service (RS)

This service handles all the resources of the Instrument through partitions and their sessions. Resources can be discovered, allocated and queried. A resource can be any hardware or software component that can be managed directly or indirectly through the network.

3.2 Information and Monitoring System (IMS)

This service uses a publish/subscribe system to disseminate monitoring data to the interested partners. More specifically each instrument, through the monitor manager publishes monitoring information to the Information and Monitoring System arranged in topics. Subscribers can register for specific messages that can be selected on the basis of the information source, the error severity level, and the type of messages or a combination of these tags. The IMS also monitors event data providing permanent recording and snapshot buffers for subscribers. Any information can also be retrieved by a client via simple query procedures.

3.3 Problem Solver (PS)

The PS subscribes to the IMS to receive the information it requires from the controlled resource it is interested in. It is positioned in the core of the Instrument Element and can collect alarms, errors, warnings, messages from all the instrument managers (more specifically from the monitor managers as described above). Artificial Intelligence techniques like rule-based experts system will be used for this task. When the PS cannot determine such an automatic recovery procedure, it simply informs the Virtual Control Room users, providing any analysis results the PS may have obtained.

3.4 Instrument Manager (IM)

The instrument managers are the parts of the instrument element that perform the actual communication with the instruments. They act as protocol adapters that implement the instrument specific protocols for accessing its functions and reading its status. Since the instruments are heterogeneous in nature there is a need to support many instrument managers in the same container, one instance for each logical set of instruments.

3.5 Data Mover (DM)

This service provides the interface with any external storage or processing elements this service that finds the 'best' mechanism to move a file from one storage resource to another. In outer words the data mover service provides a planner function that computes an I/O communication schedule minimizing the communication overheads.

3.6 Access Control

The security architecture aims at creating a mechanism by which confidentiality and non-repudiation by authentication, authorization, and encryption is provided. This service is responsible for checking user credentials and deciding whether a request should be processed by the Instrument Element. Once it is decided to proceed with the request it forwards it to one of the edge services.

4 Pilot Application

A variety of applications have been considered within the GridCC project and their use cases have had a strong barring on the design of the implementation of the software. These range from meteorology, geo-hazards, neurophysiology, and power grids to accelerator control and high energy physics experiment control.

4.1 CMS Experiment

The GridCC is the master controller of the Compact Muon Solenoid (CMS) [2] Data Acquisition (DAQ) system when the experiment is taking data. It provides physicists with a single point of entry to operate the experiment and to monitor detector status and data quality. It instructs the sub-part of this experiment to act according to the specific needs of a data-taking session. The tree main functions of the Instrument Element in this case are:

- Control and monitoring the entire instrument ensuring the correct and proper operation of the CMS experiment
- Control and monitoring of the data acquisition system
- Provide user interfaces and allow users to access the system from anywhere in the world

4.2 Power Grid

In electrical utility networks (or power grids[4]), the introduction of very large numbers of 'embedded' power generators often using renewable energy sources, creates a severe challenge for utility companies. In addition, power systems involve many geographically distributed participants: generator owners, transmission network operators, load managers, energy-market makers, supply companies, and so on. GridCC technology would allow the generators to participate in a Virtual Organization, and consequently to be monitored and scheduled in a cost-effective manner.

4.3 Intrusion Detection System

One of the main challenges in security management of large high speed networks is the detection of suspicious anomalies in network traffic patterns due to Distributed Denial of Service (DDoS) attacks or worm propagation. Anomaly sensors measure various network elements linked to Grid controlled Instrument Managers within a domain real-time Service for Monitoring & Control. The Instrument Element Problem Solver, provide algorithms aimed at fusing the collected knowledge analysing individual domain state reports, originating from heterogeneous sensors, to deduce a global view of security incidents.

4.4 Meteorology and Weather Forecasts

In this pilot application periodically a set of weather data are retrieved from the Instrument Element of the Meteorology Centres and stored in Storage Elements than, these data are processed in Computing Elements.

If an emergency condition of hazardous weather is detected the system needs a near real-time identification and allocation of available grid resources in order to quick presents the potential dangerous conduction to the users preventing a disaster.

5   Conclusions

The GridCC project is integrating instrumentation into traditional computational/storage Grids. The project started in September 2004 and the first full release of the GridCC software will be during the second year of the project.

References

1. http://www.gridcc.org
2. CMS TDR http://cmsdoc.cern.ch/cms/TRIDAS/Temp/CMS_DAQ_TDR.pdf
3. http://www.elettra.trieste.it/info/index.html
4. M Irving, G Taylor, P Hobson, IEEE Power & Energy Magazine March/April 2004, pp 40 - 44
5. The EGEE project, http://www.eu-egee.org

6. The Condor Project, http://www.cs.wisc.edu/condor/
7. The Globus Alliance, http://www.globus.org/
8. The gridengine project, http://gridengine.sunsource.net/
9. Platform LSF, http://www.platform.com/products/LSF
10. gLite - Lightweight Middleware for Grid Computing, http://glite.web.cern.ch/glite/
11. Enabling Grids for E-science, http://public.euegee.org/
12. The Open Middleware Infrastructure Institute,http://www.omii.ac.uk
13. S. Liang, The Java Native Interface, Programmer's Guide and Specification, Addison-Wesley, 1999
14. Project JXTA, http://www.jxta.org
15. Open Grid Services Infrastructure (OGSI) v1.0:
   http://forge.gridforum.org/projects/ggfeditor/document/draft-ogsi-service1/en/1
16. Web Service, W3C, http://www.w3c.org/TR/wsarch/
17. BPEL4WS.
   http://www6.software.ibm.com/software/developer/library/wsbpel.pdf
18. WS Choreography Model Overview, http://www.w3.org/TR/2004/WD-ws-chor-model-20040324/
19. W.M.P. van der Aalst, L. Aldred, M. Dumas, and A.H.M. ter Hofstede. Design and Implementation of the YAWL system. Proceedings of The 16th International Conference on Advanced Information Systems Engineering (CAiSE 04), Riga, Latvia, June 2004. Springer Verlag.
20. JSR 168: Portlet Specification, http://www.jcp.org/En/jsr/detail?id=168
21. OASIS Web Services for Remote Portlets (WSRP) TC, http://www.oasis-open.org/committees/tc_home.php?wg_abbrev=wsrp
22. P. Andreetto, S. Andreozzi, G. Avellino, S. Beco,S. Borgia, V. Ciaschini, F. Giacomini, A. Giannelle,A. Guarise, A. Krenek, D. Kuril, A. Maraschini, M. Marchi, L. Matyska, M. Mezzadri, S. Monforte, M. Mordacchini, M. Mulac, F. Pacini, M. Pappalardo, G. Patania, J. Pospisil, F. Prelz, D. Rebatto, E. Ronchieri, M. Ruda, Z. Salvet, J. Sitera, J. Skrabal, M. Sgaravatto, A. Terracina, M. Vocu, and L. Zangrando. Pratical Approaches to Grid Workload and Resource Management in the EGEE Project. In Proceedings of the International Conference on Computing in High Energy Physics (CHEP2004), Interlaken, Switzerland. 27 September-1 October 2004.
23. T.Ferrari, E.Ronchieri, gLite Advance Reservation Architecture http://edms.cern.ch/document/508055
24. OASIS UDDI, http://www.uddi.org/
25. Rob Byrom, Brian Coghlan, Andy Cooke, Roney Cordenonsi, Linda Cornwall, Martin Craig, Abdeslem Djaoui, Steve Fisher, Alasdair Gray, Steve Hicks, Stuart Kenny, Jason Leake, Oliver Lyttleton, James Magowan, Robin Middleton, Werner Nutt, David O'Callaghan, Norbert Podhorszki, Paul Taylor, John Walk, Antony Wilson, Production Services for Information and Monitoring in the Grid, in proceedings of the UK All Hands meeting 2004.
26. Globus: Monitoring and Discovery System, http://www.globus.org/mds/

Acknowledgement

This project is supported under EU FP6 contract 511382.

Collaborative Environments For The GRID:
The GRIDCC Multipurpose Collaborative Environment

Fabio Asnicar[1], Luca Chittaro[2], Luca De Marco[2], Laura Del Cano[1],

Roberto Pugliese[1], Roberto Ranon[2], and Augusto Senerchia[2]

[1] Experiments Sector, Sincrotrone Trieste S.C.p.A. di interesse nazionale
Strada Statale 14 - km 163,5 in AREA Science Park
34012 Basovizza, Trieste Italy
{roberto.pugliese, fabio.asnicar,
laura.delcano}@elettra.trieste.it
http://www.elettra.trieste.it
[2] HCI Lab
Dept. of Math and Computer Science
University of Udine
via delle Scienze, 206
33100 Udine, Italy
{chittaro, ranon, demarco, agusmag}@dimi.uniud.it

(Invited Paper)

Abstract. The GRIDCC project will extend the use of Grid computing to include access to and control of distributed instrumentation. Access to the instruments will be via an interface to a Virtual Instrument Grid Service (VIGS). VIGS is a new concept and its design and implementation together with middleware that can provide the appropriate quality of service is a key part of the GRIDCC project. The paper describes the Multipurpose Collaborative Environment (MCE) which will provide the user interface to access Virtual Instrument Grid Services, as well as tools to support group-work. After an analysis of the status of the art in collaborative environments and technologies, taking into account both the Human Computer Interaction and the architectural and technological perspectives, we present the requirements of the MCE and more specific aspects including access and control of remote instruments.

Keywords: Collaboratories, Grid technology, Quality of Serivce, Virtual Instrument Grid Service

1 Introduction

The GRIDCC project will extend the use of Grid computing to include access to and control of distributed instrumentation. Access to the instruments will be via an interface to a Virtual Instrument Grid Service (VIGS). VIGS is a new concept and its design and implementation together with middleware that can provide the appropriate quality of service is a key part of the GRIDCC development plan. The overall architecture for GRIDCC has been defined taking into account the requirements of some pilot application areas, which include distributed power systems, remote control of an accelerator and the remote monitoring of a large particle physics experiment.

The goal of GRIDCC is to build a widely distributed system that is able to remotely control and monitor complex instrumentation that ranges from a set of sensors used by geophysical stations monitoring the state of the earth to a network of small power generators supplying the European power grid. These applications introduce requirements for real-time and highly interactive operation of computing Grid resources. The overall architecture of GRIDCC is presented in figure 1. The Virtual Control Room (VCR) is the GRIDCC user interface. It allows users to build complex workflows which are then submitted to the Execution Service (ES), it can connect to resources directly and be used to control instruments in real time. The VCR is technically an instance of the Multipurpose Collaborative Environment, a groupware application that will be developed inside the project, for a specific GRIDCC application and Virtual Organization.

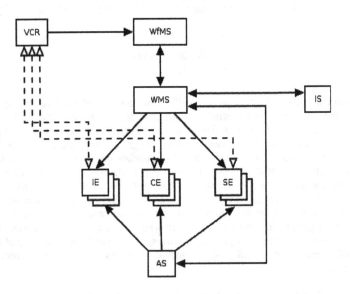

Figure 1. The GRIDCC architecture

The Instrument Element (IE) is a collection of services: the Virtual Instrument Grid Service (VIGS), the Resource Service (RS) and the Local Problem Solver (LPS).

The VIGS concept is the core of GRIDCC and is by itself made by other subcomponents. The Access Control Manager (ACM) will identify every user or entity (another VIGS for instance) and provide the proper authorization protocol. The

Instrument Manager (IM) receives commands from the VCR (or another VIGS). These are then routed to the real instruments under the control of the IM.

A Resource Service instructs the IM on the number, type and location of the instruments it should control: it knows the topology of the instruments, the connection type, their identification, etc. The Info Service Proxy collects all the information (erors, state and monitor data) from the instruments and acts as local cache before sending the data to the central Information Monitoring Service (IMS). The Data Mover Manager moves the data acquired from the instruments.

Automated problem solving in a Grid environment is a major feature of the project. Two levels of problem solver are proposed; one to solve problems related to the function of a given IE (LPS) and one to solve system wide problems (PS).

Compute and Storage Elements (CE & SE) within GRIDCC are similar in function to those in other Grid projects. However these are extended to encompass the quality of service requirements that are central to GRIDCC.

The Execution Service controls the execution of the workflows defined by the user in the VCR. It maintains the status of the jobs that make up the workflow. It controls the quality of service by making agreements with the resources available to it whether these be IEs, CEs or SEs. The ES consists of three services: the Workflow Management System, the Workload Management System and the Reservation Manager. GRIDCC requires an all pervasive Information and Monitoring System (IMS). This reports the status and availability of the resources as well as propagating errors. The Multipurpose Collaborative Environment (MCE) of the GRIDCC project is a groupware which will provide general collaboration services and allow people to control and manage remote instrumentation. This document describes the high-level requirements we identified during the course of our analysis of the state of the art, the requirements gathering and the early design stages.

The first part describes the status of the art in collaborative environments and technologies, taking into account both the Human Computer Interaction and the architectural/technological perspectives. The second part contains the requirement analysis of the MCE derived from our experience in collaborative environments and from the study of what is going on in the research.

2 State of the Art

Science is a collaborative activity and the traditional environment for scientific collaboration is the laboratory, where collaborations still rely heavily on face-to-face interactions, group meetings, individual action, and hands-on experimentation.

Electronic collaborative environments that support scientific activities, or collaboratories (a term coined by William Wulf [5]) aim at giving scientists all over the world the opportunity of remotely controlling and operating instrumentation, accessing datasets, and easily interacting with colleagues.

In particular, the core capabilities that constitute a collaboratory can be seen as technologies to link [14]:

- People to people (e.g., familiar applications, such as electronic mail, and tools for data conferencing, such as VRVS), to make remote scientists more visible to one another, and therefore, allow them to recognize common interests and concerns that can form the basis for future collaborations.
- People to information (e.g., the World Wide Web and digital libraries), to provide faster and less-restricted availability of data and results.
- People to facilities (e.g., data viewers that display the current modes and status of remote instruments as well as services that provide scientifically critical data), to enhance utilization of scarce scientific resources, by expanding access to these resources.

This section analyses the state of the art in collaborative environments and technologies, taking into account both the Human Computer Interaction in subsection 2.1 and the architectural/technological perspectives in subsection 2.2.

2.1 State of the Art: Human Computer Interaction perspective

As common in the field of Computer Supported Cooperative Work, the success of software that support group work heavily depends on psychological, social and organizational factors. This is true also for groupware to support scientific activities [2]. From this perspective, this section analyzes the state of the art of collaborative software in the context of scientific activities, describing most relevant past experiences, issues and results.

We first describe the main categories of collaboration tools used in the context of scientific activities, and then illustrate several collaborative environments (i.e., complete solutions that integrate more collaboration tools to support a specific application) that have been developed to support scientific experimental activities.

Collaboration Tools
Collaboration tools are programs that help groups of people in communicating, working together, and coordinating their activities. Most of the collaboration tools we mention are general-purpose tools; however, some of them (for example, tools to support remote operation of equipment) were developed in the context of scientific experimental activities, and are often specific to a particular scientific area or even to a specific experimental equipment and setting. In general, the effectiveness of a collaboration tools depends on many factors including the technologies used, the social environment, the goals, the level of support, and the level of need.

Communication Support Systems
Communication support tools enable people working at different places to communicate with each other. Most tools are dedicated to synchronous communication (e.g. chat, instant messenger, multi-user dungeons, video-conferencing tools), while a few support asynchronous communication (e.g., e-mail, newsgroups).

Communication may use different channels, from text (e.g., chat, instant messengers) to audio and/or video (e.g. video-conferencing tools). Some tools allow also people to talk about a shared content (e.g. a presentation or display) and to replay, archive and add notes to support continuing discussion [3].

Shared Workspaces and Applications

Shared Workspace tools support people cooperatively working on a common task. This kind of tools provides users with a virtual space in which information can be shared and exchanged. In particular, tools like whiteboards support activities like sharing information, pointing to specific items, marking, annotating, which are especially useful during a meeting or to provide expert assistance to a remote worker; shared editors facilitate the shared authoring (or co-authoring) of documents providing support for different phases of authoring, concurrent editing, annotations, comments, versions and revisions; application sharing allows existing single-user applications to be used (mostly without modification) by multiple users, simultaneously, by collating inputs from the different users (in most cases, only one at a time controls the application) and presenting output to each of the screens of the users (often, also user interface actions such as mouse movements and menu selections are shown).

These tools are used both for long-term synchronous cooperation (with fine-grained notifications giving immediate feedback about the activities of other users) and asynchronous cooperation (which can happen at different times with usually no notifications of other users' actions).

Electronic Notebooks

Paper notebooks are widely used in scientific laboratories to register important information (e.g., faults happened, important data or results) relative to the ongoing activities.

Electronic Notebooks are increasingly replacing their paper counterparts to overcome their limitations (e.g., limited space, only local availability, lack of search facilities), allowing simultaneous use by distributed researchers providing access to data, as well as automated data entry, searching, multimedia annotations and other information processing not possible in a paper notebook.

Meeting and Decision Support Systems

Decision support systems are designed to facilitate groups in decision-making and are primarily designed to facilitate meetings. They provide tools for brainstorming, critiquing ideas, putting weights and probabilities on events and alternatives, and voting; moreover, they encourage equal participation (e.g., by providing anonymity or enforcing turn-taking).

Shared or group calendars provide functionalities to facilitate meeting schedule, project management, and coordination among many people, and may provide support for scheduling equipment as well. Typical features include finding meeting times suitable for everyone and detecting when schedules are in conflict. Typical concerns are privacy, completeness and accuracy.

Remote access and Control Systems
Systems that allow to remotely access (typically) large, expensive and unique scientific and computing equipment or data repositories. The opportunity to remotely access these instruments allows reducing traveling costs (in terms of money and time losses), to integrate and share knowledge and competencies among scientists (especially in the case of data repositories, e.g., the US Protein Data Bank) and it has been used also for educational purposes (e.g., [4]).

Collaborative Environments to support scientific activities
Collaboration often involves a mixture of types of interactions, requiring an appropriate array of tools. Often, these tools are integrated into a collaborative environment devoted to a specific application area. This section presents an overview of past collaborative environments for scientific experimental activities. Following [1], we consider a collaborative environment as a network-based facility and organizational entity that spans distance, supports rich and recurring human interaction oriented to a common research area, fosters contact between researchers who are both known and unknown to each other, and provides access to data sources, artifacts and tools required to accomplish research tasks.

Upper Atmospheric Research Collaboratory (UARC) and Space Physics and Astronomy Research Collaboratory (SPARC)
The Upper Atmospheric Research Collaboratory provided (from 1992 to 1999) access to instruments located at an observatory in Greenland, and was the first non-biomedical instance of teleoperation [4]. The goal of UARC was to provide real-time control of remote instruments used to study upper atmospheric events (e.g., solar wind observation). In addition, UARC was meant to support communication among geographically distributed colleagues about shared real-time data and to provide access to archived data.

Collaboratory for Multi-Scale Chemical Science (CMCS)
The Collaboratory for Multi-Scale Chemical Science (CMCS) is a multi-institution project to develop and test an advanced collaborative community data system for chemical science. CMCS includes tools for discovering, analyzing, and visualizing data as well as for coordinating group and community processes involving the data. The main interface is the Multi-Scale Chemistry (MCS) portal. Within the portal, tools are available for exploring data collections, searching for chemical information about particular species, subscribing to receive email messages when new data appears, viewing metadata and data relationships, and visualizing data. Additional tools, such as electronic notebooks, discussion forums, task lists, and calendars, are provided for group coordination [6].
The CMCS portal is structured so that each user has a personal workspace and each team has a workspace. Furthermore, there is a public workspace that can be accessed by all users. Users have access to the ir private workspace and to all the workspace(s) for the team(s) they are members of. The functionalities provided by the CSMC portal include functionalities for management of accounts, groups and roles.

The Communication Services provided by CMCS are divided into two groups, System-level (the System announcement page and the feedback to the CMCS developing team) and Group-Level. The Group-Level Communication Services include an on-line chat, group announcement pages, a threaded discussion list, a graphical calendar and a task list management.

The CMCS portal provides a data repository that automates many aspects of data discovery, translation, and data provenance tracking. CMCS employs the Scientific Annotation Middleware (SAM) as a means to provide federated data/metadata access, extensible metadata annotation, and transformations of data and metadata [6]. Examples of provided services are a file browser to navigate the data/metadata repository, interactive visualizations of 3-D molecular structures and translations between chemical file formats using several Web service-based translators. Moreovoer, CMCS has developed an annotation tool that allows data owners and third parties to associate text, sound, images, equations, or whiteboard drawings with existing data. Annotations are stored as independent files linked via metadata to the specified data and can have their own metadata and additional relationships. Thus, annotations can be protected with access controls and threaded to organize related topics of discussion [6].

The CMCS provides also notification services (for example changes to files, to properties and submission of new data containing a specific property or chemical species). Remote users are notificated via email when events of interest occur.

Biological Sciences Collaboratory (BSC)

The Biological Sciences Collaboratory (BSC) enables the sharing of biological data and analyses through diverse capabilities such as metadata capture, electronic laboratory notebooks, data organization views, data provenance tracking, analysis notes, task management, and scientific workflow management. The overall goal of BSC is to identify and capture the various social and scientific contexts in which data sharing collaborations in biology take place and to provide collaboration tools and capabilities that can effectively support and facilitate these important data sharing contexts [7].

BSC's data sharing tools were build in conformance to a series of various miniscenarios of shared data use developed by the biologists in early participatory analysis and design sessions. These contexts helped identifying the critical features or properties of scientific data, to support data sharing (e.g., data set properties, data provenance, physical and logical organization; the full list is reported and described in [7]).

BSC presents a brand new idea in the field of collaboratories whic is a shift of the focus from a tool centric view to a data centric approach.

The Elettra Virtual Collaboratory (EVC)

The Elettra Virtual Collaboratory [8] supports scientific collaboration in experiments with ultra-bright light sources, and it has been used at the Elettra Synchrotron facility in Trieste (ITALY). The EVC is a Web based collaborative environment through which users access the different services (e.g. set-up experiments, see experimental data, run scientific programs). The Web server performs the required services by communicating with other computing equipment at Elettra (e.g., data collection workstations, data processing workstations), and translating user requests into proper commands, which are in turn executed by scripts running on specific machines. The EVC interface is Web portal that can be accessed by different categories of users (visitors, researchers, experiment leaders, staff personnel of the experimental stations), which are offered different sets of possible actions, depending on their user category and expertise level (for example, an experiment leader can add and remove collaborators from a specific experiment, while a normal user has not this possibility). Collaboration teams are led by an experiment leader (which typically, for security and safety reasons, is the one who physically carries out the experiment at the experimental station) which is able to add and remove collaborators at any time. The EVC systems provides the following collaboration tools [8]:

- a chat, which allows, besides typical messaging facilities, pasting of images (obtained from the data collection equipment) and drawing on a shared canvas (arrows, signs, or text);
- a file browser tool (this allows remote users to quickly estimate the suitability of the experimental parameters), enhanced with tools for basic processing on experimental data;
- access to scientific resources (such as the Protein Data Bank);
- monitoring of the experimental equipment both through video cameras and by a synoptic view of the equipment which is being used;
- VNC (Virtual Network Computing) to access workstations during the data analysis phase.

GridSphere-based collaborative environments

The GridSphere portal framework, developed within the GridLab EU project, provides a model for building Web portals that give access to Grid resources, as well as collaboration tools. The main idea behind the project is to adopt international standards (such as Portlets) to develop the application-specific resources of the portal, such as a job submission component or a remote file browser interface [14]. The GridSphere framework is being used to re-implement the Astrophysics Simulation Collaboratory (ASC), a collaborative environment for large scale simulations of relativistic astrophysics.

2.2 State of the Art: technological and architectural perspective

The section is focused on web technologies and portals which are commonplace and therefore will be familiar with most of users. This makes such technologies particularly interesting for the usability of the MCE and the acceptance of it by the users community.

Portals build on the same technology used for Web sites, but enhance the functionality and flexibility to cater for the demands of specific classes of user.

A portal is a presentation layer which aggregates, integrates, personalises and presents information, transactions and applications to the user according to their role and preferences. It provides a persistent state/context for each user and/or group of collaborators. This state may consist of documents, personal registries, tool configurations, calendars, or anything else related to the user's personal configuration of the their "home" at the portal. GRID portals enhance traditional portals by allowing a seamless integration with GRID and grid services [9].

A collaboration can be defined as a group of people sharing events which define state changes in the objects they are using. To implement this, an event notification service is required and a session management service by which a user can subscribe to the original data and updates via events.

A portal for interactive collaborative working is characterized by somewhat different services from a Grid application or resource discovery portal:

Mechanism to set up members (people, devices) in collaborative sessions; Generic tools like text chat, white boards; Audio-video conferencing; Visualisation with share events corresponding to changes in pixels of a frame buffer, maybe using SVG; shared maps, instruments.

The wrapping of tools and applications as Web/Grid services is a desirable way to share these objects. Several ways to set up and manage sessions could also be turned into services. Each service might have a user interface which could be a portlet. Portlets could be re-directed, e.g. to PDAs, mobile phones or other devices.

The CompreHensive collaborativE Framework (CHEF) [10], was developed by the group of Charles Severance at University of Michigan, USA. It is a collaboration tool aimed equally at course administrators and students. It offers administrators the facility to set up a course worksite, and students to set up study worksites. Each worksite can be configured with a number of included tools like chat, a discussion board, shared calendar and communal filespace. It makes use of the Jakarta Jetspeed portal framework to present these tools to the user in a collection of portlets.

The Global-MMCS v2.0 [11] and XGSP MCU from Indiana University is now being developed and deployed for test purposes. Among the open-source software being used is: OpenH323, NIST SIP stack, VIC and RAT (to link to Access Grid), Narada Brokering and Java Media Framework. Jetspeed, Java Applets or ActiveX controls are used to add non-HTML clients to HTML pages. Apache Batik could be used for SVG applications. Some generic portlets, such as a VIC portlet for unicast connection to an Access Grid session could be more widely used. A Java applet could be developed to support the full multicast protocol.

A Grid Information System simply provides information about Grid resources (computers, databases, instruments) and their capabilities. Grid information can be divided into two kinds: static data that to not change, or only do so occasionally (e.g. after a software upgrade) and dynamic data which change frequently, like current utilisation data.

Grid application portals provide a variety of mechanisms to access active services. Some of these services will invoke remote methods to generate new data, information or knowledge through simulation, data merging or mining or download from on-line instruments possibly accompanied by steering, visualisation and other control mechanisms.

The Grid Enabled web environment for site Independent User job Submission (GENIUS) [12] is an opensource project based on the commercial (but free for academic use) EnginFrame portal platform as evaluated by the European Data Grid project. It has principally been tested for EDG by the ALICE collaboration. From 2002-2003 an official grant for the INFN Grid project allowed collaboration with NICE srl to integrate in a Web portal all services offered by the DataGrid middleware. EnginFrame is built on Apache, and uses HTTPS+Java with XML+rfb (remote frame buffer) in a 4-tier architecture consisting of Web browser, EnginFrame server with Tomcat, SAP or Web services or enginFrame agents linked to services.

The Open Grid Computing Environment (OGCE) [13] Portal Project, funded by the National Science Foundation (NSF) (sometimes called the NMI portal project), represents a union of many of the American Grid related portal projects. Jetspeed, CHEF and OGSA/OGSI are currently being used with the Argonne Globus Java CoG kit. There is also an evaluation of GridSphere.

Much of the effort of the OGCE project is focused on building Grid/Web services and their access by client interfaces. Some of the work at Indiana has the goal to deploy Grid services for things like workflow tools and make it possible for portal users to discover and load the client interfaces into their portal environment or compos it into applications as components.

Extremely interesting are the frameworks for developing portals. The GridSphere [15] project is building on experience in the Java-based ASC and GPDK portal toolkits. GridSphere provides a "white-box" framework in which users can override base classes and "hook" in their own methods. It therefore requires users to become familiar with core framework interfaces which are however based on the community standard API JSR-168. The Model View Control (MVC) paradigm is used to separate logic from presentation as in other portlet frameworks.

JetSpeed is the open source portal framework maintained by the Apache Jakarta software group and released under the Apache Software License. It is written entirely in Java and uses XML based configuration files. Jetspeed will allow you to aggregate HTML pages, Java Applets, RSS (Really Simple Syndication) feeds, Java Servlets or JSPs and others. Jetspeed adheres closely to the JSR-168 standard [16].

The Access Grid (AG) [17] is an ensemble of resources including multimedia large-format displays, presentation and interactive environments, and interfaces to Grid middleware and to visualization environments. These resources are used to support group-to-group interactions across the Grid. For example, the Access Grid is used for large-scale distributed meetings, collaborative work sessions, seminars, lectures, tutorials, and training. The Access Grid thus differs from desktop-to-desktop tools that focus on individual communication. The Access Grid comes with a toolkit AGTk which can be used to develop specific functionalities and shared applications. The AccessGrid is based on Globus Toolkit and related technologies.

The Virtual Room Videoconferencing System (VRVS) [18] is a web oriented system for videoconferencing and collaborative work over IP networks. VRVS provides a low cost, bandwidth-efficient, extensible means of videoconferencing and remote collaboration over networks within the High Energy and Nuclear Physics communities. Recently VRVS also extended the accessibility to its services to other various academic/research areas.

The most promising frameworks for developing portals are the "portlet" frameworks for which a couple of standards, JSR-168 and WSRP were recently ratified. The advantage of a portlet-based architecture is that each underlying function or service can be associated with a unique portlet. This makes it easy to add new services, and many different groups can then independently contribute portlets which can be plugged into the portal. Using WSRP, they can be distributed and managed remotely on many servers and the portals composed from WSDL-like information. Each user can select and configure the portlets he/she wishes to use and selection can become part of a persistent "context". For instance certain portlets may only be useful for expert or administrative users and can be discarded by others or have their access controlled via a role-based mechanism.

The Portlet Java Specification Request JSR-168 lays the foundation for a new open standard for Web portal development frameworks. Portlets define an API for building atomic, composable visual interfaces to Web content or service providers. A portlet provides a "mini-window" which can be placed within a portal page. Multiple portlets can be composed in a single page by the developer or user through the framework.

WSRP, the Web Services for Remote Portlets API defines a standard for interactive, user-facing Web services that plug and play with portals. WSRP seeks to establish a portlet abstraction with a WSDL description for how to publish, find and bind to remote WSRP-compliant services with metadata about related things such as security mechanisms, billing, etc.

Peer to peer systems have much in common with Grid computing [19]. However, the sharing that Grid computing is concerned with is not primarily file exchange, but rather direct access to computers, software, data and other resources. One reason that Grid computing and peer-to-peer technologies have not overlapped significantly to date seems to be that P2P developers have focused mainly on vertically integrated solutions, rather than seeking to define common protocols that would allow for shared infrastructure and interoperability (a common practice for new market niches). Another is that the form of sharing targeted by P2P has been rather limited (e.g. file sharing with no access control).

However, as P2P technologies are becoming more sophisticated, it is expected that there will be a strong convergence between P2P and Grid computing.

The requirements for collaborative services, especially pertaining to order and delivery, are quite different compared to traditional distributed applications. Particularly interesting in this area are technologies like Java Messaging Framework implemented also in open source products like OpenJMS or NaradaBrokering. In particular the NaradaBrokering [20] messaging substrate enables scalable, fault-tolerant, distributed interactions between entities, and is based on the publish/subscribe paradigm. The substrate also incorporates support for Grid and Web Service. The substrate can indeed be used in scenarios where performance and scalability requirements are stringent.

3 Requirements

The Multipurpose Cooperative Environment (MCE) is a groupware which provides general purpose services to control remote instrumentation, manage experimental activity and allows the implantation of the different testbed applications through customisation and integration with application specific services. The system will be used to implement the VCR for the different applications of the GRIDCC project.

The MCE will be based on a core groupware application which solves common problems (e.g., authentication, management of the users and instrument resources, management of the Virtual Organizations, monitor of the instrument status, job control, etc) and a set of plug-ins, part of which will be general (e.g., chat, file browser, notebook, video conference), and others specific to the particular application (e.g., accelerator control, instrument control). In the following, we analyze the requirements for the general part of the MCE; detailed requirements for each GRIDCC application will be formalized in future steps of the project.

Classes of users

There are different type of users that interact with the system. With respect to the responsibilities, it is possible to define the groups of users described below. Inside some groups (e.g. MCE users) it is possible to define some sub-groups whose users play different roles with an authorizations schema specific of the application.

MCE users are the most wide and important class of end-user of the system. The ultimate usefulness of the MCE system will depend on its ability to simplify the work done by this groups by providing an easy to use, feature-laden work environment.

In particular, supporting these end-users means capturing the testbed application requirements and providing these capabilities through the MCE.

MCE users need to access all the various applications and services required via a simple, coherent common platform.

MCE administrators are responsible for the deployment and management of the portal. They are in charge of installing the software, configuring the application, providing roles and permissions to users and general user management.

A requirement of MCE administrators is a system easy to install and deploy. It is desirable that new features can be added to the system with simple changes. Another requirement is that much of its tasks can be done via the MCE itself.

The MCE will provide generic support for Grid services making it easy for a portal developer or administrator to readily make the Grid service portal accessible.

The *MCE grid service providers* are in charge of set-up and configuration of the communication between the MCE and the grid services.

The MCE could be used for remote service administration and management by a service provider. This group of users is responsible for providing application specific support. They are in charge of setting up services and features offered, end-user roles and permissions. This task might require the development of specific interfaces/modules inside MCE specification. A requirement of pilot-application administrators/developers is that the MCE system is modular and easily extensible.

General Requirements

The MCE will be a uniform and unique environment providing support for general distance collaboration and, at the same time, access to remote control and monitoring of scientific instrumentation in the context of experimental activities. The general groupware functionalities of the MCE will be extended to support different virtual instruments coming from specific pilot applications of the GRIDCC project.

The MCE should be designed considering as primary the issues of modularity and flexibility to support the many different needs and requirements of the MCE users (typically experimental scientists) and pilot applications developers. The user interface should be easy to use and accessible through standard web-browsers to enable rapid and wide adoption among all of the potential target users, thus making collaborators work as productive and efficient as possible.

It is desirable that users can authenticate themselves to the MCE services through a single sign-on. It is possible to achieve this by requiring users to login to the MCE using certificates, e.g. Grid Security Infrastructure certificates. The certificates may be retrieved from a client's local filesystem or retrieved in a secure manner from a certificate repository server (e.g. MyProxy).

Users should be able to protect their files from unauthorized access. At the same time, users should be able to easily share files and related information with other users as desired.

It would be conceivable to allow portal users to share their certificates in order to share access to resources accordingly to security policies of resource providers.

Particularly challenging is the problem of defining and managing the permissions regarding instruments exported as Virtual Instruments Grid Services.

The Resource Service handles Grid resources belonging to a given VO. Resources can be catalogued, discovered, allocated and queried. Different users can work concurrently on different set of resources (partitions). RS can also handle Grid-enabled farm and/or data storage.

The MCE should implement a basic interface to each Resource Service allowing to browse, search and eventually book resources.

Virtual Instruments are special type of resource which can also be partitioned and grouped in order to allow more advanced operations.

Another important Grid framework functionality is job management that involves at least submission and monitoring features. It is clear that users require easy-to-use and robust support tools for submitting jobs and complex workflows. After this, users should be able to monitor the status of their jobs online and receive notification when jobs complete or fail. In addition, they should be able to monitor the performance of their jobs, and to examine their job history.

Users need to store or archive their large datasets, and should be able to determine their disk or archive space allocation. Grid users require the ability to transfer large datasets between resources.

Users should be able to monitor application-specific information and receive notification about events occurring within applications.

From the literature review and analysis of the pilot applications, there is the need for tools that support remote users meetings and conversation. Such tools (e.g., chat, with integrated Instant Messaging) enable synchronous/asynchronous collaboration in the daily tasks involved in scientific research facilities and other remote environments. Presence information plays an important role in the perception that collaborators are working together. Awareness information should be notified to users to allow them recognizing the availability and the identity of other collaborators to keep in touch with them. Audio/video streaming should be provided to support one-to-one communication with remote human operators (e.g., experts), to hold remote meetings and monitor different kind of resources (e.g., remote instrumentation).

Moreover, users should be provided with the ability to easily access, share and exchange information about their activities, by means of tools such as electronic notebooks, shared repositories with annotation functionalities, formal and informal meeting support (e.g., slide presentation, whiteboards) and application sharing (e.g., through VNC).

The MCE web-based portal administration functionalities will be available online for users with administrative rights. Operations that should be provided include the creation of personal and group accounts, the assignment of resources and the management of user profiles (e.g., level of expertise, user permissions, access control and membership to different VOs). The administration of grid-based services (e.g., specific virtual instruments) and the creation of VOs will be subjected to the administrator's defined privileges (in addition to grid-services certificates).

4 MCE Design: first choices

According to the literature review, there is an increasing trend towards the adoption of web-based interfaces for the development of collaboratories. The availability and ubiquity of Internet-based network connections make web browsers one of the most lightweight, easy to use and familiar platforms to enable remote collaboration provide remote collaborators. Moreover, the scalability and portability of web browsers interfaces allows to easily extend the MCE functionalities with portlets and ad-hoc plug-ins for different platforms and devices.

As a consequence of these considerations, the MCE will mainly rely upon web-based interfaces allowing users to access its resources at any location through a standard web-browser. After the login phase, the user will have the possibility to: (i), access her/his personal workspace and information (e.g., showing urgent communications, upcoming events); (ii), see an overview of the status of the VOs he/she is involved in and to general collaboration tools (e.g., instant messenger, chat,); (iii), subscribe to different VOs to be notified about relevant events and/or messages coming from other collaborators, or, (iv) according to his/her privileges, create new VOs or modify existing ones. When the user joins a virtual organisation, she/he will have access to a shared group workspace and to the available specific resources and tools (e.g., remote virtual instruments, scientific applications). The collaborative environment will offer the opportunity to customize the web interface layout and to save user preferences, thus enabling consistency from session to session. The MCE will automatically adapt its layout, information presentation and available functions according to different user roles, abilities and client platforms.

The MCE portal will provide access to the full set of functionalities of the collaboratory through a standard web-browser. Specific pilot application tools where advanced interactive functionalities are required could be accessed through Java applets or other widely accepted cross-platform technologies.

Many mobile devices (including mobile-phones and PDAs) currently provide the ability to browse the Internet. The MCE should provide access to a (eventually limited) set of functionalities through this kind of clients, since mobile devices could offer significant support to collaborators in several scenarios. For example, mobile clients allow users to be notified of important events in a timely fashion (e.g., by receiving an SMS or even an e-mail), to keep them in continuous contact with their colleagues and to access to remote reference manuals or control panels, independently from their current location (e.g., from a maintenance area where terminals are not available).

5 Conclusions

The Multipurpose Collaborative Environment (MCE) will provide the user interface to access Virtual Instrument Grid Services, as well as tools to support groupwork. We are now designing the MCE user interface and starting to build early prototypes.

In this stage of the project, the focus is more on general design choices and collaboration tools, while more detailed aspects (including access and control of specific remote instruments, and application-specific functionalities) will be taken into account later.

More information on the GRIDCC project and on the MCE and their progress can be found at the project website www.gridcc.org.

Acknowledgement

This project is supported under EU FP6 contract 511382.

References

1: Finholt, T.A., Collaboratories, In B. Cronin (Ed.), Annual Review of Information Science and Technology 36,
http://www.scienceofcollaboratories.org/WorkshopStuff/June2001/pdfs/
Finholt_Collaboratories_03_.pdf
2: Agarwal, D., G. Olson and J. Olson, Collaboration tools for the global accelerator network, Collaboration Tools for the GAN Workshop 2002 - Report, http://www-library.lbl.gov/docs/LBNL/532/53/PDF/LBNL-53253.pdf

3: Peggs, S., T. Satogata, D. Agarwal and D. Rice, Remote operations in a global accelerator network, Particle Accelerator Conference 2003, Portland, OR, http://www-library.lbl.gov/docs/LBNL/529/47/PDF/LBNL-52947.pdf

4: G.M. Olson, D.E. Atkins, R. Clauer, T.A. Finholt, F. Jahanian, T.L. Killeen, P. Prakash, and T. Weymouth, The Upper Atmospheric Research Collaboratory (UARC), Interactions, Volume 5, Issue 3, 1998

5: R.T. Kouzes, J.D. Myers, and A.D. Wulf, Collaboratories: Doing Science on the Internet, IEEE Computers, Volume 29, Number 8, 1996, http://collaboratory.emsl.pnl.gov/presentations/papers/IEEECollaboratories.html

6: Myers, J.D., Allison, T.C., Bittner, S., Didier, B., Frenklach, M., Green, W.H., Ho, Y., Hewson, J., Koegler, W., Lansing, C., Leahy, D., Lee, M., McCoy, R., Minkoff, M., Nijsure, S., von Laszewski, G., Montoya, D., Pancerella, C., Pinzon, R., Pitz, W., Rahn, L.A., Ruscic, B., Schuchardt, K., Stephan, E., Wagner, A., Windus, T., and Yang, C., A Collaborative Informatics Infrastructure for Multi-scale Science, Proceedings of the Challenges of Large Applications in Distributed Environments (CLADE) Workshop, Honolulu, HI, http://collaboratory.emsl.pnl.gov/sam/resources/CLADE_2004_3_28.PNNL-SA-40934.pdf

7: George Chin and Carina Lansing, Capturing and Supporting Contexts for Scientific Data Sharing via the Biological Sciences Collaboratory, Proceedings of the 2004 ACM conference on Computer Supported Cooperative Work, pp. 409 - 418, http://portal.acm.org/citation.cfm?id=1031677&coll=ACM&dl=ACM&CFID=34364100&CFTOKEN=36031774

8: R. Ranon, L. Chittaro, R. Pugliese, The Elettra Virtual Collaboratory: a CSCW System for Scientific Experiments with Ultra-Bright Light Sources, Proceedings of HCITALY 2003, Symposium on Human-Computer Interaction 2003, Turin, http://hcilab.uniud.it/publications/2003-01/28.pdf

9: R. Allan, C. Awre, M. Baker, A. Fish, Portals and Portlets 2003, Technical Report UKeS-2004-06, http://www.nesc.ac.uk/technical_papers/UKeS-2004-06.pdf,

10: CHEF Architecture, School of Information + Media Union, University of Michigan, http://collab.sakaiproject.org/portal?site=!gateway&page=!gateway-400

11: Geoffrey Fox, Marlon Pierce, Wenjun Wu, Ahmet Uyar, Hasan Bulut, Portlets and Web Services for Collaboration and Videoconferencing, http://www.nesc.ac.uk/talks/261/Tuesday/Fox.ppt

12: GENIUS, , http://genius.ct.infn.it/

13: OGCE, , http://www.collab-ogce.org/nmi/

14: GPDK, , http://www.doesciencegrid.org/gridportal.html

15: Jason Novotny, Michael Russell and Oliver Wehrens, GridSphere: A Portal Framework for Building Collaborations, Proceedings of 1st Grid Middleware Congress 2003, Rio de Janeiro, http://www.gridsphere.org/gridsphere/wp-4/Documents/RioBabyRio/gridsphere.pdf

16: Jetspeed, , http://portals.apache.org

17: Access Grid, , http://www.accessgrid.org

18: VRVS, , http://www.vrvs.org

19: Stephanos Androutsellis-Theotokis, White Paper: A Survey of Peer-to-Peer File Sharing Technologies, ELTRUN, Athens University of Economics and Business, Greece , 2002, http://www.eltrun.gr/whitepapers/p2p_2002.pdf,

20: Naradabrokering Messaging System, Indiana University, http://www.naradabrokering.org

Design And Development Of A GNRB For The Coordinated Use Of Network Resources In A High Performance Grid Environment

Davide Adami[2], Nora Carlotti[1], Stefano Giordano[1], Matteo Repeti[1]

[1]Dept. of Information Engineering, University of Pisa, Italy
s.giordano@iet.unipi.it {m.repeti,
n.carlotti}@netserv.iet.unipi.it
[2]CNIT Research Unit– Dept of Information Engineering, University of Pisa
davide.adami@cnit.it

Abstract. The development and the deployment of wide-area Grid environments represent a new challenge for Next Generation Networks: not only is it necessary to implement network architectures supporting Quality of Service at IP level, but it is also essential to design and implement new network entities that monitor the network status, assure that network resources are allocated in an optimized way and interfaces Grid applications with network services, satisfying their SLSs. The paper presents the design, the development and the functional validation of a Grid Network Resource Broker (GNRB), a new architectural entity which handles the service requests coming from grid middleware (or from a single grid application) and, if necessary, allocates and reserves network resources.

Keywords: Grid, Diffserv/MPLS, GNRB

1 Introduction

Over the last years, advanced network architectures (Integrated Services, Differentiated Services, Multi Protocol Label Switching) with new capabilities in terms of reliability, Quality of Service (QoS) support and dynamic resources allocation mechanisms, have emerged. In particular, the Multi Protocol Label Switching (MPLS 1) architecture, through the dynamic allocation of Label Switched Paths (LSPs), based on the Traffic Engineering (TE) concept, allows to optimize network resources utilization.

At the same time, the great widespread and deployment of Grid technologies 4, has enabled the sharing of a wide variety of geographically distributed resources (storage systems, supercomputing clusters, databases, etc.), owned by different organizations and the creation of virtual organizations. In essence, Grids aim to integrate, virtualize and manage resources and services within distributed, heterogeneous, dynamic Virtual Organizations (VOs) across traditional administrative and organizational domains 5. Resources sharing is made easier and controlled by a set of services that allow resources to be discovered, accessed, allocated, monitored and accounted for, regardless of their physical location. These services may be seen as a software layer between the physical resources and the applications and they are often referred to as *Grid Middleware*.

Massive data-sets generated by experimental sciences can be exchanged routinely and analyzed remotely among institutions connected to high performance networks and belonging to a Grid Virtual Organization.

In this context, a new architectural model, based on the interaction and integration between *Grid Middleware* and next generation networks infrastructures, can generate a remarkable progress in the process of science.

In this paper, we first present the main features and networking issues of Grid applications. Next, we introduce an innovative architecture designed and developed to integrate networks providing QoS support and Grid applications. Finally, the experimental scenario where the functional requirements of the new architecture have been validated is reported.

2 Grid Environment Enabling Distributed System

In an ideal Grid environment, resources should be accessed in a completely seamless manner, so as to hide the physical discontinuities and differences among them. As a matter of fact, the Grid middleware turns a very heterogeneous environment into a virtual homogeneous one.

Due to its intrinsic heterogeneous nature, a Grid needs to cope with a large set of requirements and capabilities:

- *Multiple administratively autonomous domains.* Grid resources are usually geographically distributed across multiple administrative domains and owned by different organizations. The autonomy of resource owners has to be honoured along with their resource management and usage local policies.
- *Heterogeneity.* A Grid involves a lot of resources that are heterogeneous in nature and encompass a vast range of technologies.
- *Scalability.* A Grid might grow from a few to millions of integrated resources. This raises the problem of potential performance degradation as the size of Grids increases. Consequently, applications that require a large number of geographically distributed resources must be designed to be latency and bandwidth tolerant.
- *Dynamicity or adaptability.* In a Grid, resources failure is the rule rather than the exception. In fact, due to the great number of resources in a Grid, the probability that some resource fail is really high. Resource managers or applications must tailor their behaviour dynamically and use the available resources and services in an efficient and effective way.

The steps to realize a Grid which assures such capabilities include the deployment of a low-level middleware providing a secure and transparent access to Grid resources and of a user-level middleware and tools for applications development and distributed resources aggregation.

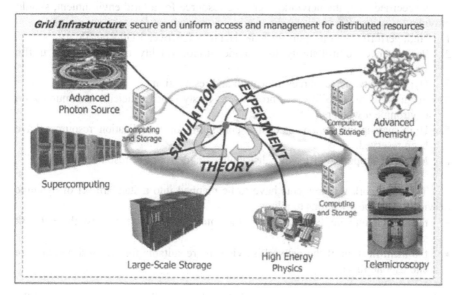

Figure 1. The invisible grid environment

From end-users point of view, Grids can be viewed as a collection of services that can be invoked with performance and accounting requirements. In this way, an *Invisible Grid* environment hides the underlying complexity to end-users, as shown in Figure 1. For example, when an application process is submitted, the grid resource manager assigns the resources to the software modules which constitute the application process, monitors the application status and returns the results.

3   Networking Issues for Grid Infrastructures

At present, Grid applications are developed for a network offering only a best effort packet delivery service. In this context, probing mechanisms are generally used to collect information and provide accurate forecasts of dynamically changing performance parameters from a distributed set of metacomputing resources, including network links. Moreover, Grid applications are usually designed in a completely independent way from the network services, even if the type of services they receive is strictly tied to the network performance.

Hence, Grid services and applications sometimes experience a quite different behaviour than expected.

Similarly, a distributed grid infrastructure with ambitious service demands stresses the capabilities of the interconnecting network more than other environments. Grid applications, therefore, often allow to identify existing bottlenecks, caused either by architectural or implementation specific problems or missing service capabilities.

As specified in 7, the network, seen as a resource for a Grid environment, should provide:
1. High performance transport for bulk data transfer (over 1 Gb/s per flow).
2. Performance controllability to provide ad-hoc quality of service and traffic isolation.
3. Dynamic network resources allocation and reservation.
4. Security controllability to provide a trusty and efficient communication environment when required.
5. High availability, when expensive computing or visualization resources have been reserved.
6. Multicast support to efficiently distribute data to groups of resources.

These network requirements have to be mapped into a standard set of dynamic Service Level Specifications to allow:
- the design and implementation of the interfaces used to request the network services;
- the identification of the network services more suitable for the Grid middleware and user applications.

Moreover, the classification, as regards high-level application requirements, of the Service Level Specifications (SLS) into different areas allows to give a formal description of the network services.
    According to 8, two areas have been identified:
- **Path-Oriented:** includes all the SLSs with special requirements in terms of traffic forwarding.
- **Knowledge-Based:** includes the SLSs requiring information about the status and the properties of the network.

In this work, as far as the Path-Oriented area is concerned, the following SLSs have been defined:
- **Premium Service (SLS-1):** low-latency communication between two end-points is assured. This service guarantees that in-profile packets (token/leaky bucket constrained) are delivered to the destination within a given delay boundary.
- **Guaranteed Bandwidth (SLS-2):** in this case, a guaranteed rate service between two or multiple grid nodes is assured.

Both SLSs need a Connectivity Service (Figure 2) with network resources allocation and reservation functions. The Connectivity Service (CS) is defined in 9 as a high-level grid network service, but its functional blocks are not specified.
    Similarly, with regard to Knowledge-Based Service, two SLSs have been defined:
- **Weighted topology discovery (SLS-3):** different metrics can be used to evaluate the "cost" of the links. In particular, the cost model may be varied depending on the Grid application and the middleware requirements.
- **SLS monitoring (SLS-4):** grid applications may require to monitor SLS parameters and to verify whether SLSs are satisfied.

To provide the latter SLSs, a further high level grid network service, referenced in 9 as Network Information and Monitoring Services (NIMS) has to be implemented. The NIMS must be designed and developed taking into account the different features and capabilities of current measurement systems integrated into heterogeneous network infrastructures.

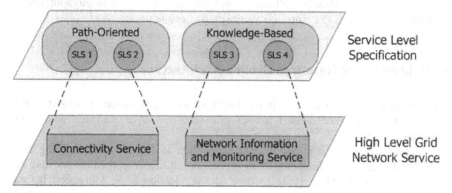

Figure 2. SLSs and Grid Network Services

4   Network Architectures Enabling High Performance Grid Environment

A reference network scenario must be defined to map Grid applications requirements into network services and assure the requested SLSs. In this work, we deal with a DiffServ over MPLS 3 network architecture with an optical transport layer, so that QoS mechanisms as well as Traffic Engineering features are available.

Indeed, a network architecture that supports Differentiated Services 2 allows providing controllable performance and QoS guarantees. The DiffServ Class-of-Services offer the possibility to define different service levels for each user application, in terms of reserved bandwidth, packet loss, maximum guaranteed boundary delay and delay variation.

Additionally, with the introduction of the MPLS architecture, several advanced network capabilities may be enabled:

- *Traffic engineering*: MPLS supports explicit paths management and traffic assignment, making it possible to dynamically plan network routes in order to optimize network resource utilization on the basis of users demands.
- *Resilience*: since mission-critical and real-time traffic is migrating to IP-based platforms, the reliability of IP networks must be enhanced. MPLS has some inherent features that can be used to protect the Label Switched Paths (LSPs). In particular, in MPLS networks, the flexible management of LSPs allows easily supporting the protection switching models.

- *Virtual Private Networks (VPNs) support*: MPLS provides efficient mechanisms to support Layer 3 VPNs, because VPN sites may be connected through dedicated LSPs.

In this network scenario, it is possible to guarantee several different network services, such as the dynamic allocation and reservation of high throughput LSPs, Differentiated Services Per Hop Behaviour (PHB) treatments, etc.

5 Grid Network Resource Broker Architecture

A particular challenge that arises in a network infrastructure aimed to support a grid environment, is the coordinated use of grid resources in compliance with Grid applications SLS.

In particular, supposing that a Grid application is capable of specifying and signalling its service requirements in an explicit or implicit way, the network may be dynamically configured in order to meet the application SLS. Moreover, before accepting the new requests coming from the application, admission control procedures should be executed to verify the availability of system and network resources. Therefore, the deployment and development of wide-area Grid environments represent a new challenge for Next Generation Networks: not only is it necessary to implement a network architecture supporting Quality of Service at IP level, but it is also essential to design and implement a network resources management system that monitors the status of the network, assures that network resources are allocated in an optimized way and matches Grid applications SLSs with network services.

Hence, to address these new issues, we have designed and developed a Grid Network Resource Broker (GNRB) which handles the requests coming from grid middleware (or from a single Grid application) and, if necessary, allocates and reserves the necessary network resources.

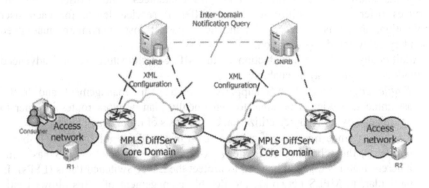

Figure 3. Hierarchical Grid Network Resource Brokers in DiffServ/MPLS networks

The network services offered by the GNRB are implemented independently of the grid environment where they are used. Moreover, the information used by the GNRB to take the decisions related to network resources utilization, reflects the real working status of the network.

When the network size or the grid extension increase, scalability issues might arise. In addition to this, if multiple administratively autonomous domains are interconnected, policies might restrict access to information concerning the status and topology of the network.

Our approach is therefore based on a hierarchical organization of the GNRBs (Figure 3). In this scenario, each GNRB is responsible for the administration and management of the network resources within its own network domain. Besides, interdomain communication allows to control and coordinate the assignment of the network resources among different domains, so as to satisfy, if possible, the Grid application SLS, even if the endpoints belong to different network domains. Finally, a grid end-user or a grid manager application may ask to the nearest GNRB the assurance of one of the SLSs defined in section 3.

The GNRB architecture consists of the following functional blocks (Figure 4):

- *Network Resources Management System* (NRMS);
- *Network Configuration System* (NCS);
- *Network Information Storage System* (NISS).

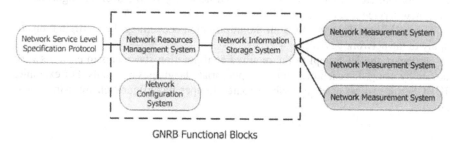

GNRB Functional Blocks

Figure 4. The GNRB architecture

Moreover, a *Network Service Level Specification Protocol* (NSLSP), on purpose defined, is used to exchange query/response SLSs messages between a consumer application and the GNRB or between two GNRBs. Both the format of the exchanged messages and their contents are specified by using XML-based schemes.

Finally, the GNRB interfaces with the *Network Measurement Systems* (NMS), which allows collecting data on nodes and links performance.

5.1 Network Resource Management System

The Network Resource Management System performs the following control functions:

1. analysis of the new service requests coming from Grid applications;

2. retrieval of the necessary information on the network status from the NISS database;
3. execution of a support decision algorithm to determine if the new service requests, with QoS constraints at network level, may be accepted, must be rejected or re-classified;
4. if necessary, network configuration operations by means of the NCS. Different actions are taken based on the requested type of Service Level Specification.

In general, the support decision algorithm module operates on weighted graphs, obtained taking into account the performance metrics required by the Grid application. The structure of the graph and the weight assigned to each branch are determined by measurements collected by ad-hoc probes as well as through the knowledge of the network topology. Such information are collected and stored in the NISS. At present, the Constraint Shortest Path First algorithm is implemented as decision algorithms.

5.2 Network Configuration System

This module allows to perform network reconfiguration operations (i.e. LSP set up and tear down, traffic control parameters setting, MPLS recovery techniques configuration, etc.) when they are considered necessary by the decision algorithm to meet new service requirements.

The NRMS sends to the NCS all the information concerning the configurations to apply to the network devices.

In particular, the NCS applies the required configuration statements to each network device by using, for each of them, the appropriate languages and tools. For example, JUNOScript can be used to perform router Juniper reconfiguration by using XML streams.

5.3 Network Information Storage System

The Network Information Storage System handles and stores measurement data. It consists of the following software components:

a) *data collection module:* allows asynchronous exchanges of information between the NISS and the measurement probes. Measurement data are typically associated with different refresh time intervals. For example, since link utilization is a quickly variable performance parameter, a small refresh time is required. On the other hand, the physical topology is a very slow varying parameter, so topology discovery operations are repeated within long time intervals.

b) *data organization module*: status information are organized hierarchically, in compliance with the *entity.characteristic.subcharacteristic* representation, defined in 10. In our GNRB implementation the storage database contains all the information concerning the current status of the network at path, link and node level.

5.4   Network Service Level Specification Protocol

The Network Service Level Specification Protocol is used by the GNRB and Grid applications manager to exchange query/response reservation messages, formatted in XML.

The following types of requests are supported:

- **Weighted Topology Discovery**: the best network paths, computed according to a predefined cost function and assuring the connectivity among the grid application nodes, are requested. In particular, when the GNRB receives a Weighted Topology Discovery query, which specifies the network metrics (e.g. bandwidth, latency, etc.) to minimize or maximize, the NRMS retrieves the information on the network resources status from the NISS database and determines the paths able to satisfy the query (Figure 5) by executing the support decision algorithm.

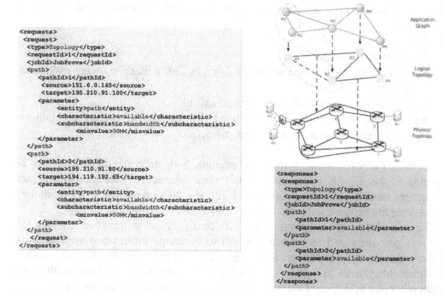

```
<requests>
 <request>
  <type>Topology</type>
  <requestId>1</requestId>
  <jobId>JobProva</jobId>
  <path>
     <pathId>1</pathId>
     <source>151.6.0.145</source>
     <target>195.210.91.100</target>
     <parameter>
        <entity>path</entity>
        <characteristic>available</characteristic>
        <subcharacteristic>bandwidth</subcharacteristic>
           <minvalue>30M</minvalue>
     </parameter>
  </path>
  <path>
     <pathId>2</pathId>
     <source>195.210.91.80</source>
     <target>194.119.192.65</target>
     <parameter>
        <entity>path</entity>
        <characteristic>available</characteristic>
        <subcharacteristic>bandwidth</subcharacteristic>
           <minvalue>50M</minvalue>
     </parameter>
  </path>
 </request>
</requests>
```

```
<responses>
 <response>
  <type>Topology</type>
  <requestId>1</requestId>
  <jobId>JobProva</jobId>
  <path>
     <pathId>1</pathId>
     <parameter>available</parameter>
  </path>
  <path>
     <pathId>2</pathId>
     <parameter>available</parameter>
  </path>
 </response>
</responses>
```

Figure 5. Weighted Topology Discovery Service: Query/Response messages

- **QoS support request**: guarantees, in terms of bandwidth, latency, delay jitter, packet loss, recovery times from links/nodes failures must be assured to the Grid application for its whole execution time. Therefore, it could be necessary to reserve network resources on-demand. The GNRB, by retrieving information on the network resources status from the NISS database, verifies if it is possible to satisfy the SLS requirements of the Grid application with the current configuration of the network and, in this case, network resources are reserved. On the contrary, if the request can not be satisfied, the GNRB searches for alternative network configurations, exploiting the knowledge of the physical topology and the availability of performance metrics of each link. If an alternative path is found, the GNRB configures the routers in order to set-up the path and reserve the needed network resources at path level, for the whole execution time of the Grid application.

- **Sorted Resource List**: a list of the sites belonging to the VO with similar storage or computing resources can be used to accomplish the mapping decision of the application modules. Since each application module is logically connected, at application graph level, to the other ones, a key to support the mapping decision can be sorting this list as regards the network resources that satisfy the execution of the job. The GNRB provides the Sorted Resource List request to solve this issue.

5.5 Network Measurement Systems

Different measurement systems may be deployed to collect the following information:
- Topology at Physical layer.
- Topology at IP layer:
- Path and Link metrics measurements (such as link utilization, delay maximum-bound, etc.)
- Performance evaluation of recovery mechanisms.
- Performance evaluation, at node level, of the traffic control components (classifier, scheduler, policer, dropper, shaper).

For these purposes, network status information, both static and dynamic, has to be reported to the GNRB by measurement probes. Moreover, network performance parameters must be evaluated and stored into the NISS database in a preliminary phase when the network infrastructure is set up

The knowledge of the physical and IP configuration information enables the topology discovery and requires a protocol to exchange information with network devices. Moreover, to deliver real-time metrics reports to the GNRB, ad-hoc measurement systems have to be deployed. For example, a network bandwidth probe might rely upon SNMP-enabled network devices, such as routers and switches.

At the end, network performance may be evaluated through experiments in the real network scenario and simulations carried out to achieve the bounds of the network performance metrics 11, 12 and to suggest the optimization strategies 13.

6 Functional Validation

The functional validation of the high performance network-aware grid environment, described in the previous sections, has been carried out in a Metropolitan High Performance Grid deployed in Pisa, Italy. In particular, a metropolitan fibre-optic ring, 22 Km long, interconnects three sites located in Pisa: the Department of

Information Engineering of the University of Pisa, the Institute of Science and Information Technologies (ISTI) of the Italian National Research Council (CNR) and the CNIT (National Inter-University Consortium for Telecommunications) National Laboratory of Photonic Networks. Each site is equipped with a Juniper M10 IP/MPLS router 15, PCs and clusters acting as grid nodes (Figure 6).

A DiffServ over MPLS network architecture has been set-up on this infrastructure, so that all the capabilities described in section 4 are available.

Moreover, we created a Virtual Organization to support a distributed encoding application, written by using ASSIST 14 and developed a first release of a GNRB prototype providing a *Guaranteed Bandwidth* SLS. The AdTech AX/4000 16 network traffic generator/analyzer was used to add background traffic into the optical fibre ring.

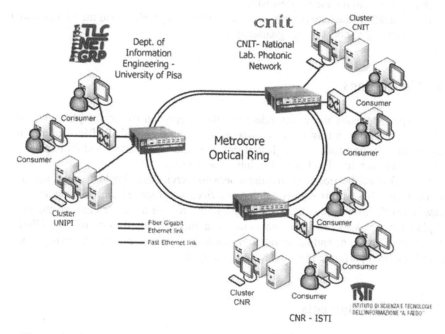

Figure 6. Metropolitan High Performance Grid Topology

The encoding application is based on stream communications among several worker modules and an emitter/collector module. Hence, when a new stream is required by the encoding application, a *Guaranteed Bandwidth* SLS request is sent to the GNRB through a QoS request message with bandwidth constraint.

The functional validation of the GNRB concerned the monitoring, support decision algorithm and configuration modules.

In order to validate the GNRB functionalities when the network utilization changes, two different scenarios were considered:
1. the network is unloaded;

2. a 950 Mbit/s background traffic must be delivered by using the Gigabit link between the Department of Information Engineering of the University of Pisa and the CNIT.

Moreover, in each network condition, the distributed encoding application requires, at run-time, to establish a 100 Mbit/s guaranteed path among each pair of the testbed sites.

When, in the first scenario, the GNRB receives the above request, it establishes three LSPs with 100 Mbit/s reserved bandwidth. Each LSP follows the shortest path (one Gigabit link) between two sites. In the second scenario, the GNRB monitoring module checks the link utilization. Hence, in order to guarantee 100 Mbit/s service rate between the University of Pisa and the CNIT, the GNRB establishes an LSPs which passes across the CNR Juniper M10 router, whereas the other LSPs remain unchanged and follow the shortest path.

In this manner, the network resources utilization is optimized and the SLS required by the encoding application are guaranteed.

6   Conclusions

This paper addresses the issues related to the integration of Grid applications and advanced network infrastructures offering QoS support and Traffic Engineering capabilities. In particular, an High Performance Grid environment and its interaction with network services has been taken into consideration.

Since Grid applications may require a network service offering higher performance levels than a best effort packet delivery service, we proposed the introduction of a new network entity, the GNRB, which plays the role of a broker between the Grid application requests and the network capabilities. The design of the GNRB architectural modules and its service functionalities has been presented. Finally, some tests have been performed to validate our prototype.

7   Acknowledgments

This work was supported by the Italian Ministry of Education and University (MIUR) under FIRB project "Enabling platforms for high-performance computational grids oriented to scalable virtual organizations (GRID.IT)".

References

1. Rosen, E., Viswanathan, A., Callon, R.: Multiprotocol Label Switching Architecture, RFC 3031, January 2001
2. Blake, S., Black, D., Carlson, M., Davies, E., Wang, Z., Weiss, W.: An Architecture for Differentiated Services, RFC 2475, December 1998
3. Le Faucheur, F., Wu, L., Davie, B., Davari, S., Vaananen, P., Krishnan, R., Cheval, P., Heinanen, J.: Multi-Protocol Label Switching (MPLS) Support of Differentiated Services, RFC 3270, May 2002
4. Foster, I., Kesselman, C., Tuecke S.: The Anatomy of the Grid: Enabling scalable virtual organization, International Journal of Supercomputer Application http://www.globus.org/research/papers/anatomy.pdf , 2001
5. Foster, I., Kesselman, C., Nick, J., Tuecke, S.: The physiology of the Grid: An open Grid services architecture for distributed systems integration, http://www.globus.org/research/papers/ogsa.pdf, June 2002.
6. Baker, M., Buyya, R., Laforenza D.: Grids and Grid technologies for wide-area distributed computing, Software-Practice and Experience Journal, 2002
7. Sander, V.: Network Issues for Grid Infrastructure, GFD-I.037, GGF Published Documents - Grid High Performance Networking Research Groups http://www.gridforum/documents/GWD-I-E/GFD-I.037.pdf , 22 November 2004
8. Clarke, P., Ferrari, T., Gaynor, M., Hasan, M.Z., Kunzst, P., Leigh, J.. Leese, M.J., Mealor, P., Monga, I., Sander, V.: Grid Network Services use Cases, GGF Meeting 13 http://www.gridforum.org/Meetings/GGF13/Documents/GGF13_Grid_Network_Services_Use_Cases_GHPN-RG.pdf , August 2004
9. Clapp, G., Ferrari, T., Hoang, D.B., Leese, M.J., Mealor, P., Travostino, F.: Grid Network Services, Draft-ggf-ghpn-netservices-1.pdf, August 2004
10. Lowekamp, B., Tierney, B., Cottrell, L., Hughes-Jones, R., Kielmann, T., Swany, M.: A Hierarchy of Network Performance Characteristics for Grid Application and Services, GFD-R.023, GGF Published Documents – Network Measurements Working Group http://www.gridforum/documents/GWD-R/GFD-R.023.pdf , 24 May 2004
11. Adami, D., Giordano, S., Maccaglia, R., Orlandini, F ., Repeti, M.: An experimental study on the EF-PHB service in a DiffServ High Speed Network, IEEE International Conference of Communications, 2004
12. Adami, D., Giordano, S., Repeti, M., Castoldi, P., Cugini, F., Valcarenghi, L.: Performance Evaluation of Recovery Techniques in a Grid Oriented Metrocore/VESPER Field Trial, Testbed Research Infrastructure for the Development of NeTworks and COMmunication, 2005
13. Adami, D., Giordano, S., Pagano, M., Secchi, R.: Optimization of Scheduling Algorithms Parameters in a DiffServ Environment, International Symposium on Applications and the Internet, 2005
14. Aldinucci, M., Coppola, M., Danelutto, M., Vanneschi, M., Zoccolo, C.: ASSIST as Research Framework for High-performance Grid Programming Environments, Technical Report, 16 February 2004
15. Juniper Home page: http://www.juniper.net
16. Spirent Home page: http://www.spirent.com/

A P2P Framework For Distributed And Cooperative Laboratories

Luca Caviglione[1], Luca Veltri[2]

[1]University of Genoa – Department of Communications, Computer and Systems Science (DIST), Via Opera Pia 13, 16145 Genova (Italy)
Phone: +39-010-3532202, Fax: +39-010-3532154
e-mail: barone@dist.unige.it

[2]University of Parma - Dipartimento di Ingegneria dell'Informazione
Parco Area delle Scienze 181/A, 43100 Parma (Italy)
Phone: +39-0521-905768, Fax: +39-0521-905758
e-mail: luca.veltri@unipr.it

Abstract. In this paper we propose a p2p framework in order to organize the cooperation and the set-up phase of interconnected laboratories for didactical purposes through the Internet. Even if the application of standard p2p algorithms to distributed cooperative laboratories is a quite new methodology, we will base our approach on a solid foundation, such as the one provided by the Session Initiation Protocol (SIP). The proposed architecture solves some general problems such as the discovering of entities, and also provides, a general framework for devices accounting. In addition, our framework offers mechanisms in order to provide an abstract namespace for adding – removing facilities on the fly and invoking them.

Keywords: p2p systems, SIP, Overlay Networks, Distributed Laboratories.

1 Introduction

Nowadays, many devices are connected through the Internet. As a matter of fact, there is an increasing trend in producing network-ready devices in order to make them remotely available. One of the oldest concepts that ignited the network technology is the one devoted to share and aggregate resources. In fact, the first network infrastructures were developed in order to share precious and expensive mainframes. With the awesome advancement of hardware technologies, this scenario is now also applicable to devices and instrumentations. The sharing of resources brings to cost reduction, and allows to set-up experiments previously unfeasible by creating chains of interconnected devices.

In another respect, also a topic under debate is how to integrate instrumentation with a Grid infrastructure. Actually, access to and control of instrumentation was fundamentally part of the original Grid vision, but it has been neglected in favor of middleware development. Recently, many projects are devoted to develop solutions to access instruments as well as reserving resource through the network [1,2,3,4].

The proposed work is aimed at introducing a framework allowing an effective cooperation among different instruments and devices in a networked environment. The basic assumption is that we will consider each laboratory as a peer and then we will build a p2p system that cooperates, in order to allow distributed experimentations with devices and resources located in different laboratories. For instance, concerning a measurement experiment, the device under test and the end user could be remotely located. In this perspective, the system is then distributed, because the resources used are not circumscribed to a local laboratory, and cooperative because different resources will be chained together in order to produce an experiment. This framework will allow to aggregate and superpose different instruments for building laboratory experiences or for providing instrument-oriented services as a Grid infrastructure.

The paper is structured as follows: Section 2 investigates the state of the art within IETF for using SIP [5] as the signaling technology to build and maintain p2p overlays, and also envisages some possible architectural blueprints, in order to exploit SIP for building a distributed and cooperative laboratory environment. Section 3 briefly analyzes the state-of-art distributed technologies and then proposes some ideas for integrating JXTA with SIP, in order to provide a "lightweight JXTA" that can be used to develop further services for distributed laboratories. Section 4 depicts how to use SIP to remotely interact with devices, and Section 5 concludes the paper.

2   On using SIP to develop p2p systems

In p2p networking all the hosts have the same capabilities and the same responsibilities. To emphasize the aspect, all the entities involved in this kind of network are named peers. However, the p2p paradigm introduces some problems: the topology is quite trivial, and the lack of a hierarchical organization brings to major difficulties in developing any kind of algorithms. There are no "well-known" nodes (in a client-server scenario, the server is the known service provider node) and the functionalities are shared across the network. This description perfectly fits for the unstructured p2p networks, such as Gnutella [6].

In this kind of network all peers interact by self-organizing themselves in order to form a network. Studies have shown that this kind of systems follow a "power-law" rule, where few nodes have a high number of connections and many nodes have a small number of connections. The main problem with this organization is due to the complete delocalization of the information and, in order to locate, content a controlled broadcast mechanism is mandatory. To cope with this, many proposals have been made in order to achieve better performance in the look-up phase both for peers and resources. The root technology is the one based on Distributed Hash Tables (DHTs), allowing to organize the overlay by mapping the peers via a hashing function. This brings to structured p2p systems, such as CHORD [7] and Kademlia [8]. On the other hand, the current IP Telephony service has been built as a phone-to-phone (peer-to-peer) system with very centric characteristics, based on server/proxy nodes (in both H.323 and SIP architectures). Of course this approach may suffer from scalability problems, and some proposals to overcome this issue and to develop p2p Internet Telephony are arising. Particularly, in [9] the authors present an architecture based on SIP. The basic idea is to view the overall Internet telephony infrastructure as a p2p architecture, where all the participants are organized in a p2p overlay network. Consequently, the overlay could be exploited to locate users for initiating a telephone call. This architectural blueprint resembles Skype, but tries to overcome some of its limitations. In fact, Skype has been developed in a closed way, it does not appear as extensible for future services and it still relies on a central server for authentication procedures. Besides, IETF has long investigated protocols and techniques for deploying Internet telephony. The most appealing candidate is the SIP architecture, and with some adjustments it could be organized as a p2p system.

In another respect, there are also some security issues that could be solved by using SIP in a p2p fashion. For instance, the architecture depicted in Figure 1 reduces the potential of failure due to a single node, resulting in increased robustness, and allows providing a preliminary protection against Denial-of-Service (DOS) attacks. This is possible, owing to the intrinsic resistance of DHT-based systems against churn, which is the continuous process of node arrival and departure.

3 Applying p2p principles on Distributed – Cooperative Laboratories

During the years, many different architectural blueprints and protocols have been developed for p2p systems, especially for file-sharing. Each developer promoted his brainchild and implemented the required protocol functionalities: from the Napster protocol to sophisticated mechanisms such as eMule, they are basically built from scratch. This fact has also produced a waste of resources, since some protocols are overlapped in functionalities. Moreover, the lack of a coordinated and standardized approach results in a variety of p2p systems that cannot interact.

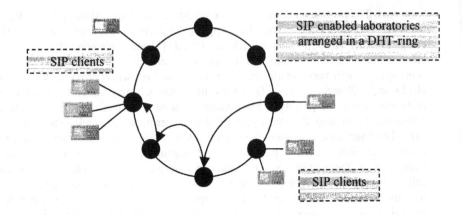

Fig. 1. SIP-based Laboratories arranged in a DHT overlay: each laboratory exports device(s).

In this perspective, even if not explicitly written for p2p, SIP could be used for different signalling purposes and, of course, also for building and managing p2p overlay systems. In addition, SIP can straightforwardly handle p2p traffic. This to demonstrate that albeit a big standardization campaign in p2p has not been undertaken, it must not be said that p2p architectures or supporting technologies are not available in existing technological pools.

For instance "*all servers in SIP are optional, allowing User Agents (UAs) to directly communicate*" [10]. Its server-less capabilities allow SIP to be one of the fundamental building blocks for p2p technologies.

To prove the maturity of the SIP technology to carry out all the signal procedures mandatory for exploiting a full featured p2p application, reference [11] tries to find out the required SIP messages in order to build a full featured Chord-based architecture. Particularly, the draft provides examples for the following operations: node registration, user registration, session establishment, node leaving the system, successful user search and unsuccessful user search. This work looks promising allowing SIP to become a killer application, raising its popularity and, at the same time, exploiting the p2p paradigm to enhance the SIP technology itself. Also [12] proposes some ideas in order to fully exploit SIP as a p2p technology.

3.1 Integrating JXTA with SIP to build services on top of distributed laboratories

Currently there is an important open project named JXTA [13], originally sponsored by SUN Microsystems, that aims to define a standard architecture and its protocols for building up p2p applications and services. JXTA comes also with a reference open-source Java implementation of its complete suite of protocols, in order to shorten the time required to develop new p2p applications. JXTA defines a series of XML-based protocols for specific p2p basic functions, such as those for self-organizing peers to into *peergroups*, publishing and discovering peer resources, communicating, and monitoring other peers.

One of the main characteristics of JXTA is that it is independent of any particular hardware platform, operating system, or programming languages, and does not require the use of any particular network transport or topology; the JXTA protocols can be implemented on top of TCP/IP, HTTP, Bluetooth, or other protocols and technologies.

The approach followed by JXTA designers was to completely re-define a new network architecture and protocols, with poor reuse of already existing ones, and this led on one hand to some duplication of network protocols and functionality, and on the other hand to a scarce interoperability with already existing p2p-like services, such as real-time multimedia communications (also known as "IP Telephony"), building up with SIP. JXTA has been used for a variety of purposes, especially for communicating, conferencing and information sharing. JXTA could also be used to provide the remote laboratories a way to communicate, in order to:

allow laboratory staff and researchers to carry out lectures or guided seminars and to help students in completing experiment and exercises;

allow students to interact, and then, to cooperate remotely as if they populated a distributed classroom;

allow using application layer techniques, such as "Application Layer Multicast", in order to exploit p2p techniques for multimedia transmission. In addition, JXTA could be used to offer virtual immersive functionalities for the interaction among involved parties as they participate in a synchronous e-learning session.

We then describe an architecture based on JXTA/SIP that tries to integrate the JXTA architecture with the SIP platform and functionality. The basic idea is to place the JXTA architecture on top of SIP, replacing all JXTA protocols and services that are already provided by the SIP architecture. The SIP infrastructure is already present, since it is a fundamental building block of our architecture. The mix of SIP and JXTA will provide a simple and efficient mechanism, allowing communication among different laboratories-devices. It will also provide an "architecture independent" layer, in order to build services on top of the networked laboratory facility. The JXTA specification originally defines six main protocols, which we recall below:

- The Peer Resolver Protocol (PRP) is the mechanism by which a peer can send a query to one or more peers, and receive responses to the query.

- The Peer Discovery Protocol (PDP) is the mechanism by which a peer can advertise its own resources, and discover the resources from other peers. Every peer resource is described and published using an advertisement.

- The Peer Information Protocol (PIP) is the mechanism by means of which a peer may obtain status information about other peers (such as state, uptime, traffic load, capabilities, and other information).

- The Pipe Binding Protocol (PBP) is the mechanism by which a peer can establish a virtual communication channel or "pipe" between one or more peers. The PBP is used by a peer to bind two or more end points of the connection.

- The Endpoint Routing Protocol (ERP) is the mechanism by which a peer can discover a route (sequence of hops) used to send a message to another peer.

- The Rendezvous Protocol (RVP) is the mechanism by which peers can subscribe or be a subscriber to a propagation service. The RVP is used by the Peer Resolver Protocol and by the Pipe Binding Protocol in order to propagate messages.

These protocols relay on a common transport layer named "Endpoint Service", which is responsible of end-to-end messaging between two JXTA peers, using one of the underlying transport protocols, such as the JXTA TCP, TLS, or HTTP.

The Endpoint Service, together with the PRP, ERP, and PBP protocols are the building blocks that a JXTA application uses to communicate with other peers. They act as signalling and messaging infrastructure allowing communications between remote peers based on peer IDs. However SIP protocol also provides a signalling platform for communicating and establishing sessions between end–systems/peers (referred as UAs). For this reason, the integration of the two technologies in the JXTA/SIP architecture seems to be very promising, providing twofold benefit: i) reuse of well-known and tested signalling platform without replicating functionality, and ii) integration with existing multimedia/IPTel infrastructure and devices.

Particularly, in the proposed JXTA/SIP architecture the upper JXTA protocols PDP and PIP, and the resolver functionality of PRP and RVP are mapped directly upon the SIP protocol and its signalling platform. In such way, all the transport and messaging infrastructure offered by the Endpoint Service, by PRP, ERP, and PBP is no longer required and is replaced by SIP.

The SIP protocol provides all addressing, messaging and routing functionality, reusing different types of underly transport protocols (e.g. UDP, TCP, TLS, and SCTP). The addressing scheme of SIP allows the unique identification of UAs and SIP users. However, in a p2p environment, it can be also useful to be able to generically address single resources, or services. For this reason JXTA/SIP extends the traditional SIP URI scheme, by providing a simple and flexible mechanism to address any kind of resource owned by or associated to a generic SIP user. The SIP URI scheme

$$\text{sip:}[user@]nameaddress[:port][;uri\text{-}parameters]$$

is still used, and the new parameter resource is introduced, obtaining the following resource address sip:*user@nameaddress*;resource=*resource-name*

where the value of the parameter resource (i.e. the *resouce-name*) is the peer URN (Uniform Resource Name), used in JXTA to identify peer entities. SIP provides JXTA also the method for delivering XML-based JXTA messages, encapsulated within the body filed of SIP messages, with Content-Type application/xml-jxta.

SIP methods used to deliver JXTA messages can vary according to the implemented upper layer service. However, the most common SIP methods used with JXTA are: MESSAGE, SUBSCRIBE and NOTIFY. Here is an example of SIP message used for sending a JXTA Discovery Query:

```
MESSAGE SIP:rdzv@zoolu.org;resource=jxta://abdcf23 SIP/2.0
Via: SIP/2.0/TCP 192.168.5.2;branch=z9hG4bK776sgdkse
Max-Forwards: 70
From: SIP:peer@neverland.net;resource=jxta://678ffr;tag=49583
To: SIP: rdzv@ zoolu.org;resource=jxta://abdcf23
Call-ID: asd88asd77a@1.2.3.4
CSeq: 1 MESSAGE
Content-Type: application/xml-jxta
Content-Length: 340
   Resolver query
     Discovery query
```

Upon SIP, there is the PRP/RVP layer that comprises the various JXTA resolver and propagation related protocols, such as PRP and RVP.

PRP is a simple query/response protocol that provides a simple addressing resolution mechanism for upper layer protocols. It can be used in point-to-point, multicast, and propagation mode. It uses the SIP MESSAGE methods for message delivery.

Although SIP provides direct peer addressing and routing, message propagation within a peer group is also required in a p2p scenario. For this scope the RVP (Rendezvous Protocol) is used within the PRP/RVP layer. The RVP provides mechanisms which enable propagation of messages to be performed in a controlled way. To most efficiently propagate messages some peers acts as Rendezvous Peer cooperating with other Rendezvous Peers and with client peers to propagate messages amongst the peers of a peer group. The PRP/RVP layer will propagate a message unless one of the following conditions is detected:

i) Loop detected: if a propagated message has already been processed on a peer, it is discarded;

ii) Time To Live (TTL) exceeded: propagated messages are associated with a TTL that is decreased each time a propagated message is received by a peer; when the TTL of a message drops to zero, the message is discarded;

iii) Message duplication: a peer discards copies of an already received message.

Upon the PRP/RVP layer there are the JXTA upper level protocols, mainly the PDP (Peer Discovery Protocol) and the PIP (Peer Information Protocol). PDP is used to discover any published peer resource. Resources are represented as advertisements.

A resource can be a peer, a peer-group, a pipe, a module, or any resource that has an advertisement. Each resource must be represented by an advertisement. The PDP enables a peer to find advertisements in its group. Custom discovery services may choose to leverage PDP. Once a peer is located, its capabilities and status may be queried. PIP provides a set of messages to obtain a peer status information. PIP is an optional protocol. Peers are not required to respond to PIP requests.

4   Direct device-to-device communications

In the previous section we showed how it is possible to build a complete p2p architecture based on SIP. Such architecture in turn can be used to realize a distributed, self-organized network of laboratories and end-devices. In this section we describe how SIP can provide also methods for direct device-to-device communications in the form of simple client-server commands or more sophisticated communication schemes. The SIP protocol has been originally defined as a mechanism for establishing sessions between two or more peers. However, the SIP architecture is very flexible and can be easily extended for building a wide spectrum of services. For example, SIP can be used to establish one-to may or many-to-many communication, to realize a presence service, as well to implement an instant messaging system.

In our scenario some mechanisms are required in order to implement direct device-to-device communications. For this goal, two main mechanisms are needed:

1) an addressing scheme able to address specific laboratories, instruments, their components and functions;

2) a set of basic methods that can be used to handle the direct interaction among these end-systems.

Regarding the first aspect, in the previous section we presented an extension of the basic SIP URI scheme that can provide the required addressing functionality to cover a distributed laboratory environment. Particularly, when the SIP URI refers to a laboratory or an instrument, the 'resource' parameter can be used to address a specific sub-component or functionality. Here is an example of how the URL of an oscilloscope within a laboratory can be

```
<sip:radio-lab@campus.net;resource=oscilloscope>
```

Concerning the second aspect, either standard SIP methods or new ad-hoc methods could be used.

SIP natively defines a number of methods for initiating a session (INVITE, ACK, CANCEL, BYE), for querying for user-agent/server capabilities (OPTIONS), and for registering with a location service (REGISTER). In addition, SIP provides means to facilitate protocol extension through the introduction of new methods and the definition of new header fields. As matter of fact, currently a number of methods have been already defined, addressing several different functions. In our scenario, two kinds of basic interactions between parties are required:

i)   a direct client/server method that can be used to issue commands or to send out-of-band data,
ii) a method for receiving asynchronous information and for being notified of specific events by a peer entity.

As far as the first kind of communication is concerned, the MESSAGE method can be used. The MESSAGE method has been defined as SIP extension for Instant Messaging (IM) [14]. This method does not establish a session and can be used to transfer single instant messages (like pager messages) carried in the form of MIME body parts within the message. In our context, the MESSAGE method can be used to send single commands or data to a peer device. The form of the transferred information is application dependent; however a common XML representation could be used. For example, here is the message that can be sent to switch on a device:

```
MESSAGE sip: tlc-lab5@campus.net;resource=generator SIP/2.0
    Via: SIP/2.0/UDP 10.10.0.5;branch=z9hG4bKdf334
    Max-Forwards: 70
    To: <tlc-lab5@campus.net;resource=generator>
    From: Alice <sip:alice@wonderland.net>;tag=1166ccae
    Call-ID: asd88asd77a@10.10.0.5
    CSeq: 1 MESSAGE
    Content-Type: application/xml
    Content-Length: ...

    <command>TURN ON</command>
```

In order to handle the second kind of communications, in which a device asynchronously receives requested data from a remote peer when a certain event occurs, the SIP event notification framework can be used [15]. The event notification framework has been properly defined as SIP extension to provide means by which SIP nodes can request notification from remote nodes, indicating that certain events have occurred. This framework uses two main methods: SUBSCRIBE and NOTIFY. Particularly, the SUBSCRIBE method should be used to request current state and state updates from a remote node, while NOTIFY messages are sent to inform subscribers of changes in state to which the subscriber has a subscription. Here is an example of how these methods can be applied for receiving asynchronous reports from a protocol analyzer in a context of distributed laboratories:

```
SUBSCRIBE sip:net-lab2@pcampus.net;resource=proto-analyzer SIP/2.0
Via: SIP/2.0/UDP 10.10.0.5;branch=z9hG4bK85c2f
Max-Forwards: 70
To: <net-lab2@pcampus.net;resource=proto-analyzer>
From: Alice <sip:alice@wonderland.net>;tag=3758f51b
Call-ID: k3143id034@10.10.0.5
CSeq: 22 SUBSCRIBE
Contact: <sip:alice@wonderland.net>
Event: trace
Content-Type: application/xml
Content-Length: ...

<filter>host 192.168.1.88 and udp and port 4444</filter>
```

```
NOTIFY sip:alice@wonderland.net SIP/2.0
Via: SIP/2.0/UDP 192.168.0.2;branch=z9hG4bKb1454
Max-Forwards: 70
To: Alice <sip:alice@wonderland.net>;tag=3758f51b
From: <net-lab2@pcampus.net;resource=proto-analyzer>;tag=763287060
Call-ID: k3143id034@10.10.0.5
CSeq: 34321 NOTIFY
Event: trace
Content-Type: application/xml
Content-Length: ...

<packet>
...
</packet>
```

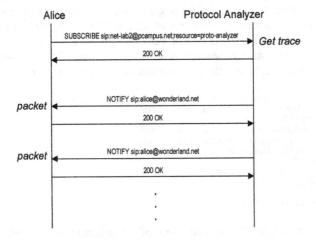

Fig. 2. Example of asynchronous event notification.

Lastly, the devices must be exported by using a proper "manager" if needed. For instance, many instruments could be managed by a standard computer without third party hardware or software. Besides, other instruments may be connected to a SIP manager running in a personal computer via a GPIB/Ethernet adapter. The rule of thumb is that the instrument data must be properly converted in order to assure the correct communication from/to the instrument.

5   Conclusions and Future Work

In this paper we proposed a p2p-based system in order to provide a framework for distributed and cooperative laboratories. We used SIP as the technological base for effectively building such a framework and, in addition, we also presented how to use JXTA to enrich the framework with valuable features. Lastly, we presented some instances of the use SIP to effectively interact with devices, providing an idea for a simple and cost-effective interoperation of instrumentation that avoids more complex systems such as one present in the Grid world. Future work will be aimed at refining this architecture and at using the SIP protocol also to issue QoS requirements to the network, in order to carry out industrial and mission critical experimentations.

References

1. L. Benetazzo, M. Bertocco, F. Ferraris, A. Ferrero, C. Offelli, M. Parvis, V. Piuri, "A Web based, distributed virtual educational laboratory", IEEE Trans. on Instrum. and Meas., vol.49, No. 2, April 2000, pp.349-356.
2. A. Bagnasco, M. Chirico, A.M. Scapolla, "XML technologies to design didactical distributed measurement laboratories", Proc. of 19th IEEE IMTC, Anchorage, AK, USA, vol.1, 2002, pp. 651-655.
3. P. Daponte, C. De Capua, A. Liccardo, "A technique for remote management of instrumentation based on web services", Proc. of IMEKO-TC4 13th Internat. Symposium on Measurement for Research and Industry Applications, Athens, Greece, 2004, pp. 687-692.
4. The Grid Enabled Remote Instrumentation with Distributed Control and Computation (GRIDCC) Project, http://www.gridcc.org.
5. Rosenberg, J., Schulzrinne, H., Camarillo, G., Johnston, A., Peterson, J., Sparks, R., Handley, M. and E. Schooler, "SIP: Session Initiation Protocol", IETF RFC 3261, June 2002.
6. Gnutella Protocol Specification, http://www.gnutella.com.
7. Stoica, I., Morris, R., Karger, D., Kaashoek, M. and H. Balakrishnan, "Chord: A Scalable Peer-to-peer Lookup Service for Internet Applications", SIGCOMM '01, August 2001.
8. Kademlia: A Peer-to-peer Information System Based on the XOR Metric, http://kademlia.scs.cs.nyu.edu/.
9. The p2p-SIP project at Columbia University, http://www1.cs.columbia.edu/~kns10/research/p2p-sip/
10. A. Johnston, "SIP, P2P, and Internet Communications", IETF Internet Draft <draft-johnston-sipping-p2p-ipcom-00>, January 2005.
11. D. Bryan, C. Jennings, "A P2P Approach to SIP Registration", IETF <draft-bryan-sipping-p2p-00>, January 2005.
12. L. Caviglione, L. Veltri, "Using SIP as P2P Technology", Invited Paper at 12th International Conference on Telecommunications (ICT 2005), Capetown, South Africa, 3-6 May 2005.
13. "JXTA v2.0 Protocols Specification", http://www.jxta.org
14. B. Campbell, J. Rosenberg, H. Schulzrinne, C. Huitema, "Session Initiation Protocol (SIP) Extension for Instant Messaging", IETF RFC 3428, December 2002.
15. A. B. Roach, "Session Initiation Protocol (SIP)-Specific Event Notification", IETF RFC 3265, June 2002.

this architecture and of using the SIP protocol also to issue QoS requirements to the network, in order to carry out in a real and [illegible] [illegible] experimentations.

References

1. [illegible] Bendazzo, M. Bertocco, [illegible] Ferraris, [illegible] Offelli, [illegible] Parvis, V. Piuri, "A Web-based distributed virtual educational laboratory", IEEE Trans. on Instrum. and Meas., vol. 49, No. 2, April 2000, pp. 349-356
2. A. [illegible], [illegible] Chirico, A.M. [illegible], "XML technologies to design didactical distributed measurement laboratories", [illegible] of IMTC 2004, [illegible] [illegible], USA, vol. 1, [illegible] pp. 651-655
3. [illegible] Grimaldi, [illegible] De Capua, [illegible], "A technique [illegible] management of instrumentation based on web services", Proc. of IMEKO [illegible] Symposium on Measurement for Research and Industry Applications, Greece, [illegible]
4. The Grid-Enabled Remote Instrumentation with Distributed Control and Computation (GRIDCC) Project: http://www.gridcc.org
5. Rosenberg, J., Schulzrinne, H., Camarillo, G., Johnston, A., Peterson, J., Sparks, R., Handley, M., and E. Schooler, "SIP: Session Initiation Protocol", IETF RFC 3261, June 2002
6. Gnutella Protocol Specification, http://www.gnutella.com
7. Stoica, I., Morris, R., Karger, D., Kaashoek, M. and Balakrishnan, H., "Chord: A Scalable Peer-to-peer Lookup Service for Internet Applications", Proc. SIGCOMM, August 2001
8. Sandonis, A. [illegible] Information System Based on the XSR [illegible] [illegible]
9. [illegible] p2p SIP [illegible] [illegible] University [illegible]
10. A. Johnston, "SIP, P2P, and Internet Communications", IETF draft [illegible]
11. D. Bryan, [illegible], "A P2P Approach to SIP Registration [illegible]", [illegible]
12. [illegible], "[illegible] SIP as P2P Technology", Invited Paper, [illegible] International Conference on Telecommunications (ICT 2003), [illegible], South Africa, [illegible] May 2003
13. JXTA v2.0 Protocols Specification, http://www.jxta.org
14. G. Camarillo, [illegible], H. Schulzrinne, C. Huitema, "Session Initiation Protocol (SIP) Extension for Instant Messaging", IETF RFC 3428, December 2002
15. A. B. Roach, "Session Initiation Protocol (SIP)-Specific Event Notification", IETF RFC 3265, June 2002

Exposing Measurement Instruments As Grid Services

A. Bagnasco, A. Poggi, G. Parodi, A. M. Scapolla

Department of Biophysical and Electronic Engineering, University of Genoa,
Via Opera Pia 11a, 16145 Genova, ITALY
{bagnasco, apoggi, parodi, scapolla}@dibe.unige.it

Abstract. This paper presents the rationale and the experience in exposing measurement instruments on the Grid. After explaining the design goals, we propose two implementation strategies targeted to access the instruments at different levels of abstraction. A performance evaluation of the realized services concludes the paper.

Keywords. Grid, remote laboratory, instruments.

1 Introduction

The Grid infrastructure allows users to access and to aggregate sophisticated resources, such as distributed computational power, large data collections and analysis codes, in a flexible, secure, and coordinated environment. So far, most of the projects taking advantage of the Grid technology focus on the distributed management and analysis of datasets after acquisition, without worrying about the instruments, which are the providers of the raw observations, used later to develop, refute, and verify scientific theories, and thus to drive scientific advances.

Now the technology for computational and data Grids has matured, and an increasing attention is directed towards instruments. The scientific research does not start from data, but rather from the process of data collection, which is not a mere repetitive procedure, and often requires strict links with interpretation and analysis, jointly to man or machine supervision. Ignoring this interaction can lead to inefficient use of computational and human resources. For these reasons, recent projects are now investigating the possibility of exposing instruments directly as Grid services.

Examples are the GRIDCC [1], the CIMA [2] and the CRIMSON [3] projects. GRIDCC, a three years project just started within the VI Framework Program, is the main European effort in this direction. It aims to integrate the interactive and real-time management of remote instrumentation (from temperature sensors to array of telescopes), storage and computational resources, and real-time data analysis and visualization services. An important part of the development plan deals with the design and the implementation of Virtual Instrument Grid Services (VIGS).

The CIMA project targets the development of a common middleware to facilitate the use of a wide range of instrument types into a Grid computing environment, exposing them as Grid services. A project goal is the description of instruments through metadata and the development of a standard methodology for grid-enabling instruments.

The recently approved CRIMSON project is currently evaluating the role played by computational grids in the interconnection of real and virtual devices. It aims at searching efficient solutions to the problems of process characterization, instrument calibration, and real-time data analysis to improve the quality of the information acquired from the field.

In the following we present our work in *GRIDifying* instruments. We introduce the approach, which we have followed in exposing diverse kind of instrument services on the Grid, and present some preliminary results in terms of performance.

2   How to Expose Instruments on the Grid

Even though the need of exposing measurement instruments on the Grid is largely recognized by the projects cited in the previous section, no precise example, describing how to do this, is still available in literature. In this section, we report our experience in turning different components of a measurement node into Grid services using the Globus Toolkit version 3 (GT3) [4].

2.1   Design Goals

A generic process of measurement requires set-up, tuning and access to diverse instruments, and these actions must be carried out in a well defined sequence. If we assume that the instruments are distributed on the Grid, the measurement program must access them remotely by efficient and secure mechanisms for remote procedure call. To improve the software reuse, and to facilitate the creation of experiments based on distributed instrumentation, it is important to have loosely coupled components. Distributed technologies already existing, such as CORBA, RMI, and EJB, result in highly coupled distributed systems, where the client and the server are very dependent on each other. These technologies works well in intranet contexts, but tend to have problems on a wider scale. For instance, they have a lot of troubles in passing firewalls, and clients require a deep knowledge about how servers work. This impinges the flexibility in assembling measurement programs. On the contrary, in a loosely coupled model, the client and the server are independent and the client can connect to and use functions from a server without the specific knowledge of the server itself. This is the model implemented by the Web Services [5].

The service-oriented approach, applied to instrumentation control [6,7], lets to encapsulate the real instruments behind general interfaces, and allows creating measurement programs as composition of services. Furthermore, the measurement program can become itself a service, offering high-level functionality. We can imagine scenarios where we select devices scattered over the Grid to compose a measurement process, or we access remote experiments without taking care of the real location of instruments and devices under test.

Grid services, which are the fulcrum of the GT3, inherit the advantages of the Web Services, and improve their possibilities adding features as lifecycle management, service data and notification, that are very promising in a distributed measurement context.

Here, we focus on a measurement node, composed by a personal computer equipped with an IEEE488 interface card that connects electronic measurement instruments. The goal is to expose the functionality offered by this setup in the form of Grid services. We start analysing which type of services can be exposed:

- the IEEE488 card communication services;
- the services offered by instruments for signal generation, data acquisition and processing;
- the services offered by the applications running on the PC. For instance, we can figure to execute an FFT on the data waveform acquired by an oscilloscope connected via IEEE488, and to expose the results as a "spectrum analyzer" service.

As the Grid philosophy is centred on optimizing the use of distributed resources, it is reasonable to limit the processing inside the acquisition node, taking advantage from other resources available on the Grid. Data can flow from the instrument service to reduction and analysis services hosted elsewhere on the Grid. Data can be archived on some medium, together with relevant metadata that can help successive automatic elaborations. When bandwidth or real-time constraints exist, it may be convenient to process data within the acquisition service. In general, less is the data processing offered by the instrument service, more are the opportunities to reuse the service itself.

In the following, we present two service prototypes, the former exposes the IEEE488 communication services, and the latter exposes the basic functionalities of an instrument, precisely an oscilloscope.

2.2 Exposing an IEEE488 Interface Card

We have a National Instruments (NI) PCI-GPIB card and we want to send SCPI (Standard Commands for Programmable Instrumentation) commands directly to the instrument connected via the card, passing through the Grid infrastructure. This is a very general approach, as the Grid service offers only the control of the communication channel, and we do not put any conditions on the instrument itself. We suppose that the application calling the service knows which kind of instrument receives the commands.

We decided to develop the service in the Linux/Java environment, because at the time we started the work, the GT3 was tested mainly on these platforms. As NI releases both the Linux driver and the Windows one for this card, we can leverage the multiplatform nature of Java to have the two versions of the service.

The card driver is a C library, thus we need to use the Java Native Interface (JNI) to wrap them in the Java code (see Fig. 1).

Fig. 2 shows the UML (Unified Modelling Language) class diagram of the CardService implementation.

The CardService exposes two main methods, which allow read and write operations on the instruments connected to the card. These methods are based on the *ibrd()* and *ibwrt()* driver APIs (Application Program Interface), and are used respectively to read and to write bytes on the interface board.

To test the CardService, we have created a simple panel, which calls the service and allows writing commands and reading answers from the instrument in the form of text strings. The service has no knowledge about the instrument itself, thus the ability to generate SCPI (Standard Commands for Programmable Instrumentation) commands for the specific instrument, task usually accomplished by the "instrument driver", resides on the client side, making it device-dependent.

In general this is not a good approach, and in the next example, we move the instrument driver within the grid service.

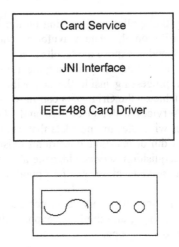

Fig. 1. The CardService stack

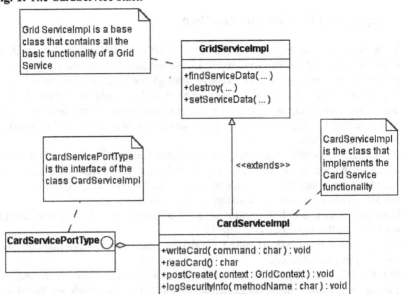

Fig. 2. CardService UML Class Diagram

2.3   Exposing an Instrument

This section extends the previous results and introduces the instrument grid service. We focus here on a grid service able to expose the functionalities of an oscilloscope, but this approach can be applied to whichever instrument.

Fig. 3, if compared with Fig. 1, presents an additional layer, named "C driver". Usually, instrument drivers are provided by vendors as a C libraries or source codes; JNI allows interfacing the C driver.

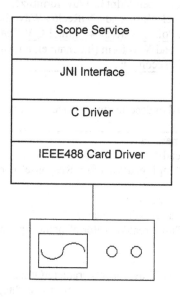

Fig. 3. Instrument Grid Service Stack

The service exposes methods for the main settings of a generic oscilloscope. Some of them are listed in Table 1.

Table 1. The Oscilloscope Grid Service Methods

Method	description
public void configureChannel (int channel, double range, double offset, int coupling, int probeAttenuation)	it configures one of the channels for acquisition
public void configureAcquisitionRecord (double timePerRecord, int minNumPts, double acquisitionStartTime)	it configures the acquisition record
public Object[] readWaveform (int channel, int maxTimeMilliseconds)	it acquires the waveform

To avoid binding the ScopeService to a specific model of oscilloscope, we work on the generic scope class stated in the IVI (Interchangeable Virtual Instruments) specifications [8]. These specifications are written for C language, thus we must convert C prototypes in Java. Table 2 reports the C and the Java prototypes for the Read Waveform method.

Table 2. C and Java Prototypes of the Read Waveform Method

C	ViStatus IviScope_ReadWaveform (ViSession vi, ViConstString Channel, ViInt32 WaveformSize, ViInt32 MaxTimeMilliseconds, ViReal64 WaveformArray[], ViInt32 *ActualPoints, ViReal64 *InitialX, ViReal64 *XIncrement);
Java	public Object[] readWaveform (int channel, int waveformSize, int maxTimeMilliseconds);

A chunk of the file used to describe the service is in Fig. 4.

```
<!-- define the response message -->
<wsdl:message name="readWaveformResponse">
    <wsdl:part element="impl:readWaveformResponse" name="parameters"/>
</wsdl:message>
<!-- define the request message -->
<wsdl:message name="readWaveformRequest">
    <wsdl:part element="impl:readWaveform" name="parameters"/>
  </wsdl:message>
<!-- define the operation -->
<wsdl:operation name="readWaveform" parameterOrder="">
      <wsdl:input                              message="impl:readWaveformRequest"
name="readWaveformRequest"/>
      <wsdl:output                             message="impl:readWaveformResponse"
name="readWaveformResponse"/>
</wsdl:operation>
```

Fig. 4. WSDL for the Oscilloscope Grid Service

An applet, reproducing the instrument virtual panel, allows testing the oscilloscope service. It has been inserted in a Grid portal, developed on the GridSphere Portal Framework [9]. ISILab-G is the portal service tab that runs the applet, displays the instrument interface, and offers a short description of the service functionalities, i.e. how to use buttons and knobs. Fig. 5 shows a snapshot of the oscilloscope panel.

The Grid portal controls user access and rights, and offer services that can be used separately or integrated to create more sophisticated processes.

The oscilloscope applet contains the necessary code, jar files, to communicate in a secure way with the Grid. The node, which runs the applet, becomes part of the Grid and establishes a private connection with the oscilloscope service.

3 Performance Evaluation

While exploring the possibilities offered by Grid technologies, it is important to consider the Grid middleware effect on the measurement process performance. An exhaustive evaluation is outside the scope of this paper; anyway, we report some results collected using the oscilloscope grid service. We have considered the "observed response time" (ORB) and the "request processing time" (RPT) [10]. The ORB is the delay, from the user point of view, between issuing a request and receiving a response, while the RPT is the time spent by the server to handle the user request. Thus ORB is calculated on the client side, while RPT on server side.

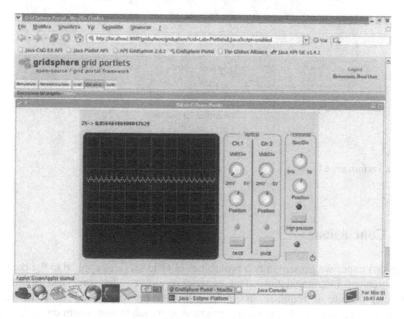

Fig. 5. Snapshot of the Oscilloscope Panel

The test bed consists of a server (1,70 GHz Pentium IV CPU and 512 MB RAM.) that hosts a plug-in PCI GPIB card from NI connected to an HP 54645D oscilloscope. It runs the Linux operating system (kernel 2.4.20-8) and the Globus Toolkit version 3.2.1. On the client side we have a PC (Linux kernel 2.4.20-8) equipped with a 2,2 GHz Pentium IV and 512 MB RAM. Both client and server are connected to the same 100 Mbit LAN.

The client invokes the oscilloscope Grid service and queries the waveform that the instrument is acquiring. The test has been repeated with 10 waveforms of different sizes, starting from 100 points to up to 4000 points. Each request has been repeated up to 1000 times with a delay of 5 seconds from the previous one. To compare the results, we calculated the ORB and RPT using the same test-bed, and a DCOM-based communication between server and client in a Windows environment. Fig. 6 shows the performance results for the scenarios described above.

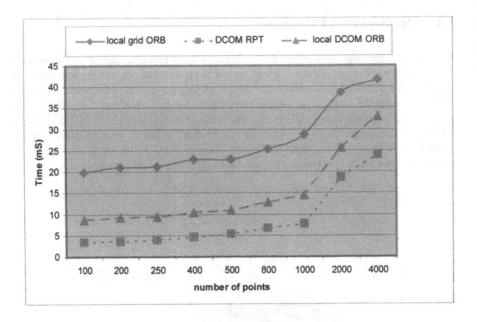

Fig. 6. Performance

4   Conclusions

In this paper we presented two prototypes of Grid services applied to the control of measurement instrumentation. The experience proved that the Grid approach is feasible, despite it requires an initial effort to be acquainted with the Globus environment and to apply the service oriented approach to instrument control.

The advantages associated to the Grid technology should justify this effort and suggest new investigations.

The results of the performance tests, which have been executed on the preliminary releases of the prototype services, show a time overhead caused by the grid middleware. This is not a surprise, but it is worth noting that the loss in efficiency becomes less important when the size of the exchanged message increases. The capability to deliver information and measurement results in real time may be a crucial feature for distributed measurement systems. Surely Grid protocols add latency, but the overhead can be considered acceptable for a lot of applications and it is irrelevant when the measurement loop is closed by a human being.

Acknowledgment

The Authors would like to thank Alberto Boccardo and Paolo Buschiazzo for their precious contribution to the software development.

References

1. http://www.gridcc.org
2. Randall Bramley, Kenneth Chiu, John C. Huffman, Kia Huffman, and Donald F. McMullen, Instruments and Sensors as Network Services: Making Instruments First Class Members of the Grid, Indiana University CS Department Technical Report 588, December 2003.
3. CRIMSON (Cooperative Remote Interconnected Measurement Systems Over Networks) – Cofin2004 project
4. http://www.globus.org/
5. http://www.w3.org/2002/ws/
6. B. Szabados, P. Hodaie, "Hybrid Architecture For Remote Instrumentation Application", IEEE Instrumentation and Measurement Technologies Conference, Anchorage AK, USA, 21-23 May 2002
7. H. Saliah-Hassane et al., "A general framework for Web Services and Grid-Based Technologies for Online Laboratories", Exploring Innovation on Education and Research, tainan, Taiwan, 1-5 March 2005.
8. http://www.ivifoundation.org
9. http://www.gridsphere.org
10. X. Zhang, J.M. Schopf, "Performance Analysis of the Globus Toolkit Monitoring and Discovery Service, MDS2", in Proceedings of the International Workshop on Middleware Performance, MP2004

Distributed Information Retrieval In GRID Environment: A Formal Approach

Igor Bisio[1], Mario Marchese[1], Maurizio Mongelli[1], Annamaria Raviola[2]

[1] DIST – Department of Communication Computer and System Science,
University of Genoa, Via Opera Pia 13 - 16145 Genoa, Italy
{Igor.Bisio, Mario.Marchese, Maurizio.Mongelli}@unige.it

[2] Selenia Communications S.p.A., Via Pieragostini 80 - 16151 Genoa, Italy
{Annamaria.Raviola}@seleniacomms.com

Abstract. A telecommunication network is composed of a number of nodes. Each node can be either a terminal host or a web cache location. Objects are downloaded throughout the network among the nodes. An object can represent either a file or any other resource to be shared (e.g. machine time). For memory saving and for safety reasons, each object is composed of a number of portions and not all the portions are located within the same node, because no node should have the complete knowledge of each object. It means that a single node can have only a part of the file. It is strongly recommendable because if a single node should be accessed without authorization, the information retrieved would not be sufficient to detect the overall content.

Concerning networking viewpoint the following main issues will characterize the performance of object exchange: Position of the information, Strategy to reach the information, Algorithm to download information, Capacity Planning.

The paper proposes a control architecture that considers the mentioned issues. It is composed of three layers: Local, Network and Planning Controller. The Local Controller acts locally to each node at object downloading time scale (seconds/minutes) and, after an "advanced flooding" signaling query to get informed about the object portions' position, decides from which node (or nodes) each portion needs to be downloaded; the Network Controller may change the distribution of the object portions among the nodes; it acts with larger time scale (hours/day) and is centralized; the Planning Controller may change the dimension of each single portion and increase/decrease the physical link and node capacities. The order of magnitude of its intervention is weeks/months.

The problem is modeled through a mathematical formulation and a specific cost function, which takes into account all the necessary details, is introduced for each controller as well as a minimization procedure.

A preliminary performance evaluation analyses the effect of the Local Controller.

Keywords. GRID, Distributed Information, Hierarchical Network Control, Resource Sharing

1   Introduction

The problem of sharing information among different locations has been widely treated both in the literature and in practical implementations. The common action is that a user issues a request (to the server, to peer elements, to the network) and the destination site returns an answer for each request. The sites may be stand-alone servers, single gateways with similar functions and, in some cases, terminal hosts.

The most common technique to access remotely located information is the client-server model, where there is a clear distinction between the consumer and the producer of services: the client asks the server to access resources (i.e. files, machine times, distributed applications) and the server processes the workload assuring, if possible, the required resources. More recently, to improve the performance of the regular client-server approach, Content Distribution Networks (CDNs) are used [1]. CDNs delocalize information and functions of interest among the main server site, which contains all the information and functions, and different surrogate server sites, where the material is duplicated. Surrogate servers are used to find new paths to retrieve the information so avoiding possibly congested routes. A third approach is represented by peer-to-peer (P2P) overlay networks, where virtual networks of many nodes, called overlays, are built over networking infrastructures. The P2P network key point is that the hosts, considered peers, are allowed to behave as clients and servers. In practice, each host may be a server and information is exchanged among peers to provide web content and to alleviate traffic burden. If a particular file is located into two remote peers, part of this file may be downloaded from one peer and part from the other peer. The advantages of peer-to-peer communication are: scalability, knowledge sharing by aggregating information, information availability. Peer-to-peer systems are used to support several network-based applications: combining the computational power of thousands of computers [2], sharing of resources [3], distributed and decentralized searching [4]. File sharing networks [5] are perhaps the most commonly used P2P applications and, at present, compose most of Internet traffic. The widespread use of such networks can be attributed to their ease of use: [6, 7], for example, have proposed the adoption of P2P as supplementary means for providing web content in order to alleviate the traffic burden on servers.

Summarizing: in client-server applications, the server is the only repository of information. It may be negative because accessing the server can create bottlenecks, both in links and machine processing, but it allows companies and entities to store important information and to make it available to others on payment. CDNs follow the same philosophy. Information multiplication allows alleviating traffic burden but increases the maintenance costs. P2P approach has many advantages but it is not safe for important information. Peer-to-peer overlay networks are the best for music clips but strategic information, bank operations, trading details may be hardly delivered by using this paradigm. Moreover, strategic networks involve many security issues. A broader framework that include all the mentioned approached and refers to information exchange in wide sense is "grid computing" (see, e.g., [8]), which allows delivering distributed contents, storing information, performing remote use of time machine and sharing files over networking infrastructures [1]. This paper takes "grid computing" networks as reference and tries to propose a generic formal approach that can take the best from the three mentioned approaches: client-server, CDN and P2P.

The idea is to have a "grid computing" network composed of N sites where each site contains important information. Each node can be considered a web cache location, a surrogate server and a peer host but, actually, it is part of an overall overlay network, which is itself repository of all needed information. In facts: there are a number of objects that are downloaded throughout the network; an object is a file or any other resource to be shared as well as machine time and distributed application. Being a possible application a strategic network, for memory saving and for safety reasons, each object is composed of portions and not all of them are located within the same node. It means that a single node has only a part of the file. It is strongly recommendable (even if it is not mathematically imposed in the proposed model for now) because, if a single node should be accessed without authorization, information retrieved would not be sufficient to detect the overall content, which needs to be composed in combination with the other nodes of the network.

The paper proposes a control architecture made up of three layers: Local, Network and Planning Controller. The former controls object downloading; the Network Controller checks the distribution of the object portions among the nodes; the latter changes the dimension of each single portion and increase/decrease the physical link and node capacities. The overall control structure is proposed through a mathematical formal model that considers the following performance optimization issues: capacity planning, position of the information, strategy to reach the information, algorithm to download it. At the best of authors' knowledge, it is the first time an overall formal model for an information exchange network (a "grid computing" structure) is proposed as well as a control architecture matching optimization and security issues together.

The reminder of the paper is structured as follows: the next section contains the description of the control architecture. The formal model for the performance optimization is reported in Section 3. Section 4 presents a preliminary performance evaluation and Section 5 shows the conclusions and possible ideas for future work.

2   Control Architecture

2.1   Preliminary Observations

The proposed approach takes the works [9, 10, 11, 12] as references concerning: the model of users' requests and of available resources [10]; the action of variable tuning dependently upon congestion measures [9, 12]; the focus on a single node making object requests [11]. The novelty of this work relies in the derivation of an optimization framework studied to get the minimization of the downloading time "seen" by each independent user. The proposed approach is decentralized and scalable, since the control laws available for both the single user and the network manager are analyzed separately.

2.2   General Framework

The considered "grid computing" network is composed of N nodes. Each node can be either a terminal host or a web cache location. I objects should be downloaded throughout the network among the N nodes. An object can represent either a file or any other resource to be shared (e.g. machine time).

For memory saving and for safety reasons, each object i is formed by $J^{(i)}$ portions and not all the portions are located within the same node. It means that a single node can have only a part of the file. It is strongly recommendable (even if it is not imposed for now) because, if a single node is accessed without authorization, information retrieved is not sufficient to detect the overall content.

Each single node k requires to download a specific object i and asks the other nodes of the network about the availability of object i portions through a specific signaling protocol. The nodes that have portions of the object i within their memory answer to node k. Node k requires to download specific portions of the object i to other nodes.

The multiple choice about which portions have to be downloaded from which nodes is performed by minimizing a cost function (local to the node) aimed at optimizing downloading time. Node k may download the same portion either from "one single node" or from "more than one node" depending on network security performance.

The performance of the overall system in terms of downloading time may be monitored also in dependence of the number of requests for a specific object by a single node. After a period of time, evaluating the obtained performance, the following actions may be taken at different time scales: portions of objects may be exchanged among the nodes; link capacities may be modified. While the former may be implemented on a reduced time period whose order of magnitude may be hours/day, the latter should be applied on planning basis.

The control architecture is shown in Fig. 1. It is composed of three layers: Local, Network and Planning Controller. The Local Controller acts locally to each node at object downloading time scale (seconds/minutes); the Network Controller may change the distribution of the object portions among the nodes; it acts with larger time scale (hours/day) and is centralized; the Planning Controller may change the dimension of each single portion and increase/decrease the physical link and node capacity. The order of magnitude of its intervention is weeks/months.

The requests are structured into X service options, characterized by a committed downloading rate ($dr^{(x)}$), where $x \in [1, X] \subset N$ is the downloading option identifier. The traffic is structured into Y classes, characterized by a specific bandwidth assigned periodically by the Network Controller for each end-to-end path. The variable $ab_y^{(hk)}$ defines the bandwidth assigned to y-th traffic class with $y = 1, ..., Y$ being Y the number of available traffic classes for the end-to-end path from node h to k. Traffic is managed through "Complete Separation with Dynamic Partitions" [13]. It means that end-to-end path allocations (called bandwidth pipes) are not shared among the traffic classes and they may be changed at each intervention of the

Network Controller. Allocations are supposed constant within the interval between two Network Controller interventions. The technology to define end-to-end bandwidth pipes for each traffic class is not specified but it may be taken from the QoS mechanisms described in the literature. The bandwidth pipe of a specific class may be obtained through ATM Virtual Paths, MPLS, IntServ and, most probably, through DiffServ.

In dependence of their priority a set of service options will be conveyed through a specific traffic class. For example, if there are: six service options and their associated downloading rate ($X = 6$, $dr^{(1)} = 1024$ *kbit/s* , $dr^{(2)} = 512$ *kbit/s* , $dr^{(3)} = 256$ *kbit/s* , $dr^{(4)} = 128$ *kbit/s* , $dr^{(5)} = 64$ *kbit/s* , $dr^{(6)} = 32$ *kbit/s*) and three traffic classes ($X = 3$) with associated allocations each end-to-end path (e.g., $b_1^{(hk)} = 50$ *Mbit/s* , $b_2^{(hk)} = 30$ *Mbit/s* and $b_3^{(hk)} = 20$ *Mbit/s, $\forall hk$*), service options 1 and 2 can be transported through class 1 bandwidth pipe, service options 3 and 4 through class 2 bandwidth pipe and service option 5 and 6 through class 3 bandwidth pipe.

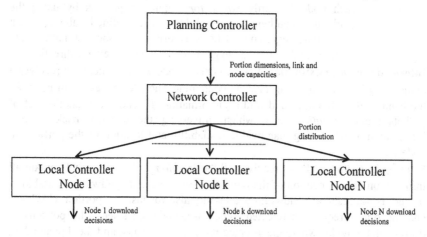

Fig. 1. Control Architecture

2.3 Operative Details

The implementation of the control structure is based on a signaling mechanism, used when a file request is issued by a node to reveal the position of the different portions composing the desired file. The most popular search mechanism (in use in P2P networks) blindly floods a query to the network. To avoid useless multiplication of signals, this paper uses an "advanced flooding" algorithm, where, differently from "blind flooding", a node does not forward all the requests already received but performs this operation having some knowledge about the "cost" (measured in terms of residual bandwidth) of the paths traversed. The signaling scheme is topical to get good performance. Future work will be dedicated to its study but, for now, it is out of the scope of this proposal. The steps of the object request algorithm are briefly summarized in the following including also some observation about CAC and QoS routing schemes.

1. **File Request:** when a file request is issued by generic node k, *exploratory* signaling packets are generated by node k and forwarded to each node of the network through the advanced flooding scheme.

2. **Exploratory Phase:** *Exploratory* signaling packets:

   a.   traverse the network node by node;

   b.   check the bandwidth availability of each link along the back route (or, more precisely, of the proper bandwidth pipe of each link); the minimum bandwidth availability (b_{min}^{hk}) over the path from h to k "defines" the cost of the path as $\left(\dfrac{1}{b_{min}^{hk}} \right)$;

   c.   each node forwards *exploratory* signaling packets by using the minimum bandwidth availability cost defined above; each *exploratory* signaling packet memorizes the Shortest Path Route from h to k, followed to get to node h in the reverse direction.

3. **Intermediate and Destination Nodes:** generic node h of the network receives a number *exploratory* signaling packets containing: the file request from node k, the bandwidth availability and routing information. It sends back another set of packets (called *location_info*), which, traversing the network back, reports information about routing, bandwidth availability and location of the portions to node k.

4. **Routing Decision:** node k: receives all the *location_info* packets that contain all information about: location of the portions, best route and bandwidth availability; if there is not residual bandwidth availability also for just one portion, CAC acts and the object request is rejected, otherwise, node k: selects the portions to download, the nodes where the selected portions are located and the downloading rates (i.e. either the committed ones, if possible, or assigning the residual bandwidth on the path) by minimizing the cost function proposed in the next section within the Local Controller. The node k selects the best route to get to the selected locations and forwards a *resource_confirmation* packet, which reserves the resources over the selected shortest path, informs the nodes about portions to download and rates and authorizes file downloading. If the resources should not be available any longer over the path, CAC acts again and the file request is blocked.

3   Performance Optimization Model

The following definition should help formalize the problem. Reported definitions are aimed at focusing on the main content of the paper (file downloading and distribution), so service option and traffic class indexes are neglected for the sake of clarity. The formal description, in practice, assumes just one service option defined by its downloading rate and just one traffic class, whose bandwidth is the same of the physical link capacity. The same assumption is kept in the performance analysis.

3.1 Local Controller

Definitions:

I, number of objects to share;

i, object identifier $1 \le i \le I$, $i \in \mathbb{N}$;

N, number of nodes of the network;

L^i, dimension of the $i-th$ object;

J^i, number of portions that compose the $i-th$ object;

D^{ij}, dimension of the $j-th$ portion of the $i-th$ object;

R^{ki}, number of requests from node k regarding $i-th$ object (traffic matrix);

C^{lm}, physical capacity of link *(lm)*;

b^{lm}, residual bandwidth available over link *(lm)*;

pd^{lm}, propagation delay over link *(lm)*;

dr, committed downloaded rate;

Path(hk), sequence of links (defined as couple of nodes) composing the path from h to k;

$C_{hk}^{min} = \min\limits_{(lm) \in Path(hk)} C^{lm}$, physical capacity bottleneck for *Path(hk)*;

$b_{min}^{hk} = \min\limits_{(lm) \in Path(hk)} b^{lm}$, minimum residual bandwidth available for *Path(hk)*;

$\tau^{hk} = \sum\limits_{(lm) \in Path(hk)} pd^{lm}$, propagation delay for *Path(hk)*;

M^k, storage capacity of node k;

$\phi_k^{ij} = \begin{cases} 1, \text{ if } j\text{-th portion of } i\text{-th object is present at node } k \\ 0, \text{ otherwise} \end{cases}$;

$\varphi^i = \begin{pmatrix} \phi_1^{i1} & \cdots & \phi_1^{iJ^i} \\ \vdots & \ddots & \vdots \\ \phi_N^{i1} & \cdots & \phi_N^{iJ^i} \end{pmatrix}$, distribution matrix of $i-th$ object;

$A_k^i = \begin{pmatrix} A_{1k}^{i1} & \cdots & A_{1k}^{iJ^i} \\ \vdots & \ddots & \vdots \\ A_{Nk}^{i1} & \cdots & A_{Nk}^{iJ^i} \end{pmatrix}$, matrix defining the decisions of node k concerning the

portions of $i-th$ object;

$$A_{hk}^{ij} = \begin{cases} 0, & \text{if } \phi_k^{ij} = 0 \\ \{0,1\}, & \text{otherwise} \end{cases}$$

The estimation of the capacity b_{min}^{hk} is topical to get an estimation of the downloading time of each single portion. In practice, b_{min}^{hk} is verified through the signaling protocol at the beginning of the operation. Being the approach followed within a bandwidth reservation framework and knowing the downloading rate that is considered fixed for the overall downloading operation, when the signaling packets flow through the network, they can verify exactly which is the residual bandwidth on each link and take the minimum value. The approach might be used also in best effort networks with self-regulating TCP connections. In this case signaling should perform a measure of the average bandwidth still available. Details of that should include the format of signaling packets, the presence of time stamps and any other information that can help estimate the bandwidth available. The topic is very interesting and it will be the object of future research.

Being:

$$f^{hk} = \begin{cases} dr, & \text{if } b_{min}^{hk} \geq dr \\ b_{min}^{hk}, & \text{otherwise} \end{cases} \tag{1}$$

the bandwidth that can be assigned to the file download request on the *Path(hk)*.

The contribution of node h to downloading time of the $i-th$ object "seen" by node k is defined by the function $T_{hk}^i(A_k^i)$ in (2).

$$T_{hk}^i(A_k^i) = \sum_{j=1}^{J^i} \left[\frac{D^{ij}}{f^{hk}} \cdot A_{hk}^{ij} \right] + \tau^{hk} \cdot \Psi(A_k^i) \tag{2}$$

where

$$\Psi(A_k^i, h) = \begin{cases} 1, & \text{if } \sum_{j=1}^{J^i} A_{hk}^{ij} \geq 1 \\ 0, & \text{otherwise} \end{cases} \tag{3}$$

It means there is at least one portion of the $i-th$ object downloaded from node h.

The downloading time of the $i-th$ object "seen" by node k is defined as in (4).

$$T_k^i(A_k^i) = \max_h \left[T_{1k}^i(A_k^i), ..., T_{hk}^i(A_k^i), ..., T_{Nk}^i(A_k^i) \right] \tag{4}$$

Node k takes decisions about which portions to download from where by minimizing $T_k^i(A_k^i)$ under the variable A_k^i. In other words, it defines the matrix A_k^i, which minimizes the cost $T_k^i(A_k^i)$ under the constraint in (5).

$$\sum_{h=1}^{N} A_{hk}^{(ij)} \geq \Lambda, \forall j \tag{5}$$

where Λ defines the redundancy, i.e. the minimum number of nodes from which each portion needs to be downloaded. If $\Lambda = 1$ (as done in the performance analysis), it means that each portion of the $i-th$ object must be downloaded at least from one node. The Local Controller acts at "object request" time scale.

3.2 Network Controller

Supposing a time interval of consecutive network operations, denoted by C, it is feasible to have an overall centralized network optimization, acting over C, that "decides" both file portion exchanges on the basis of the performance obtained in the period C and the new bandwidth portions for traffic class. It important to give some details about portion exchange in this context. Defining:

$P^{k,ji}(C)$, number of requests from node k regarding $j-th$ portion of $i-th$ object within the observation interval C (traffic matrix in the period C);

$dl_r^{hk,ji}$, identifier of the $r-th$ download from node h to node k regarding $j-th$ portion of $i-th$ object within the observation interval C;

$\beta^{hk}(d_r^{hk,ji})$, bandwidth allocated by the Local Controller (or average capacity measured in case of best effort) for $r-th$ download from node h to node k regarding $j-th$ portion of $i-th$ object within the observation interval C; in case of best effort network, the quantity is the ratio between the portion dimension and the really measured downloading time.

The average bandwidth availability from node h to node k for the observation interval C is:

$$\bar{\beta}^{hk} = \frac{1}{I}\sum_{i=1}^{I}\frac{1}{J^i}\sum_{j=1}^{J^i}\frac{1}{P^{k,ji}(C)}\sum_{r=1}^{P^{k,ji}(C)}\beta^{hk}(d_r^{hk,ji}) \tag{6}$$

Similarly as the previous case, the average downloading time for node k and object i can be obtained from (7) and (8):

$$\bar{T}_{hk}^{i}(A_k^i(\varphi^i)) = \sum_{j=1}^{J^i}\left[\frac{D^{ij}}{\bar{\beta}^{hk}}\cdot A_{hk}^{ij}(\phi_k^{ij})\right] + \tau^{hk}\cdot\Psi(A_k^i) \tag{7}$$

$$\bar{T}_k^i(A_k^i(\varphi^i)) = \max_{h}\left[\bar{T}_{1k}^i(A_k^i(\varphi^i)),...,\bar{T}_{hk}^i(A_k^i(\varphi^i)),...,\bar{T}_{Nk}^i(A_k^i(\varphi^i))\right] \tag{8}$$

The average downloading time for the overall network may be written as:

$$\overline{T} = \sum_{i=1}^{I}\sum_{k=1}^{N}\overline{T}_k^i(A_k^i(\varphi^i)) \tag{9}$$

under the constraints (5) and (10), which states the limitation on the memorization capacity of each node.

$$\sum_{i=1}^{I}\sum_{j=1}^{j^i}D^{ij}\cdot\phi_k^{ij} \le M^k, \ \forall k, \ 1\le k\le N \tag{10}$$

If the Network Controller acts also on the bandwidth pipes, a possible choice is to assign, for each *Path(hk)*, a bandwidth that considers both the number of object requests and the number of rejected calls. The computation is similar to (6) but also not accepted connections should be included as well as the constraint. If $\overline{\xi}^{hk}$ is the computed bandwidth value under the physical constraint $\overline{\xi}^{hk} \le C_{hk}^{min}$, the object portion relocation can proceed as envisaged above from (7) on, but substituting $\overline{\beta}^{hk}$ with $\overline{\xi}^{hk}$.

3.3   Planning Controller

A complete optimization of the network may be reached at Planning Layer where also the dimensions of each portion may be modified. In this case the evaluation period Q should be much larger than at network Control Layer. Q may have the scale of weeks or months. The cost function to be used is similar as in (9) but also the variables D^{ij} are object of the optimization process under the additional constraint $\sum_{j=1}^{j^i}D^{ij} = L^i$. To complete the analysis, it is also possible to extend the optimization process at the channel and capacity acting on the constraints C_{hk}^{min} and $M^{(k)}$.

Concerning the performance evaluation, the authors will focus on the Local Control leaving the Network Controller and the Planning Controller to future work.

4   Performance Evaluation

The performance evaluation is limited to check the behavior of the Local Controller. In practice, the aim is showing the performance of the object download decision process. The considered overlay network, which implements the advanced flooding signaling to get information about the position of the object portions, is reported in Fig. 2.

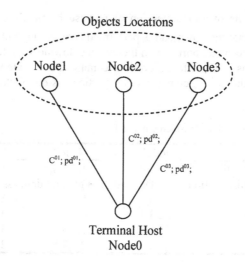

Fig. 2. Overlay Network Considered

Only Node0 is considered as terminal host. The objects are located in Nodes from 1 to 3. The following data have been used in the tests:

- number of objects to share: $I = 1$;
- number of nodes in the network $N = 4$;
- dimension of the object: $L^1 = 5$ *Mbit*;
- number of portions that compose the object: $J^1 = 5$;
- dimension of the object portions: $D^{1j} = 1$ *Mbit* $\forall j \in [1,...,J^1]$;
- number of requests from node 0 regarding object 1 (traffic matrix): $R^{01} = 1$.

The physical capacity of links *(10)* and *(30)*, equivalent to the physical capacity bottleneck of *Path(10)* and *Path(30)*, is $C_{10} = C_{30} = (1 + Bandwidth\ Increase)$ *Mbit/s*, where the parameter "*Bandwidth Increase*" is varied in the tests performed. The physical capacity of link *(20)* is set to $C_{20} = 1$ *Mbit/s*.

The capacities of the links are always considered fully available and coincident with the residual bandwidth available over the network links. The committed downloading rate is supposed to be equal to the overall bandwidth available over the paths. The propagation delay over each link is fixed and equal to *10 ms*.

Three local control strategies are compared in the tests: the algorithm proposed in Section III and called "Opt" in the following figures; a "Blind" method, where all the portions of the object, if present, are downloaded from all the nodes; and a "Heuristic" method where downloading of portions, when present, is performed from the node reporting the largest value of the expression:

$$\left[D_h \left(C^{hk}_{min} \right)^{-1} + \tau^{hk} \right]^{-1} \tag{10}$$

Where D_h is the overall dimension of requested object's portions available at node h. It means that the download is performed considering "intelligent" distinction of the impact of the single portions on the average downloading time. In practice, the decision is taken as if the different portions contained in a node were just one single unit. In the tests, the object, composed of 5 portions, is distributed as reported in the following table:

Table 1. Object Distribution Considered in the Tests

Type	Node1	Node2	Node3
Full Distribution	1,2,3,4,5	1,2,3,4,5	1,2,3,4,5
Random Distribution	Defined through a random assignment		
Case 1	1,2,3,4	5	5
Case 2	1,2,3,4	4,5	4,5
Case 3	1,2,3,4	3,4,5	3,4,5
Case 4	1,2,3,4	2,3,4,5	2,3,4,5

All the figures show the Downloading Time (in seconds) versus the Bandwidth Increase parameter, acting over links *(10)* and *(30)*, expressed in *Mbit/s*.

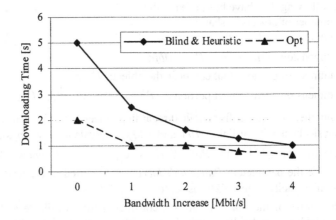

Fig. 3. Downloading Time versus Bandwidth Increase [Full Distribution case]

The results of the Full Distribution case (Fig. 3) highlight the main advantage of the performance optimization model proposed in this paper. The procedure defined allows the simultaneous download of different portions of the object without duplications and, in particular, uses the fragmentation to optimize the downloading time. Actually, the Downloading Time versus the Bandwidth Increase, when "Opt" is used, is much lower than the other two cases, which offer exactly the same performance in this case. Also in the Random Distribution case (Fig. 4) the "Opt" solution provides the lowest Downloading Time.

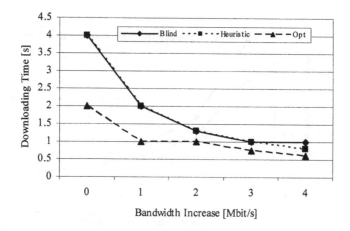

Fig. 4. Downloading Time versus Bandwidth Increase [Random Distribution case]

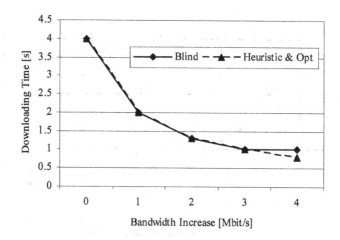

Fig. 5. Downloading Time versus Bandwidth Increase [Case 1]

In the Case 1 (Fig. 5), the Downloading Time is practically the same for all the mechanisms, even if, when the "Bandwidth Increase" is high, "Heuristic" and "Opt" (undistinguished in this case) have better performance than "Blind" method. This result is due to the distribution of the portions (see Table 1 – Case 1) that are concentrated in node 1, so removing the advantage introduced by the new method. Actually, there is only one choice and it is followed by all the schemes except for "Blind". When the information is more distributed through the network all the features of "Opt" are used and it offers a very satisfying performance (Figures 6 and 7).

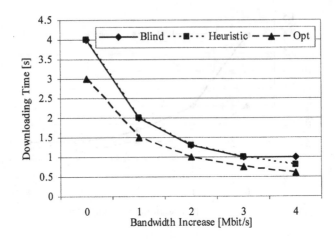

Fig. 6. Downloading Time versus Bandwidth Increase [Case 2]

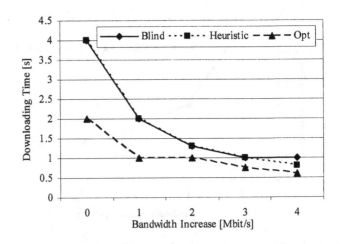

Fig. 7. Downloading Time versus Bandwidth Increase [Case 3 and Case 4]

5   Conclusions

The paper has proposed a control architecture composed of three layers within the framework of "grid computing" networks: Local, Network and Planning Controller.

The Local Controller acts locally to each node and decides which portions to download from which nodes. It is based on an "advanced flooding" signaling algorithm that transmits information about the object portions' position and the bandwidth availability along the paths. The problem is modeled through a mathematical formulation. A specific cost function, which takes into account all the necessary details, is proposed for each controller, as well as a minimization procedure. The performance evaluation has highlighted the basic mechanisms used by the Local Controller and has given a first idea about the advantages and drawbacks of the proposal. Network Controller and Planning Controller may change the dimension of each single portion and increase/decrease the physical links and node capacities. Its detailed description is put off for future research.

References

1. Verma, D. C.: Content Distribution Networks, An Engineering Approach. John Wiley & Sons, Inc., New York (2002)
2. SETI@Home, URL: http://setiathome.ssl.berkeley.edu
3. Freenet, URL: http://freenetproject.org/
4. NeuroGrid P2P Search, URL: http://www.neurogrid.net/
5. Parameswaran, M., Susarla, A., Whinston, A.B.: P2P Networking: An Information Sharing Alternative. IEEE Computer (2001) 31-38
6. Stavrou, A., Rubenstein, D., Sahu, S.: A Lightweight, Robust P2P System to Handle Flash Crowds. IEEE Journal on Selected Areas in Communications (2004) 6-17
7. Stading, T., Maniatis, P., Baker, M.: Peer-to-Peer Caching Schemes to Address Flash Crowds. Proceedings of IPTPS'02, Cambridge, MA, March 2002, 203-213
8. Ripeanu, M., Foster, I., Iamnitchi A.: Mapping the Gnutella Network: Properties of Large-scale Peer-to-Peer Systems and Implications for System Design. IEEE Internet Computing (2002) 50-57
9. Ng, T. S. E., Chu, Y., Rao, S. G., Sripanidkulchai K., Zhang H.: Measurement-Based Optimization Techniques for Bandwidth-Demanding Peer-to-Peer Systems. Proceedings of IEEE INFOCOM 2003, San Francisco, CA, March 2003, 2199-2209
10. Ge, Z., Figueiredo, D. R., Jaiswal, S., Kurose, J., Towsley, D.: Modeling Peer-Peer File Sharing Systems. Proceedings of IEEE INFOCOM 2003, San Francisco, CA, March 2003, 2188-2198
11. Yang, X., de Veciana, G.: Service Capacity of Peer-to-Peer Networks. Proceedings of IEEE INFOCOM 2004, Honk Kong, March 2004, 2242-2255
12. Kalogeraki, V., Chen, F.: Managing Distributed Objects in Peer-to-Peer systems. IEEE Network (2004) 22-29
13. Ross, K.: Multiservice Loss Models for Broadband Telecommunication Networks. Springer Verlag, Berlin (1995)

Chapter V

Architectures And Techniques
For Tele-Measurements

Chapter V

Architecture And Techniques
Performance Measurement

Architectures For Remote Measurement

M.Bertocco

Padova University, Dept.Information Engineering (DEI)
Via Gradenigo 6, 35151, Padova, Italy
matteo.bertocco@unipd.it

(Invited Paper)

Abstract. This discussion paper present issues in the design of a networking measurement system. Key requisites are summarized at first. Then, architectures will be analyzed in order to let a designer aware of advantages and efforts of presented choices.

Keywords: Remote and Distributed Measurement, Distributed Laboratories, Instrumentation and Measurement.

1 Introduction

There are many reasons beyond the increasing interest in remote measurement. In some cases the main reason is just *a need*. Key applications comes for instance from the field of telecommunications, where a transmitter, a receiver and a measuring instrument should be controlled in such a way that useful information is obtained. Usually the three above items are far located each other, hence some mechanism is required in order to let a test engineer to accomplish the task. To quote a less trivial example, when quality of service of a network is under investigation, a common procedure consists in setting up many test points at different locations. Measurements are continuously collected and summarized into a few meaningful parameters that are stored in a data base for further investigation. In this latter case again, there is the need for an architecture that enable an easy set up of remote test points that should be controlled from a centralized location, i.e. where the data is collected and analyzed.
Environmental monitoring is another interesting field where distributed measurements are not an option. In this case a rather large number of sensors is distributed in a geographical area. Procedures should be set up in order to interact with the sensors both for setting sensor parameters, and for collecting data coming from the sensors themselves. At a higher degree of abstraction such data is used to extract some more meaningful information. For example one may detect alarm conditions on a given region from the geological point of view. The majority of a town may understand wether a traffic restriction action provided significative reduction in the mean pollution level, and so on. From time in the

control of industrial plants dedicated networks are used, now with the emerging trend of using ethernet also for fieldbus operation a new opportunity emerges of considering from a unique point of view "classical" instrumentation and in-the-field sensors or actuators. Although not necessary, applications can benefit from such unified view, where services kept far apart can now share resources.

There are other meaningful cases where remote measurement system may add a significative value.

A first one consists in the cooperative usage of expensive instrumentation or devices. A full-compliance certified lab for electromagnetic compatibility testing easily has a value greater than 1,000,000 $: some of the expensive resources of such a lab, such as the anechoic chambers with built-in motorized antennas, cameras, analyzers and waveform generators could be shared in a networking environment between test engineers and design engineers of a given product speeding up the whole design process of electronic devices.

In any case, in an integrated factory environment, with the usage of remote measurement procedures a better interaction is possible between design engineers and test-engineers, where remote usage of testing facilities enables people a closer interaction, while gaining a quicker developement of test procedures. In fact, software code developed for testing purposes can be more easily shared providing a greater efficiency.

Another case where remote access to electronic instrumentation and devices provides new challenges is in the field of continuous education. Electronic systems are becoming complex, expensive and quickly changes, so that a single school or university many not have sufficient resources to keep pace with technology update. Moreover the "students" (university or college students as well as graduated continuos-education attendee) often use such complex resources for a limited amount of time. Once that a student has gained a sufficient skill level, instrumentation is no longer necessary. Hence the costs are payed back only if a large number of people could share in a time-scheduled way physical resources. This means that specialized test lab can be set-up each offering a limited number of state-of-the-art instruments and experiments via a remote access policy. This last example shows that remote measurement architectures could probably receive a considerable improvement from grid computing technology. However, this latter issue and discussion of the various services that an e-learning organizazion should offer to students and teachers is beyond the scope of this paper.

In any of the above examples, remote measurement architectures are needed. Despite the strong difference between the above applications, some interesting common needs can be stated, and hence some general architectures can be summarized in order to giving networking capabilities to a measurement application. In this paper, an overview of the main requirements and off-the-shelf solution that a designer should consider are recalled. Than, a brief overview of solutions for networking measurement applications is presented: from the instrument up to the whole system.

The aim of this discussion paper is to help the pratictioner to understanding the challenges beyond a networking measurement procedure, and hence to obtain guidance in efficient design of the final application or in a clever selection of an already-available one.

2 System Specifications

Many vendors now claim to be selling network capable instruments, that the development of a distributed measurement system may seem a solved problem. However, some critical pre-requisites should be kept in mind when choosing an measurement system from some third-part system integrator or when developing a new one from scratch.

A first critical requirement is *security*. In the context of networking measurement systems security does not mean just prevention against hackers attacks, but the guarantee that unauthorized use of devices should be forbidden. Otherwise, non-trained users could innocently and unpredictably change the setup of instruments just by randomly clicking at a web page exported by the instruments themselves. Security means in turn to consider whether *user accounting* should be performed or not. Probably in most applications this is not a crucial requirement, but for e-learning or when a remote lab is rented it may be, because access to the system via a login/password paradigm and logging of user activity in practice means money charge.

Another critical issue is the possibility that *multiple users* are simultaneously enabled to interact with remote instruments or devices. This feature has to be planned in advance, and proper policies should be implemented. In some cases exclusive allocation of resources may be preferred: hence suitable schemes must be implemented, while mechanisms that avoid starvation or deadlocks should be taken into account.

Another important requisite is *device-independence*. While this is desirable for the control of devices in general, it becomes even more important for networked instruments. It is advisable that the various code written need not be changed at all, or at least except some well defined points, whenever a remotely available test bench is updated by changing some device or instrument. In this view, device-independence means saving of investments. Generalizing the requirement, one may argue that well-agreed protocols for the exchange of data/commands, object oriented techniques and re-usable software objects should be adopted whenever possible.

Finally, the *learning curve* for users of a remote measurement system should be kept as short as possible. One should note that networking implies a number of features and configuration parameters previously unknown by technicians in the measurement context. In principle, network configuration should not be entrusted to the end-user. Moreover, user autentication and accounting is a task that should be left to a system-administrator of a computer network, and not to a test engineer designing a distributed measurement application.

3   System components

Various items contitute a distributed measurement system: hardware components, interfaces toward computers or the network, software modules variously interacting each other in order to provide the required services via suitable protocols, and so on.

The designer has a number of options to determine or choose among the available ones. Some among the various options are here summarized at different abstraction layer in order to provide some guidance in such choices.

3.1   Hardware Solutions

Many instruments offer a IEEE-488 interface for the dialog with a computer. This well-known standard allows the control of instrumentation via an exchange of suitably formatted strings. The format of such command/data strings may depend on the specific instrument at hand. Some manufacturers have announced that IEEE-488 interface will be no longer supported in the near future, as the cheaper and faster Ethernet and USB connections will take over as the main instrumentation interfaces. Instruments optionally provided with 802.11 wireless lan or Bluetooth connections can also be found. Command/data strings exchanged between instrument and computer can remain substantially the same as those exchanged via a IEEE-488 bus. The use of an Ethernet link, however, opens the opportunity for direct remote control of the instrument via a TCP channel, i.e., through the Internet. Hence, remote control of the device becomes as simple as opening a network connection and exchanging data packets.

Devices provided with the more recent connection schemes often can directly act as simple web-servers. A web-capable instrument directly publishes web pages that enable a user to change its setup by using a browser. Likewise, measurement data are seen from other pages published by the web-server running inside the instrument. No further coding is needed to remotely interact with the remote instrument. For those who do not have such recent instrumentation, but are looking for a simple solution to networking, protocol converters are currently sold at a reasonably low price. For instance, a full IEEE-488 bus can be mapped into one Internet address. Moreover, drivers for the IEEE-488 bus (e.g. the VISA driver) that map the physical interface port into a network address are currently sold as a standard package with IEEE-488 interface modules, so one can say that the "old" 488 bus is already network-aware. Anyway, it should be noticed usually the above solutions do not satisfactorily solve security issues, do not cope with multiuser management and are not device-independent. Moreover, exported web-pages are not a real solution to system integration. In summary, they represent straightforward and easy solutions for a very limited number of practical applications.

3.2 Middleware Solutions

A well-designed networked measurement system should implement mechanisms that separate the actual control of devices from the logic that coordinates them in order to obtain the final results. In other terms, tools should be provided with that enable a designer to see the whole system in a generalized form, so that attention is concentrated on the sequence of input stimuli to be provided and processing of the corresponding answers from the Device Under Test (DUT). Approaches proposed with this aim might be called *middleware solutions*. Some among the available variety are discussed below.

IEEE-1451. The IEEE-1451 series of standards include among their purposes the definition of architectures that allow remote interaction through the network with sensors and actuators [1], [2]. The study group associated with this standard is currently rather active and interesting extensions, for instance for the management of a large sensor network, are under consideration. As a matter of fact, the logical gap between the term "sensor" and "instrument" is becoming ever thinner, so that many ideas of the 1451 study group can be adopted for networking measurement applications. For instance, a relevant aspect of this standard is its independence from both the adopted communication infrastructure (for instance, Ethernet, Profibus, etc.) and the physical characteristics of devices.

IVI. The Interchangeable Virtual Instruments (IVI) Foundation has, among others, the aim to develop software modules, called "IVI device drivers" [3], realized in order to enable dialog with instrumentation in a device-independent manner. These drivers export a standard set of functionalities for a given instrumentation class; within the IVI driver each function call is translated into the right commands for the actually available instrument.

The main advantage of IVI is the abstraction of communication from the physical device. No reconfiguration at all is required when some IVI-compliant instrument available in the test system is replaced with another IVI-compliant model, as long as it belongs to the same IVI class. On the negative side, it should be considered that, at present, there is no IVI driver for every instrument, although really important IVI instrument classes (oscilloscope, multimeters, function generators, switches) are already implemented and work is progressing to expand the coverage.

SCPI. The Standard Commands for Programmable Instrumentation (SCPI) Consortium gives a generalized description of the functionalities of a generic instrument, in a unified and device-independent manner in order to guarantee interoperability [4].

Besides, it defines the syntax of the commands that a generic instrument has to support. This mechanism is also independent from the specific interface the instrument uses for the dialog with a host computer. The SCPI standard is somewhat more abstract than IVI: it doesn't take care both of the problems concerning networking and of the programming language to be used. In fact, it is just a way by which commands exchanged with instruments are standardized, regardless of any other issue. SCPI may be a better choice with respect to IVI because it gives a unified and more easily extensible representation of instruments and is not related to any protocol or specific operating system.

A drawback of SCPI is that it is not supported by every instrument or, even worst, it is partially supported; some commands might be available in a way that make them subtly incompatible with the standard, thus compelling the programmer to write suitable software modules dependent on the specific instrument at hand.

In the end, one should note that the technologies presented here do not conflict with each other; these environments should be seen as complementary and co-operative tools, not necessarily as alternative ones. For instance, an IVI driver can interact with a SCPI instrument. Moreover, one may propose an IVI driver for a smart sensor compliant with IEEE 1451. Another example consists in the possibility for a web server application like a Servlet to use an IVI driver in order to control the instrumentation.

It is highlighted the fact that device independece in a key factor in applications where hardware components (e.g. instrument, interfaces) are expected to be changed during the life cycle of the system itself. For instance this aspect must be carefully considered for the case of expandible distributed measurement system, or when plug & play capabilities would be offered, in a continuous education scenario, or when experiments are being shared in a grid computing–like environment. Moreover, at the present time vendors are releasing devices with proprietary drivers that sometimes require a specific operating platform or even worse require call to vendor proprietary library functions for device access. This imply simple warns to the designer:

- a clever project must include in its specification a way for separating low-level interaction with devices or interfaces with the adopted network protocol for exchange of data/commands in such a way that communication between the adopted protocol and low-level drivers is accomplished by a "translator". The above protocol should be kept as far as possible general, i.e. not tied to a particular device, or vendor solution, while the translator should provide techniques for an easy upgrade to new hardware components.

- devices/drivers offering direct networking capabilities should be considered intrinsecally unsure in a multiuser environment, hence masking techinques (sub-networking, firewalling, ...) should be considered at the design level.

- proprietary data/commands protocols that are not technically relased to the public domain (e.g. DataSocket in LabVIEW) should not be used at all in a project that claims to gather resources from many sites, like in a e-learning project aiming in sharing resources from many institutions.

3.3 Development environments

Some commercial environments offer interesting features for the development of networking measurement systems. Two well-known examples are LabVIEW from National Instruments [5], and VEE from Agilent Technologies [6]. Both products provide similar facilities. For instance, they can publish on the Internet a graphical front panel of a virtual instrument so that an end-user can interact with remote (virtual) instruments through a browser, by seeing physical devices in the form of web pages. Another interesting feature is that it is possible to design graphical applications that share software objects through the Internet: as an example, an application can share a knob used to control some instrument setup with other certified applications running elsewhere. The simplicity of use of these environments is paid for in terms of execution speed. Hence, if speed is a critical issue, standard programming languages should be preferred. It should be recalled that vendors supply interesting sets of libraries, whose features are quite similar to those provided by the environments mentioned above. Moreover, rapid development platforms for Java or C languages provide a number of standard libraries that allow the easy implementation of networking applications, the connection with data base engines and so on (for further details see for instance online documentation of reference [7]). One issue to be kept in mind for the choice of the platform is the learning curve of the designer. While LabVIEW/VEE are very easy to use, Java/C platforms require a higher skill level, and more time for the designer to let him keep pace with updates of the platform itself.

To help the reader in the choice, and on the author experience, if a new network protocol is under consideration for the remote dialogue with instrumentation, or issues like security and user accounting are major issues, or a whole distibuted system composed of many interacting components is being set-up, then a graphical enviroment (LabVIEW/VEE) is probably not the right choice. Opposite, for a quick development of applications that interact with "famous brand" instruments and exports features like a web page or a simple panel for remote control of some sub-functionality, then LabVIEW/VEE are very nice development environments.

4 Architectures

Different architectures with an increasing level of complexity can be adopted in a networking measurement system. From the software point of view interesting technologies are ready to be advantageously adopted to build up distributed measurement applications. Each of them suggest a corresponding architecture, hence some among the most widely adopted are recalled in the following [8] – [27]. First of them is the so called client-server architecture [20–22]. In practice, a server computer runs a "daemon" application that listens to network requests coming from client computers in the form of string messages. Then, the server translates messages into execution of some measurement task, and finally sends back to the client packets containing the results of the operation accomplished. One should note that a very simple implementation of the above client-server scheme may exploit well-established web server engines, like Apache or Tomcat [28]. Any web server allows an extension of its functionalities via Common Gateway Interface (CGI) techniques. In practice, CGI makes possible the remote invocation of simple programs via http requests coming from a browser. Since a generic program can be started upon client invocation via a web transaction, one may easily translate requests coming from the client browser into corresponding execution of measurements [24, 25]. Noticeably, a Java Applet, directly downloaded from the web server, can be started at the client side. In this way a more user-friendly or sophisticated interaction between the user and remote programs controlling instrumentation is possible. An interesting improved version of the client-server paradigm is the Remote Procedure Call (RPC) technique. The RPC layer runs on top of the TCP/IP protocol, and enables a computer program to obtain the execution of code functions directly performed by another computer as if they were locally executed. Due to its nice features, the VXI plug & play alliance [29] suggested a few years ago to adopt RPC for distributed measurement systems. Anyway, RPC alone is too general to say that it constitutes a standard for measurement systems, because it does not specify any set of mandatory or recommended functions specifically tailored for a measurement application (e.g. "trigger" "data acquisition" and so on). Besides, the RPC protocol is usually filtered out (i.e. dropped) by Internet firewalls and does not guarantee inter-operability if different versions of RPC are run in different machines. Some limitations of RPC have been addressed by more recent protocols, such as in the Distributed Component Object Model (DCOM) technology, that extends the idea behind remote invocation of functions to the idea of remote software objects distributed in a network and run by different computers on demand. Noticeably, the IVI proposal above quoted is based on intensive usage of DCOM objects, hence intrinsically providing support for distributed measurement applications. A drawback of DCOM is that it is not supported, neither does it run, on non Windows-based systems. Moreover, DCOM objects have unlimited control on operating system library invocation: this is an open door to serious security failures, even though work is currently under development to take care of this issue. An interesting alternative approach to DCOM is provided by Common Object Request Broker Architecture (CORBA), which is an open standard supported by a few hundred Information Technology (IT) firms [30]. CORBA defines a language for the description of software interfaces and a "communication bus" (Object Request Broker) that implements the distribution, allocation, creation and activation of objects in a transparent way. Detailed discussion of CORBA is beyond the scope of this paper. Anyway, it should be noticed that complex distributed measurement systems based on this technology have already been produced (e.g. [31]). Another interesting approach is provided by Java Remote Method Invocation (RMI) that offers distributed computing services to Java applications. In particular, an extended functionality version of RMI, called JINI, allows the creation of network distributed applications in a very simple yet effective way. In fact, JINI implements the idea of a community of service providers (Jini community) that announce themselves through a Lookup Server. An application can advantageously exploit mechanisms embedded in a Jini community concerning "service discovery", "leasing" of services and "transactions" [32, 33].

The fig.1 shows a first simple example of client-server configuration, where routing issues are taken into account. In the figure, the "bridge" represents a component that takes care of protocol handling in such a way that client applications are simplified. For instance, the client "sees" a server as a simple socket connection, while the bridge encapsulates requests into http packets and takes care of opening and closing connections to the remote server. The "proxy/firewall" avoids exporting laboratory resources directly to everybody. A simple practical implementation of a proxy access may consists in encapsulating requests/data to/from instrumentation inside http packets. The advantages of this scheme come from the availability of robust off-the-shelf firewall and proxy servers that can be used for filtering out most of malicious packets. Moreover, if a proxy is used, intranet addresses can be used for the instrument server, that is a server address can be hidden behind an URL symbolic name instead of being announced as a public IP.

Fig. 1. Client-server routing.

A shared devices laboratory, can be seen in figure 2. In the figure three main components can be recognized. Client computers accessing to remote devices through a network connection. "Measurement servers" providing access to physical resources, and repository servers providing all various information needed to the clients to accomplish the task of a remote measurement. In practice, one user first of all finds useful information concerning the laboratory, the available experiments and so on via dynamic web pages provided by the web server. Then, the user downloads applets or suitable software programs from the web server that will be run on his client computer. Such downloaded software in its simplest configuration design scheme directly interacts with some measurement server. Each maeasurement server takes care of multiuser operation, that is it grants access to its resources depending on the data stored in te SQL server, and keeps trace of the number of users requesting exclusive usage of physical resources. In a more complicate scheme the measurement servers publish available resources and services (i.e. software chuncks) in a lookup server, the downloaded client software first of all performs a "discovery query" in order to obtain from the lookup server a list of available services, and then collects run-time modules (the same provided from instrument servers to the lookup server) in order to perform useful tasks. In this second case again measurement servers takes care of exclusive access to physical resources, while user grants can be stored directly in the lookup server by suitable "daemon" processes running in the lookup server itself. Once more, the SQL server may store measurement procedures together with all the information needed for the management of the whole laboratory itself: dynamic contents of web pages, user privileges, access policies, and so on. On the other side, the measurement servers benefit from this organization in that they can store in the SQL server up-to-date drivers for instruments that enable device-indepedent features, measurement procedures, and object-oriented components for a reusable interaction with physical devices.

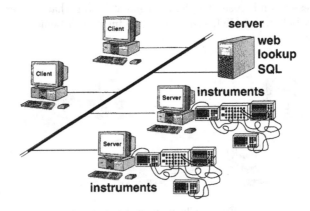

Fig. 2. Simple multiuser configuration.

The scheme of fig.2 can be easily extended to the wiev of a distributed laboratory, where multiple test-benches and multiple users cooperates (fig.3). The web, SQL and lookup engines still are present but store data of multiple laboratories where many experiment are made available. Despite the figure depitcs one web/lookup/SQL server such services are distributed through the net, as usually. Moreover, a client application can simultaneously interact with more far located servers in order to accomplish a truly distributed (and more complex) experiment.

One should note that the dialog with remote devices is context-dependent. Namely the setup of instrumentation, and the retrieval of measurement data must be performed in some pre-determined order without any interference from

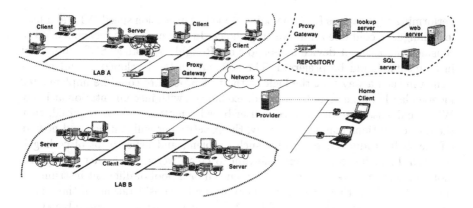

Fig. 3. Example of distributed laboratory.

others. For instance, a CGI procedure must be atomic and exclusive; that is, it should interact with the instruments without letting another procedure try to use the same devices too. This fact is simple to understand while looking at a standing-alone measurement bench, but it is sometimes forgotten in the literature when remote measurement procedures are proposed. Nearly none of the previously quoted environments ensure an exclusive use of instrumentation and takes into account that an instrument remembers setup commands in most cases even when switched off. Most of the above architectures do not have native leasing functionalities and do not provide lock/unlock services, but requires further coding. In this perspective, CGI services are a bad solution, Ethernet instruments are almost useless, while client-server approaches should be carefully designed. Instead, CORBA and JINI technologies offer native services to implement transactions in the use of remote instrumentation. From a "grid computing" point of view of distributed measurement, where also the use of centralized repositories of data and procedures should be considered, JINI and CORBA seem to offer more promising features. The client-server approach is instead a good solution to solve some practical localized task, but it does not offer a unified view of the whole problem, while the use of Ethernet interface converters or Ethernet instrumentation is a somewhat raw solution that quickly solves a specific measurement task, and no more.

An alternative point of view to be considered is the one of the practitioner that could be a little confused by the number of available solutions. On the basis of the author' experience, a practical suggestion is to carefully plan the intended usage of the distributed measurement system before taking a decision. This means that, if the main goal is to interact with a given remote coputer-based device in single-user mode and using a trusted Internet or Intranet connection, then devices with Ethernet interfaces or Ethernet converters may suffice as an adequate solutions. When some software program should be developed to let an end user to interact with remote devices, then LabVIEW or VEE may quickly provide all the functionalities needed at a comparatively low effort. Moreover, for better

performing applications from the point of view of execution speed IVI and a high
level language could be an interesting choice. In any case, it is recommended to
consider the issue of device-independence. Roughly speaking, SCPI-compliant
instruments should be preferred. The software to be developed should be designed in such a way that the logic of the test procedure could be implemented
independently of the actual commands exchanged with instrumentation at hand.
When multi-user use of some computer-based remote system is required, then
the above solutions may fail. Anyway, the practitioner will just not find now an
off-the-shelf product able to solve such problem, and hence a lot of coding is
required. In such case, the next question to be answered is whether a single remote measurement bench should be managed or a pool of different instruments
(hopefully cooperating with each other) are under consideration. In the former
case a client-server approach could be the right answer, at a reasonable effort,
while in the latter situation JINI/CORBA technology should be considered, but
a lot of coding will be required.

5 Acknowledgment

The author is grateful to prof. L. Benetazzo for the valuable discussions during
the preparation of this paper. Some concepts reported in this work can be also
found in reference [35].

References

1. IEEE-1451 website at NIST, [Online] Available: http://ieee1451.nist.gov
2. Potter,D., Smart plug & play sensors, IEEE Instrumentation and Measurement Magazine, vol.5, (March 2002) 28–30
3. Interchangeable Virtual Instrumentation (IVI) consortium documentation, [Online] Available: http://www.ivifoundation.org
4. Standard commands for programmable instruments, SCPI Consortium,
   1997. [Online] Available: http://www.scpi.org
5. LabVIEW website at National Instruments, [Online] Available:
   http://www.ni.com/labview/
6. VEE documentation at Agilent Technologies website, [Online] Available:
   http://www.agilent.com. (note: use the search box with the keyword
   "VEE")
7. Eclipse website, [Online] Available: http://www.eclipse.org
8. Wheelwright,L., The network centric system, in Proc. IEEE Internat.
   Conf. AUTOTESTCON/2003, (22-23 September 2003), 44–47
9. Benetazzo, L., Bertocco, M., Cappellazzo, S., Pegoraro,P., Platform independent architectures for cooperative remote control of automated test
   laboratories, in Proc. IEEE Internat. Conf. AUTOTESTCON/2003, (22-
   23 September 2003), 52–75
10. Bertocco, M., Cappellazzo, S., Carullo, A., Parvis, M., Vallan, A., Virtual
    Environment for Fast Development of Distributed Measurement Appli-
    cations, IEEE Trans. On Instrum. And Meas, (2003), 681–685

11. Cristaldi, L., Deshmukh, A., Lazzaroni, R., Monti, A., Ponci, F., Otto-boni, R., A Monitoring System Based on a Multi-Agent Platform, in Proc IEEE conf. IMTC/2004, (2004) 907–911
12. Ferrero, A., Salicone, S., Bonora, C., Parmigiani, M., ReMLab: a Java-based remote, measurement laboratory, IEEE Trans. On Instrum. And Meas, (2003) 710–715
13. Grimaldi, D., Nigro, L., Pupo, F., Java-based distributed measurement systems, IEEE Trans. On Instrum. And Meas, (1998) 100–103
14. Spoelder, H., Virtual instrumentation and virtual environments, IEEE Instrum. Meas. Mag., vol. 2, (1999) 14–19
15. Winiecki, W., Karkowski, M., A New Java-Based Software Environment for Distributed Measuring Systems Design, IEEE Trans. On Instrum. And Meas, (2002) 1340–1346
16. Trullols, E., Sorribas, J., del Rio, J., Samitier, C., Manuel, A., Palomera, R., LabVir: A virtual Distributed Measurement System, in Proc. IEEE IMTC/2003, (May 2003) 776–779
17. Cicirelli, F., Furfaro, A., Grimaldi, D., Nigro, L., Pupo, F., "Management Architecture for Distributed Measurement Services," in Proc. IEEE IMTC/2004, (May 2004) 974–979
18. Bertocco, M., Cappellazzo, S., Flammini, A., Parvis, M., A multi-layer architecture for distributed data acquisition, Proc. 19th IEEE Instrument. and Meas. Technol. Conf, (2002) 1261–1264
19. Wienicki, W., Mielcarz, T., Internet-based methodology for distributed virtual instrument designing, in Proc. IEEE IMTC/2003 Conf., (2003) 760–765
20. Bertocco, M., Ferraris, F., Offelli, C., Parvis, M., A client-server architecture for distributed measurement systems, IEEE Trans. on Instrum. and Meas. vol.47, no.5, (1998) 1143–1148
21. Panegiani, F., Macii, D., Carbone, P. Open Distributed Control and Measurement System Based on an Abstract Client-Server Architecture, IEEE Trans. On Instrum. And Meas, (2003) 686–692
22. Ewald, H., Page,G.F., Client-server and Gateway-Systems for Remote Control, in Proc. IEEE IMTC/2003, (2003) 1427-1430
23. Chen, S., Open Design of Networked Power Quality Monitoring Systems, IEEE Trans. On Instrum. And Meas. (2004) 597-601
24. Pyszlak, P., Remigiusz, R., Rak, J., Majkowski, A., The next approach to the design of a web-based virtual laboratory, in Proc. IEEE IMTC/2003, (2003)1083–1086
25. Giannone, P., Pecora, E., Pitrone, D., "Development in Web-based Laboratory Sessions," in Proc. IEEE IMTC/2004, (2004) 1155–1158
26. Dricot, J.M., De Doncher, P., Diericks, M., Grenez, F., Bersini, H., Development of distributed Self-Adaptive Instrumentations Networks using JINI technology, in Proc IEEE Internat Workshop VIMS/2001, (2001) 22–27
27. Bertocco, M., Cappellazzo, S., Narduzzi, C., Parvis, M., A distributed sensor network based on JINI Technology, in Proc IEEE Internat Workshop VIMS/2002, (2002) 68–71
28. Apache organization website, [Online] Available: http://www.apache.org
29. VXI Plug & play alliance website, [Online] Available: http://www.vxipnp.org

30. The Common Object Request Broker: Architecture and Specification, [Online] Available: http://www.omg.org (1998)

31. net-7 monitoring system, [Online] Available: http://www.tek.com/site/ps/0,,2F-14219-INTRO_EN,00.html

32. Edwards, W.K., Core Jini. 2nd ed. Prentice-Hall, (2001)

33. Jini community website, [Online] Available: http://www.jini.org

34. ADA project website, [Online] Available: http://www.ada-ist.org

35. Benetazzo, L., Bertocco, M., Narduzzi, C., Networking Automated Test Equipment Environments, IEEE Instrumentation and Measurement Magazine, (2005) 16–21

Traceability Assurance Of Tele-Measurements

Alessio Carullo, Simone Corbellini, Marco Parvis

Dipartimento di Elettronica - Politecnico di Torino
Corso Duca degli Abruzzi, 24 - 10129 Torino, Italy
alessio.carullo@polito.it

(Invited Paper)

Abstract. This paper deals with the traceability-assurance problems that are faced when a distributed measuring system is employed. These problems are highlighted for different distributed-system architecture, then a set of guide-lines are provided in order to correctly manage such systems from a metrological point of view. The proposed solution is based on a network-assisted calibration procedure, which requires suitable traveling standards that are sent to the nodes of a distributed system under calibration and are remotely controlled through the Internet. A prototype of a traveling standard is also described that has the capability to be synchronized with standards sent to the other nodes.

Keywords: Measurement, calibration, quality assurance, interconnected systems, large-scale systems

1 Introduction

The basic concept of measurement, which consists in the comparison between the unknown quantity and a known standard, is commonly implemented by means of electronic devices that are frequently based on a digital architecture. Such devices usually transduce the unknown quantity into an electrical-voltage signal, compare it with an internal standard and, after some form of processing, provide an indication that is suitable for the end user. These devices, which are hereafter referred to as "stand-alone instruments", are commonly employed to carry out measurements of a single quantity. An electronic switch, which can be either internal or external to the instrument, may allow the sequential measurement of several quantities to be performed.

An important milestone in the measurement field was the development of communication buses devoted to the instrumentation, such as the IEEE-488 bus, which give the possibility to assemble a set of stand-alone instruments in an "automatic test-bench". In this case, a software program that typically runs on a Personal Computer (PC), is responsible for instrument configuration, data acquisition and data processing.

In the last decade, the availability of both wired and wireless networks has permitted to extend the size of the communication bus, that can now cover very

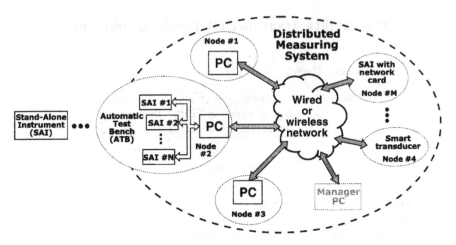

Fig. 1. Evolution of the measuring systems: from the stand-alone instrument to the distributed measuring system, which can either include a manager PC or rely on a grid architecture.

large areas. Different automatic test-benches, as well as smart transducers [1] and instruments that embed a network card, can be interconnected through this extended bus, thus arranging a distributed measuring system. These systems can either be based on a manager PC, which remotely configures the different subsystems, acquires from them the corresponding measurements and, after some form of processing, makes the final results available to the users, or on a grid architecture, where each node has management capabilities and co-operates with the other nodes on a peer-to-peer basis.

This evolution of the measuring systems, which is summarized in figure 1, has enormously increased the usefulness of such systems and has brought to the new concept of "tele-measurement", i.e. the measurement carried out by an instrument at a certain site and shared with other users by means of a telecommunication network. Thanks to the capillarity of the Internet, a distributed measuring system allows the measurement of several quantities distributed over a large area to be obtained at a reasonable cost. Typical applications are the monitoring of the environmental pollution and of the power quality.

On the other hand, the described evolution has made the traceability assurance a very difficult task, because the obtained results are the combination of hardware and software processes that take place over a wide-distributed environment. Furthermore, other problems arise due to the data exchange over the network, such as net latency and lack of synchronization among units [2]-[5].

In the next sections, these issues are tackled by revising the concept of traceability and a set of guide lines are provided in order to correctly manage a distributed measuring system from a metrological point of view.

2 Metrological management of a distributed measuring system

2.1 Traceability-related problems in a distributed measuring system

The concept of traceability is defined in [6] as *the property of the result of a measurement or the value of a standard whereby it can be related to stated references, usually national or international standards, through an unbroken chain of comparisons, all having stated uncertainties*. Traceability is a fundamental aspect of any measurement that has to be compared with other measurements or with acceptance limits.

When a measurement is carried out by means of a stand-alone instrument, the traceability can be assured by periodically calibrating the instrument against devices that are directly or indirectly related to the primary standards.

In an automatic test-bench, more attention has to be paid in order to assure the traceability of the obtained results, that typically depend on the measurements the different instruments carry out and the software process such measurements are subjected to. This implies that:

- the involved instruments have to be periodically calibrated;
- the software program has to be managed within a quality-assurance model (see an example in [7]).

A distributed measuring system presents traceability-related problems that are similar to the ones encountered in an automatic test-bench. However, the management of such a kind of systems is made more difficult by their distributed nature. The instruments could be spread over a large area and their number could be very high. In this situation, the common calibration approach, which requires the instruments to be moved to a laboratory for the periodical calibration, is not a sustainable solution, because of its high cost for calibration and transportation. Furthermore, during the calibration time, the different nodes are unavailable, thus making the distributed measuring system not completely operative.

The verification of the software programs that are involved in the measuring process is another tricky problem, since the software process could be either split between the manager and the peripheral PCs or, in the case of a grid architecture, distributed over the different nodes. It is therefore more convenient to verify the whole "software chain" than verify each program independently from each others. For this reason, a good solution consists in performing a whole calibration procedure, that is intended for verifying both hardware and software sections of the system.

Other problems could arise that depend on the kind of measuring task a distributed measuring system performs, as described in the next section.

2.2 Measuring tasks performed by a distributed measuring system

The measuring tasks a distributed measuring system can perform can be subdivided into three main categories:

1. tasks whose results are the set of data each node provides and collected -as is- by the manager PC;
2. tasks whose results are provided by a manager PC as the processing of data the different nodes provide;
3. tasks performed at each node by processing local measurement and data obtained by other nodes, according to the grid paradigm.

In the first case, the manager PC acquires the measurement results from each node and makes them available to the authorized users. Therefore, the nodes perform independent measurements and the manager PC acts as a simple data collector. An example of this situation is a distributed measuring system for the monitoring of industrial plants, where each node has the role to find anomalous situations in the different units of a plant. Another example is a system for the monitoring of the environmental pollution over a geographical area, whose main goal consists in comparing the results of each node with alarm thresholds in order to highlight dangerous situations. The traceability assurance of the tele-measurements obtained with these kinds of systems does not present particular problems, since each node can be managed independently of the others from a metrological point of view. In addition, the out-of-service of a single node does not impair the functionality of the whole system.

In the second case, the manager PC is no longer a simple data collector, but is part of the measuring chain, since it processes the acquired raw data and provides the final results. In the field of the environmental monitoring, the manager PC could process the data acquired from different nodes in order to interpolate pollutant concentration between adjacent units [8], or to predict pollutant levels by means of models that take weather forecasts into account. In the field of the power-quality, a typical application consists in monitoring the harmonic pollution in the mains. In this case, each node processes the acquired data and estimates different quantities, such as power, energy and quality indices; such quantities are sent to the manager PC, where the harmonic power flow is estimated and harmonic sources are localized [9]-[12]. The effectiveness of such estimations often requires a strict synchronization among the different nodes, usually obtained by means of the Global Positioning System (GPS) [9]-[10]. When the measured quantities are instead averaged over long intervals, the synchronization is not required provided that the time shift among nodes is lower than a limit value [3]. It appears that the traceability assurance of tele-measurements is definitely a not trivial operation, since the distributed measuring system has to be considered as a large instrument, whose inputs are located at the different nodes, while the outputs are available at the manager PC. The calibration of such an instrument requires to contemporaneously stimulate the different nodes with synchronized signals, so that the robustness of the whole system with respect to lack-of-synchronization can be evaluated.

A distributed system that performs the third kind of task implements a Grid Measuring System (GMS), since each node co-operates with the other ones in order to carry out the final results. In this situation, the presence of a manager PC is not mandatory, even though it can be used for grid management or for other purposes, such as data and results storing in a central database and the control of user access to the stored data. The GMS is therefore made up of interconnected nodes that usually embed a PC, which provides both networking and processing capabilities. A node of a GMS could be an automatic test-bench or a unit based on data-acquisition boards plugged into a PC. A distributed measuring system with this architecture is useful, as an example, to predict the pollutant concentration at each node by employing the measurements that the adjacent nodes perform. Another typical example is in the power-quality field, where each node monitors a section of the mains and localizes the harmonic source by processing the own results and those of the adjacent nodes. One should node that these tasks could be also performed by means of a distributed system based on a manager PC. However, the grid architecture allows an high reliability to be obtained, since in 'a centralized architecture, a crash of the manager PC could make the whole system unavailable.

The concept of GMS further makes the traceability assurance a very complicate problem, since the system is now a multi-inputs multi-outputs distributed instrument. The calibration of a GMS therefore implies that the output at each node has to be verified while stimulating the other nodes with known inputs.

A network-assisted calibration procedure is proposed in order to tackle the mentioned problems, with the aim to reduce the metrological-management cost of a distributed measuring system and the out-of-service time of each node, that affects the whole system.

3 Network-assisted calibration procedure

The proposed network-assisted calibration procedure is based on traveling standards that are sent to the different nodes and are remotely controlled through the Internet. Such traveling standards, which have to be equipped with a network interface, can be commercially-available equipment, e.g. multifunction calibrators and signal generators, or specifically-designed devices [13].

As previously mentioned, the calibration of the first kind of distributed measuring system can be performed by calibrating each node independently of the other ones. It is therefore sustainable the application of the commonly employed calibration procedure, provided that the software programs that manage the acquisition and the processing at each node are suitably validated. However, significant advantages could be obtained by employing the network-assisted procedure, which requires to stimulate a node by means of a remotely-controlled traveling standard and acquire the corresponding results at the manager PC. In such a way, each measuring chain is calibrated at a glance, since hardware and software sections are involved in the verified path.

If the distributed system implements the second kind of measuring task, the network-assisted calibration procedure seems to be the optimal solution both from a technical and an economical point of view. In this case, the calibration procedure is remotely exercised by the manager PC, which manages the traveling standards, in order to apply known stimuli at the node inputs, configures

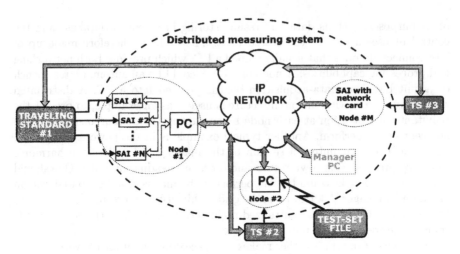

Fig. 2. Network-assisted calibration procedure of a distributed measuring system.

the involved nodes and acquires the corresponding measurements (see figure 2). The obtained results are compared with the expected ones in order to state the "in-calibration" or "out-calibration" condition of the whole system. If the system performs measuring tasks that require a strict synchronization among the different nodes, synchronized stimuli have also to be employed during the calibration phase.

The calibration of a GMS can be viewed as the set of calibrations of the different nodes. Thanks to the network-assisted procedure, a set of traveling standards is employed in order to stimulate different nodes, one of which is the unit under calibration. Since a GMS could not include a manager PC for the management of the calibration procedure, the best choice seems to be the configuration of the different traveling standards as a grid architecture, i.e. the calibration of a grid system is performed by means of a grid of standards, as shown in the figure 3. An important implication of such a choice is that each standard has to have enough capabilities to speak with the other standards and to manage the calibration procedure. One of the traveling standard is therefore configured as the calibration manager, in order to remotely control the other standards and to process the calibration results. Another problem arises that is related to the number of nodes that made up a GMS, which can be very high thus requiring a great number of traveling standards. However, from a measurement point of view, the results a node provides are typically affected only by the adjacent nodes. For this reason, it is possible to define a co-operation area, which includes all the nodes that mutually affect each others. As an example, in the figure 3 is highlighted the co-operation area for the nodes 2, 3, 5, 6 and 8. The calibration of a node that belongs to such a co-operation area therefore requires

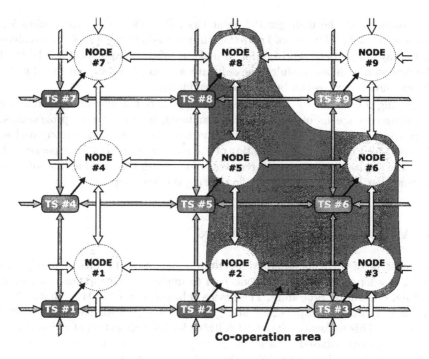

Fig. 3. Network-assisted calibration procedure of a GMS.

five traveling standards. This is a common situation in GMSs for the monitoring of the environmental pollution, since far nodes have negligible mutual effects.

3.1 Calibration Test-Set

The calibration of a generic distributed measuring system is performed in a group of test points, hereafter referred to as test-set. The test-set has to be representative of the possible operative conditions of the system, thus making effective the calibration. The test-set has therefore to be suitably chosen, by taking several information into account, such as range and nature of the measured quantities, operating principle and configuration of the instruments, operative conditions of the different nodes. Furthermore, it is important to consider the presence of acquisition and processing programs in the measuring path and the multi-input nature of the algorithm that could be implemented at each node and at the manager PC. The verification of the software programs could lead to a test-set with a very high size, thus making unsustainable the proposed procedure. A modified procedure can be therefore implemented, which is based on a two-step calibration procedure.

During the first step, the nodes that implement complex algorithms, are fed with a great amount of data by means of a test-set file. The corresponding results are processed at the manager PC or at the PC configured as the calibration manager in a GMS. One should note that the processing programs at the involved nodes have to be designed in order to make a software input port available. If the first step ends successfully, the software section of the system that follows the measuring devices can be considered validated.

During the second step, the inputs of the distributed measuring system are stimulated by means of the traveling standards, which are now programmed to provide a reduced test-set. In this situation, the measurements acquired at the manager PC allow the verification of the measuring devices to be performed, since the results the manager PC receives follow a path that includes the software section, that has been already verified during the first step.

4 An example of traveling standard

A prototype of traveling standard has been designed by the authors and is now under development [14] that is conceived to implement the proposed network-assisted calibration procedure. The device is based on a PC-104 module, which provides networking capabilities and gives the full potentiality of a personal computer. This makes the standard suitable for the calibration of a distributed measuring system based on a manager PC, as well as for the calibration of a GMS, since it can be configured as the calibration manager.

The prototype has been specialized for the calibration of distributed measuring systems that monitor the mains power-quality. The traveling standard is basically an arbitrary-signal generator with three voltage outputs and three current outputs. Furthermore, the device can be synchronized with other standards, since the power-quality monitoring is usually based on the synchronous acquisition of the measured quantities at the different nodes. For this reason, the time base of the traveling standard employs a local oscillator disciplined through a GPS-based clock. In such a way, different traveling standards can be synchronized within a few microseconds in order to generate the stimuli with a precise timing. A first prototype of traveling standard, which provides two voltage outputs and a current output, is shown in figure 4.

5 Conclusions

The increasing diffusion of distributed measuring systems in different fields has introduced the new concept of tele-measurement. At the same time, the difficult in assuring the traceability of tele-measurements has emerged from the diffusion of such systems. Besides the problem that is faced in the estimation of network-related effects on the measurement uncertainty, the traceability assurance of a tele-measurement could become a tedious and very expensive activity if performed in the traditional way. For this reason, an innovative approach has been proposed in this paper for the calibration of a distributed measuring sys-

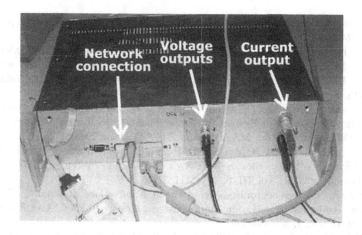

Fig. 4. The prototype of traveling standard based on a GPS-based clock.

tem, with the aim to make such an activity both economically and technically sustainable.

The main advantages the proposed procedure offers are: the minimization of the out-of-service time of a distributed measuring system; a low management cost, since its implementation does not require the presence of skilled technicians at the different nodes; the possibility to calibrate multi-inputs multi-outputs distributed measuring systems; the possibility to store the calibration results in a centralized database and automatically process them in order to dynamically manage the calibration interval of the whole system, thanks to the available history.

References

1. IEEE Std 1451.1-1999, IEEE Standard for a Smart Transducer Interface for Sensors and Actuators - Network Capable Application Processor (NCAP) Information Model, The Institute of Electrical and Electronics Engineers, Inc., 3 Park Avenue, New York, NY 10016-5997, USA.
2. D. Potter, "Using Ethernet for industrial I/O and data acquisition," Proc. *IEEE Instr. and Meas. Tech. Conf.*, Venice, Italy, pp. 1492–1496, May 1999.
3. L. Cristaldi, A. Ferrero, C. Muscas, S. Salicone, R. Tinarelli, "The Impact of Internet Transmission on the Uncertainty in the Electric Power Quality Estimation by means of a Distributed Measurement System", *IEEE Trans. on IM*, August 2003, vol. 52, n. 4, pp. 1073-1078.
4. H. Eren, W.J. Nichols, I. Wongso, "Towards an Internet-based virtual-wire environment with virtual instrumentation," Proc. *IEEE Instr. and Meas. Tech. Conf.*, Budapest, Hungary, pp. 817–820, May 2001.
5. P. Daponte, D. Grimaldi, M. Marinov, "Real-time measurement and control of an industrial system over Internet: implementation of a prototype for educational

purpose," Proc. *IEEE Instr. and Meas. Tech. Conf.*, Budapest, Hungary, pp. 1976–1980, May 2001.

6. VIM - International Vocabulary of basic and general standard terms in Metrology.

7. General Principles of Software Validation; Final Guidance for Industry and FDA Staff - U.S. Food and Drug Administration - Center for Devices and Radiological Health, Ed. January 2002.

8. W. Tsujita, S. Kaneko, T. Ueda, H. Ishida, T. Moriizumi, "Sensor-Based Air-Pollution Measurement System for Environmental Monitoring Network ," Proc. *Transducers '03*, Boston, MA, USA, pp. 544–547, June 2003.

9. H. Ukai, K. Nakamura, N. Matsui, "DSP- and GPS-Based Synchronized Measurement System of Harmonics in Wide-Area Distribution System", *IEEE Trans. on IE*, December 2003 , vol. 50, n. 6, pp. 1159-1164.

10. H. Saitoh, "GPS Synchronized Measurement Applications in Japan", *IEEE/PES Transmission and Distribution Conference and Exhibition 2002: Asia Pacific*, 6-10 Oct. 2002, vol. 1, pp. 494-499.

11. L. Cristaldi, A. Ferrero, S. Salicone, "A Distributed System for Electric Power Quality Measurement", *IEEE Trans. on IM*, August 2002, vol. 51, n. 4, pp. 776-781.

12. C. Muscas, "Assessment of electric power quality: Indices for identifying disturbing loads", *ETEP*, 1998, vol. 8, n. 4, pp. 287-292.

13. A. Carullo, M. Parvis, A. Vallan, "A Traveling Standard for the Calibration of Data Acquisition Boards," *IEEE Trans. on IM*, April 2004, vol. 53, n. 2, pp. 557-560.

14. A. Carullo, M. Parvis, A. Vallan, "A GPS-Synchronized traveling standard for the Calibration of Distributed Measuring Systems," Proc. *IEEE Instr. and Meas. Tech. Conf.*, Ottawa, Ontario, Canada, pp. 2148-2152, May 2005.

Exploring The Capabilities Of Web-Based Measurement Systems For Distance Learning

D. Grimaldi[1], S. Rapuano[2], T. Laopoulos[3]

[1] Dept. of Electronics, Computer and System Science, University of Calabria,
87036 Rende (CS), Italy
grimaldi@deis.unical.it
[2] Dept. of Engineering, University of Sannio, Corso Garibaldi 107,
82100 Benevento, Italy
rapuano@unisannio.it
http://lesim1.ing.unisannio.it
[3] Electronics Laboratory, Physics Dept., Aristotle University of Thessaloniki,
Thessaloniki, Greece
laopoulos@physics.auth.gr

(Invited Paper)

Abstract. The paper presents a brief state of art of distributed laboratories for teaching of electric and electronic measurement and proposes an innovative platform for distance learning in such field. The platform consists of a learning management system (LMS) and a remotely accessible distributed laboratory to carry out experiments on real instrumentation without requiring specific software components on client side. In particular, the user authentication and management and the tracking of learning process are provided by the LMS. The experimental phase is carried out by means of Virtual Instruments executed on computers connected to the instrumentation included in a remote laboratory system distributed on geographical scale.

Keywords. E-learning, distributed laboratory, measurement instrumentation, computer network, software architecture.

1 Introduction

E-learning has been a topic of increasing interest in recent years. It is often perceived as a group effort, where content authors, instructional designers, multimedia technicians, teachers, trainers, database administrators, and people from various other areas of expertise come together in order to serve a community of learners [1].

In order to simplify their joint work, a lot of software systems have been developed. They are generally referred to as *Learning Management Systems* (LMSs) and *Learning Content Management Systems* (LCMSs).

The primary objective of a LMS is to manage learners, keeping track of their progress and performance across all types of training activities. The LMS manages and allocates learning resources such as registration, classroom and instructor availability, instructional material fulfilment, and online learning delivery.

The LCMS usually includes an LMS and adds an authoring system providing an infrastructure that can be used to rapidly create, modify, and manage content for a wide range of learning to meet the needs of rapidly changing business requirements. The LCMS can retrieve detailed data on learner scores, question choices, navigation habits and use them to give content managers crucial information on the effectiveness of the content when combined with specific instructional strategies, delivery technologies, and learner preferences.

These software systems also provide support for interaction between learning space participants, a mechanism to deliver course materials over the Web, administrative components to allow instructor tracks student records and monitors their progress and collaborative components like Bulletin Board, Chat, E-mail etc. [1].

Most existing web-based learning environments are based on basic instruction models. Their main functionalities are centred on the management and distribution of learning materials, synchronous and asynchronous communication, and progress tracking and reporting [2]. Some existing learning environments address collaborative learning. However, most of them do not support effectively innovative pedagogical approaches based on models of collaboration used in modern working life [2].

One of the most interesting new teaching strategies proposed for distance learning systems is Project Based Learning (PBL) pedagogy [2-3]. The approach provides teaching through the development of a project that involves the learners. The most relevant step in learning process, in fact, especially in scientific and technological fields, is the application of the acquired theoretical knowledge.

The basic and high instruction as well as the adult training have been recognized to be at the centre of the growth, innovation and integration processes in the democratic societies. The European Union officially announced the mission of improving the education systems in Europe with the declaration of Lisbon in 2000. Among the objectives to realize such mission, two of the main ones are: to give to all citizens the same opportunities of improving his/her degree of instruction and to promote the institution of a life-long learning system to update the competences and to encourage new specializations of the adult people, thus increasing their capability of finding or changing their work. E-learning seems to be the best way to reach these objectives, as it removes the physical, geographical and cultural barriers to the education and enables the learners in choosing their own learning path and time.

A huge practical training is absolutely essential to assure a good knowledge transfer from teacher to students and so to educate good professionals. Laboratory activity is an open challenge for on-line teaching applied to scientific domains. The remote control of instrumentation and the execution of real experiments via Internet are topics of interest for many researchers [4-10].

In particular, in teaching of electric and electronic measurement topics, from academic courses to continuous training in industry, learners should achieve an accurate practical experience by working in real conditions and on real instruments. However, electric and electronic measurement laboratories, both public and private ones, due mainly to their costs, are not widespread, and this complicates the life-long learning of specialized technicians especially in the field of process control, quality control and testing engineering.

How much the E-learning technologies can be useful to teach electric and electronic measurement topics is an open question, currently dealt with by the Working Group (WG) "e-tools for Education in Instrumentation & Measurement" (etEI&M), operating into the IEEE Instrumentation and Measurement Society, Technical Committee TC-23 Education in Instrumentation and Measurement.

None of the most recent proposals of didactic laboratories for electric and electronic measurement includes the noticeable advantages that a LMS could give for supporting a learner-centric approach to the teaching. In particular, concerning the students, it is not possible to self-design their own learning process, or carry out a collaborative or project-based learning. Concerning the teachers, it is not possible to track the activity of the students, or carry out an interactive experiment in a virtual classroom.

In this paper, a distributed platform based on a LMS is proposed in order to provide full courses of electric and electronic measurement including theory as well as practical experiments on actual instrumentation. The proposed solution integrates the advantages provided by an off-the-shelf LMS, compliant with international standards for web based training, and a new approach for providing remote experiments on measurement instrumentation, based on web services and the thin client paradigm. The proposed approach relies on Virtual Instruments (VIs) developed in LabVIEW and ensures that the students access the instrumentation without downloading heavy plug-ins.

Based on the proposed approach has been projected and realized a national measurement laboratory that will operatively provide the students of electric and electronic measurement courses with the access to remote measurement laboratories that deliver different didactic activities related to real-time measurement experiments. The project has been financed by the Italian Ministry of Education and University (MIUR) within the National Operating Program (PON) 2000-2006 [11]. The initial infrastructure is composed of the laboratories at the University of Sannio and at the University of Reggio Calabria "Mediterranea" under the patronage of the National Research Association on Electric and Electronic Measurement (GMEE), and the collaboration of about twenty Italian universities and some specialized instrumentation, e-learning and publishing companies such as National Instruments, Tektronix, Agilent Technologies, Yokogawa, Keithley, Rockwell Automation, Didagroup, Augusta publishing. Moreover, the work carried out until now led to a second project, financed by the Italian Space Agency, aiming to design a distance learning system that uses satellite networks as a backbone for providing web based training to mobile as well as home/office learners located in the whole Europe [12].

The paper is organized as follows. Section 2 describes the main functions and the trends in development of LMS. Section 3 summaries the most recent proposals for realizing remote laboratories for didactic purposes. Section 4 describes overall architecture of the proposed platform and the delivered functionalities. Section 5 presents the system architecture and its hardware and software components. Section 6 illustrates some performance evaluations of remote visualization of instruments panels based on the proposed solution.

2   Remote laboratories for didactic aims: the Virtual Laboratories

In order to understand the measurement procedures and measurement system design, it is necessary to repeat the same experience of actual measurements of physical phenomena many times in order to make all learners able to operate the measuring instrumentation [13-15].

Some drawbacks make it difficult to provide a complete set of updated workbenches to every learner. The most relevant are: (i) the high cost of measurement equipments and in general of the experimental laboratories in educational sites and industry, (ii) the growing number of students and specialized technicians, (iii) the reduced number of laboratory technical staff, and (iv) the continuous evolution of measurement instrumentation involved, that makes it difficult and very expensive to keep the technical staff up-to-date.

The potentiality of remote teaching [16] and, in particular, the use of the Internet as a channel to reach the students or workers at their homes was soon recognized [17-19]. Therefore, currently a lot of teaching material can be found as (i) Web based lectures and seminars, sometimes interactive, provided by hardware or software producers, mainly directed to professionals that want to reduce the time to market for a new application [20]; (ii) Web support to University courses, including slides of lectures and exercises [21, 22]; (iii) simulation of actual experiments to be executed either remotely or on student's PC [23, 24], and, more rarely, (iv) remotely accessible laboratories, where the learners can access real instrumentation through a Web page [9, 10, 25-27]. These solutions are often referred to as Virtual Laboratories (VL).

2.1   The Virtual Laboratory architecture

By following the innovation in the software and in the hardware market, the VL has assumed different structures and has increased its complexity and versatility in the years. Initially the VL was based on a sort of Virtual Instrument (VI) only. The VI was able to control the single measurement instrument by means, in the most cases, of an IEEE 488 standard interface designed to permit the instrument control from different manufactures to be interfaced in the same way. A Graphical User Interface (GUI) permitted to achieve this aim in friendly way.

Successively, the availability of the local area networks and software innovations made possible to manage a set of measurement instruments [32]. Nowadays, the internet technology permits (i) to implement the VL over wide area networks, and (ii) to use the advantages of the web technology [5], [10], [32]. Consequently, the VL architecture can be based on (i) stand-alone VIs operating stand-alone as the fundamental components, and (ii) both the VI and the connecting network as components of a Distributed Virtual Laboratory (DVL).

There are basically two ideas to be considered when the VI is taken into account. The first is the possibility of controlling a real measurement instrument by a friendlier GUI. To achieve this aim the IEEE 488, the serial or the VXI interfaces can be used. In this way a set of different measurement instruments, co-operating among them could be presented like a single, more sophisticated, instrument.

The second idea is to present a program simulating the real instrument's behaviour as a VI.

By taking into consideration both ideas, it is possible to realize mixed systems, more sophisticated and more flexible, that provide two modules (i) one for the simulation of the experiment, and (ii) another to carry out the real measurement process.

The DVL complicates the VL approach by using also Internet or Ethernet as medium or long-range communication bus for the instruments. By means of a DVL, the student can access the laboratory by means of a remote client from his/her home.

There are many available networks designed to meet differing requirements of the measurement applications, (i.e. Profibus, DeviceNet, Inter-bus, Ethernet). The Ethernet technology for restricted areas and Internet for geographic area are very attractive and cost-effective

Even if the interaction between the student and the instruments take place through the network, the main idea is still valid: the student is supplied with a graphical interface that makes the instrumentation easier and friendlier. How the real measurement procedure takes place, what kind of instruments and algorithm are used and how the network communication is implemented can be made transparent to the student. Differently, the student can know how the DVL works if necessary from the education point of view.

2.2   Development of DVLs

Two different approaches to develop the measurement software of the VL can be taken into account: the first one refers to the use of the commercially available integrated development systems, i.e. LabVIEW and related software tools as Measurement Studio. The second approach, also used in the literature, is based on the Object-Oriented Programming (OOP) as using Java and related technologies. By using both the approaches, the main aim of the VL can be achieved, but it is also true that both have advantages and disadvantages that should be considered before the architecture design.

In the case that the VL is part of distance learning platform, it is necessary that the GUI is integrated into the Learning Management System (LMS). The objective of the LMS is to manage the learners, keeping track of their progress and performance across all types of training activities. The LMS manages and allocates learning resources such as registration, classroom and instructor availability, instructional material fulfilment, and online learning delivery [9].

As the off-the-shelf LMSs are usually closed, proprietary software, often not customizable at all, the research carried out by scientists to provide teaching of electric and electronic measurement including experiments didn't take in account the possibility of integrated remote laboratories and LMSs.

The focus of their activity, instead, was the development of the remote laboratory itself, and adding scheduling and user account management modules. From the realization point of view, the solutions found in literature, except for [25], require that the software enabling the remote control of the instrumentation is developed almost from scratch, in C, C++, Java language. When a standard communication structure is not used, the reusability and the interoperability of a VI is greatly limited to the specific laboratory application and the expandability of the system is bound from the availability of skilled technicians able in developing new VIs to be included that system. The project [25], following the research trend in [8, 9, 28], reverses the problem relying on the use of LabVIEW from National Instruments, a standard language for VI development, for producing the VIs and the software AppletView, from Nacimiento, to produce Java applets that constitute a remote interface of LabVIEW VIs. In such a way it is possible to reuse the wide number of already developed VIs for integrating existing instrumentation in a remote laboratory without developing new software. Java applet ensures the compatibility of the laboratory with every operating system and doesn't oblige the student to download heavy plug-ins from Internet. Moreover, most of the solutions found in literature require the development of the accessory software applications enabling the sharing of the laboratory instrumentation, such as the scheduling and security policy. Following the same approach found in the LMS oriented research, in the last years a new trend started for ensuring the VI interoperability and reusability using the XML (eXtensible Markup Language) and SOAP (Simple Object Addressable Protocol). The VIs are realized and could be accessed as web services [8, 10]. The main advantage of these solutions is that they can be easily integrated in LOs for existing LMSs. However, in these cases, the language used to develop the VIs is C++ or J2EE, so the existing LabVIEW VIs cannot be reused.

3   The proposed solution

At the Universities of Benevento and Reggio Calabria, Italy, it has been realized a virtual learning environment devoted to the teaching of electric and electronic measurement that integrates an off-the-shelf LMS and a geographically distributed laboratory, accessible from the Web by using a simple browser. The distributed laboratory is accessed through the LMS executed on a central server that delivers such functionality to users by means of Web Services technologies [29] and the VIs controlling the instrumentation are written in LabVIEW.

3.1 Delivered services

The developed platform delivers the typical functionalities of a common LMS described above, including user authentication and management and tracking of learning process at user level. Moreover, it provides several innovative functionalities encapsulating in specific LOs the remote control of measurement instrumentation. This objective has been achieved by developing an additive module for the LMS Inform@ from Didagroup [30]. The module ensures the integration of VIs written in LabVIEW in the LMS as LOs, thus enabling remote users to transparently get the control of a real measurement instrument and to display the measurement results within the normal learning activities.

The remote laboratory is distributed on geographical scale since the measurement instruments are physically located in laboratories belonging to different universities. At now, there are two laboratories involved, one is located at the University of Sannio in Benevento, and the other is located at the University of Reggio Calabria "Mediterranea", Italy [31]. The access requests to the measurement instruments are handled by a scheduling system which, transparently through specific scheduling policies, connects the user to a specific physical laboratory in which the required measurement instruments are available.

Different users profiles are managed by the system: student, teacher and administrator.

The services delivered by the remote measurement laboratory module to the student are mainly the following:

(1) Experiment Visualization, this service allows the student following on-line a laboratory activity hold by the course teacher. The student obtains the displaying on his computer of the server desktop used by the teacher to control the measurement instruments involved in the experiment. The experiment is carried out by operating on the front panel of a LabVIEW VI controlling all the involved instrumentation.

(2) Experiment Control, this service allows the student to perform an experiment by controlling remotely one or more actual measurement instruments. The student can choose a specific experiment in a set of predefined ones and he can run it only if the required measurement instruments are currently available.

(3) Experiment Creation, this service allows the student to remotely create an experiment interacting directly with specialized software executed on the servers used to control the measurement instruments. This feature enables the adoption of PBL as didactic model. Under the supervision of the teacher, the students can develop a specific project by producing, in an individual or collaborative manner, a VI to control a set of real instruments.

The services delivered to a teacher are related to the remote handling of the available experiments (remote creation, modification and removal of experiments, etc.).

Finally the administrator is responsible of the correct working of the overall distributed system and of handling of user profiles.

3.2   Architecture of the distributed laboratory

In order to allow a student to access a remote and geographically distributed didactic laboratory, the paper proposes a Web-based multi-tier distributed architecture centered on the LMS that can be considered as the core component of the overall system.

The module designed to manage the remote laboratory is based on a scheduling system that manages the catalogue of available measurement instrumentation, and re-directs the user request to the measurement laboratory, chosen among the partner laboratories, in which the required instrument is currently available. Moreover, it enables the requestor to gain the control of a measurement instrument by the LabVIEW software which has not to be installed on client-side.

The proposed multi-tier architecture is composed of the following three tiers (Fig.1, Fig.2):

(1) the presentation-tier: it manages the experiment visualization on client-side. It is based on standard Web browsers with no need of specific software components (no specific operating system is required).

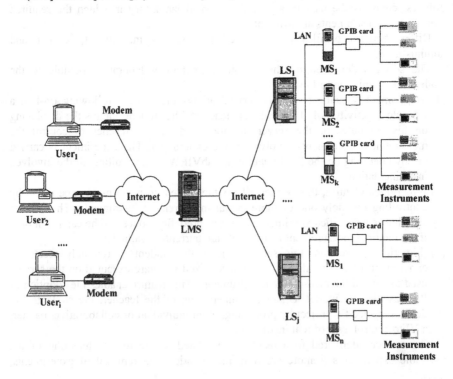

Fig. 1. Hardware components of the proposed architecture.

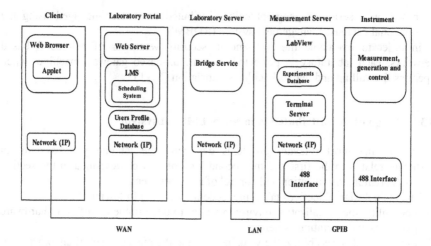

Fig. 2. Software layers of the components of the laboratory architecture.

(2) the middle-tier: it manages the system logic on server-side. It is composed of the following distributed server components:

- The LMS is executed on a central server of the overall distributed laboratory, called Laboratory Portal. The LMS interfaces to the users through a Web Server that is hosted on the same machine.
- A Laboratory Server (LS) is used to interface a real measurement laboratory with the rest of the distributed architecture. There is a LS for each measurement laboratory of the universities involved in the project. It delivers the access and the control to the laboratory measurement equipments through a service, called Bridge Service. Moreover, the Laboratory Server is the only machine in a measurement laboratory directly accessible through the Internet, while the other server machines are not accessible and typically constitute a private local network. For this reason the LS can also be used for security purposes in order to monitor the accesses to the measurement laboratory and to protect it against malicious accesses.
- A Measurement Server (MS) is a server located in a measurement laboratory that enables the interaction with one or more instruments. A MS is physically connected to a set of different electronic measurement instruments through an interface card. The GPIB interface has been used to connect the MS to the instruments.

The used VIs are stored in a database of the MS, where the LabVIEW environment is installed. No adjustment is necessary to include a VI in the virtual learning environment, therefore the wide number of existing VIs can be reused without requiring additive work.

(3) the storage-tier: it performs the data management, related for example to the management of the user profiles and the distributed management of the data related to the available experiments at the different measurement laboratories. It is

based on a series of geographically distributed databases, managed using the relational database management system (RDBMS).

In order to overcome the well-known security weakness of Microsoft based networks, each laboratory is protected by a Linux-based gateway machine which operates firewalling and network address translation (NAT).

3.3   Design of the distributed laboratory LMS subsystem

One of the most relevant problems designing the laboratory subsystem, in particular related to the first and the second kind of service described in the section 4.1, is the remote visualization and control of an experiment.

The main design objectives taken into account were the following:
(1) portability: the visualization environment has to be portable on different hardware platforms and operating systems;
(2) usability and accessibility: the visualization and the management of an experiment have to be easy to understand and to perform, even for users that are not expert of information technologies, and the system features have to be accessed easily and homogeneously by students operating at university laboratories or at home;
(3) maintenance: reduced maintenance costs. This can be made possible through a client-server approach that eliminates the need for installing and upgrading application code and data on client computers;
(4) client-side common technologies: students have to access to the system using their desktop computers based on common hardware and software technologies, with no need of very powerful processors or high central or secondary memory capacity, and connecting to Internet through low speed dialup connections;
(5) security: the remote access of the students through the Internet must preserve the integrity of recorded and transmitted data and of the system;
(6) scalability: the system performance has not to be degraded with the growing of the number of connected users.

In order to achieve these objectives it has been chosen to adopt a solution in which a student obtains the remote displaying on his home computer (thin client model) of the application executed on the MS and used to control the instrumentation scheduled for his request. Moreover, it has been chosen to use the Web and Java technologies integrated with the thin-client model. These technologies, in fact, can be opportunely used to allow a student, using exclusively a common Web browser and Java 2 Runtime Environment, the remote visualization and control of an experiment, ensuring a high system portability and usability and the fulfillment of the other objectives described above.

A thin-client model is based on a distributed computing paradigm in which the network separates the presentation of the user interface from the application logic. It is a client-server architecture in which the application execution and data management is performed on the server, called Terminal Server.

The user interacts with a lightweight client, called Presentation Client, which is generally responsible only for handling user input and output, such as sending user input back to the server and receiving screen display updates over a network connection. As a result, the client generally does not need many resources and can have a simple configuration, so reducing support and maintenance costs.

By using thin client technologies students are also able to use limited hardware devices, the so called Thin Client devices equipped only with a processor and a flash memory (no hard disk or other storage units). This solution extends the class of possible learners to mobile users, owning a smart phone, a personal data assistant or a notebook with a modem or a wireless LAN adapter.

The thin client paradigm allows the platform to visualize and control a remote device through the interaction flow among the distributed system components described in the following points:

(1) the student executes the authentication phase on the LMS platform using a login and password, interacting with the Web Server used by the LMS platform;

(2) the student chooses a service (Experiment Visualization, Experiment Control, Experiment Creation);

(3) the student visualizes on his/her desktop the VI front panel on the Measurement Server which he/she is connected to.

If the chosen service is the Experiment Visualization:

a. the student is connected to the Laboratory Server where the teacher is performing the experiment;

b. the Bridge Service of the Laboratory Server finds the Measurement Server that is currently used by the teacher, and allows the student to connect to the related Terminal Server in order to visualize the experiment on his/her own computer.

If the chosen service is the Experiment Control or Experiment Creation:

a. the student chooses an experiment among a list of available ones;

b. on the basis of the required experiment, the student is connected to a Laboratory Server of a partner university which is equipped with the required measurement instruments that available for the deployment of the experiment, chosen by the scheduling system of the LMS platform;

c. the Bridge Service of the Laboratory Server finds the Measurement Server that is connected to the required measurement instruments, and allows the student to connect to the related Terminal Server in order to visualize and manage or develop a new VI from his/her own computer.

It is worth to note that in the Experiment Creation phase the student has to develop a VI whose functionalities have already been foreseen by the teacher. Therefore, once a project has been assigned to a student, the related instrumentation is also known. The main difference between the Experiment Control and Creation is the student access to a VI front panel only or to the whole LabVIEW development environment.

3.4 System security

Security concerns are an essential feature of distributed multi-user scenarios. Service code downloaded from the remote site requires being trusted along with the remote site itself. User credentials must be verified and the usage of system resources granted, and resource accesses controlled.

In distributed laboratory the authentication and authorization mechanisms are implemented which permits user identification and acknowledgment of his roles (e.g. administrator or normal user), privileges and grants [32].

The concept of user groups is introduced and users may become members of one or more groups. A finer authorization control is achieved by specifying service method/function name(s) in the permission.

The authentication and authorization concerns are transparently managed with respect to service functionalities and service implementation.

Users can freely propose new groups. However, only signed user and accepted groups can be effectively used. A system administrator signs a new user and establishes its expiration time. Signed users cannot be modified by users: the system is able to recognize when a user is corrupted, modified or just invalid (e.g. it expired). A user submission form is offered and group membership is achieved by choosing items from a list of already accepted groups. Inspection of group properties is allowed. Likewise to user, a group submission form is also available. Submitting a new group requires the group name and a list of group permissions.

A reserved area, offering an overall vision of existing users and groups, is under the control of the system administrators. New UserACKs/groups can be signed/accepted and single permissions can be added or removed in a group as well as in a submitted UserACK.

Access control mechanism is regulated. Like a reservation, a valid lock permits service to be used according to a certain time window. Accesses can be allowed or denied depending on different criteria. Load balancing aspects, instrument availability and instrument exclusive use are often considered during a lock acquiring phase. If a measurement service is already locked or unavailable when requesting a lock, a different time window can be negotiated on the basis on existing reservations, user priority and so forth. Each service is responsible of its own lock administration.

4   Evaluation of thin-client technologies

Many approaches have been researched by academia and industry to support the distributed computing paradigm based on the thin-client model. In the paper [32] the following protocols have been taking consideration for the integration of the remote VIs and the LMS: RDP, ICA, X-Window and VNC.

RDP (Remote Desktop Protocol) (version 5.2) [33] is a proprietary protocol developed by Microsoft and based on T.120 protocol family standards. T.120 protocol is an international-standard protocol for collaborative work that Microsoft uses in NetMeeting. RDP allows the direct connection of a client executing RDP to the Windows Server of several Microsoft's Windows operating systems. In particular the first thin-client solution of the Microsoft was the release of the operating system Windows NT 4.0 Terminal Server Edition, in 1998, in which RDP was the default communication protocol. Successively, with the introduction of Windows 2000 Server, the Terminal Server become a component of the operating system and has been identified with Microsoft Terminal Service [34]. Terminal Service is available for Windows 2000 Advanced Server, Windows XP Professional and Windows 2003 Server. RDP supports different mechanisms to reduce the amount of data transmitted over the network connection such as compression and caching of bitmaps in RAM. In addition, RDP adds persistent bitmap caching, so improving performance over low-bandwidth connections, especially when running applications, such as LabVIEW, make extensive use of large bitmaps.

ICA (Independent Computing Architecture) was released by Citrix [35]. It allows the connection to the Windows Server through a server on which the software Windows Metaframe, that leverages the Microsoft Terminal Services, is executed. ICA is the protocol used to communicate between client and application server. ICA was designed to provide remote displaying with emphasis on the minimization of network traffic. It is extensible and includes optional protocol drivers for functionalities such as session control, encryption and compression hooks. Even if from the point of view of the features delivered, ICA is more mature than RDP, a main limitation in the use of the Citrix system is of economic nature. To activate a Citrix environment, the Citrix MetaFrame license (that is very expensive) is necessary, whereas RDP adds no additional cost to those related to Microsoft Windows Server.

X-Window is based on the so called *X protocol* [36]. It was developed in the mid 1980's to provide a network transparent graphical user interface primarily for UNIX operating system. It is significantly different from the ICA and RDP protocols. ICA and RDP implement mechanisms that grab an application GUI and send it to the client for displaying it. The X protocol in contrast specifies a client/server relationship between an application and its display at the application level. Furthermore, the X protocol is based on vector graphics whereas ICA and RDP are bitmap-oriented. In the X protocol, the software that manages a single screen, keyboard and mouse is known as an *X server*. A client is an application that displays on the X server and is usually termed as *X client*. This protocol was not chosen because, from the performance point of view, it lacks efficiency and clients require much more storage and processing power compared to other thin client technologies. The X protocol allows separating the presentation logic of an application but it was built for high-end UNIX workstations and not for thin-clients. The X server puts a heavy load on the network and the X protocol offers no sophisticated data compression or transmission optimization technologies. There are compression and caching extensions to the X Protocol such as LBX (Low Bandwidth X) but they are effective at the cost of GUI responsiveness. This renders X applications unusable over low bandwidth connections and therefore not suitable to be used in our system.

VNC (Virtual Network Computing) was released by AT&T [37]. VNC is also referred as *Remote Frame Buffer* (RFD) *protocol*, a very simple protocol based on the concept of remote frame buffer. VNC server merely sends its frame buffer contents to the client for display. Frame buffer updates are requested by the client only, so the client determines how often its display is updated. This allows for a flexible adaptation to the available bandwidth. The absence of multi-user support and security disqualify VNC for any other than single user usage on a private LAN, and so it is not suitable for a geographically distributed system. Moreover, some performance analyses have been carried out concerning the bandwidth occupation on a 56k modem connection. These evaluations showed that the amount of exchanged data between the client and the server are very huge with respect to the RDP protocol, causing some consequences such as high response times, slowness in the remote visualization, and also poorly displaying quality.

The RDP protocol has been chosen for implementing the remote laboratory services thanking to its lower cost with respect to ICA, and higher efficiency with respect to X-Window and VNC.

5   Realization and first performance evaluation

The proposed learning environment, called LADIRE [31], has been realized in its software components and is currently being deployed at the Laboratory of Signal and Measurement Information Processing (LESIM) of the University of Sannio [22]. Referring to the architecture described in section 4.2, the LADIRE has been realized with 1 LS, 1 firewall/NAT, 10 MSs, 40 measurement instruments going from simple Agilent 6.5 digit multimeters to the most up-to date Tektronix TDS 7704 oscilloscope and LeCroy SDA 6000 serial data analyzer, including spectrum analyzers, impedance meters, power meters, signal generators, power supplies. The enhanced LMS has also been installed on a Dell PowerEdge server, located in the same LESIM and a Web portal has been realized to enable the authoring of didactic contents from other universities. The enhanced LMS is the brain of the whole system and manages the resource allocation among the different physical laboratories constituting the didactic system. The PowerEdge server has been put in a private LAN behind a Linux based PC providing the security from external intrusions and the network address translation for the units in the LAN that have to access the Internet. In the current configuration, in fact, the LS is connected to the same LAN of the LMS and has to access the Internet. The LS manages the resource allocation in each physical laboratory and routes the traffic coming from the MSs to the Internet.

Once the MS resource is granted, the communication involves only the remote client and the MS by means of the LS. Therefore the data flux concerning the remote access to the VIs doesn't reduce the available bandwidth of the LMS. The MSs are inserted in the same private LAN of the LS and cannot be accessed from the Internet, so that the LS only should be protected from unauthorized access.

The collaboration of electric and electronic measurement professors coming from about twenty Italian universities and the National Research Association on Electric and Electronic Measurement is providing the didactic contents of the LMS as well as several measurement experiments. Some others are currently being developed at the LESIM.

The bandwidth required to the laboratory and student connections to the Internet depends on the specific implementation of the thin client paradigm.

In order to allow a client to access the system without the need of pre-installed software, it has been chosen to use a RDP client that allows the remote desktop to be displayed on the client-side using a standard Web browser. In particular, it is possible to use on client-side the *Terminal Server Advanced Client* (*TSAC*) released by

Microsoft. It is a Win32-based ActiveX control, which can be used to run Terminal Services sessions within Microsoft Internet Explorer. On the other hand the fulfilment of the main goals of our remote measurement laboratory required the design and implementation of a specific RDP client. In particular an important goal is that the student, after the authentication phase on the LMS platform, has to obtain the remote visualization and control of the required experiment executed in the LabVIEW environment on the appropriate Measurement Server through the Bridge Service component of the respective Laboratory Server. Moreover, in order to avoid malicious attacks to the system, the student has to obtain the remote visualization of the LabVIEW application without the knowledge of a user account on the Measurement Server, that is, instead, necessary to exploit the functionalities of the Terminal Server executed on that machine. It has been chosen to achieve this goal through the insertion, made by the Bridge Service component, of a valid username and password in the connection request made by the RDP client. Because of the legacy nature of the TSAC, it is not suited to modify at run-time the RDP connection request in order to insert the valid username and password. For this reason it has been decided to develop a specific RDP client using the Java Applet technology, an open technology that permits to develop client components executed into a Web browser and to leverage all the advantages of the Java language, particularly suitable for programming in distributed and heterogeneous computing environments, providing a direct support to many programming aspects such as multithreading, code mobility, and security. The applet permits the connection and the automatic authentication on the Terminal Server of the chosen Measurement Server.

To develop the RDP client, first the *ProperJavaRDP* [38] applet has been analyzed. It is an open source Java RDP client for Windows Terminal Services. It is based on *Rdesktop*, a SourgeForge project [39]. ProperJavaRDP runs on Java 1.1 up (optimized for 1.4), and works well on Linux, Windows and Mac. ProperJavaRDP was developed at Propero Ltd.. In order to evaluate the performance capability of the approaches based on ActiveX and on Java Applet technology, the bandwidth occupation of the client-server interactions has been measured, during the execution of an experiment. Because thin-client platforms are designed and used very differently from traditional desktop systems, quantifying and measuring their performance effectively is a very difficult task [40]. Many factors, in fact, can influence a performance comparison. Some of these are the use of optional mechanisms of the thin-client protocol (such as persistent caching and compression for the RDP protocol), the application executing on the server, the network bandwidth and the kind of traffic that shares the bandwidth, etc. In order to obtain the performance comparisons, for the purposes of the present work, it has been decided to trace the network bandwidth occupation in the interaction of the client and the Terminal Server during the visualization of a reference experiment (the Spectrum Measurement VI from National Instruments) executed on server side in the context of LabVIEW. The configuration of the server on which the Terminal Server is executed is: CPU Pentium IV at 2,80 GHz, 512 MB of RAM, Windows 2003 Server. Its network bandwidth towards the Internet is of about 768 kbps. The configuration of the client is: CPU at 2,0 GHz, 256 MB of RAM, Windows XP Professional. Its network bandwidth towards the Internet is that of a normal dial-up connection made through a modem at 56 kbps.

As it can be seen from Fig. 3, the ActiveX-based client occupies nearly the entire bandwidth during the time in which the user works on the VI. After this time interval, thanks to the use of the compression and persistent caching mechanisms of the RDP protocol, the bandwidth occupation decreases, up to about 700 bps.

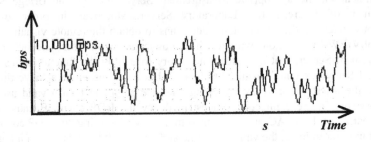

Fig. 3 Bandwidth occupation using ActiveX-based client.

The use of ProperJavaRDP client (Fig. 4), implies high bandwidth occupation (included between 4600 and 8950 bps), even when the user does not perform any action on the VI. This is due mainly to the fact that this client does not use compression and persistent caching. The reported bandwidth measurements have been obtained by using the tool integrated in the Windows operating systems and have been used to obtain a qualitative evaluation of the occupation in order to select the best trade-off.

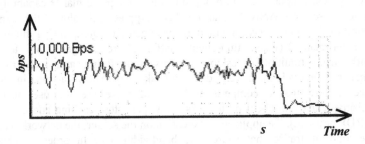

Fig. 4 Bandwidth occupation using ProperJavaRDP client.

6   Optimizations

Because of the high bandwidth occupation of ProperJavaRDP client, a specific RDP client Applet has been developed, called *LaboratoryApplet*. This RDP client

extends the solution of Propero by adding some functionalities and improving communication performances. Some of the added functionalities are the following:
- *compression* of transmitted data in order to minimize the network traffic;
- choice of the *cache dimension* on client side (not available for TSAC client that limits the cache dimension to 10 MB);
- choice of *load balancing* option: enabling to use of a load-balancing solution for Terminal Server based on farms of servers. So the client can be connected to the least-loaded available server, in terms of bandwidth occupation, number of opened sessions, CPU load and memory occupation.

By using the LaboratoryApplet, new occupation bandwidth measurements have been carried out during the execution of the reference experiment (see Fig. 5). As it can be seen this solution permits a bandwidth occupation lower than those obtained using the ProperJavaRDP client or the Microsoft ActiveX-based client. This is possible thanking to the persistent caching mechanism of RDP protocol and to the possibility of opportunely changing the cache dimension on client-side. In particular this solution permits to come to a bandwidth occupation of about 650 bps, after a time interval when the user doesn't perform actions on the VI, during which the bandwidth occupation is included between 4000 and 7000 bps, when the user acts on the VI.

Some optimizations have been made with respect to the Visualization Experiment service. This service allows a group of students to follow a laboratory activity hold by the teacher.

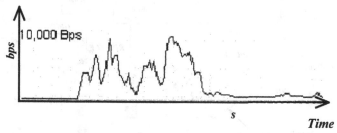

Fig. 5 Bandwidth occupation trend using the LaboratoryApplet client.

The Microsoft implementation of RDP protocol is layered on the TCP/IP protocol stack, which permits only a one-to-one communication. As a consequence, for the Visualization Experiment service, it requires the handling of a RDP session for each student of the group. This limitation of the RDP protocol implementation is the reason of the non-scalability of the system, since a bandwidth occupation proportional to the number of connected students that can lead to the saturation of the available bandwidth.

The solution proposed to solve this problem is to allow the Terminal Server to handle a unique data flow with all the connected clients. This can be made through the implementation of RDP on UDP/IP protocol stack, in order to leverage the multicast communication of the IP protocol. In particular, the proposed solution is shown in Fig. 6.

A first phase permits to initialize the multicast channel between the clients and the Bridge Service of the chosen Laboratory Server, whereas a default RDP session based on the TCP protocol is established between the Bridge Service and the Terminal Server of the required Measurement Server. The output data flow sent by the Terminal Server for the remote displaying to the clients is handled as a unique data flow and in particular the Bridge Service intercepts the RDP packets and encapsulates them in UDP packets, which are then sent to the clients using a multicast channel.

Another optimization has been made with respect the workload of the servers [41]. The workload of the server can be evaluated by means of the CPU occupancy parameter. This parameter is defined on the basis of the delay time occurring to the system CPU to answer at the system call. Obviously, higher the occupancy of the system CPU, longer is the delay time of the answer. Low CPU occupancy parameter means that the server can support other measurement tasks. Therefore, the estimation of the CPU occupancy parameter is based on the evaluation of the time response of the CPU by using some system library tied to the operating system.

The technique is based on the intelligent probe packet sent by the PC client. As an example, the PC#1 can be the object of the CPU occupancy evaluation and it must be configured as server. Consequently, another PC, as an example the PC#m, must be configured as client.

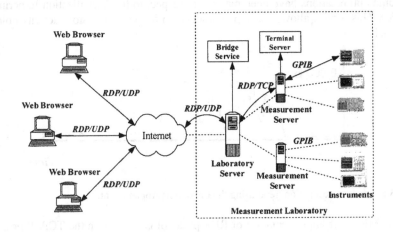

Fig. 6 Interactions for the Experiment Visualization service based on RDP/UDP.

Each probe packet operates according to the following steps: (i) n consecutive requests to the system clock are sent, (ii) the first and the n-th answer containing the timestamps of the clock are taken into account and stored, and (iii) the CPU occupancy parameter is evaluated as difference between the two timestamps stored. In order to make the CPU occupancy parameter independent from the clock frequency of the system CPU, the final value is furnished as the percentage ratio between the time interval defined from the two stored timestamps and the maximum time delay of the CPU response. The reliable solution is accomplished by stepping outside Java and writing few C code lines able to be (i) used in the different operating systems like Windows 2000/XP and Linux, and (ii) integrated with Java application via Java Native Interface.

7 Conclusions and future work

The paper described the hardware and software architecture an innovative platform for distributed e-learning that integrates a classical LMS with remotely accessible measurement laboratories through the Web. To this end, a solution based on the thin client model has been described, and the performances of different implementations of the model have been analyzed.

The work goes on in the following directions: (i) to develop a set of reference experiments including real instrumentation; (ii) to finish the development of the resource scheduler; (iii) to carry out more extensive performance evaluations in order to quantify the limits of the current solution in a geographically distributed environment; and (iv) to start a field test of the laboratory prototype with a group of students collecting their feedback opinions.

References

1. S.S. Ong, I. Hawryszkiewycz: Towards personalised and collaborative learning management systems, Proc. of the 3rd IEEE Int. Conf. on Advanced Learning Technologies (ICALT'03), Athens, Greece, (2003) 340-341.
2. H. Batatia, A. Ayache, H. Markkanen: Netpro: an innovative approach to network project based learning, Proc. of the Int. Conf. on Computers in Education (ICCE'02), Auckland, New Zealand, (2002) 382-386.
3. M. Guzdial, A. Palincsar: Motivating project based learning: Sustaining the doing, supporting the learning, Educational Psychologist, vol.26, no. 3-4, (1991) 369-398.
4. M. Albu, K. Holbert, G. Heydt, S. Grigorescu, V. Trusca: Embedding remote experimentation in power engineering education, IEEE Trans. on Power Systems, vol.19, no.1, (2004) 144-151.
5. P. Arpaia, A. Baccigalupi, F. Cennamo, P. Daponte: A distributed measurement laboratory on geographic network, Proc. of IMEKO 8th Int. Symp. on New Measurement and Calibration Methods of Electrical Quantities and Instruments, Budapest, Hungary, (1996) 294-297.
6. P. Arpaia, A. Baccigalupi, F. Cennamo, P. Daponte: A measurement laboratory on geographic network for remote test experiments, IEEE Trans. on Instrum. and Meas., vol. 49, no.5, (2000) 992-997.
7. P. Daponte, D. Grimaldi, M. Marinov: Real-time measurement and control of an industrial system over a standard network: implementation of a prototype for educational purposes, IEEE Trans. on Instrum. and Meas., vol. 51, no.5, (2002) 962- 969.
8. Bagnasco, M. Chirico, A.M. Scapolla: XML technologies to design didactical distributed measurement laboratories, Proc. of 19th IEEE IMTC, Anchorage, AK, USA, vol.1, (2002) 651-655.
9. G. Canfora, P. Daponte, S. Rapuano: Remotely accessible laboratory for electronic measurement teaching, Computer Standards and Interfaces, vol.26, no.6, (2004) 489-499.

10. P. Daponte, C. De Capua, A. Liccardo: A technique for remote management of instrumentation based on web services, Proc. of IMEKO-TC4 13th Int. Symposium on Measurement for Research and Industry Applications, Athens, Greece, (2004) 687-692.

11. Ministero dell'Istruzione, dell'Università e della Ricerca, Piano Operativo Nazionale 2000-2006, Avviso 68, Misura II.2b, project "Laboratorio Didattico Remoto Distribuito su Rete Geografica".

12. P. Daponte, A. Graziani, S. Rapuano: Final report on "Progetto preliminare di teleducazione", Agenzia Spaziale Italiana, (2004) http://progetto-teleducazione.cres.it

13. K. Mallalieu, R. Arieatas, D.S. O'Brien: An inexpensive PC-based laboratory configuration for teaching electronic instrumentation, IEEE Trans. on Education, vol. 37, no. 1, (1994) 91-96.

14. L. Finkelstein: Measurement and instrumentation science - an analytical review, Measurement, vol. 14, no. 1, (1994) 3-14.

15. S.S. Awad, M.W. Corless: An undergraduate digital signal processing lab as an example of integrated teaching, IEEE IMTC-97, Ottawa, Canada, (1997) 1314-1319.

16. M.Cobby, D.Nicol, T.S.Durrani, W.A.Sandham: Teaching electronic engineering via the World Wide Web, IEE Colloquium Computer Based Learning in Electronic Education, London, U.K., (1995) 7/1-11.

17. G.C. Orsak, D.M. Etter: Connecting the engineer to the 21st Century through virtual teaching", IEEE Trans. on Education, vol.39, no.2, (1996) 165-172.

18. M.A. Pine, L.A.Ostendorf: Influencing earth system science education: NASA's approach", Int. Geoscience and Remote Sensing Symposium, IGARSS'95, Florence, Italy, vol. I, (1995) 567-569.

19. S.M. Blanchard, S.A. Haie: Using the World Wide Web to teach biological engineering, Frontiers in Education 1995 - 25th Annual Conf. Eng. Education for the 21st Century, Atlanta, GA, USA, vol.2, (1995) 4c5.9-14.

20. TechOnLine - Educational Resources for Electronics Engineers. http://www.techonline.com .

21. NETwork per l'UNiversità Ovunque. http://www.uninettuno.it .

22. Laboratory of Signal and Measurement Information Processing - Engineering Faculty of University of Sannio. http://lesim1.ing.unisannio.it.

23. Agilent Technologies Educator's Corner. http://www.educatorscorner.com/experiments/index.shtml

24. V.M.R. Penarrocha, M.F. Bataller: Virtual instrumentation: first step towards a virtual laboratory, IEEE Int. Workshop on virtual and intelligent measurement systems, Annapolis, Maryland (MD), USA, (2000) 52- 56.

25. A. Bagnasco, A. Scapolla: A grid of remote laboratory for teaching electronics, 2nd Int. Learning Grid of Excellence WG Workshop on e-Learning and Grid Technologies: a fundamental challenge for Europe, Paris, France, (2003). http://ewic.bcs.org/conferences/2003/2ndlege/index.htm

26. L. Benetazzo, M. Bertocco: A distributed training laboratory, Eden annual conf. on open and distance learning in Europe and beyond, Granada, Spain, (2002) 409-414.

27. L. Benetazzo, M. Bertocco, F. Ferraris, A. Ferrero, C. Offelli, M. Parvis, V. Piuri: A Web based, distributed virtual educational laboratory, IEEE Trans. on Instrum. and Meas., vol.49, no.2, (2000) 349-356.

28. W. Winiecki, M. Karkowski: A new Java-based software environment for distributed measuring systems design, IEEE Trans. on Instrum. and Meas., vol.51, no.6, (2002) 1340 – 1346.

29. Web Services. http://www.w3.org/2002/ws/

30. http://www.didagroup.it

31. http://www.misureremote.unisannio.it

32. F. Cicirelli, A. Furfaro, D. Grimaldi, L. Nigro, F. Pupo: Management architecture for distributed measurement services, IEEE IMTC 2004, Instrumentation and Measurement Technology Conference, Como, Italy, (2004) 974-979.

33. N. Ranaldo, S. Rapuano, M. Riccio, E. Zimeo: A platform for distance learning including laboratory experiments, 42nd Annual Conf. of the Italian Association for Informatics and Computing (AICA), Benevento, Italy, (2004) 503-516.

34. Remote Desktop Protocol.
http://www.microsoft.com/windows2000/techinfo/howitworks/terminal/rdpfandp.asp.

35. Windows 2000 Terminal Services.
http://www.microsoft.com/windows2000/technologies/terminal/

36. Citrix. http://www.citrix.com.

37. X Protocol. http://www.x.org/X11_protocol.html.

38. VNC. http://www.realvnc.com.

39. ProperJavaRDP, Propero Ltd. http://properjavardp.sourceforge.net/.

40. Rdesktop, a Remote Desktop Protocol Client. http://www.rdesktop.org/.

41. J. Nieh, S.J. Yang, N. Novik: Measuring Thin-Client Performance Using Slow-Motion Benchmarking, ACM Trans. on Computer Systems, vol.21, no.1, (2003) 87-115.

42. A. Condarcuri, D. Grimaldi, A. Panella: Workload measurements and synchronization into a distributed measurement laboratory, 13th IMEKO TC-4 International Symposium, Athens, Greece, (2004) 360-366.

A WEB-Based Architecture Enabling Cooperative Telemeasurements

Alberto Roversi, Andrea Conti, Davide Dardari, and Oreste Andrisano

WiLAB, IEIIT-BO/CNR-CNIT, DEIS, University of Bologna,
V.le Risorgimento 2, 40136 Bologna, ITALY
{aroversi, aconti, ddardari, oandrisano}@deis.unibo.it

(Invited Paper)

Abstract. This paper addresses the concept of cooperative telemeasurements in which resources such as instruments and programmable devices are distributed in a network of different laboratories and can cooperate to set up augmented experiments. In particular, both the end user's benefits and the technical implementation challenging issues in enabling cooperative distributed laboratories are discussed. An example of possible network architecture design guidelines able to enable cooperation and remote control of instrumentation is provided with particular regard to the access of a user to distributed resources. Finally, an implementation of a real cooperative telemeasurement platform at WiLAB, Bologna, Italy, is described.

Keywords: Tele-measurement systems, cooperation, WEB user interface, architecture design.

1 Introduction

Recently, different works have been presented in the literature to describe hardware and software architectures enabling the remote access to complex instrumentation and measuring test bed concerning different application fields, that is enabling the so called telemeasurement (see, e.g., [1]). In 1994, the CNIT (*Consorzio, Nazionale Interuniversitario per le Telecomunicazioni*) [2] has been founded in Italy to create a research network among telecommunications groups of italian universities; one of the main projects of CNIT has been the set up of Networked Laboratories (LABNET, i.e., *Laboratorio Nazionale di Comunicazioni Multimediali*) [3] made up in Naples starting from and in conjunction also with previous experiences at the WiLAB, Bologna, Italy. One of the goals of LABNET is the definition of methodologies for remote access to complex instrumentation and a test bed concerning communications systems and networks.

In the last years, both in literature and in practice, the classical telemeasurement concept has been enhanced by another research task concerning with distributed measuring systems based on sensors and acquisition subsystems or distributed educational workbenches for measure, for which different solutions have been presented [4],[5],[6].

In [7] the new *wide sense telemeasurement* paradigm, where not only the instruments but also devices under measurement and interconnections can be remotely defined and controlled, has been proposed taking advantage of current flexible hardware platforms related to the Software Defined Radio (SDR).

In this paper, the concept is further elaborated by introducing another level of extension belonging to the *cooperative distributed telemeasurement*. This would allow a user to participate to a complex telemeasurement experience based on distributed laboratories that cooperate transparently to the user. This requires the set up of a proper communication network and the management of distributed resources.

This activity is performed in the context of the PRIN CRIMSOM and Summit projects, funded by MIUR and Regione Emilia-Romagna respectively, that consists in the study and development of new advanced telemeasurement architectures and is based on a basic hardware test platform developed in the VICOM project (*Virtual Immersive Communication*) [8]. This is an immersive platform that enables virtual laboratory experiences working in a distributed fashion with advanced 3D user interfaces.

The paper is organized as follows: in Section 2 we introduce the concept of cooperative measurement giving some examples of scenarios and highlighting outcome technical challenges, in Section 3 we treat the architecture design issue, while in Section 4 an example of WEB based cooperative telemeasurement implementation is reported. Finally, our conclusions are given in Sections 5 and 6.

2   Cooperative distributed measurements

The increasing necessity of cooperative actions in various measuring processes has been highlighted in recent works, where different definitions, levels of cooperation, challenges and solutions have been proposed [9, 10].

The general trend is to underline the importance to conceive a distributed measurement system as a single aggregated system, even if it is composed by several resources (measure instruments, sensors, programmable platforms, etc.), with different roles and functionalities, namely nodes. A large variety of applications ranging from industrial measure processes, environmental monitoring to educational engineering could take advantage from the strength of distributed resources aggregation. The real-time dialogue and exchange of measured data between nodes, the aggregation of information obtained by different measure processes, supported by high-level information, communication and elaboration networks will enrich results and methodologies of measurements.

To this purpose a new cooperation level can be added in the node's network architecture with the purpose to acquire information from distributed resources and offer, in a centralized way, functionalities like management, centralized pro-

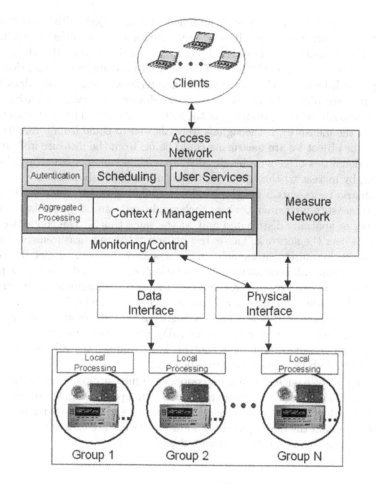

Fig. 1. Philosophy design of a cooperative distributed heterogeneous platform

cessing or other operations on the data acquired by distributed nodes. Another way could be the introduction of cooperation at first level, between nodes. This way of cooperation is possible only when each node in the network is able to exchange data or information directly with other nodes as could happens, for example, in wireless sensor networks where data are aggregated in a decentralized way (data fusion) by nodes before they are delivered to the final user [11]. In this case both the network throughput and the node energy consumption are reduced.

By considering complex measure instruments as nodes, we can in a similar way aggregate distributed resources in a such way a remote user could have, in a unified integrated interface fashion, all the functionalities and services offered by the distributed remote instruments.

Normally a measure instrument is intended like an object with specific characteristics and functionalities. It is in general composed by different functional blocks: Analog Section (analog signal processing, analog to digital and digital to analog conversions), Digital Section (elaboration unit, digital information processing) and Interfaces (User and computer interfaces, control unit, data bus). These units are physically placed in only one device (normally expensive) that needs to be positioned in proximity of the source of the signal to be characterized (object of the measure). If the signals were allowed to come from a remote site, not only the object we are measuring could be far from the measure instrument, but functionalities of geographically distributed instruments could also be aggregated, by increasing the added value of a network of laboratories.

This scenario represents a geographical extrapolation of the basic structure of a typical measure instrument, where each single unit can be placed in the same laboratory or spatially distributed and, at the same time, it can cooperate with other units over the network. This extrapolation of the telemeasurement concept increases the measurement capacity since it aggregates several distributed resources and avoids all necessary instrumentation being located in the same place. In this way users can experiment the access to a sort of "augmented-laboratory" composed by different cooperative distributed instruments. The sensation to work on a measure workbench even configuring remote resources is increased by the realization of a three-dimensional (3D) interface offering the aggregation of distributed resources and their control. For example, the 3D user interface realized at WiLab Bologna is reported in Fig. 2.

The design of a such distributed cooperative measure platform requires the analysis of different technical issues related to signal processing, the development of a web architecture able to manage heterogeneous and distribute resources, the user interface and the development of a library of measure experiences and, first of all, a proper communication network design.

3   Architecture design guidelines

In this session we approach the design architecture philosophy. A general cooperative system is composed by different groups of devices. The groups are geographically distributed, the devices are heterogeneous and with various functionalities. Hence, a hierarchic architecture approach is preferable.

The general cooperative and distributed architecture scheme proposed is depicted in Fig. 1. It is composed by different functional blocks that perform different tasks here described in detail:

-   *Groups*: the groups (e.g., laboratories) are distributed on the territory and are composed by different devices (instruments, programmable platforms or sensors). Each group has the possibility of local pre-processing like, for example, compressions of big quantities of measured data to reduce the network throughput, basic data fusion functions on distributed sensor acquisitions or other advanced operations.

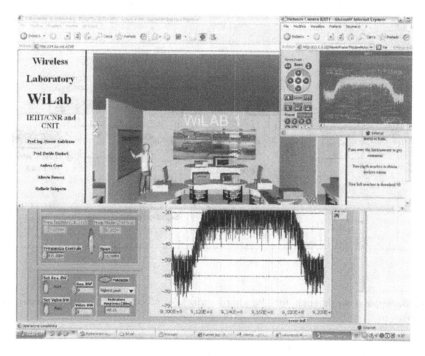

Fig. 2. An example of a 3D-Telemeasurement user interface

- *Access Network*: the access network is composed by the network and the interfaces used by clients to access to the platform. The evaluation of the impact of different access networks (wired, wireless, radio, satellite) on the QoS perceived by user is one of the planned task of the project.
- *Measure Network*: this is the network responsible for the data transfer between different groups. Note that the measure network can be used to transport the results of distributed measure processes and also to transport the signals to be measured or elaborated from one device belonging to another group. Physically, it can be composed by dedicated links or coincide with the access network. In the second case, the quality of service becomes a critical issue that has to be investigated.
- *User Services*: this block is the entire set of services offered by the system to the user. It collects data from connected functional blocks, elaborates them and presents the result to the user through opportune interfaces.
- *Authentication*: this block has to recognize the user by giving him rights to access and control the entire groups of devices (advanced user) or only a limited access to the platform (educational experience). The authentication block for example can be based on Lightweight Directory Access Protocol (LDAP) [12].
- *Scheduling*: In a multi-user and distributed environment, the presence of a resource management system makes strictly necessary to avoid conflicts

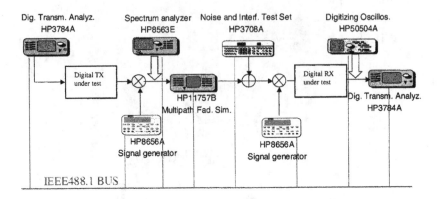

Fig. 3. Typical workbench for the characterization of communication transmission chain.

among different users contemporary accessing to the same resources. To this aim, we need a scheduling system in order to allow users to register himself and then to guarantee the reservation of the requested set of resources.

- *Aggregated Processing*: this block has two main functionalities. Firstly, it is a centralized processing support to devices with no significant intelligence on board; for example, if we have a distributed wireless sensor network, due to complexity constraints, each sensor will perform a minimal data elaboration, leaving this task to this block.

The second functionality is to manage the cooperation between devices and groups by allowing, for example, parallel measurements thus reducing the measure time. For example, suppose each group is composed by instruments for the characterization of the bit error rate (BER) for different values of the signal-to-noise ratio (SNR) in a digital communication transmission chain. It is well known how this kind of measure would be particularly time consuming if it were performed locally (i.e., using only the instrumentation present in the same laboratory). Fig. 3 shows a typical workbench configuration necessary for the BER measurement which is composed by a digital transmission analyzer, mixers, signal generators, multipath channel fading simulators, noise and interference test set, plus oscilloscopes and spectrum analyzers for signals observation in the time and frequency domains. Using collaborative laboratories equipped with the same set of instruments and making use of SDR platforms for system definition [7], it is possible to start parallel sessions of measure, one for each SNR value, and to get only one final result consisting in the curve reporting the BER as a function of the SNR with a considerable reduction of time. The architecture will be responsible for controlling, collecting and aggregating the results produced by distributed groups. The user will have the sensation to control only one "big" laboratory and will receive from the user services block only the final results.

Another approach is related to the case where not all the required resources are present in the same laboratory and the whole measurement process has to be split into several sub-processes each of them performed in a different laboratory. In this case the measure process starts in one laboratory, equipped with a part of all the necessary instruments, and it is completed in another laboratory with the rest of needed instruments. In the serial cooperative measurement approach, the presence of embedded hardware to acquire the signal in one laboratory and reconstruct it in another laboratory is necessary, as far as the software resources to packet and transport the signal over the measure network. Note that in the serial scenario the real-time transmission of the signal over the network may be or not a requirement depending on the application.

- *Context/Management*: when an advanced user approaching the cooperative telemeasurement platform is interested in using some particular devices or group of devices, he must know which of them are available, choose and reserve the resources and finally obtain the exclusive remote control. That is the task of the Context/Management block that sends queries to the Scheduling block asking for context information about the current use of resources. By interfacing to the Monitoring/Control blocks, it verifies through the Data Interface block the effective availability of groups and resources, upgrades his own dynamic database and return to the user, through the User Services block, the actual availability of groups and resources. In the next Section we will give a description of the architecture realized in our laboratories and we will provide an example of implementation of managing of distributed measure resources.

- *Data Interface*: in the proposed architecture each group is interfaced to the higher levels by a local server that plays the role of bridge between the devices interface and the monitoring/control unit. Normally, each commercial device product, like GPIB Enet [13], for example, providing a "bridge" between the IEEE 488.1 [14] and TCP-IP protocols, is available with his own interface. Commercial DSP (digital signal processor) boards are normally connected to a PC host via serial link (standard EIA232), PCI bus (Peripheral Component Interconnect) or, in some cases, there is the direct connection to the data network through commercial daughter cards [17]. Recently there is the availability of programmable FPGA devices with integrated network interface [16] whereas wireless sensor base stations are normally interfaced with Ethernet or serial interfaces [15]. This large variety of interfaces make the global integration one of the most critical implementation issues.

Physical Interface: it is mainly composed by platforms applied to the acquisition and reconstruction of analog signals under analysis. Each physical interface is controlled and managed by upper functional blocks monitoring/control and management. In this way the system can start the acquisi-

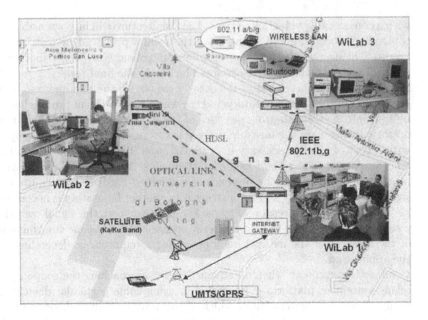

Fig. 4. Networked laboratories at WiLab in Bologna.

tion and reconstruction processes in a synchronized way among distributed groups and can control parameters like sampling-rate, filtering, transport network protocol, etc.. In our architecture the physical interface is supported by two software packages; the first one is at physical level with the task to configure the acquisition and reconstruction processes and to compensate eventually clock mismatches in the A/D and D/A conversion; the second one is finalized to the transport of the samples on the measure network. The technical challenges and the research interests to adopt these features are expected to be increased if the measure process is expected to work with real-time characteristics (for example real-time BER measurement, evaluation of subjective quality of services for real-time video streaming).

4   Example of a WEB based cooperative telemeasurement platform implementation

A first approach to the realization of the general architecture, described in Section 3, under development at the WiLAB laboratories at the University of Bologna, Italy, is now described. A sketch of the network composed by three different laboratories is illustrated in Fig. 4. Each laboratory is equipped by different instrumentation and programmable devices also accessible via wireless technologies. All laboratories are connected by networks of heterogeneous nature, like a wireless bridge IEEE802.11 b/g with speed up to 54 Mbit/sec, an optical link with speed up to 100 Mbit/sec and HDSL wired link with a guaranteed speed of 512Kbit/sec. The network architecture is also accessible by UMTS/GPRS and satellite networks technologies. The following architecture blocks have been implemented: user services, scheduling, context/management, monitoring control, data and physical interfaces. One of the main difficulty in implementing a general architecture depends on the fact that each component is equipped with different interfaces, software platforms and drivers. So that a great effort has been paid to integrate everything in only one architecture.

We addressed the study, the design and the realization of a server-based architecture able to manage the instruments and programmable devices on the measurement workbench at the WiLAB. When a remote user approaching the telemeasure platform is interested in using some particular instrument, he needs the knowledge of which instruments are available, choose and reserve the resources and, finally, obtain the remote controls. From the point of view of a remote user, the complete availability of a resource means that it must be turn on and nobody else got its control.

Here we describe the server-based architecture, represented in Fig. 5 and developed in our laboratories, which is able to handle multiple remote user access to the instruments and programmable devices present in the network.

The architecture is characterized by a first level composed by different laboratories (groups) containing two different classes of resources: the instruments are connected each others by the GPIB bus and interfaced to the second level through a GPIB-ENET. The programmable devices are connected to the second level by an integrated network interface or by the use of a dedicated Host PC.

The second level is composed by several slave servers, one for each laboratory based on Apache software [18]. They role is to test if the local instrumentation and devices are turned on and, in that case, to keep the status on their availability. The third level is represented by one resources reservation manager, composed by RRserver server Apache [18], My SQL [19], Labview Server [20]), that can be physically located in one of the laboratories. The resource reservation server informs the remote user about the availability of the instrumentation, reserves and gives to the user the remote control of the chosen set of instruments and keeps continually upgraded a special database containing the state of the resources. When the client needs to know the availability of the distributed resources, the master server queries the slave server which instruments are turned on, the slave server runs locally a particular Common Gateway Interface (CGI) that inquires the state of instrumentation to the GPIBEnet and, through TCP sockets, obtains the state of programmable devices from each device network interface. The resource manager is based on a parallel architecture composed by three servers:

- *WEB server Apache [18]*: his role is to dialog with clients. It contains both static (HTML language to configure programmable devices) and dynamic (PHP language) pages which can be visited by the remote user;
- *Database server MySQL [19]*: his role is to keep, manage and upgrade information about the instrumentation and the programmable devices present

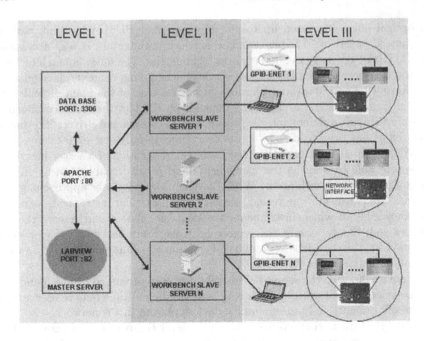

Fig. 5. Hierarchical Management/Control platform for distributed resources.

in the laboratory every time there is a query about their availability. The MySQL database server dialogs only with the Apache web server;

– *LabVIEW server[20]*: his role is to keep and send to the clients the executable Virtual Panel to control the chosen instruments.

Hence, the client chooses the resources, reserves them for a specific time slot and downloads from the LabVIEW Server the Virtual Instrument Panels, embedded in a HTML page as an ActiveX object, with the permission to control the selected instrumentation. In addition, a HTML page is provided to run different configuration files on the software radio based devices if present and available. In this way the number of users that can access the resource reservation system is not limited. A constraint is on the number of clients connected to different Virtual Instrument Panels depending on the number of Labview licenses available on the server (in our platform this number is 20). The reservation system is multi-user whereas the control of each virtual panel and programmable platforms is obviously possible only by a single user. It is foreseen the development of a resource scheduler in order to give to the user the possibility to reserve a set of instrument for a certain period. The implemented system has atomic characteristics so that when multiple users want to reserve more instruments involved in a particular experience, only one user will be able to reserve all desired resources in one shot.

Since different hardware technologies and platforms are used, the choice of particular programming language is not unique, but it is strictly related to the constraints imposed by the hardware technology. In particular, the C language has been used to develop the CGI interfaces taking advantage from the API libraries offered by National Instrument for the GPIB-ENET device. The realization of the Virtual Instruments panels using Labview language is due to its wide use in the community and the wide range of instrument drivers already available for this platform. However, these implementation aspects are completely hidden to the final user which interface is entirely based in HTML pages representing the chosen experience and the results of telemeasure process.

5 Preliminary tests and future activity

In Fig. 6 an example of measured consumed network bandwidth vs frame rate of the Virtual Panel display, is reported for the digital oscilloscope in a typical measure experience. In this way we can have an idea of the throughput requirements of the application. Obviously some data-rate consuming functionalities, such as the WEB cam, cannot work through low data rate access networks (e.g., GPRS). A rigorous measurement of the response time is a very hard task because of the variety of hardware, software and network technologies involved. However, qualitative experiments show that using wired and wireless local area network accesses to open one virtual instrument panel, the response is practically immediate (less that 1sec.). The timing response increases when the traffic generated by the opened virtual instrument panels saturates the GPIB-ENET/100 interface (maximum rate of 100kbyte/sec of TCP/IP traffic).

Actually we are working on to increase the value of this platform. In particular, as a first step, we are developing the aggregating processing block with the aim to parallelize the same measurement on different laboratories, as explained in the previous Section.

Secondly, we will integrate a wireless sensor network in the general architecture to perform a first test-bed on real-time detection and collaborative processing of distributed signals.

6 Conclusions

In this paper, the concept of cooperative telemeasurements for a network of distributed laboratories has been described. In particular, the end user benefits and technical implementation challenging issues in enabling cooperative use and aggregation of distributed measurement resources have been discussed. An example of possible network architecture design guidelines, to enable cooperation and remote control of instrumentation, has been provided. Finally, some implementation issues of a real telemeasurement platform have been described with particular regard to the set-up realized at WiLAB, Bologna, Italy.

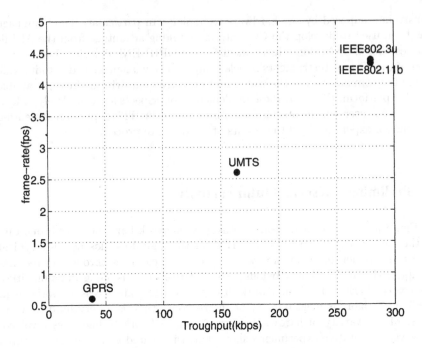

Fig. 6. Bandwidth requirements characterization for oscilloscope remote management.

Acknowledgments - This work has been performed under contract with the projects CRIMSON and Summit, funded respectively by MIUR and Regione Emilia-Romagna, Italy.

References

1. L. Benetazzo, et al., "A Web-Based Distributed Virtual Educational Laboratory", IEEE Trans. on Instrumentation and Measurement, vol. 49, no. 2, April 2000, pp. 349-356

2. CNIT site, url: *http://www.cnit.it.*

3. LABNET site, url: *http://www.labnet.cnit.it.*

4. D.Grimaldi, L.Nigro, F.Pupo, "Java-based Distributed Measurement Systems", IEEE Trans. on Instrumentation and Measurement, vol. 47. no 1, pp.100-103, Feb 1998.

5. L.Cristaldi, A.Ferrero, S.Salicone, "A Distributed System for Electric Power Quality Measurement", IEEE Trans. on Instrumentation and Measurement, Vol. 51,

no. 4, pp.776-781, Aug 2002.

6. M. Bertocco, F. Ferraris, C. Offelli, M. Parvis, "A client server architecture for distributed measurement systems", IEEE Trans. on Instrumentation Measurement, vol. 47, pp. 11431148, Oct. 1998.

7. O.Andrisano, A.Conti, D.Dardari, A.Roversi, "Telemeasurements and Circuits Remote Configuration through Heterogeneous Networks: Characterization of Communications Systems" Submitted to IEEE Trans. on Istrumentation and Measurements.

8. VICOM site, url: *http://www.vicom-project.it.*

9. F.Amigoni, A.Brandolini, G.D'Antona, R.Ottoboni, M.Somalvico, "Artificial Intelligence in Science of Measurements: From Measurement Instruments to Perceptive Agencies", IEEE Trans. on Instrumentation Measurement, vol. 52, no. 3, pp. 716723, Jun. 2003.

10. D.Castaldo, D.Gallo, C.Landi, "Collaborative Multisensor Network Architecture Based on Smart Web Sensor For Power Quality Applications", Proceeding of IMTC2004, Como, Italy, May 2004.

11. D. Dardari, A. Conti, R. Verdone, "Process Estimation through Self-Organizing Collaborative Wireless Sensor Network", IEEE Globecom 2004 - Wireless Communications, Networks, and Systems, Dallas, Texas, Dec. 2004.

12. "Authentication Methods for LDAP (RFC2829)" IEFT May 2000 available at http://www.ietf.org/rfc/rfc2829.txt.

13. J. Humphrey, V. Malhotra, V. Trujillo, "Developing Distributed GPIB Test Systems Using GPIB-ENET/100 and Existing Ethernet Networks", National Instruments, Application Note 103, www.ni.com.

14. ANSI/IEEE sdt 488.1, "IEEE standard: digital interface for programmable instrumentation", 1987.

15. Crossbow site, url: *http://www.xbow.com.*

16. COM5001 datasheet, Comblock site, url: *http://www.comblock.com.*

17. DSK.91C111 - 10/100 MBit Ethernet Daughter Card for DSK boards, reference site url:*http://www.dsignt.de/products/peripherals/dsk91c111.htm.*

18. Apache HTTP Server Project, www.apache.org/

19. MySQL Database Server, www.mysql.com/.

20. Labview User Manual, National Instrument, 1998.

The MOME Workstation As A Platform For Automatic Analysis Of Measurement Data

Marek Dabrowski[1], Jaroslaw Sliwinski[1], Felix Strohmeier[2]

[1]Warsaw University of Technology, Institute of Telecommunications, Poland
[2]Salzburg Research, Austria
E-mail: [1]{mdabrow5,jsliwins}@tele.pw.edu.pl, [2]fstrohmeier@salzburgresearch.at

Abstract: The paper describes the Data Analysis Workstation developed by the IST project MOME (Monitoring and Measurement Cluster). The MOME project maintains the repository of information about publicly available measurement data, like e.g. packet traces, flow traces, routing data, and others. The collected information includes the detailed description of an assumed measurement scenario, as well as selected statistical parameters obtained by data analysis methods. The MOME repository aims at providing assistance to the research community in finding and retrieving raw measurement data, which is most appropriate from the point of view of different research objectives. In this paper we discuss the design and development of the MOME Data Analysis Workstation. It extends the MOME Database with capabilities for automatic statistical analysis of submitted measurement data. The included exemplary results of analysis performed on packet traces captured in a measurement site associated with MOME illustrate that the MOME Database provides the researchers with useful information about available measurement data.

Keywords: measurements and monitoring, meta-database, data analysis

1   Introduction

Traffic analysis and modelling is considered as important topic in research on IP and IP QoS (*Quality of Service*) networks. Despite of the significant work performed recently in this area (see e.g. [3][4]), the Internet traffic is still not well understood and considered as difficult to be described by mathematical formulas. Broad range of Internet research domains rely on well-suited measurement data. Therefore effective research on Internet traffic requires access to large repositories of real and representative measurement data, still being able to find appropriate data for the specific research task. One can observe that a number of research projects were started with the aim of collecting such data (see e.g. [5][6][7][8]), but typically each one does it on its own way, using different measurement tools. Due to the lack of standard formats for data storage and annotation, finding appropriate data among various uncoordinated repositories is often quite difficult.

Therefore, the IST MOME (*Monitoring and Measurement Cluster*) project [1] has developed a meta-database. The MOME meta-database does not perform measurements by itself. Rather, the main goal is to collect and disseminate information about measurement data available from different repositories. The defined meta-data model (i.e. format of stored data) allows the submitters to describe the measurement scenario and environment by a carefully selected set of attributes [2]. Notice that the information about measurement scenario can be complemented by certain statistical parameters, which may be important for assessing the contents of the stored data. In this paper we discuss the attributes for storing the analysis results, which have been included in the MOME meta-data model. We claim that these parameters are the most adequate from the point of view of describing the characteristic features of the captured traffic.

The meta-database approach has been previously studied by several projects dealing with measurement data. For example, the SIMR architecture [9] and CADIA Trends project [10] aim at collecting and annotating various measurement data. The MOME database not only stores the information, but can also perform automated analysis of submitted measurement data. The MOME Data Analysis Platform allows for executing selected analysis tools. New analysis tools can be quite easily integrated in the MOME Workstation, by implementing appropriate filters, which parse their output files and insert results into the database. The access to the results is available for all visitors browsing the MOME database. Together with meta-data describing the assumed measurement scenario, the analysis results should help them in assessing, if a particular raw data set is appropriate for given research targets.

The remainder of this paper is structured as follows. In section 2 we introduce the MOME meta-data model, with special focus on the attributes defined for storing the results of data analysis. Section 3 describes in detail the architecture of MOME Data Analysis Platform. The exemplary case study, which demonstrates the capability of MOME system for performing selected analysis tasks, is presented in section 4. Finally, section 5 concludes the paper.

2   The MOME meta-data model

In this section we describe the meta-data model, which was used as a base for implementation of the MOME meta-repository. The meta-data model consists of a set of attributes, which annotate the most important features of various measurement data. These attributes describe in detail the measurement scenario, like e.g. the type of measurement, type and format of stored data, type of capture platform. Additionally, the meta-data model comprises some statistical parameters, which represent the internal characteristics of measured traffic.

Notice that the statistical parameters selected for implementation of the MOME database correspond only to a small subset of known analysis methods, applicable for a given type of data. However, recall that the main goal of the MOME repository is to assist the researchers in searching for measurement data, which is most appropriate for their needs. Then, they can retrieve the data and perform further processing, using more advanced methods (like e.g. compare the captured traces with various analytical traffic models). Therefore, the attributes of MOME meta-data model should represent only the basic and most important information, which adequately describes the characteristic features of the actual data. In particular, it should allow assessing, if the data is representative for the conditions assumed for particular investigated traffic models. However, we stress that the MOME database is flexible, i.e. it allows for adding new fields, if the results of some new analysis methods will be required. Currently five different types of measurement data are considered and described below: packet traces, flow traces, QoS results, routing data and HTTP traces. In addition, the MOME database supports a generic type for web repositories having mainly a pointer to a base URL.

2.1   Packet traces

The packet traces consist of collections of time-stamped records of packets, captured in certain measurement point. Obviously, the availability of representative packet traces is crucial for developing and validating packet-level traffic models. Such traces are also essential in the relatively new area of research on multi-level and multi-timescale traffic models (see e.g. [11]).

Notice that each trace is captured at a certain point of a network, only within a finite time interval. The trace can be regarded as representative, if its relevant features sufficiently reflect the general characteristics, and can be confirmed by a trace collected at a different time, or in different network. For example, different traffic models are proposed for the access and core networks, which can differ essentially from the point of view of level of traffic aggregation. For evaluating the traffic model, developed for core or access network, the used packet traces must be representative for the assumed type of network. Additionally, the duration of the trace should be sufficiently long to allow evaluating the stationary behaviour of captured traffic (or, to reveal its non-stationary behaviour).

Table 1. Packet traces analyses

Analysis name	Data type
Average packet inter-arrival time	Real
Average packet size	Real
Average packet rate	Real
Packet size histogram	Graph
Average bit rate	Real
Average bit rate per protocol	Graph
Average bit rate per application	Graph
Time-series of bit rates, calculated over non-overlapping intervals	Graph
Stationary variance of bit rate	Real
Required value of effective bandwidth	Graph
Required traffic descriptor parameters	Graph
Value of Hurst parameter	Real

To allow assessing if the trace is representative and appropriate, the information stored in the MOME meta-database should include such information, as: type of measured network, measurement time, location of capture device, speed of monitored link, capture method, filtering and anonymisation rules, capture platform, and the storage format of collected data. This information gives us detailed description of the assumed measurement scenario. In addition, several attributes for storing *analysis results* were defined, as presented in Table 1.

The first group of attributes corresponds to calculation of the empirical statistical parameters. The input data consists of a statistical sample of observed packet lengths, or packet inter-arrival times. Based on this sample, one can easily calculate the following parameters: **average packet inter-arrival time, average packet size, packet size histogram, average packet rate** and **average bit rate**. Additionally, **average bit rate per protocol** and **per application** correspond to the split of the captured traffic between different transport protocols (TCP, UDP) and different applications (identified by the port numbers). These statistical parameters (and graphs) give us basic information about the volume and type of traffic carried on the monitored link.

The **time-series of bit rates** shows the plot of bit rate (calculated as total number of bits transmitted within the time interval, divided by the length of this interval), as a function of time. It gives us information, how the observed traffic fluctuated throughout the trace duration. Notice that the rates can be calculated over intervals of different length. In MOME database, a few typical values are considered: 10ms (to reveal very-short time scale traffic changes), 1 second (short time-scale), and 3 minutes (medium time-scale). From the set of obtained bit rate samples we can calculate the **variance of bit rate**. This parameter attempts to quantitatively estimate the level of variability of traffic observed within trace duration. The set of bit rate samples can be also used as input data for more advanced analysis tasks, like e.g. fitting the parameters of assumed analytical traffic models, and validating their correctness. Currently, such methods are not considered in the MOME database, but can be added in the future, if needed.

The **effective bandwidth** gives us estimation of the resource requirements of captured traffic. In IP QoS networks, such information is necessary for performing the admission control decision. The algorithms for measurement-based estimation of effective bandwidth were discussed for example in [3][12][13].

The **traffic descriptor** parameters provide an envelope, which describes the worst-case behaviour of the traffic arrival process, in a deterministic way. Typically, the *token bucket* mechanism is used for this purpose. The information about required values of token bucket parameters can help in assessing the worst-case variability of the traffic captured in particular trace.

The last considered attribute is the **Hurst parameter**, which evaluates the level of self-similarity of captured traffic. Traffic is regarded as self-similar, if it is highly variable on multiple time scales. Several known methods for estimating the value of Hurst parameter are based on analysing the plot of Index of Dispersion for Counts (IDC), analysing the variance-time plot, the R/S analysis, periodogram-based analysis in frequency domain, or wavelet-based analysis [3][14].

2.2   Flow traces

The flow traces consist of a collection of records of distinct flows, observed in certain measurement point. Single flows can be defined as by IETF-IPFIX in [19]. The flow record contains the timestamps of flow arrival and departure, as well as the number of packets and bytes transmitted within its lifetime. The flow traces are useful in developing and validating the traffic models on the flow (or connection) level. Similarly to the case of packet traces, the capture location and duration should be considered for assessing the usefulness of the trace. Therefore, the following meta-data attributes have been defined: type of measured network, measurement time, location of capture device, speed of the monitored link, capture method, filtering and anonymisation rules, type of capture platform, and the storage format. The proposed list of adequate parameters corresponding to statistical analysis of flow records is presented in Table 2.

Table 2. Flow traces analyses

Analysis name	Data type
Average flow inter-arrival time	Real
Average flow duration	Real
Average number of packets per flow	Real
Average number of bytes per flow	Real
Average flow arrival rate	Real
Average traffic bit rate	Real

The **average flow inter-arrival time** and **average flow duration** describe basic flow-level characteristics of observed traffic. In the case of packet-based networks, the "size" of the flow must be described not only by its duration, but also by the total number of transmitted bytes and packets. Therefore, both the **average number of packets per flow** and **average number of bytes per flow** are included as information about typical size of observed flows. The **average flow arrival rate** corresponds to the intensity of new flow arrivals. Finally, the **average bit rate** gives us long-term estimation of utilised resources.

2.3 QoS results

In the case of QoS measurements, the raw data consists of a set, or time-series, of measured values of QoS metrics, like e.g.: one way delay, delay variation, packet loss ratio, etc. The singleton values correspond to the measurement taken on a single packet (probe packet in the case of active measurement, or data packet in the case of passive measurements).

The meta-data attributes should give precise information on the network, where the measurement was performed, and the time, when it was performed. Therefore, the defined meta-data model includes such information, like: the name and type of network, location of measurement equipment, method for time synchronisation between the transmitter and receiver, calculated metrics, and the storage format of the measurement results. The proposed additional attributes related with data analysis tasks are presented in Table 3.

Table 3. QoS results analyses

Analysis name	Data type
Minimum packet delay	Real
Average packet delay	Real
Maximum packet delay	Real
10-percentile of packet delay	Real
90-percentile of packet delay	Real
Histogram of packet delay	Graph
Minimum packet delay variation	Real
Average packet delay variation	Real
Maximum packet delay variation	Real
Packet loss ratio	Real

The analysed statistical sample consists of the set of singleton measurements of packet delay, or packet delay variation. The basic statistics calculated from this sample include: **minimum**, **maximum**, and **average** value, or certain **quantiles** of the distribution of the considered quantity (packet delay, or packet delay variation). Additionally, the **histogram can** give us estimation of the distribution of the packet delay or delay variation. Finally, the **packet loss ratio** completes the assessment of QoS delivered to the measurement traffic.

2.4   Routing data

The routing data consists of the snapshots of routing tables, or collections of routing update messages captured within particular measurement interval. The routing data may be useful for investigating the changes of logical inter-domain topologies, observing trends in number of advertised prefixes, or detecting abnormal behaviours of the inter-domain routing. The selected statistical attributes for routing data are presented in Table 4.

Table 4. Routing data analyses

Analysis name	Data type
Routing table size	Integer
Number of update messages	Integer

The **routing table size** (expressed in the number of prefixes) allows us for assessing the global state of network at a time of taking the capture. It can be useful for investigating the trends in routing table size and growth of the total number of advertised prefixes. The **number of updates** can be evaluated in the case of capturing the routing protocol messages.

2.5   HTTP traces

The HTTP (*Hyper Text Transfer Protocol*) is the application layer protocol, designed for the web application. The HTTP trace contains records of observed user HTTP requests and server HTTP responses typically collected by log-files in web- or proxy-servers. The information stored in such traces corresponds to the application (or session) layer traffic model. The statistical parameters selected for describing the contents of HTTP traces in the MOME database are shown in Table 5.

Table 5. HTTP traces analyses

Analysis name	Data type
Average HTTP request inter-arrival time	Real
Average response size in packets	Real
Average response size in bytes	Real
Average HTTP request arrival rate	Real

Several statistical parameters can be useful for evaluating the activity of web application users. The arrival process of user request is described by **average request inter-arrival time** and **average request arrival rate**. The basic parameters characterizing the server responses are **average response size**, measured **in packets** and **in bytes**.

3   Automated analysis of measurement data

Based on the proposed meta-data model, the MOME meta-repository has been implemented and deployed (see Fig. 1). It is available publicly via the MOME project webpage [1]. The main part of the system is the **Database**, which stores all the MOME meta-data. Users can browse the MOME database using the **Graphical User Interface (GUI)**, implemented on the MOME **Web-server**.

The user can request analysis of raw data, stored in some public repository or available directly from the measurement site associated with MOME. The data analysis is performed on the **MOME Data Analysis Workstation** ('data analysis' block on Fig. 1). The raw data file is retrieved from its original location and temporarily stored it on a local disk. After performing the analysis, the results are inserted into appropriate fields in the MOME Database.

The MOME system assumes that external, freely available data analysis tools can be used for performing actual operations on the measurement data. Each tool must be adapted to the MOME Workstation by implementing appropriate filters, which are able to read and parse the analysis results. At the beginning stage of the project, only few exemplary analysis tools were selected. They allow performing basic analysis tasks for packet-level traces, stored in the popular format of *libpcap* [15] However, new analysis tools can be quite easily added, depending on the needs.

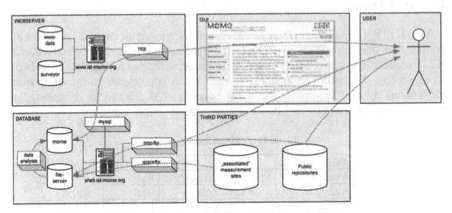

Fig. 1. Deployment of the MOME meta-repository

3.1   Executing data analysis tools

Fig. 2 illustrates step-by-step the process of executing data analysis task. When an authorised user browses the database, the **Analysis Request Manager (ARM)**, which is a part of MOME GUI, displays the data analysis tasks available for particular measurement data. If the user requests the analysis (step 1), the ARM inserts a new pending request into the MOME database.

The software modules, which constitute the MOME **Data Analysis Workstation,** are implemented as independent processes and can run concurrently on the MOME server. The **Autonomous Process Manager (PM)** is implemented as a PHP CLI (*Command Line Interface*) script and is periodically executed on the MOME Database server. The PM searches the database for new pending analysis tasks. If PM finds out that new analysis task has been requested, it starts the **Download Manager (DM)**, providing it as parameter the link to the file, which should be analysed (step 2).

The DM, which is also implemented as a PHP CLI script, retrieves the file identified by the provided link and saves it on the local storage (step 3). Notice that multiple DMs can be started at the same time, to download different remote files. After the DM has retrieved the file, the PM can change the status of the meta-data entry in the MOME database to *"analysis pending, local copy"* and execute the **Analysis Manager (AM)** (step 4). The parameters, which he provides to the AM, are: the type of requested analysis, and the name of the local file with raw data. The status of the appropriate entry in the meta-database is now changed to *"analysis in progress"*.

The **Analysis Manager (AM)** is implemented as a PHP CLI script. Only one instance of the AM runs at a time. The AM executes the external analysis tool (step 5), providing the name of the local file to be analysed. Therefore the tool needs to be installed with appropriate privileges for running on the MOME server. After finishing the actual analysis work, the analysis tool saves the results into the output file (step 6). The AM parses this file and inserts appropriate values into the MOME database (step

7). Finally, it sets the data status to *"analysis done"* and finishes the data analysis procedure.

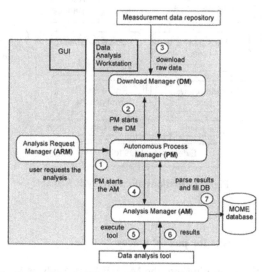

Fig. 2. Operations of the MOME Data Analysis Workstation

3.2 Adding new analysis tools

Fig. 3 shows the internal structure of the AM module. Recall that this module manages the execution of data analysis tools integrated in the MOME Workstation. The *'check available tools'* function finds available analysis tools and selects the tool that is capable to perform the requested analysis. The *'execute analysis tool'* function starts the actual analysis tool, providing as parameter the name of the local file with raw measurement data. The analysis tool performs the requested operations and saves results to the local output file. This file must be read by the function *'parse results, plot graphs'*. This function parses the results, extracts appropriate data and, if necessary, generates graphs using the *GNUPlot* tool. Finally, the results are inserted into the appropriate fields of the MOME database (*'update database'* function).

The actual analysis operations are performed by the external data analysis tools. Any freely available analysis tool can be quite easily integrated in the MOME system, if it meets the following requirements: can be executed on the Linux system, can read input data from the file and write output data to the text file. Notice that particular analysis tools may be dedicated for specific types of measurement data (e.g. packet or flow traces), and specific input data formats (e.g. *pcap* in the case of packet traces, or *NetFlow* in the case of flow traces).

Adding new analysis tool to the MOME Data Analysis Workstation only requires implementing two new functions in the AM: *"execute analysis tool"* and *"parse results, plot graphs"* (see Fig. 3). Those functions act as an interface to the external tool and are implemented separately for each tool supported by the system.

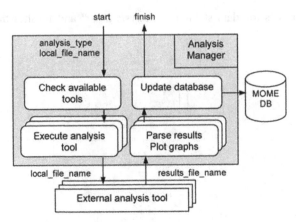

Fig. 3. Operations of the Analysis Manager

3.3   Basic analysis tasks for packet traces

As a first tool for integrating into the MOME Data Analysis Workstation, a simple utility called *tcpdstat* [16] has been chosen. *Tcpdstat* performs basic analysis of packet-level traces captured by tools like *tcpdump* [15] or *tcpdpriv* [17]. As input data, it reads the binary files saved in the *libpcap* format. The statistics calculated by *tcpdstat* include: file size, measurement start-time and end-time, total number of captured bytes and packets, average and peak bit rate and standard deviation, number of observed distinct flows, average number of packets per flow, top 10 flow sizes, number of observed distinct IP addresses, packet size distribution, average rate, split between different protocols and applications.

The *tcpdstat* saves the results in a form of human-readable ASCII file (see Fig. 4). Some of the values calculated by *tcpdstat* can be directly used for filling appropriate attributes of the MOME meta-data model (see section 2.1). For example, the **average bit rate** can be directly extracted by parsing the *tcpdstat* output file. Notice that the **average packet rate** can be easily calculated, based on other values produced by *tcpdstat*. Additionally, based on the textual data parsed from the *tcpdstat* output, MOME Workstation creates the graphs of **packet size histogram**, and **average bit rate per protocol** and **per application**. The *GNUPlot* tool, executed on the MOME server, produces the actual graph, which is then stored as a PNG picture in the MOME database.

```
TotalTime: 899.82 seconds
TotalCapSize: 166.31MB  CapLen: 96 bytes
# of packets: 3103414 (1490.21MB)
AvgRate: 13.89Mbps  stddev:1.10M   PeakRate: 18.61Mbps

### IP flow (unique src/dst pair) Information ###
# of flows: 279098  (avg. 11.12 pkts/flow)
Top 10 big flow size (bytes/total in %):
  7.6%  2.9%  2.8%  2.1%  1.3%  1.3%  1.2%  1.2%  1.1%
 1.1%

### IP address Information ###
# of IPv4 addresses: 190536
Top 10 bandwidth usage (bytes/total in %):
 29.7% 13.2%  9.1%  7.8%  4.9%  4.5%  3.4%   3.1%  2.9%
 2.9%
### Packet Size Distribution (including MAC headers) ###
<<<<
 [   16-   31]:      8285
 [   32-   63]:    948296
 [   64-  127]:    835986
 [  128-  255]:    165792
 [  256-  511]:    158287
 [  512- 1023]:    137072
 [ 1024- 2047]:    849696
```

Fig. 4. Fragment of exemplary results file of tcpdstat

4   Exemplary results of analysis performed by the MOME Data Analysis Workstation

In this section we demonstrate the capabilities of the MOME Data Analysis Workstation for analysing the data obtained in a measurement site "associated" with MOME. By "associated" measurement site we mean that the results of performed measurements are automatically uploaded and annotated in the MOME meta-repository. We show that the results provided by simple analysis tools integrated in the MOME system can provide interesting information for the researchers visiting our meta-repository.

4.1   Capturing measurement data at the MOME associated sites

As a reference and for demonstration, a monitoring and meta-data storage process for packet traces has been implemented in the premises of Fachhochschule (FH) Salzburg. The information from the packet headers is captured, made anonymous, compressed and uploaded together with necessary meta-data to the MOME database. This approach provides an automatic way to continuously get up-to-date measurement data to the MOME database. The procedure is divided into the following steps:

(1) Header capture + anonymisation (*tcpdpriv* [17])
(2) Data compression (*bz2*)
(3) Meta data construction
(4) Data upload (trace + meta data), by *ftp*

(5) Meta data entry to MOME database

The capturing in step (1) can be done according to different schedules and filters. A first scenario assumes a full packet capture process, starting daily at 02:00 until 23:59, with traces stored in hourly intervals. Data transfer to the analysis station is scheduled for the time period between 00:00 and 01:59. Note that the data upload is never performed during the capturing process, to avoid interferences with the monitored traffic. Due to the low access link speed (4Mbps), the amount of a full data capture can be handled without losses. Additionally, header extraction and compression reduce the accumulating amount of data usually below 500MB per day. This scenario will be adapted based on the produced analysis results.

While steps (1) - (4) are performed by the monitoring host, the last step is done at the MOME database server in regular intervals. The following figure shows the network set-up as it was installed at the reference capture point at the FH Salzburg. Due to the simple installation without major changes in the operational network, a set-up with a mirror port on the operational edge router has been preferred against other installations, e.g. with Linux routers or splitters. Only outgoing traffic is monitored.

The automatically generated meta-data is listed in Table 6. Exemplary values are given for a specific monitoring scenario at FH Salzburg.

After the meta-data entry is available in the MOME database, analysis requests can be set on these traces by using the web interface. In the case that all automatically uploaded monitoring traces are dedicated for analysis, the analysis requests can be set in the database automatically.

Table 6. The meta-data values generated for the trace collected in measurement site "associated" with MOME

Common Attributes		Packet Trace Specific Attributes	
Attribute	**Generated Value**	**Attribute**	**Generated Value**
StartTime	local start time of the measurement trace	NetworkType	"WAN Access Network"
EndTime	local end time of the measurement trace	CollectorLocation	"AT, Salzburg, FH Salzburg"
FileSize	comressed file size in Bytes	TrafficType	"operational unidirectional outgoing traffic, educational institution"
Description	"Trace FH Salzburg"	LinkProtocol	X.21
DataLocation	link to the uploaded file	LinkSpeed	4 Mbps
FileCompression	bz2	CaptureMode	mirror port
MD5Sum	the md5sum	FilterRules	None
SubmissionDate	local time of data upload	NumberPackets	counted number of packets
LastUpdate	database insertion date	TraceAnonymisation	-P99 -A50
Tool	tcpdpriv		
DataAvailability	local copy	CapturePlatform	Linux 2.6.9
SubmitterID	a dedicated id	DataFormat	libpcap
		AdditionalInfo	2x2Mbps X.21 connections

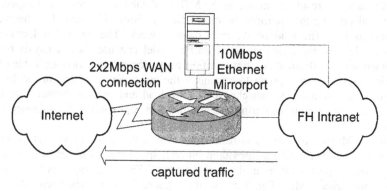

Fig. 5. Network Set-up

4.2 Analysing packet traces captured at MOME associated sites

Below we present the results of analysis performed on the 1-hour long traces from the measurement site in FH Salzburg, captured during three days on the 20th February (Sunday), 28th of February (Monday), and 1st of March (Tuesday) 2005. The traces were automatically uploaded and annotated in the MOME meta-repository. Then, the trace analysis was performed by the MOME Data Analysis Workstation, as described in section 3. The analysis results can be accessed directly via the MOME repository GUI (see Fig. 6). The average values of bit rate, packet rate, packet size and inter-arrival time are directly displayed on the screen, together with other detailed information about the trace. The plots of packet size histogram, as well as per-protocol and per-application distribution of bit rate, are displayed after clicking on the *'graph'* button.

Traffic rate, averaged over entire trace duration, in bit/s	2442547
Average packet inter-arrival time in sec	0.003186
Average packet size in bytes	972.7
Average packet arrival rate in pkt/sec	313.89
Histogram of packet sizes	graph
Bandwidth use per-protocol	graph
Bandwidth use per-application	graph

Fig. 6. The representation of analysis results for exemplary packet trace annotated in the MOME meta-database

The presented case study validates the design and implementation of the MOME Data Analysis Workstation. In addition, it demonstrates how MOME meta-repository can provide the research community with relevant, pre-processed information about the gathered traces. For example, Fig. 7a and Fig. 7b present the daily plots of average bit rate, and packet rate, respectively. This kind of plots can be quite easily obtained from the analysis results available in the MOME database. As one could expect, the traffic peaks during the afternoon hours are clearly visible. One can also observe the fluctuations in traffic load on different days of week. The researcher looking for traces needed for validation of some traffic model can use these analysis results, combined with the detailed description of the measurement scenario (see Table 6), for selecting the most appropriate trace among the collection available in the MOME database. For example, he can choose to download and process further only those traces, which were captured during higher utilisation period of the FH Salzburg WAN connection.

Another MOME data analysis results are demonstrated by the graphs on Fig. 9. They show the distribution of measured bit rate, split between different applications (recognized by port numbers in the packet header). Fig. 9 corresponds to one of the FH Salzburg traces, while Fig. 9 to one of the traces from the public repository [18]. As a simple example of usage of such information, consider a researcher looking for traces needed for his work on modelling traffic related with the audio streaming application (in particular, *realaudio*), or with peer2peer applications (*Kazaa*, or *Napster*). Thanks to the graph from Fig. 9, he can conclude that the FH Salzburg trace is not useful for him, because such applications are not represented in this trace. On the other hand, one can find the traffic related with the considered applications in the MAWI trace (see Fig. 9b).

Summarising, the MOME repository can assist the researchers in finding and retrieving the traces, which are most appropriate for their needs. Although the currently integrated analysis tasks are quite simple, the demonstrated scenarios show that they can achieve the main goal, which is providing the researchers with high-level information about the contents of the trace.

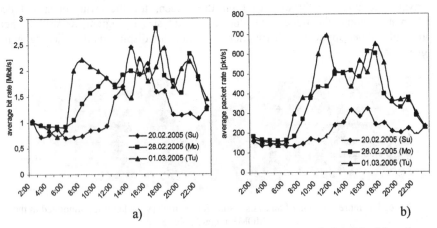

Fig. 7. The plots of: a) average bit rate and b) average packet rate, calculated for 1-hour long traces collected during 20 Feb, 28 Feb and 01 Mar 2005

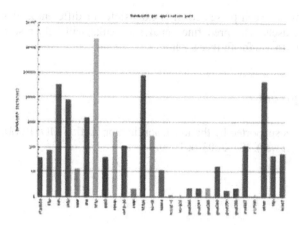

Fig. 8. The average bit rate of traffic related with different applications in the FH Salzburg trace from 20.02.2005, 14:00-15:00

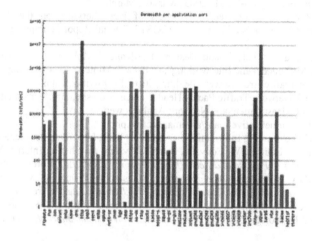

Fig. 9. The average bit rate of traffic related with different applications in the MAWI trace from 31.01.2005, 14:00-14:15

5   Conclusions

In this paper we presented the automated measurement data analysis approach assumed by the MOME Workstation. The goal of this approach is to enhance the existing MOME measurement data catalogue, storing general information about measurement data, like: time, duration or file sizes of measurement traces, with additional attributes describing the contents of the measurement trace itself. A framework providing the measurement data handling tasks, like downloading and the database access has been presented. It is extensible for different analysis algorithms on different datasets. The predefined analysis results will help researchers to find traces they want to use for further studies.

Acknowledgment

The work was supported by the IST Coordination Action MOME (*Monitoring and Measurement Cluster*), IST-2003-001990.

References

1. IST-MOME web site, http://www.ist-mome.org
2. P. Aranda-Gutierrez et al., „MOME: An advanced measurement meta-repository", in proceedings of *3rd International Workshop on Internet Performance, Simulation, Monitoring and Measurements*, IPS-MoMe 2005, Warsaw, March 2005
3. P. Tran-Gia, N. Vicari (eds.), "Impacts of New Services on the Architecture and Performance of Broadband Networks", COST257 Final Report, compuTEAM, Wuerzburg, 2000
4. M. Menth (ed.), "Analysis and Design of Advanced Multiservice Networks Supporting Mobility, Multimedia and Interworking", COST279 Midterm Report, Aracne, Roma, 2004
5. The Internet Traffic Archive web site, http://ita.ee.lbl.gov/index.html
6. The NLANR Network Traffic Packet Header Traces web site, http://pma.nlanr.net/Traces/
7. The NLANR Measurement and Network Analysis web site, http://watt.nlanr.net/active/maps/ampmap_active.php
8. The RIPE NCC Routing Information Service Raw Data, http://data.ris.ripe.net/
9. M. Allman, E. Blanton, W. Eddy, "A Scalable System for Sharing Internet Measurements", in *Proceedings of Passive and Active Measurements*, PAM 2002, Fort Collins, March 2002
10. The CAIDA "Correlating Heterogeneous Measurement Data to Achieve System-Level Analysis of Internet Traffic Trends" project, http://www.caida.org/projects/trends/
11. P. Salvador, A. Pacheco, R. Valadas, "Multiscale Fitting Procedure using Markov Modulated Poisson Processes", *Telecommunication Systems Journal*, 23(1-2):123-148, June 2003
12. C. Courcoubetis, V.A. Siris, G.D. Stamoulis. "Application of the Many Sources Asymptotic and Effective Bandwidths to Traffic Engineering", *Telecommunication Systems*, 12(2-3):167-191, 1999
13. M. Dabrowski, F. Strohmeier, "Measurement-based Admission Control in the AQUILA Network and Improvements by Passive Measurements", *LNCS 2698*, W. Burakowski, B. Koch, A. Beben (eds.), Springer 2003
14. D. Veitch, P. Abry, "A wavelet based joint estimator of the parameters of long-range dependence", *IEEE Transactions on Information Theory*, vol. 45 no. 3, pp.878-897, 1999
15. The *tcpdump* web site, *http://www.tcpdump.org/*
16. The *tcpdstat* web site, *http://staff.washington.edu/dittrich/talks/core02/tools/tools.html*
17. The *tcpdpriv* web site, *http://fly.isti.cnr.it/software/tcpdpriv/*
18. The MAWI working group traffic archive, *http://mawi.wide.ad.jp/mawi*
19. J. Quittek, T. Zseby, B. Claise, S. Zander, "Requirements for IP Flow Information Export", RFC3917, IETF, October 2004

LATERE: A Remote-Access Laboratory For Experiences On Management And Control Of Networking Devices

Renato Narcisi, Antonio Pantò, Giovanni Schembra
CNIT-University of Catania Research Unit, Italy
V.le A. Doria 6, 95125 Catania (ITALY)
Tel 0039-095-7382370 Fax 0039-095-7382397
{renato.narcisi, apanto, giovanni.schembra}@diit.unict.it

Sergio Armenia, Giuseppe Tropea*
Netsense s.r.l.
{sergio.armenia, giuseppe.tropea}@netsenseweb.com

Abstract. A networking laboratory contains expensive equipment and offers a small grade of scalability. Remote laboratories allow for much more efficient use of laboratory equipment, and give users the opportunity to batch-schedule their experiments and manage them from the comfort of their local clients. In this paper, LATERE, a new interactive laboratory which allows remote testing of complex satellite scenarios and wireless indoor networks on real equipment, is detailed. Users can select arbitrary network topologies; characteristics of each link and transport layers, used by the traffic generators, can be freely set. Moreover, LATERE offers a user-friendly web interface to define each experiment parameters and to asynchronously view all results, which are displayed and also emailed to the students; an experiments' database is kept, allowing per-user profiles and history, plus logging and test reproducibility. LATERE is scalable, and built upon a lightweight middleware distributed and embedded on the interface with the equipment. LATERE is based on standard inexpensive hardware running open source software.

Keywords. Laboratory, remote control, web-application, middleware, satellite

1 Introduction

During the last years we have witnessed the growing and widespread interest in telecommunication and computer networks; as a consequence, academia strongly feels the necessity of effective tests on network topology design and on router configuration, as well as transport protocol analysis. Students are more motivated and can learn more effectively if they have the opportunity to conduct experiments. Together with theoretical courses, a practical experience is very important for a student, almost fundamental for his/her future job. At the same time, limitations of a

---

* Corresponding Author

traditional laboratory structure are well-known. Some of those limitations are: high price of the equipment (routers, meters); small physical space available inside the laboratory rooms, limiting the number of students at a time; laboratory availability limited to office hours.

Remote laboratories allow for much more efficient use of laboratory equipment and give students the opportunity to conduct experiments from the comfort of an Internet browser.

LATERE remote laboratory is a simple, cheap and scalable laboratory, designed to give students a better understanding of computer networks. In this paper we describe the design and the ongoing implementation of this interactive laboratory that enables experiments in:

- complex satellite networks environment, including network topology and satellite link characteristics control;
- wireless network performance evaluation.

These two topics have been chosen because of current evolutions in telecommunication technology: satellite networks are more and more used for IP data traffic and applications' behaviour over satellite links is a key issue in the development of future technology; indoor wireless networks have seen a rapid growth, both in consumer and businesses scenarios, because they are cheap and easy to deploy.

A networking laboratory containing expensive routers needs considerable funding and offers a small grade of scalability. We have realized a very scalable laboratory, based on standard inexpensive hardware running open source software solutions: routers, satellite links and traffic generators are implemented through GNU/Linux operated routers, variable-link-parameters emulation tools and open source traffic-injection tools. The laboratory monitor and management system is provided by a highly scalable Java lightweight distributed middleware.

Remote access is given to the students via a user-friendly web application, which manages concurrent accesses, per-student history, laboratory booking and service notifications also. The server-side GUI allows:

- to choose between two platforms (satellite system or wireless local network) for the tests;
- to configure the network topology and the network elements' parameters;
- to select the traffic generators and characteristics of data flows;
- to choose the performance values the user wants to observe.

The remainder of this paper is organized as follows. In Section 2 most interesting related works are presented; Section 3 gives an overview of the architecture and system components, which are described in detail in Section 4. Section 5 describes all experiences offered by the LATERE system; finally in Section 6 some concluding remarks on the activity carried out and on the future work are drawn.

2   Related Work

Despite the number of so-called e-learning applications has been increasing in recent years, remote laboratories that give even partial control of real network devices to their users, have had limited development. Nevertheless, several well-documented implementations do exist.

An example of a remote laboratory is VITELS [1], the result of a collaboration between four Swiss universities and an Engineering School. The VITELS system offers to students (after their proper authentication), seven different modules coming from different departments and institutions, but evenly integrated under a common web interface. The modules are called: simulation of IP network configuration, configuration and performance evaluation of real IP network, management and configuration of a virtual network, firewalls, protocol analysis, Linux system installation and configuration, client/server programming.

Another example is LABNET [2], a remote telecommunications laboratory developed under a joint CNIT (Consorzio Nazionale Interuniversitario per le Telecomunicazioni) CINI (Consorzio Interuniversitario Nazionale per l'Informatica) project. LABNET is made of a number of different projects all having the common, main goal of lab automation. Under this common paradigm, all equipment of technical-scientific interest is kept under constant monitoring and sampling, to enrich students' knowledge and training, when they are remotely located at sites different than the equipment itself. LABNET offers interactive, remote experience of laboratories' equipment to its users, coupled with web-based learning and real-time interaction with teachers.

In [4], authors present the evolution of a real telecommunications laboratory towards a remote laboratory, to overcome the well-known limitations of a traditional lab. They have designed a networking lab that solve these problems, since it can be accessed through the Internet, and it makes use of old PCs converted to IP routers running the free router software Zebra[6].

In [3], a remote laboratory in the field of Control and Automation is presented, which can be easily adapted to different technical areas. Its main features are: server-side-only code execution; special attention to didactical issues; simple web-based interface; experiments and tests run on the real equipment.

The aim of our LATERE project is to overcome a traditional "fixed model" approach to the remote tests. Tests run on real hardware and the complex web interface allows for the maximum possible flexibility, both in configuring the equipment, the traffic flows, and the network topology.

Batch processing and asynchronous multiple-user access allows for high efficiency, so that the system is always running at full rate when tests exist and are queued. Post processing and analysis of the results is also offered to the users, as well as per-user status and history of past experiences.

3   Architecture Overview

The aim of our work has been to develop a complex system to carry out tests concerning both wireless and satellite communications including network topology creation. The main characteristic of this system is its full remote accessibility and control; users can configure it concurrently via its interactive web site in order to execute one specific experience among those provided. The features of the system are reported as follows:

- automatic setup and handling of fault situations, including an auto-recovery mechanism;
- concurrent access to the platform;
- user accounts management including management of user history;
- sessions management with respect to the ongoing experiences;
- output production in human readable format, including remote notification to the user via email.

The system architecture, whose scheme is shown in Fig 1, is composed by the following parts:

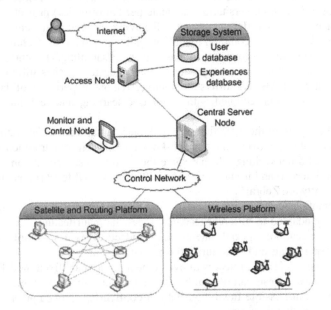

Fig 1: scheme of the overall system architecture

- an **Access Node,** which is devoted to offer remote access via an interactive WEB interface. The Graphical User Interface (GUI) is offered through a Web Server; it is usable with any Web Browser without stringent requirements. Additionally this node performs user and session management, including registration, authentication and maintenance of user sessions. In order to accomplish this, the access node has full access both to a **User database** and an **Experience database.** The former is dedicated to the storage of user information, while the latter stores all the information regarding the user's ongoing experience and all the previous experiences that he has already carried out or configured;
- an internal network, namely the **Control Network,** which connects all the hardware components to the Central Server Node. This network is designed

to offer a full control of the entire platform in a way that is highly robust. As an example regarding its robustness, it is important to deny the access to the users in order to avoid configuration mistakes which could lead to node isolation.

- a **Central Server Node**, which is devoted to the management and control of all those hardware components which form the platform. It receives input from the access node in the form of text files, each file corresponding to a different experience to be carried out. The Central Server Node processes the inputs and sends suitable commands to the **Satellite and Routing Platform** and the **Wireless Platform,** in order to carry out the specific experience. These commands, sent through the Control Network, have a general form (i.e. activate/deactivate network interfaces, start a server or a client application, send me the results, etc.); this is a strategic choice to make this node unaware of the characteristics of the two platforms, for example operating systems running on the hardware, tools installed, syntax and semantic, etc. Additionally a concurrent process, which runs in the background, takes care of handling all the possible fault situations. If a fault is recognized, this process activates an automatic recovery mechanisms, otherwise the current experience will be reset and the hardware components of the platform are rebooted.

- a **Satellite and Routing Platform**, composed of three routers and four traffic generators/receivers which form a fully meshed network. Each hardware component executes the commands received from the Central Server Node independently of others component in the same platform. Since they form a fully meshed network, a user can create a complex network topology by suitably activating a set of links and configuring the routing tables of the routers. Additionally, a user can set up satellite link choosing its characteristics in terms of bit error rate (BER), round trip delay and bandwidth. Finally a user can choose a certain number of traffic flows to be analysed depending on network parameters.

- a **Wireless Platform**, composed of six wireless traffic generators/receivers and four IEEE 802.11 b/g access points. Similarly to the Satellite and Routing Platform, each wireless node executes the commands received from the Central Server Node independently of others nodes in the same platform. A user can freely configure up to three independent ad hoc networks and up to four managed networks. Finally it is possible to choose a set of traffic flows to be analysed depending on the condition of the wireless environment, including topology, interference and load of access points.

- a **Monitor and Control Node**, which is used by a system administrator to have a local access to the platform. Through the Control Network a system administrator has full access both to any hardware component and to the databases of the system. Additionally he can be aware of the platform state through a merging of the log files produced by each node.

4   Implementation Choices

The system-wide glue that keeps all pieces together is composed by a proper combination of hardware and software elements which enables robust communication between the test equipment and the management equipment.

The hardware part is essentially built upon a **control LAN,** shown in Fig 2 which connects all test equipment with the Central Server Node. Each test equipment if thus provided with a dedicated Ethernet port and IP address, insulated from the data paths used to carry on the tests, thus providing a convenient means for all "out-of-band" signalling.

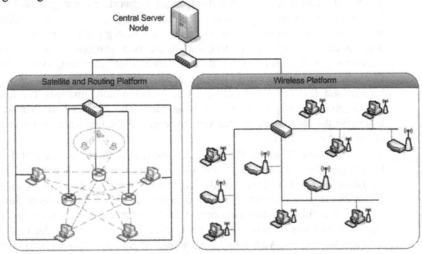

Fig 2: control LAN structure

The software is essentially composed by a lightweight **middleware** distributed on each system element and a custom **communication protocol**. Our middleware offers the following important functions:

- it defines the interface between the Central Server Node and each specific, hardware-dependant feature of the different test boxes, by means of a translation layer between our standard protocol commands and the several test hardware commands;
- it passes, queues and processes our standard configuration and control commands between all system nodes;
- it constantly monitors test equipment status, inducing appropriate actions when alarms do trigger.

The communication protocol is based on messages, passing between nodes, and acknowledgement from the middleware demons running on each test box, in order to maintain, at each step, the whole platform in an overall known status. The Central Server Node serves as a centralized place to collect asynchronous responses and dictate system status.

4.1 Hardware and Software details

The whole middleware is developed using Java. This design decision has been taken because of the availability of lightweight embedded Java VM that can be deployed inside test equipment, and thus serve as a standard container for the translation layer which runs on the equipment, and bridges to specific hardware configuration commands. All platforms' test hardware is composed of machines which are configured via low-level configuration command sequences (normally fed through a remote console), which we feed through our translation layer running on the embedded VM. The satellite platform is composed of the following elements:

- 3 routers running Linux based operating system; each router has 7 ethernet ports, one connected to the Control LAN, two connected to the remaining routers and four connected to the traffic generators/receivers.
- 4 traffic generators/receivers using open source applications. Each element has 4 ethernet ports, one connected to the Control LAN and the remaining connected to the routers.

All platform nodes are connected through a full meshed network.
The wireless platform is composed of the following elements:

- 4 802.11b/g wireless Access Point Zyxel G-1000, with their ethernet port connected to the Control LAN;
- 6 wireless generator/receiver nodes, with an ethernet port and a wireless port. The former is connected to the Control LAN while the latter is used to send/receive data flows.

4.2 System scalability

Obtaining a scalable test environment has been the constant issue during all system design. This goal is reached on a two-level scale. The LATERE system allows a seamless extension of a single platform; this is simply done by acquiring new test equipment for the platform, connecting to the control LAN, to the data paths, and adding entries to the system configuration files. Moreover, the whole system can be extended by instantiating a whole new platform without much effort; this is accomplished through a robust, object-oriented and multithreaded (multi-platform) design of the middleware.

5 Tests Available within LATERE

The set of experiences provided by the LATERE system has been designed to deal with the main critical factors involving traffic on wireless or satellite networks, as well as the creation and analysis of complex network topologies. At the same time these experiences are proposed in a simple and user-friendly way, in order to maximize the learning phase, instead of driving users' attention towards system's inner details.

The main critical factors in wireless networks are: their scalability in terms of supported number of users [5] and their robustness to radio channel interferences. In satellite networks the BER, bandwidth and delay offered by the satellite links play a crucial role. In Table 1 an overview of the experiences a user can carry out within LATERE system is reported; then a brief description of them is given.

Experience name	Platform	Target of the experience
Connectivity	Wireless/Satellite	create a complex network topology and perform a connectivity test bounded to the created topology
Traffic Flows versus BER	Satellite	analyse the impact of a satellite link's BER on both the throughput of a TCP connection and the packet loss rate of a UDP connection
Bandwidth-Delay Product	Satellite	analyse the impact of bandwidth-delay product on the throughput of a TCP connection
Packet Size	Satellite	analyse the impact of changing packet size on the throughput of a TCP connection in a satellite environment when the satellite link offers a certain value of BER
WiFi Scalability	Wireless	analyse the WiFi scalability, evaluated in terms of the number of users that a wireless LAN can support
WiFi Interferences	Wireless	analyse the effect of inter-channel interferences on wireless TCP and UDP connections

Table 1: an overview of the experiences offered within LATERE

5.1 Connectivity

In this experience, which is available on both platforms, a user is first asked to create a complex network topology. The way to do so depends on the chosen platform:

- within the Satellite Platform it is possible to switch on and off all links, to create routing tables and to assign a gateway to each traffic generator/receiver;
- within the Wireless platform a user can create a set of ad hoc and managed networks simply setting the SSID and the radio channel of the wireless nodes.

The user is then asked to choose a node, which will be the sender of standard ICMP "ECHO REQUEST" packets addressed to remaining nodes; results of this experience are built upon all replies to the aforementioned echo requests and show whether the routes from sender node to all network endpoints are correctly set up. These replies are collected in a text file which is sent back to the user via email,

5.2 Traffic Flows versus BER

As reported in Table 1, with this experience a user can analyse the impact of the satellite link's BER on both the throughput of a TCP connection and the packet loss rate of a UDP connection. A student is first asked to set up the satellite emulator by choosing the involved links and a value for the BER, including values for uplink and downlink bandwidth and delay. A network topology can then be configured, similarly to the "Connectivity" experience. Finally a maximum of three traffic flows can be configured in order to analyse them versus different BER values. The output of this experience is collected in a table, which contains all values of BER and correspondent values of throughput/packet loss rate for all configured flows. The user receives this table via email and he can easily plot the values using his favourite plotting tool.

5.3 Bandwidth-Delay Product

In order to fully exploit link capacity, TCP must use a window size equal to the bandwidth times the round-trip latency of the link, referred to as the "Bandwidth Delay Product" (BDP). A large BDP typically exceeds the default TCP buffer size used by the different operating systems running on the endpoints. This experience aims to clarify this critical factor to the student. The steps that a user is asked to follow are similar to the "Traffic Flows versus BER" experience, as well as its output, which is a text file containing a table of values.

5.4 Packet Size

The goal of this experience is to improve students familiarity with the relation between the link BER and the payload size of a TCP connection. For a certain value of BER the packet loss rate is given by the following formula:

$$PLR = 1-(1-BER)^{payload\_size}$$

where `payload_size` is expressed in bit. The analysis carried out within this experience emphasizes the impact of payload size on throughput of a TCP connection. A user is first asked to set up the satellite emulator by choosing the involved links and the BER value; in this test the bandwidth and delay assume their default value. Finally a set of maximum three traffic flows can be configured. The output table reports the values of throughput related to the different payload sizes.

5.5 WiFi Scalability

Goal of this experience is to show the effects of multiple wireless users loading the same wireless LAN segment. A user is first asked to select the access point to be included in the test, as well as the wireless receiver node, which will collect all concurrent flows sent by other nodes. Configuration of this test is concluded with the choice of a traffic flow (TCP/UDP) to be monitored. The output table reports the values of throughput/packet loss rate versus the number of concurrent wireless nodes loading the chosen access point.

5.6 WiFi Interferences

This experience aims to show the trend of throughput/packet loss rate versus inter-channel interference, the latter being generated by a concurrent wireless host. The experience is composed of seven single tests. In each of them the competing node uses a different radio channel, starting from the same channel used by the monitored node (case of maximum interference); the last test is performed with zero inter-channel interference. The configuration steps are:

- set up either an ad hoc or a managed network involving two wireless nodes, so that a target flow is established and can be monitored;
- set up either an ad hoc or a managed network involving two concurrent wireless hosts, so that a competing flow is established.

The output table will report values of throughput/packet loss rate versus the distance between the radio channels of the two networks.

6   Conclusions

We have presented an economical and scalable remote networking laboratory, accessible through the Internet. As said, system scalability is one of the strongest LATERE features; we plan to extend the whole system by including an entirely new platform, devoted to wireless sensor networks analysis and testing. Moreover, in the near future we are going to evaluate the LATERE remote laboratory "on the road", as a support to an undergraduate telecommunications course, in order to assess its strengths and weaknesses, and to integrate it in our learning environment for students.

References

1. Virtual Internet Telecommunications Laboratory of Switzerland, http://www.vitels.ch
2. Labnet, http://www.labnet.cnit.it/
3. C. Chiculita and L. Frangu. A Web Based Remote Control Laboratory. *The 6th World Multiconference on Systemics, Cybernetics and Informatics*, July 14-18, 2002 Orlando, Florida.
4. S.Yoo and S.Hovis. Remote Access Internetworking Laboratory. *Proceedings of the 35th SIGCSE technical symposium on Computer science education, pages 311--314*, 2004.
5. http://www.networkcomputing.com/1113/1113f2.html
6. http://www.zebra.org

The LABNET-Server Software Architecture For Remote Management Of Distributed Laboratories: Current Status And Perspectives

Stefano Vignola, Sandro Zappatore

National Inter-university Consortium for Telecommunications (CNIT),
National Laboratory for Multimedia Communications,
Via Diocleziano 328,
80123 Naples, Italy
stefano.vignola@cnit.it, sandro.zappatore@cnit.it

Abstract. The availability of a wide range of instrumentation that can be remotely controlled, and the possibility to access the Internet with cheap and fast xDSL lines have encouraged the design of tele-measurement platforms by means of which researchers, instructors and students can exploit the facilities offered by complex and expensive laboratories. Although the market offers several solutions to remotely manage equipment, little attention has been paid to the hardware and software architectures devoted to distance learning experimental environments. The paper presents the architectural approach, and the related implementation, which was proposed and developed within two projects funded by the Italian Ministry of Education, University, and Research (MIUR). Both projects were aimed at building the experimental test-beds, which include very heterogeneous equipment, and at developing an open system architecture for their actual managing. Specifically, the attention is focused on the supervising central unit, which represents a crucial and critical element of the system. The software implementation of the central unit is described in detail, as concerns both the data structures adopted and the protocol machine. The results of some performance tests are shown to highlight the efficiency of the proposed solution.

Keywords. Multimedia systems, tele-measurement systems, computer networks

1 Introduction

The increasing interest in distance learning and the availability of cheap lines (e.g., xDSL) to access to the Internet have been strongly encouraging the design and development of platforms, which can provide the users with powerful tools, devoted not only to personal communication, but also to training with complex and expensive laboratories. This topic does not strictly involve the field of distance learning: indeed, many sectors, ranging from medicine and biology to mechanics and telecommunications [1-4], can potentially benefit of effective suites, which allow to remotely manage and control experiments.

Distributed Cooperative Laboratories

The possibility of remotely driving experiences and measurements represents an issue of great relevance for both commercial, research and educational purposes. To this aim, the Italian Ministry of Education, University and Research (MIUR) funded the LABNET project (2000-2003), whose main goals were the building of the CNIT National Laboratory for Multimedia Communications in Naples (Italy), and the development of applications for the management of laboratories, devoted to distance learning and training [5, 6]. The significant results achieved after three years of activity encouraged to promote a follow-up, in order both to integrate and upgrade the existing equipment and to add new functionalities and facilities to the software system architecture. In this framework, a new project, called "Technological Network for Instrumentation and Measurements in Telecommunications", has been started in July 2003: besides the exploitation of the results of the LABNET, the project attempts to create a common interface for integrating the services offered by two different training platforms: the first one is represented by the LABNET itself, the second one was designed and implemented by the University of Catania within the Multicampus Project [7].

Commonly, a tele-measurements system architecture includes a supervising central unit (SCU), whose task is to control and monitor the instrumentation, thus coordinating the data exchange among instruments and the stations of users interested in the execution of the experiment. Furthermore, the SCU must provide data and service abstraction, so that end users (scientists, R&D technicians in the industry, teachers and students of educational institutions, etc.) can effectively receive data collected by the instruments, and send commands and possibly other data to them. In other words, the SCU represents the core of a system operating at the boundary between the client/user space and equipment/laboratory space. Its main tasks are: i) to maintain the data-base of the instruments and their related status; ii) to get requests from clients and dispatch them to the set of devices involved in the specific experience; iii) to gather data from instruments and to push them to the appropriate group of clients.

In the literature, several architectures have been proposed to remotely control and manage measurement instrumentation. For each architecture, the structure and the functionalities of the SCU are also outlined. For instance, in [8] the authors present a platform for supervising and monitoring a power industry process, based on Java and on a CORBA oriented middleware. The attention is here focused on the remote monitoring rather than the remote control and management of the measurement equipment. Java is at the base of many other software platforms for distributed measurement systems [9-11]. Bertocco et al. in [11] propose an intriguing client-server architecture: the server/SCU simply acts as a sort of instrument broker, while the clients, accessing the (virtualized) instruments, actually process and integrate the collected data. In such a model, the clients can be thus regarded as cognitive, reasoning and self-determining computing entities. The issue of tele-laboratories for control engineering and robotics is discussed and faced in further recent works, such as [12, 13].

In spite of the wide variety of the proposed architectures, few appear to be oriented to the remote measurements for a distance learning experimental environment [14-16] and, for the most part, no support (e.g., multicast) is provided for an efficient and simultaneous dissemination of the measured data to a potentially large group of users.

Moreover, the systems proposed are mainly oriented to manage "classical" measurement instruments (such as oscilloscopes, function generators, etc), with little attention to the modern network measurement devices. Finally, several platforms exploit proprietary software at the SCU level, thus possibly limiting the implementation of new functionalities and constraining the software developers to operate according to some rigid (and not completely known) models.

The present paper mainly focuses on the architecture and protocols at the basis of the design of the SCU, in the following referred to as LABNET-Server (LNS), initially developed within the LABNET project.

The paper is structured as follows: in Section II the designed system architecture is discussed. The third Section illustrates in detail the LNS software platform, describing both the data structures adopted and the protocol machine. In Section IV the results of some performance tests are presented that prove the efficiency of the LNS. Finally, in the last Section, conclusions are drawn.

2 System Architecture

The system architecture design was started in 2000 and carried out through three years of activity provided by the LABNET project; during the last two years (2003-2005), the deployment of new measurement instrumentation, as well as the upgrade of the software suite (necessary to manage the laboratory), were carried on within the new project.

The preliminary phases of the both projects were devoted to highlight some concerns, which were successively taken into account during the actual implementation of the devised system and, especially, of the SCU. The latter must be able to address several crucial issues embedded in the system: i) the possible heterogeneity of the application environments, as well as of the instruments, ii) the software portability and scalability (in order to support different types of terminals and access networks), iii) the level of flexibility (to allow the user to configure, manipulate and interface every kind of equipment in a simple way), and iv) the capability of multicasting the data gathered from the measurement instrumentation for an efficient use of the transmission resources. All these aspects, although quite relevant, are not sufficiently well focused, and often neglected, in some products available on the market.

Obviously, the system must provide mechanisms for instruments sharing and reservation and it must support, especially in the case of education and training applications, "ad-hoc" graphic user interfaces, which do not necessarily generate a virtual representation of the entire control panels of the instrument set involved in the experiment. Indeed, in many cases it would advisable to show only a subset of possible controls actually required by a particular experiment, thus focusing the attention on the experience itself, rather than on the measurement devices. All the previously mentioned concerns represented a sort of guidelines for the designing of the entire LABNET laboratory system, whose "hardware" consists of, besides some ancillary PCs, two test-beds (see [6] for more details): the first one is devoted to telecommunication systems and the second one is centered on telecommunication networks.

The facilities offered by the laboratory can be exploited by means of a simple common Internet browser, which communicates with an ad-hoc server (LNS), which represents the SCU of the system. The latter, in a sense, manages the border line between the *outer* laboratory space (user/client space, the domain where a user can access "unified instruments" by means of "standard protocols") and the *inner* laboratory space (i.e., "the domain of the instruments and of their specific rules and protocols"). More in detail, the clients download, through an HTTP connection, a proper Java archive, which contains all the applets (related to a specific experience) needed for communicating measured data and displaying them on a proper graphic panel. Each applet exchanges two main kinds of data with the LNS: the first one consists of commands toward the server, and then to the instrumentation on the field, the second flow bears data collected (by the LNS) from the instruments toward the clients. In this manner, the LNS hides the languages/environments (e.g., Labview, Openview,…) specific to the system under test and to the test-bed used to carry out the experiment.

On the side of the *inner* laboratory space, the LNS communicates with an experience manager that controls the low level data exchange with the physical devices, through a set of drivers. Obviously, the manager depends on the selected experience and the drivers are strictly related to each device or equipment and implement specific communication protocols, such as IEEE 488 for measurement instrumentation, SNMP or command lines for network devices, and so on.

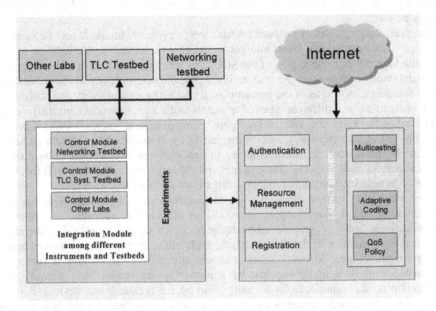

Fig. 1. Overall software architecture of LABNET.

The controlling processes are themselves organized in two sublevels. The first one only depends on the specific device class and provides the first level of abstraction: each device belonging to the class can be monitored and controlled without respect to the particular instruments actually in use. This abstraction enables to see device classes as the mere set of services they offer. On the contrary, the second sublevel depends on the experiments to be issued via the network: in fact, each experiment should require a specific set-up for the various devices involved and has to gather and process data in some desired suitable order and format. A higher-level module performs the integration of data (coming from the heterogeneous instruments) before making them available to the LABNET Server.

In order to better detail the role of the various components, let us fix an operative context, e.g. a distance learning one, and suppose that an instructor (whose workstation acts as master and, initially, holds the so-called Control Token (CT) that grants writing rights) has chosen a particular experiment (among those available in a menu) and that a number of students, remotely attending such a training event, are present in one or more classrooms. At first, the instructor's request to initialize the experience is sent to the LNS, which consults its internal database and launches the appropriate experience manager. The latter sends the initialisation commands to the drivers of the devices involved in the experiment, and waits for their acknowledgement. If all is in order, the actual data exchange can start.

It should be highlighted that commands to the LNS can be sent only from the station currently holding the CT. The latter is uniquely managed by the master station (i.e., the instructor's workstation) that can freely assign and revoke the CT to any student attending the experiment. In this way, after a demonstration by the instructor (where the multicast capability of the system plays an important role), a student can directly control the instrumentation and perform his/her own tests. In this phase, the actions exerted and the subsequent result can be made visible to the other students (in which case, multicast would again be useful) or not.

User commands received by the LNS are routed to the specific experience manager, responsible for the subsequent dispatching to the appropriate drivers. Data produced by the instruments are gathered by the drivers, routed to the experience manager, and (after possible integration with other data, if necessary) "published" in a data repository, maintained by the LNS. At this point, all the participants in the experiment can receive the data, either upon client request (GET mode) via a unicast connection (this request typically occurs whenever a new client station enters the virtual classroom and needs knowledge of the experience status), or by server initiative (PUSH mode). In the latter case, two different modes are available: if the network supports multicast, a multicast update is sent to the group associated with the experience, otherwise a unicast update is delivered to each client station participating in the training event.

3 Labnet-Server architecture

The LABNET-Server represents the core of the software platform and, in a sense, it can be viewed as a middleware that provides elements to offer services through a common interface, in order to establish a contact between who asks for a service and who offers it.

A relevant feature of the LNS is its capability of dispatching data in a multicast fashion (when required and allowed by the network), thus reducing the overall bandwidth needs.

It mainly acts as a repository where labelled data can be referenced, stored, updated and retrieved. The LNS is diagrammatically depicted in Figure 2, where the principal functional blocks and data structures are shown.

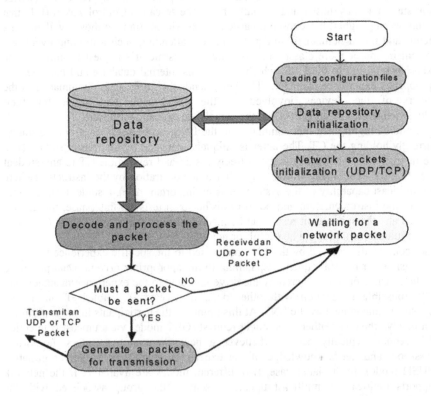

Fig. 2. LNS principal functional blocks and data structures

During the starting phase, the LNS initializes its internal structures and opens a pool of Internet sockets to communicate to both the inner and outer space. Then, the LNS waits for packets from the network. When something arrives from the outer space, the packets are related to commands used to register a new client as a participant, to select the desired experience, to set a value of a variable inside the repository, to assign/revoke the CT and to notify a disconnection. Vice-versa, a packet from the inner space may consist of an acknowledgment at the end of an experience initialization phase and of updates from the instruments involved in the experiment.

If the packet received refers to some variable stored in the repository, the value of the variable is changed according to the contents of the packet's payload. Internally, data are stored by means of a chain hash table.

As regards the format, each packet consists of two main sections: the first one (header section), common to all types of packets, presents a fixed length and it contains general information (timestamp, sequence number, total packet length). Further fields permit to uniquely identify the experience the data are related to, and the specific type of packet: the type indicates whether the packet is a command or it is bearing data gathered from the instrumentation through an appropriate experience manager. The second section is constituted by zero or more data structures, called containers. Each of them refers to a scalar or array variable(s) and conveys either measured data or parameters needed to set up the various devices in the test-bed. The container is organized in two portions, a header with the type and name of the variable, and an information field with the actual data to exchange. Figure 3 sketches the packet structure.

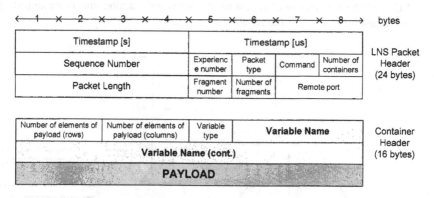

Fig. 3. Structure of an LNS packet. Each packet consists of a header and of a variable number of containers, according to the corresponding field in the packet header. Whenever the container refers to a bi-dimensional array, the first two fields specify the number of rows and columns of the array, respectively.

Table 1 presents the principal packet types provided for the LNS and their synopsis. The first part of the list regards the packets within the outer space, the second concerns the interactions within the inner space and, finally, the last row refers to DataManagement packets, which are common to both spaces.

A more detailed description is need to better understand how a DataManagement packet is processed by the LNS. Such a type of packet can be issued by both the client stations and the experience managers. Upon receiving this type of packet, the LNS checks the field command and acts as follows.

- If a STORE command is specified, the action performed depends on the component which issued the request. In case of an incoming packet from an experience manager, the LNS updates the referenced variable in the repository and forwards the packet to all client stations involved in the experience. If the related packet comes from the client station holding the CT, it is simply forwarded to the appropriate experience manager.
- If a READ command is specified, the LNS sends a DataManagement STORE packet, whose payload is the value of the referenced variable, to the client station that originated the request.
- If a DELETE command is specified, the LNS deletes the referenced variable from the repository.
- If a CREATE command is specified (usually by an experience manager), the LNS allocates the variable(s) specified.

The protocol machine shown in Figure 4 can help to better understand the actual mechanism, which controls the LNS. Whenever the SELECTEXP packet is received from an instructor station, the machine leaves the *idle state* and reaches the *experience start* phase. If all instruments in the set involved in the specific experience are ready within a timeout, the LNS enters the *experience running* state, during which the DATAMANAGEMENT packets (most of them carrying the data related to the measurements) are exchanged, until a RELEASEXP packet is received. Otherwise, a SELECTEXP NACK messages is sent to the instructor's station thus communicating the unavailability of some of the required elements.

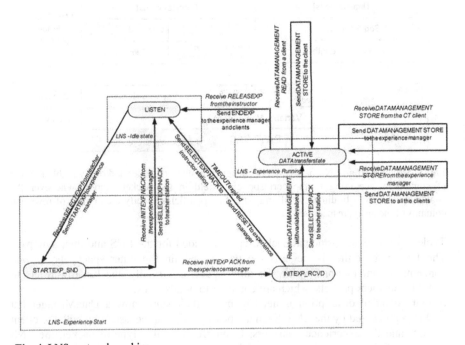

Fig. 4. LNS protocol machine.

Table 1. List of the main packet types managed by the LNS with the corresponding synopsis.

Packet type	Synopsis
NEWCLICONNECTION	Connection request generated by a client station in order to access the experience selected.
NEWCLICONNECTIONRESP	Simply an ACK (or NAK) to the above request.
SELECTEXP	Request sent by the instructor's workstation to initialize one of the available experiences. Upon receiving this packet, the LNS performs a STARTEXP request (see below).
SELECTEXPRESP	Just an ACK/NAK of a SELECTEXP.
RELEASEEXP	Command issued by the instructor's workstation to release (terminate) the current experience. Upon receiving this packet the LNS sends an ENDEXP packet to the appropriate experience manager.
GETCLILIST	Command to get the list of all student stations participating in the current training event. Only the instructor's station can perform this command.
CLILISTRESPONSE	A data packet (possibly consisting of one container) with a list of the IP addresses of all the participants.
PASSTOKEN/REVOKETOKEN	Command issued to assign/revoke the CT to a selected client station.
DISCONNECT	Disconnection request sent by a student station to leave the training event. This is the last packet delivered by the client station before terminating the applets involved in the experience and closing the Internet sockets used for data exchange.
STARTEXP	Command issued by the LNS to the experience manager to request a specific experimental set-up. Upon receiving this command, the manager checks the availability of all the devices involved in the experience. If all is in order, the manager invokes the related instrument drivers to initiate the actual operations.
ENDEXP	Command issued by the LNS to the experience manager to shut-down the experience and to the student stations to terminate the applets

	involved in the experience, thus closing the Internet sockets used for data exchange.
INITEXP	A packet issued by the experience manager to notify the LNS the successful initialization of a test-bed. Together with this packet, the manager also delivers as many DATAMANAGEMENT as the number of variables used to represent all the measured data involved in the experience. Upon receiving the INITEXP packet, the LNS sends a positive SELECTEXPRESP to the instructor's station.
DATAMANAGEMENT	Packet sent by any element (clients, LNS, experience manager) which carries the current value(s) of one or more variables defined in the LNS repository. The field command specifies the operation required: STORE, READ, DELETE or CREATE.

4 Preliminary results

In order to evaluate the performance of LNS, some tests were performed: the main aim was to estimate the possible delay and jitter that the LNS introduces while archiving data in its repository and dispatching them to the user stations involved in the experience. To this goal, a suitable experimental set-up, sketched in Figure 5, was employed.

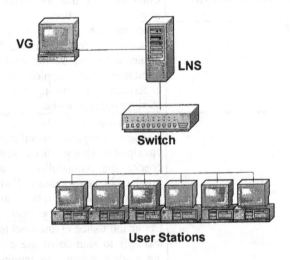

Fig. 5. The experimental setup.

The "variable generator" (VG) plays the role of an experience manager: it (quasi-) synchronously produces, every D seconds, a set of LNS packets, conveying a group of n variables. The LNS receives the packets, arriving from the VG, decodes them, updates the values of the referenced variables in its repository, and, finally, for each variable, it generates as many LNS packets as the number of user stations interested in the experience. More precisely, if v is the number of variables referenced by the VG, $v * n$ is the total number of packets that the LNS has to send every D seconds.

Since the VG is implemented by means of a PC running a standard Linux kernel, the generation of packets is itself affected by a certain level of jitter, due to the intrinsic not real-time nature of the operating system. Therefore, the generic instant t_k when the k-th packet is produced can be written as

$$t_k = kD + \varepsilon_{tx} \tag{1}$$

where D is the (theoretic) packet time (i.e. the time interval between two successive packet departures), and ε_{tx} is a random variable expressing the uncertainty about tk. The algorithm used by the VG assures that ε_{tx} has a zero mean.

The user station receives the k-th LNS packet at the instant T_k given by

$$T_k = t_k + o + \varepsilon_{lns} = kD + \varepsilon_{tx} + o + \varepsilon_{lns} \tag{2}$$

where o is a fixed time offset due to the physical transmission, and ε_{lns} is a random variable expressing the uncertainty related to the time spent by the LNS perform its tasks (packet decoding, data archiving and dispatching). At the (k+1)-th LNS packet, T_{k+1} is given by

$$T_{k+1} = (k+1)D + \varepsilon'_{tx} + o + \varepsilon'_{lns} \tag{3}$$

Then, the time interval between the arrivals of two consecutive packets is

$$\Delta T = D + \alpha_{tx} + \beta_{lns} \tag{4}$$

where

$\alpha_{tx} = \varepsilon'_{tx} - \varepsilon_{tx}$ is a zero mean random variable, whose variance σ_α^2 is twice that of ε_{tx}, and

$\beta_{lns} = \varepsilon'_{lns} - \varepsilon_{lns}$

As α_{tx} and β_{lns} are independent and α_{tx} has a zero mean, then

$$E[\beta_{lns}^2] = E[(\Delta T - D)^2] - E[\alpha_{tx}^2] = E[(\Delta T - D)^2] - \sigma_\alpha^2 \tag{5}$$

Since

$$E[\beta_{lns}^2] = E[(\varepsilon'_{lns} - \varepsilon_{lns})^2] = E[(\varepsilon'_{lns})^2] + E[(\varepsilon_{lns})^2] - 2E[\varepsilon'_{lns}\,\varepsilon_{lns}] \tag{6}$$

under the assumption that ε'_{lns} and ε_{lns} are independent,

$$E[\beta_{\ln s}^2] = E[(\varepsilon'_{\ln s})^2] + E[(\varepsilon_{\ln s})^2] - 2E[\varepsilon'_{\ln s}]E[\varepsilon_{\ln s}] =$$

$$= 2\left\{E[(\varepsilon_{\ln s})^2] - E[\varepsilon_{\ln s}]^2\right\} = 2\sigma_{\varepsilon \ln s}^2 \qquad (7)$$

where $\sigma_{\varepsilon \ln s}^2$ is the variance of $\varepsilon_{\ln s}$.

Combining Eq. (5) with Eq. (7) gives

$$\sigma_{\varepsilon \ln s}^2 = \tfrac{1}{2}\left\{E[(\Delta T - D)^2] - \sigma_\alpha^2\right\} \qquad (8)$$

Hence, the RMS of time delay introduced by the LNS tasks, can be estimated by

$$RMS(\varepsilon_{\ln s}) = \sigma_{\varepsilon \ln s} = \sqrt{0.5(E[(\Delta T - D)^2] - \sigma_\alpha^2)} \qquad (9)$$

In practice, by evaluating the variance of the packet time at each user station, viz $E[(\Delta T - D)^2]$, and by computing the variance of the packet time of the flow produced by the VG, viz σ_α^2, it is possible to estimate the standard deviation of $\varepsilon_{\ln s}$, that is the root mean square of the time a packet spends to "pass-through" the LNS.

Two main groups of tests were performed by means of the experimental set-up above described. In the first group, the VG synchronously produced a set of 30 variables, 26 of them scalars (4 bytes each) and 4 one-dimensional arrays (1024 bytes each). In the second group of tests, the VG generated a set of 60 variables (52 scalars and 8 arrays). For both groups, the packet times selected, i.e. the time intervals between two consecutive set-of-packets generations, were 1000, 500, 250, 125, 60, and 32 ms. The VG, the LNS and the 15 user stations were hosted by a set of PCs (Fujitsu-Siemens Scenico P300 VKM266), running the Linux kernel v. 4.6.11. The switch employed was a CISCO Catalist 2900xl.

The results of the tests are graphically reported in Figures 6, 7 and 8. Specifically, Figure 6 and 7 show the behavior of $\sigma_{\varepsilon_{tx}}$ (dotted line) and of $\sqrt{E[(\Delta T - D)^2]}$ (solid line) when 30 and 60 variables are involved in the experiment, respectively. The former RMS was estimated at the VG, the latter one was obtained by averaging the $(\Delta T - D)^2$ measured at every user station. Finally, Figure 8 plots the behavior of the standard deviation of the time spent by a packet to pass through the LNS. It should be highlighted that, in the 30 variables case, the spread of delay introduced by the LNS is negligible for (theoretical) packet times of 1000, and 500 ms, and it assumes very low, about constant values for the other packet times.

Some more comments on the case of 60 variables managed by the LNS are in order: the standard deviation of delay monotonically increases, and within the range 500-60 ms is about twice that of the previous case. When the packet time is 32 ms, the delay spread drastically increases, becoming about 4 times that observed with 30 variables. This fact can be motivated by noting that the time required to transmit 15 packet sets (each of them conveying 60 variables) at 100Mb/s is about 15 ms, which equals about one half of the packet time. In other words, the transmission process taking place in the kernel strongly affects the user space, where the LNS is running. It should be noticed that this spread value, though significantly higher than the others, represents only the 0.34% of the related packet time, thus confirming the overall good efficiency of the designed system.

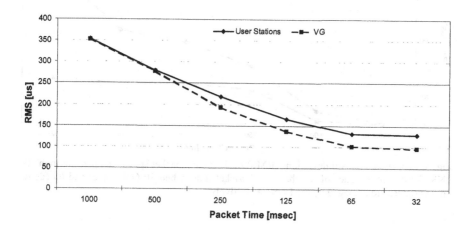

Fig. 6. Behavior of the standard deviations (RMS) of packet time observed at the VG (dotted line) and at the user stations (solid line), for different values of the theoretical packet time, when 30 variables are managed by the LNS.

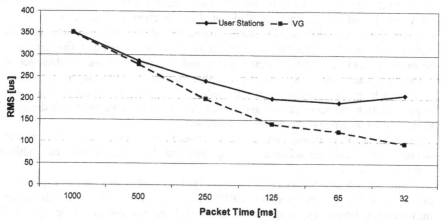

Fig. 7. Behavior of the standard deviations (RMS) of packet time observed at the VG (dotted line) and at the user stations (solid line), for different values of the theoretical packet time, when 60 variables are managed by the LNS.

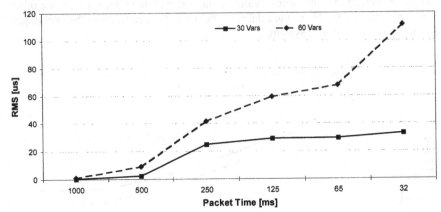

Fig. 8. Estimated standard deviations (RMS) of the time a packet spends to "pass-through" the LNS, for different values of the theoretical packet time, when 30 (solid line) and 60 (dotted line) variables are managed by the LNS, respectively.

5 Conclusions

The paper has presented the tele-laboratory system designed and developed within two projects funded by MIUR. The attention has been focused on the description of the software platform devised in order to allow accessing the experimental testbeds for research and remote training. This platform is characterized by a number of original features, which include addressing a broad spectrum of heterogeneous instrumentation, the exploitation of multicasting capabilities, and the separation between equipment-specific functionalities and the ones related to data access, transfer and management. Specifically, the paper has described the supervising central unit, which plays a very important role in the entire system. Its software implementation has been outlined and some results have been presented and discussed to prove the efficiency of the proposed solution.

Further work is being carried on to investigate whether some performance improvements could be achieved by implementing the protocol machine within a multi-thread context.

Moreover, another intriguing topic under investigation is the possible exploitation of the LABNET platform within a GRID structure, thus creating a real distributed measurement laboratory, where both users and instruments could be spread in different physical locations.

Acknowledgments

This work was supported by the Italian Ministry of Education, University and Research (MIUR).
The authors thanks Caterina Calcagno for writing part of the software used during the performance tests.

References

1. IEEE Commun. Mag., Special Issue on Communications and Tele-Learning, vol. 37, no. 3 (1999)
2. Hackbart, S.E., Ed.: Web-based Learning, Educational Technol., vol. 37, no. 5 (1997)
3. Wu, Y.L., Chan, T., Jong, B.S., Lin, T.W.: A Web-based Virtual Reality Physics Laboratory, Proc. 3rd IEEE Internat. Conf. on Advanced Learning Technologies (ICALT'03), Athens, Greece (July 2003) 455
4. Gillet, D., Latchman, H.A., Salzmann, Ch., Crisalle, O.D.: Hands-On Laboratory Experiments in Flexible and Distance Learning, J. Engineering Education, vol. 90, no. 2 (April 2001) 187-191
5. Davoli, F., Maryni, P., Perrando, M., Zappatore, S.: A General Framework for Networked Multimedia Applications Enabling Access to Laboratory Equipment: the LABNET Project Experience, Proc. IEEE Internat. Conf. on Information Technology: Coding and Computing (ITCC 2001), Las Vegas, NV (April 2001) 359-365
6. Davoli, F., Vignola, S., Zappatore, S.: A Multimedia Laboratory Environment for Generalized Remote Measurement and Control, Proc. 2002 Tyrrhenian Internat. Workshop on Digital Communications (IWDC 2002), Capri, Italy (Sept. 2002) 413-418
7. Multicampus – Task W1-A3. Configurazione HW e SW di una aula multimediale per teledidattica, http://www.diit.unict.it/users/gmorabito/multicampus, Catania (2003)
8. He, F., Wang, F., Li, W., Han, X., Liu, J.: Object Request Brokers for Distributed Measurement, IEEE Comp. Appl. in Power, vol. 14, no. 1 (Jan. 2001) 50-54
9. Grimaldi, D., Nigro, L., Pupo, F.: Java-based Distributed Measurement Systems, IEEE Trans. on Instrumentation and Measurement, vol. 47, no. 1 (Feb. 1998) 100-103
10. Nigro, L., Pupo, F.: A Modular Approach to Real Time Programming using Actors and Java, Proc. 22nd IFAC/IFIP Workshop on Real Time Programming (WRTP-97), Lyon, France (Sept. 1997) 83-88
11. Bertocco, M., Ferraris, F., Offelli, C., Parvis, M.: A Client-Server Architecture for Distributed Measurement Systems, IEEE Trans. on Instrumentation and Measurement, vol. 47, no. 5 (Oct. 1998) 1143-1148
12. Overstreet, J. W., Tzes, A.: An Internet-Based Real-Time Control Engineering Laboratory, IEEE Contr. Syst. Mag., vol.19, no. 5, (Oct. 1999) 19-34

13.You, S., Wang, T., Eagleson, R., Meng, C., Zhan, Q.: A Low Cost Internet-Based Telerobotic System for Access to Remote Laboratories, Artificial Intelligence in Engineering, vol. 15, no. 3 (July 2001) 265-279

14.Yuan, X. F., Teng, J. G.: Interactive Web-Based Package for Computer-Aided Learning of Structural Behavior, Comp. Appl. in Engineering Education, vol. 10, no. 2 (Oct. 2002) 79-87

15.Murphy, T., Gomes, V. G., Romagnoli, J. A.: Facilitating Process Control Teaching and Learning in a Virtual Laboratory Environment, Comp. Appl. in Engineering Education, vol. 10, n. 3 (Jan. 2003) 121-136

16.Ferrero, A., Salicone, S., Bonora, C., Parmigiani, M.: ReMLab: A Java-Based Remote, Didactic Measurement Laboratory, IEEE Trans. Instr. Meas., vol. 52, no. 3 (June 2003) 710-715

A New Perspective In Instrumentation Interfaces As Web Services

Alfonso Vollono[1], Andrea Zinicola[1]

[1] CNIT – Consorzio Nazionale   Interuniversitario per le Telecomunicazioni
National Multimedia Communications Laboratory, Via Diocleziano 328, Napoli, Italy
Alfonso.Vollono, Andrea.Zinicola@cnit.it

Abstract. Both educational and industrial institutions have a growing need to test experiments in highly complex, expensive structures. On the other hand, high-speed networking opens up the possibility of accessing remote sites and performing experiments with a high level of quality of service. The main goal of the LABNET ("LABoratories on the NETwork") Project [1] was the realization of an Integrated Learning System that allows students and instructors, located in geographically dispersed areas, to access technological resources, such as sophisticated laboratory equipment, measurement devices and, in general, complex test systems, through a scalable networking infrastructure and a number of supporting multimedia technologies. The paper introduces a new perspective in the definition of a suitable interface to the instrumentation and proposes a web-service based instrument interface, as an evolution of the LABNET remote measurement framework.

Keywords. Multimedia systems, tele-measurement systems, computer networks, virtual instrumentation, web services, SOAP

1   Introduction

Both educational and research activities in the scientific and industrial world often demand sophisticated laboratory instrumentation with specific characteristics. Such features may render the required instruments relatively expensive, difficult to replicate in many sites and, as a consequence, hardly accessible.

The classical distance learning paradigms should therefore be complemented with the support of practical applications, such as the execution of laboratory experiments, without the need of the physical presence of the users in the laboratories. The capability of performing remote operations on measurement instrumentation allows sharing a large number of valuable and expensive resources among various structures. In this way, students can choose among a number of laboratory experiments, to complement their instruction in a particular subject [1-4]. Moreover, distributed laboratories open up new possibilities for access to sophisticated instrumentation for research and industrial measurements, especially for Small and Medium Enterprises (SMEs).

In an educational or training context the typical scenario sees two or more customers taking part, with different rights (lecturer-instructor/student), to a laboratory experience. A measure is therefore an intrinsically collaborative application.

Collaborative systems are able to group customers geographically distributed, but connected by a network. Given the hierarchy among the customers of a group, the management of privileges becomes necessary, at least as regards the distinction between a master of the experience and the normal users. In the cooperative environment, it must be possible to acquire and release the control of the instrumentation: the master (lecturer-instructor) must be given the possibility to enable another user (student) to gain control of the experiment, when necessary.

Therefore, a functionality is needed for the management of both accounting and access privileges to the laboratory resources. In the LABNET framework, a server is dedicated to the management of these tasks, and the remote users connect to it through the Web, by different access modes, to use services available. Therefore, the software architecture is basically a client/server one.

The paper introduces a new perspective in the definition of a suitable interface to the instrumentation and proposes a web-service based instrument interface, as an evolution of the LABNET remote measurement framework.

The paper is organized as follows. The next section describes the LABNET Framework software architecture. In section 3 a new definition for the measurement instrument interface is introduced. In the last section this concept is extended in a web-service point of view.

2   Software Architecture

On the server side, building Web pages on the fly is useful, above all because the generated Web pages depend on data submitted by the user. Java Server Page (JSP) is the technology chosen to dynamically generate HyperText Markup Language (HTML) code. JSP frees the server choice from a particular vendor and guarantees:

- efficient improvements (in comparison with traditional CGI, Common Gateway Interface),
- high performance (in comparison with PHP, Personal Home Page tools PHP needs to be compiled into byte code every time a *.PHP page is invoked, while a *.JSP is compiled only once, then it is saved and accessible by the server),
- high power level (again, in comparison with PHP –Personal Home Page tools PHP has access only to PHP libraries, JSP has access to all the Java libraries)
- high portability (with JSP, if one wants to use the code inside a page elsewhere, one can cut it out, put it into a Java class, and invoke it from anywhere in one's application, even not from a page, whereas with PHP one is stuck inside the HTML box).

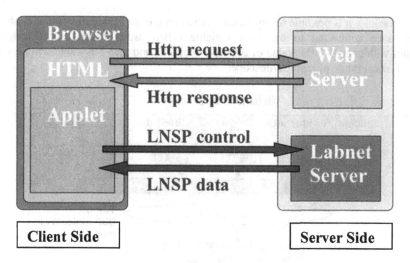

Fig. 1. LABNET framework: Client/Server data flow.

The generated HTML pages contain *Java Applets*, which are, in turn, containers of graphical basic objects, called *Java Beans*. Each one of them, integrated in a Web page, is the "virtualization" of a real instrument component. The "virtual components" realized are the most commonly used ones in real devices under test: from knobs to sliders, from buttons to displays, from text areas to combo-boxes. These components are designed to be instrument-independent; therefore, they are fully reusable and readaptable. Furthermore, the Java programming language, is platform-independent and ensures the highest level of application portability over different operating system. On the client side, it is sufficient to have a generic browser, with a (free) Java Virtual Machine installed, without the need of paying for any other licensed software.

Once a valuable experience has been designed and realized, exporting the needed capabilities of the instruments involved is enough, in order to make it remotely controllable. We do not want to reproduce a mere group of instruments, but we want to provide the users an ergonomic and easy to use *Web graphical interface* (Graphical User Interface -GUI), where the attention should be focused on the specific features of the experiment being performed. So, the Web interface actually consists of a *meta-instrument*, being able to group capabilities possibly referred to different real instruments among those involved in the experiment.

At the first abstraction level the *experience* becomes a set of variables (Ids, variablesExpId), attached to the data flow to and from the remotely controllable capabilities of real instruments.

These experience variables reside in the server, which we will call *LabNetServer* (LNS), and "virtual components" on the client side are connected to them, thanks to a custom communication protocol (LabNetServer Protocol LNSP).

In fig.1 the client/server data flow is shown: once the GUI web pages have been downloaded via the HTTP protocol, instrumentation and control data are exchanged through the LabNetServer Protocol LNSP.

As an example, it is possible to say that a rotation on the client side of "virtual knob", for instance connected to the LNS variable called *oscilloscope_vertical_range*, matches with changing the value of this LNS variable, and then this value will be set as the vertical range of real oscilloscope.

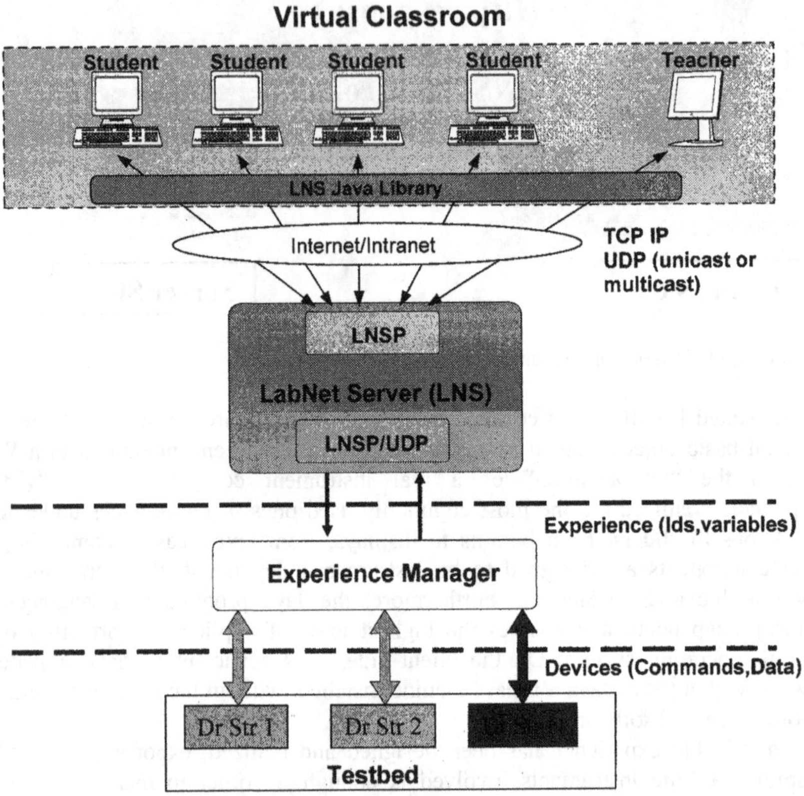

Fig. 2. LABNET Framework Architecture.

On the basis of the above experience definition, being able to perform instrument control is equivalent to being able to change the values of the respective LNS variables. A *super user* grants the *user* privileges, for exclusive use, to write the LNS variables (*pass/revoke TOKEN*). The *super user* can start and close the experience, that is to say, allocate and de-allocate the LNS variables.

So, among LNS tasks are the data providing, but also accounting and resource management, scheduling the interaction with different client types (*super user, user*) and allocating the proper resources for each requested experience, thus avoiding conflicts and collision among users.

User data providing relies upon *UDP*, in *unicast* or *multicast* fashion. This connectionless communication protocol is light, but also unreliable. Therefore, the LNS has to deal with lost packets and QoS problems. Educational sessions often involve a large number of user stations. For this reason, wherever it is supported by the network, multicast transmission should be chosen, for a more efficient use of the available bandwidth. On the other hand, for each kind of user, there is a reliable control connection to the server over the TCP communication protocol. It is used both for TOKEN exchange and for starting or taking part in an experience. TCP is heavier than UDP, but it guarantees stability and control of parameters that are critical for the correct working of the system.

In fig.2 the LABNET Framework architecture is shown: the most important software component and their communication protocol are pointed out.

The LNS knowledge is limited to the experiences and to their allocation, based on different types of user access (*super user, user*), but it does not concern the instruments being used.

So, a logical element is needed to establish a link between the LNS and the heterogeneous and variegated instrumentation world: the *Experience Manager* (EM).

The EM knows the link between the LNS experience variables and the commands of the instruments to be remotely controlled, which are accessible by means of instrument drivers. In this way, instrument implementation details are hidden from the LNS. The EM manages also the allocation of instruments in different experiences. In fact, before starting an experience, the EM must verify if there are all the instruments needed, if they are not busy, and if it is possible to connect them with one another. Only if these conditions are verified, then the instruments are allocated, the LNS variables are initialized and the experience can start. All experiences involving busy instruments cannot be started.

Once an experience is started, an action, like a rotation of a "virtual knob" on the client side, causes a succession of operations that involve all the components of the illustrated software architecture.

The graphical user interface, as a consequence of the rotation of a "virtual knob", sends the LNS the value to be set. The LNS stores this value in the variable related to the moved "virtual knob"; then, it sends a packet to the EM. The EM resolves from the packet the instrument involved, the command and the value to be set. Through the appropriately developed instrument driver, the command reaches the instrument and the EM waits for a response. The instrument involved replies by sending back the same value received, if the setting comes to a successful conclusion, or a different value, corresponding is the actual parameter setting, if an error has occurred. This value goes to EM, then to LNS and eventually arrives at the graphical user interface to set the "virtual knob".

In this way, the software and hardware architecture guarantees the coherence between the real instrumentation world and the virtual one and so the consistency of the proposed laboratory experiences.

In fig.3 the exchange of values through the link between GUI and instrumentation driver in the LABNET Framework architecture is shown.

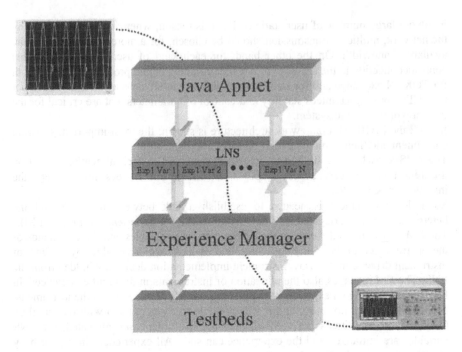

Fig. 3. LABNET Framework: the link between GUI and Instrumentation.

3   Instruments as their parameter sets

The heterogeneity of the instrumentation and devices to be controlled and driven, both in terms of electrical interfaces and communication protocols, calls for an integration level among the different environments, in order to hide the implementation details in the interaction with the EM. As an example, consider interfacing equipment for measurements of physical layer characteristics of telecommunication systems (e.g., spectrum analyzers, oscilloscopes, signal and noise generators, ...), usually accessed via LabVIEW proprietary software and the GPIB bus, and networking equipment (e.g., routers, traffic generators, traffic meters, ...), which may require interaction via SNMP and the use of C++ or Java code.

By hiding such details by means of some level of abstraction, the Integrated Learning System will be able to manage remote classes and laboratories, where, in principle, some instruments may be controlled neither via GPIB, nor via SNMP. In this context, the user accesses the real device and/or the instrument (really present in the Lab), through an abstract instance of the latter. Such instances may be even utilized to represent cooperating groups of instruments, in order to provide an "extended virtual instrument".

It is, in the first place, necessary to supply an abstraction, able to describe whichever measurement instrument, which can be led back to the definition of web-service. The definition that we have reached sees an instrument as constituted by the parameters that characterize it (e.g., making reference to a spectrum analyzer, Video BandWidth - VBW, SPAN, acquired signal - AS, visualized signal - VS). For each parameter, two kinds of access operations can be present:

- Read
- Write

For each operation, two access modes are defined:

- Synchronous
- Asynchronous

As an example, for the "visualized signal" parameter, the access mode "asynchronous reading" is foreseen: as soon as a new waveform is available, we might want to visualize it (this is an intrinsically instrument-driven operation). The SPAN, instead, is a parameter for which both reading and writing operations in synchronous way may be of interest: the customer or the script that supervises the measurement experience sets up the parameters of interest for the experience in the most suitable way. This measurement instrument abstraction defines an "instrument" object through the parameters that characterize it and the operations available on them. In Figure 4, an exemplary reduced class diagram for a Spectrum Analyzer is shown.

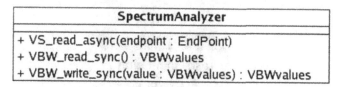

Fig. 4. Spectrum Analyzer simplified class diagram.

4 Instruments as web-services

Given the definition of an instrument as a set of access operation on the parameters that represent it, we look at an extension of this concept in a web-service view.

Because of the need for distributing instrumentation resources to and sharing them with one or more Experience Managers, the need arises of a *remote procedure call* interface for device drivers.

The SOAP (Simple Object Access Protocol) specification [5] defines a model for exchanging messages (as XML, eXtended Markup Language , documents). It relies on three basic concepts:

- messages are XML documents;
- these documents travel from a sender to a receiver;
- a sender and a receiver are chained together through a Transport.

Working with just these three concepts, it is possible to build sophisticated systems that rely on SOAP. The SOAP Specification contains a convention for performing remote procedure calls (Remote Procedure Calls -RPC) using SOAP messages.

Each instrument is a web-service, whose interface is composed by the operations defined on the parameters that characterize the instrument itself.

In our architecture we have chosen HTTP as Transport mechanism. As represented in fig. 5, instrument driver functionalities can be invoked by means of SOAP-RPC transported over HTTP. At the moment, we are investigating the use of different Transports in order to choose the most efficient one. We implemented asynchronous operations in a way similar to a video streaming server with a SOAP interface: what is exported is the capability to start/stop the acquisition of data from the instrument and the transmission of the acquired data via the light UDP protocol to an endpoint (e.g., Experience Manager, or another instrument). In this way, we can transfer compressed acquired raw data, without the overhead determined by the SOAP header. This is a critical issue in bandwidth consuming applications, as might be the transfer of acquired signals.

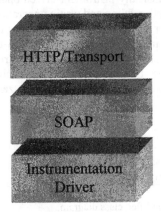

Fig. 5. Virtual Instrument Block Diagram

5   An Operative example

Within the framework of the project "Technological Network for Telecommunication Measurement Instrumentation", funded by the Italian Ministry of Education, University and Research (MIUR), as a LABNET follow-up, a number of new measurement devices have been acquired, to be connected in the Telecommunications Systems Measurement Testbed, available in the CNIT National Laboratory for Multimedia Communications in Naples. In particular, we concentrated on highly specific and valuable equipment for wireless network measurements, an area of growing interest in the recent years.

A specific testbed, including some pieces of the above-mentioned equipment has been set up for the Tyrrhenian International Workshop on Digital Communications in Sorrento, Italy, from July 4 to 6, 2005. The goal of the testbed is to show the application of the concepts outlined in the paper, by allowing remote users to test the operating conditions of an IEEE 802.11b wireless LAN (WLAN), in the presence of an adjacent interfering channel. The network segment under examination is that of a single Access Point (Cisco Bridge 350), whereas the interfering signal is created by an Agilent E-4438C vector signal generator. The resulting sum traverses a splitter, whereby a part of the signal power is directed to another Access Point, to be then transmitted over a satellite link to the remote site; another part reaches a Spectrum Analyser (Agilent E-4404B), where the interference phenomenon can be visualized. Fig. 6 shows the complete scheme of the testbed.

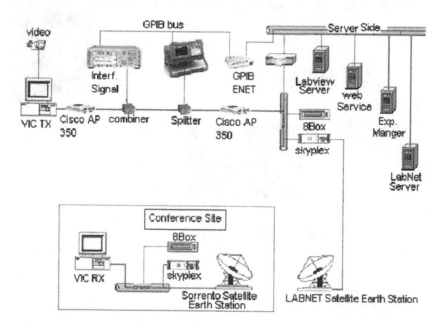

Fig. 6. Testbed

The signal under test is H.261-coded video, generated by a multicast VIC application. The interfering traffic is, for the time being, a deterministic constant bit rate signal. Different patterns, in particular on-off ones with various statistical characteristics are under realization on the signal generator. The current experiment, which is mainly aimed at showing the roles of the LABNET Server and Experience Manager and, in general of the whole architecture that has been discussed, allows the qualitative analysis of the channel throughput, under different conditions:

- When the two transmission are on the same channel (CH 7),
- When the two transmissions are on adjacent channels (CH 7, CH 6),
- When the two transmissions are on non-overlapping channels (CH 1, CH 7).

The experiment can be controlled at the remote site, and the throughput behaviour can be appreciated by observing the received video sequence and, at the same time, monitoring the output of the spectrum analyzer.

Fig. 7 shows a snapshot of the web interface, with a badly interfered video frame, the signal spectrum and the windows visualizing the packet loss statistics.

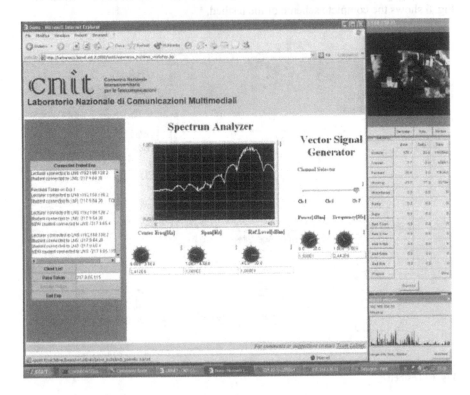

Fig. 7. Web interface for the WLAN interference measurement.

6   Conclusion

The architectural framework of the LABNET system for remote control and management of telecommunication measurement experiences has been discussed in the paper. In particular, we have highlighted the characteristics of instruments as web services, and the advantages of decoupling the low-level drivers to the maximum possible extent from the system architecture. An operative example of a testbed, containing all the described software elements has been briefly sketched. Based on the current design, the system architecture is evolving smoothly toward an integrated Grid platform for measurement instrumentation [6].

References

1. Davoli, F.; Maryni, P.; Perrando, M.; Zappatore, S.: A General Framework for Networked Multimedia Applications Enabling Access to Laboratory Equipment: the LABNET Project Experience. In Proc. IEEE Int. Conf. on Information Technology: Coding and Computing, Las Vegas, NV, (2001) 359–365
2. Hoyer, H.; Jochheim, A.; Rohrig, C.; Bischoff, A.: A multiuser virtual-reality environment for a tele-operated laboratory: IEEE Transactions on Education, Vol. 47 (1), (2004) 121–126
3. Nedic, Z.; Machotka, J.; Nafalski, A.: Remote laboratories versus virtual and real laboratories. Frontiers in Education, 2003. FIE 2003. 33rd Annual Volume 1, (2003) T3E-1 - T3E-6 Vol.1
4. Asumadu, J.A.; Tanner, R.; Fitzmaurice, J.; Kelly, M.; Ogunleye, H.; Belter, J.; Song Chin Koh: A Web-based electrical and electronics remote wiring and measurement laboratory (RwmLAB) instrument. IEEE Transactions on Instrumentation and Measurement. Vol. 54 (1), (2005) 38-44
5. SOAP Version 1.2 specification: http://www.w3.org/TR/soap12
6. The GRIDCC project – http://www.gridcc.org.

Chapter VI

Virtual Immersive Communications And Distance Learning

Architectures For Context Aware Services

Federico Morabito[1], Roberto Stomeo[1], Massimiliano Lunghi[1], Giovanni Cortese[1],
Fabrizio Davide[1]

[1] Telecom Italia Learning Services, Viale Parco de' Medici 37, ,
00100 Rome, Italy
{federico.morabito, ue006123,
massimiliano1.lunghi, fabrizio.davide}@telecomitalia.it,
g.cortese@computer.org
http://www.tils.it

(Invited Paper)

Abstract. In this paper, we describe our application framework (Multi-Site Navigator) for delivering context-aware services. The application framework is based on our middleware solution, named conteXt Data Manager (XDM), using DHTs technologies as underlying substrate for integrating different contextual data generators or sensors.

Keywords: context-awareness, distributed data management, overlay networks, localization

1 Introduction

When we speak of support infrastructures for context-aware systems, we refer to a scenario in which mobile users use wireless communications to interact with surrounding networks, sensors, applications, services, and other users present in the environment. The enormous number of wireless network solutions currently available means that we will require an infrastructure capable of supporting services that can adapt to diversity at many different levels including: data transmission, Quality of Service, user expectations and user profiles. In this paper, we analyze issues related to the requirements and design of infrastructures for context-aware services.

For the system to adapt to a rapidly changing environment, the technological infrastructure must itself be context-aware and adapt constantly to ambient conditions. Adaptation must take place as far as possible without the intervention of the user. It is the task of the "infrastructure" to provide these capabilities.

Also the seamless communication feature is required, intended as an interplay between layers in the architecture, from the physical up to the applications level.

1.1   Scenario

An architecture for context-aware services [2] provides services that are enhanced by knowledge of the physical environment such as proximity, environmental characteristics, and specific knowledge on the local and user. A physical environment seamlessly integrates the services provided by all the computing devices it contains – including embedded systems, network connected appliances and portable devices of mobile users – as well as those supplied by more traditional application servers running in a separate computer room or remotely in a computing grid.

Users moving in such pervasive computing environment [1] access all service available with ease; in turn, users' devices provide services to and share data with other components in the environment.

Context awareness add a level of intelligence to the architectures in the management: the integration between environmental context and service environments is related to the ability of the architectures to 1) collect data, 2) provide information ,3) adapt the services to the user profiles.

Environmental Context is information describing the current state of the environment. This includes dynamic context data, such as position, presence and identity of people, specific situations occurring within the space, as well as static data such as spatial model of environment, inventories of objects, and so on. Service environments should gather, organize and 'publish' context information so that applications can use them.

Similarly, users are expected to share with the environment a wide range of information about themselves, including their position, preferences, profile information, tracks of past interactions with the services in the environment, skills.

In this way, the environment is able to provide services that are enhanced by knowledge of the physical environment itself such as proximity, environmental characteristics, and specific knowledge on the local and user. A physical environment seamlessly integrates the services provided by all the computing devices it contains – including embedded systems, network connected appliances and portable devices of mobile users – as well as those supplied by more traditional application servers running in a separate computer room or remotely in a computing grid.

Users moving in such pervasive computing environment access all service available in a easy (transparent) way; users' devices provide services to and share data with other components in the environment.

2   Context Awareness Architectures for large-scale services

To deliver context-aware applications to a vast mass of users, several application and information engineering issues have to be addressed. This research line addresses issues in software engineering, information management, network management, data management, sensor network, ontologies, user interface, etc. in order to support the design of large scale, context-aware applications [3].

2.1 Requirements

A generic context-aware system includes different main components that can be schematized as follow:
1. Information providers (sensor networks, personal users information space, network parameters, multimedia data providers etc)
2. Middleware to collect and distribute contextual data
3. Applications that use context data to adapt offered services to users

Our research is focused on information management architecture able to capture data from sensor networks and use them for building context aware applications.

As sensor networks capture data from the user and from the environment, there is a need for a highly sophisticated information management infrastructure that transforms sensed data into information, that can be used by the user and applications, and is able to store and distribute it.

Most previous research on architectures for context aware services focused on a 'toolkit' approach, targeted to simplify engineering of services which use information originated from sensors.

A context infrastructure should give, to users and applications, access to information originated from the physical world with comparable ease to that provided by the current Web for the sharing of 'traditional' content.

However, there are a number of specific properties that an ubiquitous, Internet-like infrastructure should have:
- Extreme scalability. The infrastructure is expected to manage a high volume of rapidly changing data where the publishing rate is application dependent andcan range from seconds to hours and the data published are represented in few bytes. It should be able to implement mechanisms for storage and retrieval of large amounts of data
- Shared access to data
- Support for sophisticated data models for representing information e.g. supporting ontology languages
- Support for mobile user
- Flexibility about available resources (for storage, processing, bandwidth etc) and ability to distribute computations (e.g. filtering) accordingly
- Ability to cope with the barriers which derive from existence of different communication protocol ant technologies and data ownership domains
- Privacy and security

2.2 conteXt Data Manager (XDM): a DHT-based solution

We propose a solution for the context data management architecture called conteXt Data Manager (XDM) based on Distributed Hash Table technology, that addresses collecting, (interpreting), filtering, storing, and sharing information which is dynamically generated by sensors in the environment (Fig. 1) and that permits to build applications using the contextual information for delivering context aware services.

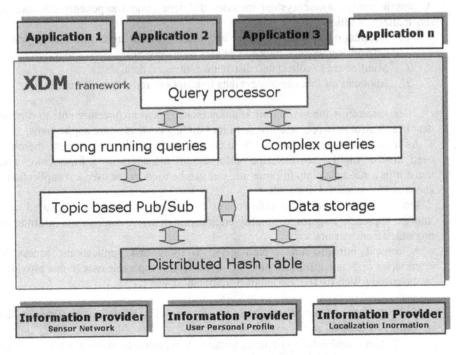

Fig 1 conteXt Data Manager (XDM) architecture

DHTs ([4],[5]) have emerged in the P2P community as a promising architecture on which to base a wide variety of large scale applications, in open and dynamic systems. DHTs address to some extent most of the requirements we listed above for ubiquitous context awareness infrastructures. DHTs provide mechanisms to implement large scale information retrieval, such as support for data replication and support for locality-aware accessing to data, security/ privacy through encryption, host independence, indirection to address problems of accessing data in multiple domains, name-based communication, and autonomic load-balancing. Our solution uses DHT technologies (based on Pastry [9]) as routing substrate to implement a lookup services into a distributed and dynamic environments, where nodes generate, require or forward data. The use of DHT technologies as routing substrate and lookup service guarantees properties of scalability in terms of nodes and data: this property is relevant for context data management architecture where a large number of sensor data are published within the network and a lot of information providers are distributed across the networks.

Over the DHT substrate, XDM includes the data storage and retrieval functionality with the aims to maintain the information in the system with a level of persistency depending on the type of data.

For example, personal information (user profile) remain persistent in the system since the information owner decides to change while sensors generated data have expiration mechanism based on aging. XDM permits to build distributed and collaborative storage containing data published by participating nodes.

The communication model (based on Scribe [10]) used in XDM is optimized for supporting the communication-intensive applications. The publish/subscribe paradigm permits to implements a multicast tree related to a topic creation: for example, when the application or the user requires a localization service, the "subscribe operation" to the service permits to obtain the position information through an optimized multicast tree [4]. The many-to-many communication model is obtained by defining subscribers and publishers and the optimized graph for the communication is maintained by p2p message passing model.

The ability to perform spatial queries is a fundamental issue in context-aware systems, since a lot of the information will include spatial references. The objective here is to implement such queries efficiently over a widely distributed collection of data sources. For example, position information is collected by a variety of sensors and is stored in the context data grid. We have done initial work in multi-attribute range queries on DHTs, which is required to query position information, using XML modeling of data. A simplified version of query processor has been built using XML data representation to perform query in XML-based language (for example XPath).

The prototype of the architecture shown in Fig. 1 provides the following features (see figure 1):

1   Topic-based Publish/Subscribe, to allow clients to register for information of interest and to collect and distribute information.

2   Distributed Persistent Data Storage, with appropriate support for indexing and complex query mechanisms. To make this service flexible to support a variety of practical deployment scenarios, also appropriate strategies for placement of data in the network should be supported (for example, an agent should be allowed to store the generated data locally or to ask another node for storing it. Similarly, desired levels of replication should be controllable depending on the requirements of the scenario).

3   Query processor (based on XPath language [14]) supporting also Long Running queries, by extending the topic-based publish/subscribe component.

2.3   User Context in XDM framework

The XDM framework purpose is to seamlessly extend the context of the user mobile stored into his/her personal computing environment (often hosted in her home or enterprise) through his/her mobile device.

User's Data (such as profile, documents, contacts, etc.) belonging to the user workspace, that are relevant to 'current' activities, will be made always accessible in the XDM framework to the mobile user through a Shared Context space.

The Shared Context space in XDM implements a common, distributed "blackboard" offering publish-subscribe style coordination to a number of peers .

Providers of information to this space include:

- sensors in the physical environment (cameras, environmental sensors, etc.);
- sensors in the user's Personal Area Network (GPS, motion sensors, cameras, speakers etc);
- inference components for high level, interpreted context.

The XDM framework, supported by the DTH infrastructure, enables to maintain the consistency of Shared Context space as a data storage distributed among several different nodes. This distribution of the data information permits to maintain up-to-date the dynamic information collected by a broad class of sensors whit less traffic on the network.

The topic-based publish/subscribe communication model permit to collect all information related and similar (for example, all information related to "position") providing them to some service (for example, "localization").

In the XDM framework sensors can be standalone devices (e.g. temperature, pressure, motion sensors, camera, microphone etc.). or wireless user devices (e.g. PDAs with GPS, etc.).

2.4   Case study: Multi-Site Navigator

We developed an application prototype (see figure 2), conducted as part of the Virtual Immersive Communications (VICOM) project, funded by the Italian Ministry for School, Universities and Research [15], so called *Multi-Site Navigator*, to test the XDM middleware.

This application provide user to access contextualized services provided by the environment, based on the following list of Context Data:

- users profile
- localization sensors (indoor and outdoor)
- simulated sensor (Temperature; Light)
- shared device (Video Wall; Printer)

through a simple web interface.

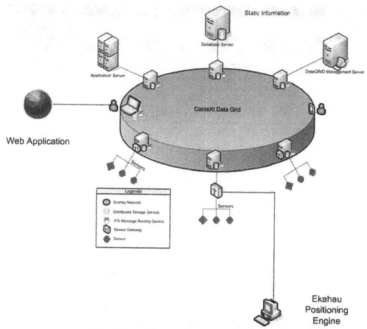

Fig. 2. Multi site navigator architecture

We deployed and tested the Multi-Site Navigator in a lab environment, with a network architecture based on a 802.11b WLAN [8]. The sensors (real or simulated) generate dynamic context information captured and stored in a Dynamic Shared Context Space trough the XDM middleware.A database models, was also implemented, to describe the static object the environment (e.g. building, floor, room, etc.) sensors, devices and persons and their context information (position, url, owner, etc).

The localization services, instead, are provided by the following third-part technology provider:

- **indoor**: Ekahau Positioning Engine [7]
- **outdoor**: GPS system.

In the Multi-Site Navigator the contextual information is graphically represented in the web interface (see figure 3) that provides the following kind of services:

- Management of contextual information using a 2D-Map;
- Access the Directory of distributed sensors.
- Searching and query the data from the sensor (using XPath [14] based on XML standard);

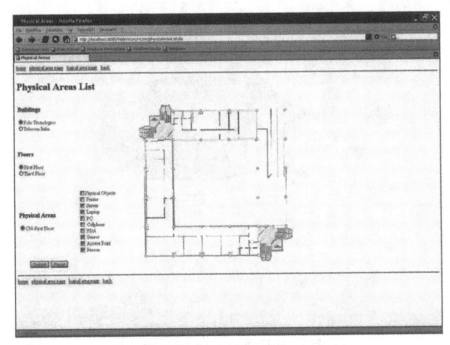

Fig. 3. Multi site navigator: the visualization portal

3 Conclusions

We have discussed some of the architecture requirements for a generalized deployment of context aware services, which heavily depend on emergence of standard architectures for pervasive computing. We are actively researching some of these themes. We have also developed a prototype of context aware application based on the proposed architecture solution.

3.1 Future works

A relevant feature to be investigated is related to raising the expressiveness of queries over context information. The use of semantic web models such as RDFS (Resource Definition Framework Schema) to represent information and their relationships is a promising way that can be obtained using DHT infrastructures and XML data representation.

Also, mechanisms that allow the context information to be shared even if the individual peers do not have consistent ontologies are to be considered in the infrastructure design (although some degree of standardization of ontologies is required, it is desirable that peers can interoperate even if they do have partially different ontologies).

In this field, semantic interoperability is a specific challenge for adding context awareness to pervasive computing environments.

4   References

1. A. Ferscha "Coordination in Pervasive Computing Environments" Proceedings of the Workshop on Enabling Technologies: Infrastructure for Collaborative Enterprises, Proceedings of the Twelfth IEEE International, Linz - Austria, IEEE, 3-9, June 2003.
2. A.K. Dey et al, "Towards a Better Understanding of Context and Context-awareness," GVU rep.GIT-GVU-99-22, College Comp., GA Inst. Tech., 1999.
3. "ISTAG Scenarios for Ambient Intelligence in 2010", ftp://ftp.cordis.lu/pub/ist/docs/istagscenarios2010.pdf
4. DKS(N, k, f): A Family of Low Communication, Scalable and Fault-Tolerant Infrastructures for P2P Applications. Luc Onana Alima, Sameh El-Ansary, Per Brand, and Seif Haridi. In 3rd International Symposium on Cluster Computing and the Grid - CCGRID2003, Tokyo, Japan, May 2003.
5. Chord A scalable peer-to-peer lookup service for internet applications, Ion Stoica and Robert Morris and David Karger and M. Frans Kaashoek and Hari Balakrishnan, Proceedings of the 2001 conference on applications, technologies, architectures, and protocols for computer communications, ACM Press, 2001
6. "Skipnet: A scalable overlay network with practical locality properties" "N. HARVEY and M. JONES and S. SAROIU and M. THEIMER and A. WOLMAN", In Proceedings of USITS (Seattle, WA, March 2003), USENIX.", 2003
7. "Ekahau Positioning Engine™ 2.0", http://www.ekahau.com
8. "eCASA : an Easy Context-Aware System Architecture", G. Cortese, F.Davide, A. Detti IEEE Vehicular Technology Conference
9. A. Rowstron and P. Druschel, "Pastry: Scalable, Distributed Object Location and Routing for Large-Scale Peer-to-Peer Systems," Proc. Int'l Conf. Distributed Systems Platforms (Middleware), ACM Press, 2001, pp. 329–350.
10. A. I. T. Rowstron, A.-M. Kermarrec, M. Castro, and P. Druschel. SCRIBE: The design of a large-scale event notification infrastructure. In Networked Group Communication, 2001.
11. M.Lunghi; G.Cortese; F.Davide; Ambient Intelligence – the evolution of technology, communication anc cognition towards the future of human-computer interacion: Context-Awareness for Physical Service Environments. Ios Press, 2004, Amsterdam, The Netherlands
12. M.Lunghi, G.Riva, P.Loreti, F.Vatalaro, F.Davide; Presence 2010: The Emergence of Ambient Intelligence on Being There: Concepts, Effects and Measurements of User presence in Synthetic Environments. Ios Press, 2003, Amsterdam, The Netherlands

13. M. Lunghi, F. Davide, P. Loreti, G. Riva, F. Vatalaro, Communications through Virtual Technologies, Advanced Lectures on Networking, Springher 2002, 2497, 124 – 154
14. XPath, www.w3.org
15. www.vicom-project.it

The Teledoc2 Project: A Heterogeneous Infrastructure For International E-Learning

Gianluca Mazzini[1], Andrea Ravaioli[2], Cesare Fontana[2], Paolo Toppan[2],
Luca Simone Ronga[3], Stefano Vignola[4] and Oreste Andrisano[2]

[1] CNIT Research Unit of Ferrara - gianluca.mazzini@cnit.it
[2] CNIT Research Unit of Bologna - {andrea.ravaioli, cesare.fontana,
paolo.toppan, oreste.andrisano}@cnit.it
[3] CNIT Research Unit of Firenze - luca.ronga@cnit.it
[4] CNIT Research Unit of Genova - stefano.vignola@cnit.it

(Invited Paper)

Abstract. In this paper we describe the Italian online learning project, called Teledoc2, financed by the Italian Ministry of Education, Universities and Research (MIUR) and realized by CNIT (National Inter-University Consortium for Telecommunications)[1]. The Project is running through 2003-2005 and aims to build a complete, multimedia, interactive and fully-featured online-learning service for ICT researchers and PhD students of Italian Research Centres, being branches of CNIT. The paper gives an overview of all the technical, didactical and organizational aspects implemented in order to achieve the successful activation of the Teledoc2 integrated services[2]. CNIT chose avant-garde and technologically updated solutions in all these aspects, in this way building an innovative and efficient didactic service and giving also a research tone to the Project. The main attention of the paper, in particular, is devoted to one of the key elements of the service: the technical aspects regarding the heterogeneous network architectures, designed, implemented and operated by CNIT to distribute multimedia information among its terminal sites, enabling the Teledoc2 learning service. A series of objective and subjective measurements are still in progress to evaluate the reliability of the different network solutions and the Quality of Service guaranteed by the different architectures for real-time transmission of audio-video and data contents. The pervasive use of avant-garde technological solutions, like multicast, satellite Ka band and NAT, further increases the efficiency of the widespread distribution of Teledoc2 learning service in all the Italian territory. Finally, the Project give special attention to the research and experimental activity, called "Network of Laboratories", developed in particular at the WiLab (Wireless Communication Laboratories) of the University of Bologna.

Keywords: e-Learning multicast, e-Learning architecture, e-Learning satellite, e-Learning measurement, e-Learning management

1   Introduction

In recent years, two different and important factors are completely changing the features of the educational market inside Research Centres: first we register a growth of knowledge and information necessary to increase professional life; at the same time, these capabilities are valid only for a short time because they are changing very quickly. Therefore we feel the need of continuous training during our lifetime. Unfortunately, the real possibility to learn is often obstructed by factors such as distance, time, money, difficulty to realize shared meetings, etc. So the importance of distance learning is increasing and, in few years, we have seen a fast evolution, from the books sent by mail (first generation of distance learning) to the integrated use of TV transmission, video and audio recordings (second generation) to the interactive real time data transmission (third generation)[3].

CNIT gave a significant contribution to fill these needs: during the year 2000-2001, a first learning project, called Tele- doc1[4], was realized addressing its attention to the PhD stu- dents of Italian Research Units, being branches of CNIT; with the current Teledoc2 Project, CNIT completely upgrade the technical approach to initiatives related to e-Learning, using the most avant- garde technologies in the field of ICT. Teledoc Projects are planned for the dif- fusion of scientific and technological culture in the ICT field, allowing students to attend courses broadcast by different Italian Research Centres. The Teledoc2 project, in particular, aims at building an efficient service of distance learning of third generation: the courses, offering educational and specialist contents in the forefront of research, can be attended in real-time by students just connecting to the CNIT proprietary packet communication network and using simple Personal Computer (PC) running a custom multimedia application. The successful acti- vation and permanence of the Teledoc2 service need a perfect integration of all technical, didactical, organizational aspects that characterize the final service. In the sections of this paper we describe all these elements and we point out the importance and critical aspects of each of them.

2   The Teledoc2 Learning Service: Needs and Requirements

The Teledoc2 service aims at providing a fully-featured learning service, made up of real-time learning functionalities and the possibility to record lectures for offline studies: all the students from all the Italian CNIT Research Units have to be able to attend the full services. The learning strategy, therefore, aims at recreating a live virtual classroom environment, with a real-time face-to-face relationship and high level of interactivity among the users: more, the concept of virtual classroom has to be extended to "ubiquitous distributed service", with no kind of limitation for the user's position. With this "direct" contact and control, students are encouraged, in spite of the distance, to pay attention as happens in a real classroom. In order to lead a successful project and stimulate a large and enthusiastic user participation, the whole learning system must be complete, efficient, user-friendly and characterized by having fixed and suitable QoS level. CNIT characterized the Teledoc2 service with the following composing elements, needs and features:

- *network:* a widespread data communication network is necessary to connect all the CNIT Italian Research Centres. The Teledoc2 learning service imposes requirements to the CNIT network and, in particular, a minimum guaranteed network bandwidth for all the connected centres. Besides this, CNIT chose to adopt an innovative technological way of transmission, particularly useful for real-time multimedia transmission towards a multitude of users: the multicast. A communication network supporting the multicast allows reducing the required bandwidth and obtaining scalability, when the number of student increases. CNIT fixed the multicast support as a basic feature for its terrestrial and satellite networks.

- *e-Learning software:* The second fundamental element is a valid and user-friendly e-Learning software application for the transmission of audio, video and data on the network and for the management of all the teaching aspects (accounts' management, subscription to courses and lectures, download of teaching materials, etc). Keeping in mind the CNIT network infrastructure and the objectives of the Teledoc2 service, we fixed the main requirements for this software: real-time transmission of audio, video and Power Point slides (also with animations); support for multicast; shared pointer on the slides; synchronization of the streaming; interactivity via chat, audio and video among teacher and students, under the control of the teacher (hand-rising and floor-control functionalities); possibility to record the lectures; fixed quality of transmission; reliability and simplicity of use; system of users' authentication; management of lectures' calendar and users' registration to lectures.

- *courses contents::* The Teledoc2 Project aims at offering a high level of didactic contents; a formative route for PhD Students was planned, made up of interesting courses, held by Professors affiliated to CNIT and very well known in the international Information and Communication Technologies field. The improvement of knowledge among PhD students is fulfilled by 31 courses covering a broad range of subjects, concerning specialized and avant-garde topics all related to ICT. To obtain an international relevance, all the courses are in the English language.

Finally, in order to build a successful Learning Service, another important element must be considered beyond the technical and didactical ones: network, software and courses' contents contribute to create a complete learning service only if they are perfectly integrated and if the access to the service is user-friendly, authenticated and well managed. A Web site[2], in particular, seems the easiest way to access the service; a client-server system of authentication and users' management need to be created and integrated with the Web-server and with the real-time learning software.

3   The Software Solution

CNIT conducted a long and deep evaluation phase to choose a complete tool, a
so-called "Software Platform", able to provide synchronous service (multimedia,
interactive, real-time distribution of lectures) and asynchronous service (recording and playing of recorded lectures) with all the fixed features described in
Section 2. To minimize the possible combined drawbacks due to the software
platform and the communication network, a commercial choice was decided for
the software platform. It has to be underline that the performance of this software has to be suitably checked on a limited bandwidth network. CNIT researchers analyzed and tested commercial software platforms and, in parallel,
promoted a research activity, aiming at building a proprietary e-learning software tool (Mbone-CNIT-Tool), obtained by the integration of the open-source
Mbone tools (like Vic, Rat)[6] used during the Teledoc1 Project (tools have
been recompiled, modified and optimized) with new software created by CNIT
researchers.

3.1   Analysis of the Software Platform

After a research on the market, CNIT selected a series of commercial tools
and tested them by means of demos, involving a lot of CNIT Research Centres
and giving the opportunity to deeply evaluate all the aspects (transmission,
functionalities, reliability, stability). After this preliminary selection, we analyzed
the following software packages:

- Learn-Linc[7]
- X-Learning[8]
- X-Stream[9]
- Centra Simposium[10]
- Vip-Teach[11]
- Paiper-Platform[12]
- Breeze-Live[13]

The software analysis has been developed in two steps, as shown in Figure 1.

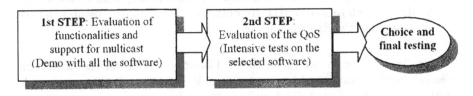

Fig. 1. *Steps of the software analysis*

During the first step, the software providers gave us trial versions of the
tools, letting us test them on the CNIT multicast network. In these tests we

simulated real lectures and we evaluated the general level of quality and the functionalities at both application and network levels. Tests aimed to evaluate a lot of parameters: multicast-enabled transmission, quality of audio and video, slide managements, interactivity, simplicity of utilization, synchronization, delays, stability, presence of functionalities like floor-control, lectures' recording, shared pointer. The results of these tests are summarized in the Figure 2.

	LearnLinc	X-Learning	X-Stream	Centra Simposium	Vip Teach	Mbone-CNIT-Tool
Way of transmission	Multicast server + unicast	Unicast + multicast	Unicast + multicast via satellite	Unicast + multicast	Complete multicast support	Complete multicast support
Audio & Video Quality	Good	Good	Good	Fair	Good	Good
Slides' managements	Jpeg + pointer + html + application sharing	Streaming slide + pointer	Jpeg + pointer + html	Jpeg + pointer + html + application sharing	Power Point slides' progress	Power Point slides' progress
Interactivity	Excellent	Poor	Poor	Good	Good	Fair
Floor Control	Good	Absent	Good	Good	Good	Absent
Simplicity of utilization	Good	Fair	Poor	Good	Good	Good
Synchronization and delays	Fair	High delays	High delays	Good	Good	Good
Shared Pointer	Ok with Jpeg no with HTML	Yes	Yes	Ok with Jpeg no with HTML	Yes	Yes
Lessons' recorder	Yes	No	No	Yes	Yes	No
Stability	Excellent	Good	Fair	Instable	Fair	Fair

Fig. 2. *Comparison of e-Learning software platforms*

The most critical aspect we found in the tested software was the complete support for multicast.

3.2 Platform Selection

Anyway, we have finally selected the two platforms that gave the better performance according to the Teledoc2 service requirements:

- VIP-Teach (produced by LightComm)
- MBone-CNIT-Tool (open source tools + original software by CNIT researchers)

On these two platforms we made a lot of tests in order to evaluate the provided Quality of Service; for these evaluations we used subjective judgments, expressed by the participants in the tests who completed forms at the end of every test. According to the ITU-T Recommendations[14], the forms consist in multiple-choices questions, regarding various parameters of Quality of Service; data has been collected in databases and statistically elaborated, using the Mean-Opinion-Score (MOS) method. Analysing the values of MOS, regarding the different QoS parameters, we noticed that:

- The two platforms show the same qualitative level, as regards: experience level, simplicity in use, audio and video quality
- MBone-CNIT-Tool gives better quality in the video and let the students interact easier via audio (no "floor control" or "hand-rising" mechanisms)

- Vip-Teach offers additional functionalities ("floor control" and "hand-rising"), better synchronization, facility in use and functional graphical interface

Considering the final Teledoc2 service, we evaluated the software Vip-Teach giving more complete functionalities, so we decided to focus our attention on it, collaborating with LightComm to customize the tool for the Teledoc2 requirements. The development of the MBone-CNIT-Tool became a research project. Looking at these emerged results, Vip-Teach platform offers:

- Good quality of audio, thanks to the introduction of a new 64 Kb/s codec;
- Transmission of video with reasonable quality and fluidity in movements;
- Transmission of Power-Point commands to have synchronized progress of slides, with animations and shared pointer;
- Various interactivity instruments like audio, video and chat;
- Added functionalities (hand-rising, floor control) that are suitable for a didactical environment where one user (teacher) should control the lecture and the interactivity with students: the professor is the chairman and the students need to ask for the talk with a very simple mechanism not compromising the real-time relationship;
- Good stability reached with numerous tests on the CNIT network;
- Simplicity of use and graphic user-friendly interface that were obtained by the feedbacks of CNIT people during the demonstrations;

In Figure 3 we report an example of an online lecture's screenshot:

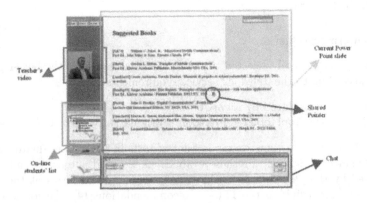

Fig. 3. *Vip-Teach screenshot*

4 Main Network Infrastructures

An efficient online learning service is based on a communication network with suitable trade off between bandwidth and cost: large investments have been made by CNIT to build its own terrestrial and satellite systems that can cover all the Italian territory, using IPs assigned to the CNIT autonomous system. Particular attention has been paid also to possible integration with European networks, like Geant, and European projects developed in the framework of Unesco.

In order to provide a good online learning service, it must be considered that reliability, answering the service requirements and integration with all the technical and didactical aspects are critical features for the network; CNIT used all its experience in the ICT field to guarantee these features both in the backbone connections and the local connections. One of the key network requirements fixed by CNIT has been the support for multicast[15][16], a strong element of innovation and originality, compared with most of the other distance learning systems. CNIT decided to build multicast-enabled networks, because this way of transmission seems particularly appropriate for online learning services, like Teledoc2. These applications, in fact, need the co-presence of two basic conflicting network requirements: on one hand they need communications one-to-many and many-to-many to reach all participants and to promote interaction; on the other hand they need high bitrates, as they have to transmit audio and video of fixed quality. During a Teledoc2 lecture the teacher does indeed transmit multimedia streaming to all the students connected; with the classical unicast transmission the traffic on the network would be too intense and cause problems of quality degradation. Using the multicast technology, the data transmitted to multiple users are not duplicated in more copies of the same data for every single destination (like in unicast modality) but just one copy is sent on the network and it's duplicated only where it's necessary (fig.4).

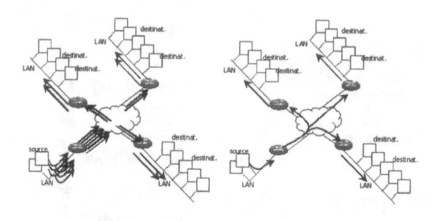

Fig. 4. *Unicast vs Multicast*

This offers a lot of advantages for e-learning services: extreme bandwidth economizing, scalability when the number of participants increase, good quality of service also for users with not high bandwidth availability. On account of these considerations, the CNIT network architectures have been characterized by:

- Minimum guaranteed network bandwidth for all the connected centres (not lower than 384 Kbps);
- Complete support for multicast

For the backbone connections among the national CNIT Research Units, CNIT adopted heterogeneous solutions:

1. A wideband terrestrial network, during the year 2004
2. A satellite Ka-band network during the year 2005

Other solutions to distribute the Teledoc2 formative services are now under study:

3. Integration between CNIT satellite network and terrestrial GARR network
4. Use of Ku band satellite link to re-transmit the Ka band lectures' streaming, thus extending the service to European level

4.1 Scenario 1: CNIT Terrestrial network

CNIT built, during the 2004, a main terrestrial network and a backup network. CNIT organized a tender for contract for the building of these networks: Albacom won it for the main network; Wind won it for the backup network.

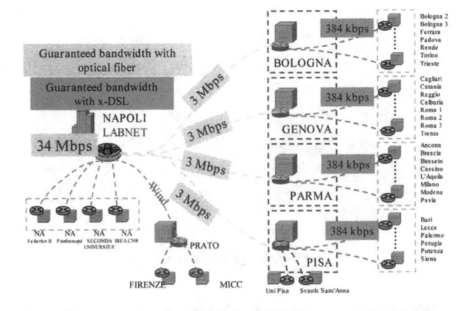

Fig. 5. *CNIT main terrestrial network*

The typology of the fixed network is a three level hierarchical tree, reflecting the different role each site has in the overall network architecture (fig.5). The tree root is the CNIT National Laboratory of Multimedia Communications located in Naples. The second level nodes are in Bologna, Florence, Genoa, Parma and Pisa; these sites have to distribute the data packets over the territory and they are directly connected to Naples by ATM technology and/or Frame Relay (layer 2, reference model OSI), so there is full transparency to the upper levels. All the other sites are the tree leaves, i.e. they are located at the third (and last) level in the tree.

In Bologna and Parma an optical fiber connection with carrier at 34 Mbps has been installed, while in Genoa and Pisa an HDSL link on copper lines at 8 Mbps (4*2Mbps) has been developed, delivered by IMA trunking technology. The virtual circuits coming from Naples and from third level sites converge on these circuits. The third level sites are connected to the second level sites by a virtual circuit based on Frame Relay with minimum guaranteed bandwidth of 384 kbps (512 kbps for Bologna2 and Bologna3); the links are realized by means of HDSL on copper at 2 Mbps. Every site got a router connected to the backbone CNIT through provider by Albacom (Wind for Florence) and a RJ45 plug-in able to give Ethernet or FastEthernet connectivity. Each site got some different public CNIT IP addresses (typically 32) that could be used for Teledoc2 purposes or for other research projects and collaborations between CNIT centres.

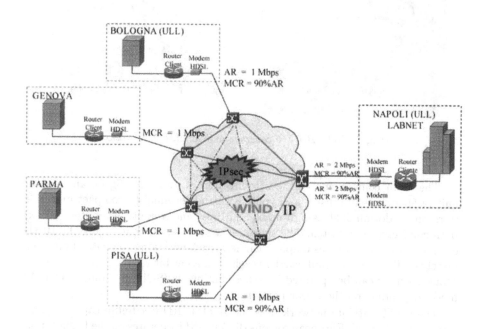

Fig. 6. *CNIT terrestrial backup network*

In order to guarantee the continuity of the Teledoc2 Service, even when some failures in the main terrestrial network occur, CNIT decided to build also a backup network, assigned to the provider Wind (fig.6). The backup network involves the 4 sites belonging to the second level (Bologna, Genoa, Parma, Pisa) and the site of first level (Naples). This network uses a methodology type Frame relay with V.35 interface and a public IP address is provided for each interface. This network gives the possibility to create IP tunneling between involved sites.

4.2   Scenario 2: CNIT Ka band satellite network

In the framework of the Teledoc2 Project, CNIT has set up and operates an advanced proprietary satellite network that connects, in a mesh typology, 24 of its Research Units and Laboratories. The satellite network, based on the Skyplex technology, operates in Ka band and provides a guaranteed bandwidth of 2 Mbps shared among the connected terminals. In our typical application most of this bandwidth is reserved for audio and video multicasting, so that a significant improvement of their quality is achieved. CNIT selected EUTELSAT as space operator partner.

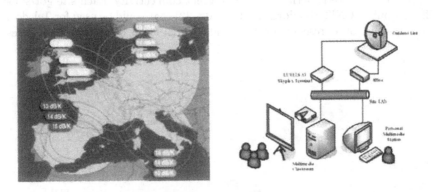

Fig. 7. *HOT BIRD 6 coverage and satellite user station*

The main components of the satellite user station are (fig.7): an outdoor unit (a 90cm satellite dish); a Skyplex Ka-band terminal (a compact unit able to receive a downlink stream of up to 38 Mbps) and a multicast router (8-Box) both connected to a switch. The stream is assembled on board the HOT BIRD 6 satellite from as many as 6 uplink carriers, each with a net bit rate of approximately 6 Mbps, or 18 uplink carriers each with a net bit rate of approximately 2 Mbps. Skyplex can be operated in either continuous (SCPC) or burst (TDMA) mode, depending on the traffic pattern.

The CNIT Satellite network uses the HOT BIRD 6 satellite capacity. This satellite has 4 uplink European regions (fig.7): CNIT operates on the Italian spot. The uplink traffic from each single CNIT station is multiplexed onboard through by the Skyplex payload. This satellite system will also enable the pan-European broadcast due to the downlink Ka coverage of HOT BIRD 6.

4.3 Scenario 3: Integration between CNIT satellite network and GARR terrestrial network.

As not all the CNIT Research Units have a connection to the satellite network, a series of alternative solutions are under study in order to let these sites participate in the Teledoc2 services anyway. In a future perspective, for example, CNIT is testing the integration between CNIT satellite network and GARR (the main academic network in Italy), using a solution proposed by LightComm, the same company providing the Vip-Teach software platform. In this configuration, a PC, running suitable software, works as a gateway between a multicast-enabled network (the CNIT satellite network) and a unicast one (GARR); it should be emphasized that GARR is now completing the implementation of the multicast technique. The real-time multimedia streams, coming in multicast from the satellite network, are multiplied in several unicast streams addressed to the connected users of the GARR network. All this process runs at the kernel level of the Vip-Gateway workstation, increasing its operating speed and reducing the delays introduced in the real-time transmission.

4.4 Scenario 4: Re-transmission of Teledoc2 lectures on Ku Satellite Network

The use of Ku-band satellite network is another scenario that CNIT is testing, within the Didanet Project, in order to spread the Teledoc2 lectures in wider areas. The main idea is the integration between Ka-band CNIT satellite network and a broadcast network over the Ku band, whose access has been provided by Aersat. The Teledoc2 lectures, transmitted over the Ka band network, are retransmitted in real-time to a hub, located by Skylogic in Turin, and from there they are relayed in Ku to the Eutelsat W3 satellite for broadcasting to simple DVB-IP receivers. In the current configuration, the available Ku bandwidth allows a transmission speed of 256 kbps, which is less than the one used over the Ka band. Therefore, a transcoder is used to adapt the audio/video/control flows before retransmission. The rationale behind such solution lies in the increased scalability that is achieved, at the expense of renouncing real-time two-way interactivity for the Ku receivers. The latter may be in principle a very large number, over the W3 coverage area (which includes Europe, the whole Mediterranean and North Africa), without affecting the bandwidth need. With this solution, in fact, users located in the coverage area of the Eutelsat W3 satellite can attend the Teledoc2 lectures in no-interactive modality: they need a small parabolic antenna, a DVB-IP card and the software Vip-Sat-Player, a proprietary application aimed at receiving the lectures' transmission on Ku band network.

5 Local Network Solutions

In section IV we heve described the heterogeneous connections among the national CNIT Research Units, built to receive and transmit the Teledoc2 lectures. Once the learning service has been received in a single point of every Centre, it's important to evaluate how to distribute it ubiquitously inside the campus, so that every interested user can participate from his/her own workstation. In this section we present a series of local network solutions: some of them have been tested in CNIT sites and others are under study.

In the terrestrial architecture, every CNIT site has a router with a RJ45 connector offering Internet connectivity with a classical network cable; the connection can be used in different modalities for the local distribution in the Research Centre:

1. Connect a PC directly to the CNIT router by a cross cable; this configuration is proper for Research Centres having a classroom (equipped with projector and audio-diffusion system) where students attend the lectures in groups;
2. Connect a switch to the CNIT router to realize a small local CNIT network. PCs connected to this switch must be configured with public CNIT IPs.
3. Connect the CNIT router to the department switch. By configuring the ports of the switch in order to realize a local CNIT VLAN, all the workstations connected to these ports have direct link to the CNIT network;
4. Connect an access point to the CNIT router, so that wireless users can be served. If we connect more access points, they are not able to distinguish between broadcast and multicast traffic, so multicast packets are sent at the slowest bitrate[17]
5. Use NAT (Network Address Translation) technology to translate local IPs to private CNIT IPs; to support multicast, anyway, NAT must be one-to-one.

The most critical aspect to be taken into account in the local distribution of the Teledoc2 service refers to the management of multicast traffic. The multicast routing is useful if you need to offer real time services to different hosts at the same time, but the routing must be enabled and correctly configured in all the network routers: this implies a global administration of the network. There are a lot of possibilities for the realization of a multicast network; in the planning you must consider which is the best intradomain multicast protocol for your needs and only after this choice you can design the structure of the network. CNIT network administrators have considered PIM-SM protocol (Protocol Indipendent Multicast Sparse-Mode), with static RP (Rendezvous Point); multicast protocol in IP environment does not support the transport of real-time data with a specified QoS (it uses UDP connection), it is a best effort transport network. There are some approaches for the measurement of the reliability on IP multicast networks in order to put some quality to IP, but they are more complicated than a normal TCP/IP connection, in particular in case of widespread networks.

As example of CNIT Local Research Unit, the architecture of Bologna CNIT site, is presented in Figure 8.

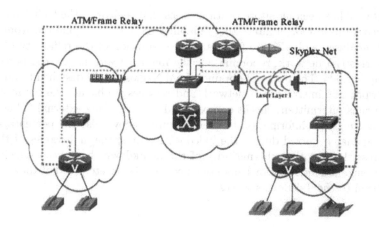

Fig. 8. *CNIT Research Centre at the University of Bologna*

The node is divided into 3 macroareas, each one of them being a laboratory. In the central area (Bologna1), a Cisco 3745 router is the core of the network; it is connected to the rest of CNIT network with ATM 34Mbit/s link and with satellite Skyplex Net. Bologna2 is connected to the centre with LASER layer1 link, and for backup with HDSL Frame/Relay PCR 2 Mbit/s MCR 1Mbit/s. Bologna3 is connected to the centre with Wi-Fi Bridge (802.11g) and for backup with HDSL Frame/Relay PCR 2 Mbit/s MCR 1Mbit/s. The three places are connected also with VoIP technology. This heterogeneous structure permits to study the different features of every link and to find the correct configuration to distribute the lectures of the Teledoc2 project into the tree.

Referring to the local solutions to redistribute the multicast streaming coming from the Ka-band satellite network, it's possible to connect PCs directly to the satellite switch or connect a terrestrial LAN, but it's generally necessary to change into the switches configuration. The "IGMP snooping" must be deactivated, if not the switches prune ports that are not joined to IGMP, causing the stop of the multicast traffic from the "passive" satellite receiver.

6 Measurements of reliability and network utilization

Once the Teledoc2 service has been implemented, integrating all the composing elements and the heterogeneous network solutions, it's interesting and necessary to monitor and measure the reliability and the global Quality of the Service, in particular for the aspects related to network; tests concerning these aspects have be done on the terrestrial network and are still in progress on the satellite scenarios.

During the whole of 2004, the CNIT terrestrial network worked finely, allowing a continuative and good level formative service to all the connected centres.

This network has been monitored with utilities that registered the traffic load on the network links; a lot of interesting reliability data can be found out from the network analyser tools. Figure 9 gives a representation of the reliability of the CNIT terrestrial network, reporting, for the main network nodes, the percentage of up periods. The data are obtained from Nagios, a tool designed to inform the network administrators of network and services problems; the monitoring daemon runs intermittent checks on hosts and services using external "plugins" which return status information to itself. Figure 9 shows also the percentage of up period that occurred during the Teledoc2 lectures; the high values of these data stress the fact that the majority of the breakdowns have been scheduled (electric current's interruptions, maintenance operations, etc.) so that they cause no problem to the Teledoc2 services.

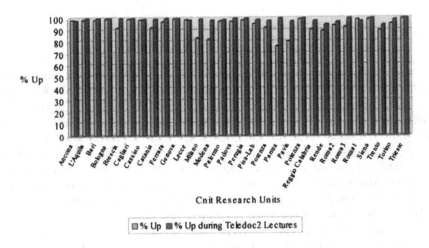

Fig. 9. *Reliability of the CNIT terrestrial network*

In Figure 10, for example, we report the particular trend of state breakdowns for a third level node in Bologna2, from the 1st of January to the 31st of December 2004.

Other interesting data can be extrapolated using another tool, called MRTG (Multi Router Traffic Grapher) that uses SNMP (Simple Network Management Protocol), protocol designed to give a user the capability to remotely monitor the network events. Analyzing the MRTG trends concerning the CNIT terrestrial network during the 2004, we can see the different level of network's utilization in the research centres: some links are used only for the Teledoc2 services, while other sites used the CNIT connection to generate normal Internet traffic too. This different behavior, for example, can be seen Analysing the MRTG log files and the graphs of traffic generated during the last months of 2004 into 2 different third-level nodes, Padova and Bologna2, both connected to Bologna1:

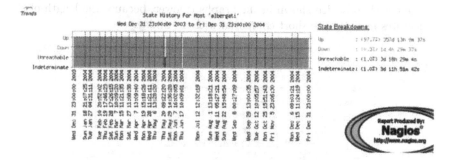

Fig. 10. *Network breakdowns in the CNIT node of Bologna2*

- Interface Padova-Bologna1 (Figure 11): the traffic, from May to December 2004, is present only during the Teledoc2 lectures, with concentration in the months of service.
- Interface Bologna2-Bologna1 (Figure 12): the traffic is better distributed, with higher bitrates.

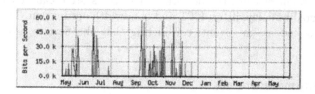

Fig. 11. *Interface Padova-Bologna1*

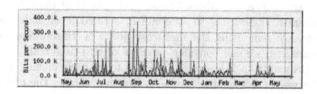

Fig. 12. *Interface Bologna2-Bologna1*

- Interface Bologna1-Naples (Figure 13): we notice very high bit rates because all the links from the third level sites connected to Bologna converge into the second level node of Bologna.

The MRTG measures are averaged in the traffic of one day; this implies that the max bandwidth shown in the graphs is lower, because the length of the lectures in a day are short compared with a full day.

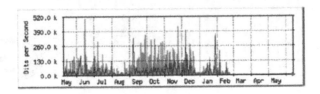

Fig. 13. *Interface Bologna1-Naples*

7   Measurements of Quality of Service

The global Quality of Service provided by Teledoc2 multimedia online learning service can be evaluated, using both objective and subjective measurements. On an objective point of view, the performance of the system can be characterized by parameters like delay, jitter, RTT and bandwidth occupation. The subjective evaluation is just as important, considering that the Service is mainly user-oriented.

7.1   Objective measurements

A testbed has been built in the CNIT Research Centre of Bologna to simulate a real-time Teledoc2 lecture, using the CNIT terrestrial network and the Vip-Teach e-Learning software tool. These tests want to evaluate how QoS parameters change, in online learning services, with different fixed values of guaranteed bandwidth. In Figure 14 the scenario of the testbed is shown. We have considered the teacher in a second-level node (Bologna 1) of the CNIT terrestrial network, transmitting lectures to students in a third level node (Bologna 2); since the data are sent using the multicast protocol, the number of receivers is non influential, so we can consider one flow of real-time traffic between two nodes as a representation of a service extended to multiple clients. The available bandwidth between the two sites has been changed, setting the traffic shape at different values of bitrate on the router of Bologna2; two hosts with monitoring functionalities are connected to the switches in each site.

Using the software Vip-Teach of the Teledoc2 lectures, we have simulated lectures in 6 scenarios with different guaranteed bandwidth in the receiver's site: 1 Mbps, 512 Kbps, 328 Kbps, 256 Kbps, 128 Kbps and 56 Kbps. The effective link capacity has been verified in every scenario using a free measure network tool, called Iperf. We have simulated a lecture of 2 minutes in every scenario: in the first minute the teacher has the control (transmission of teacher's audio and

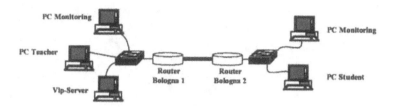

Fig. 14. *Testbed for QoS measurements*

video); in the second minute the student interacts with the teacher (transmission of teacher's audio, student's audio and video). During the lecture we have started a sniffer in the monitoring PCs.

The Figures 15, 16 and 17 report some outcomes of these tests, comparing the throughput measured at teacher and student's sides, during the second minutes of the test in three scenarios with different guaranteed bandwidth. Y-axis reports the values expressed in byte/sec, while X-axis the time expressed in seconds (in particular from 60 sec to 120 sec, corresponding to the second minute of every test).

- *1 Mbps guaranteed bandwidth* (Figure 15): the audio and video's graphs of teacher and student are nearly identical, when they are measured at transmitter or receiver's side. All the audio and video streaming are received at the same quality level they have in transmission.
- *384 Kbps guaranteed bandwidth* (Figure 16): the audio's graphs of teacher and student are similar, when they are measured at transmitter or receiver's side: audio streaming are fair and understandable. The video graphs highlight some lost packets from transmitter to receiver; anyway the video streaming is still acceptable
- *128 Kbps guaranteed bandwidth* (Figure 17): the graph of the audio at the receiver's side presents very low values and it's perceived as incomprehensible. The video streaming, thanks to the TFRC system (a LightComm technology that estimates the available bandwidth and adapts the bitrate of the multimedia transmission), is transmitted at very low bit rates; besides, a lot of video packets are lost, so the quality of the video streaming results bad.

In conclusion, these experiments show that multimedia distance learning services need a minimum guaranteed bandwidth in order to provide a suitable Quality of Service. Planning its network's architecture, in fact, CNIT fixed values of minimum guaranteed bandwidth so to obtain a fixed Quality of Service[5].

7.2 Subjective measurements

The subjective Quality of Service perceived by students attending the lectures can be evaluated using method of measure like the Mean Opinion Score. According to the ITU-T Recommendations[14], we have realized a list of multiple choice

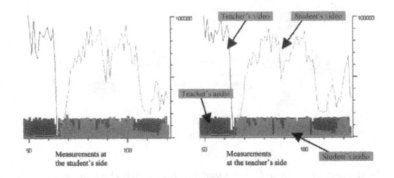

Fig. 15. *Results for 1 Mbps guaranteed bandwidth*

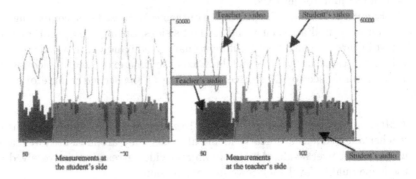

Fig. 16. *Results for 384 Kbps guaranteed bandwidth*

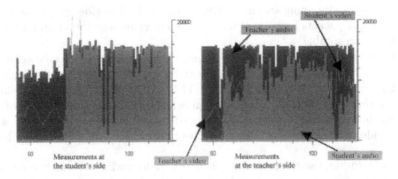

Fig. 17. *Results for 384 Kbps guaranteed bandwidth*

questions, with which the students can express opinions on parameters like quality of audio-video, usability of the system, synchronization, level of interactivity, level of satisfaction on functionalities etc.); we have collected these questions in forms and students have to compile them at the end of every lecture. The results are then assembled in databases and elaborated with the method Mean-

Opinion-Score (MOS). Figure 18 shows the outcomes of these tests, realized with Teledoc2 users attending Teledoc2 lectures on terrestrial and satellite networks. Y-axis reports the values of Mean Opinion Score (MOS), going from the worst value (0) to the best (5); X-axis shows the parameters of the evaluation. The judgments of the users are over the fair level for all the QoS parameters, with similar values in both the network architectures: this shows that the online learning service built in the Teledoc2 Project can be implemented in heterogeneous network architectures, granting always a suitable Quality of Service. Better performances on the terrestrial network are detected as regards the audio quality and the video fluency. Moreover, the judgments about the parameters "Easy to interact via audio" and "Easy to chime in" show that the interaction during the lectures is easier on the terrestrial network; this is probably due to the delays occurring in the real-time transmissions of audio and video on the satellite network.

With these tests we focus our attention on the user's satisfaction of the online service; beyond the technical considerations, in fact, the subjective judgments of the participants are the most important element to consider and to evaluate.

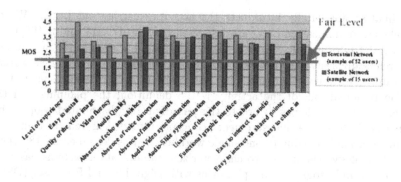

Fig. 18. *Results of the subjective QoS evaluations*

8 Courses Contents and Web-Access

The 31 multimedia courses of the Teledoc2 Project offer a specialized and avant-garde educational route in the field of the Information and Communication Technologies; each course is made up of 5 lectures of two hours which will be held between May 2004 and December 2005, for a total of more than 300 hours of online lectures. The lectures, in particular, aim to increase the formation of PhD students of Italian Universities; the courses belong to 5 main thematic areas: Communication Networks, Optical Communication, Coding and Transmission Systems, DSP & Compression Techniques, ICT Related Technologies. The Teledoc2 educational strategy will plan final exams at the end of each course, aimed at the evaluation of the students' understanding's level. The Teledoc2 Service seems particularly valuable also on a didactical point of view: participants can submit questions, send feedbacks and receive answers and clarifications in real time. The direct contact and interaction among the participants increases the attractiveness of the lectures and the motivation of both the teacher and the students.

The Teledoc2 Web-site[2], finally, is an important communication channel to give visibility to the Project, to keep all the users up to date with the latest news and to facilitate the users' access to all the services. From this site, in fact, students can create their accounts, see the courses' calendar, register themselves to the scheduled lectures and download the didactic material. All this information is synchronized with the management of real-time lectures, thanks to the interface between the Teledoc2 Web-Server and the Vip-Server of the learning software. The global management of users' authentication, accounting and registration is perfectly integrated and user-friendly.

9   Network of Laboratories

This final session briefly introduces the activities carried out by the WiLab (Wireless Communication Laboratories) research unit of CNIT and IEIIT/CNR (Institute of Information, Electronic and Telecommunication - Italian National Research Council) at the University of Bologna, within the context of Networked Laboratories for telemeasurement, in the framework of the Teledoc2 project.

The principal aim of the activity is the definition, the planning and the realization of technologies to create the paradigm of "distributed cooperative telemeasure". The "Telemeasurement" concept (intended as remote control of instrumentation belonging to one single workbench) was described in [18] where the methodology was applied to characterize communication systems based on instruments and programmable platforms with Digital Signal Processor (DSP). In the companion paper [19] the concept of telemeasurement has been enhanced with the introduction of cooperative telemeasurement in which various resources are distributed in a network of different laboratories and can cooperate to set up augmented experiments. This extension of the telemeasurement concept is finalized to increase the measurement capabilities since it concentrates several distributed resources giving the user the possibility to access remote laboratories and to use remote devices without having all needed instrumentation locally. The definition and the realization of this platform was partially performed analyzing different technical challenges relating to signal processing, management of distributed resources, development of aggregated user interfaces, transport of signals for measure, finding innovative and efficient operative solutions concerning remote controls, protocols to gain access and control of specific laboratory instrumentation, building prototypes to test the schemes designed on the field, realization of an integrated didactic and researching environment developing a collection of experiences of measure and the design of a proper communication network. Some of these topics and examples are discussed in the cited paper [19].

10   Conclusions

In this paper a summary of different activities carried out in the framework of the Italian research Project Teledoc2 has been reported. It's worthwhile emphasizing the efficient cooperation among different Research Units of the National Inter-University Consortium for Telecommunications which has been necessary in order to achieve the ambitious experimental objective of this research Project.

References

1. National Inter-University Consortium for Telecommunications, http://www.cnit.it
2. The Teledoc2 Project, http://www.teledoc2.cnit.it
3. O. Conlan, V. Wade, C. Bruen, M. Gargan, "Multi-model, Metadata Driven Approach to Adaptive Hypermedia Services for Personalized eLearning", Springer-Verlag, Aug. 2003.
4. The Teledoc Project, http://www.teledottorato.cnit.it
5. First eLearning Regions & Cities Quality Awards, http://www.eife-l.org/news/pressrelease/elrcaward2004
6. Mbone Conferencing Applications, UCL Network and Multimedia Research Group, http:// www-mice.cs.ucl.ac.uk/multimedia/software/
7. Web Conferencing Software and Audio Conferencing Solutions, iLink communications, http://www.edtlearning.com
8. Coolaborative tools, Altosistemi, www.altosistemi.it
9. Multimedia, Telespazio, http://www.telespazio.it
10. Online Business Collaboration for the Real-Time Enterprise, Centra, http://www.centra.com
11. VIP Conference, LightComm, http://www.lightcomm.it
12. Videoconference, e-learning and streaming, Paiper, http://www.paiper.it
13. Macromedia Breeze, Macromedia, http://www.macromedia.com
14. ITU-T Recommendations: SERIES P: TELEPHONE TRANSMISSION QUALITY, TELEPHONE INSTALLATIONS, LOCAL LINE NETWORKS - Audiovisual quality in multimedia services" - P910, P911, P920 - http://www.itu.int
15. D. Estrin, D. Farinacci, A. Helmy, D. Thaler, S. Deering, M. Handley, Van Jacobson, C. Gung Liu, P. Sharma, L. Wei, "Protocol Independent Multicast-Sparse Mode (PIM-SM): Protocol Specification.",draft-ietf-idmr-PIM-SM-spec-10.ps, March 15 1997
16. B. Cain, S. Deering, I. Kouvelas, B. Fenner, A. Thyagarajan, "RFC-3376: Internet Group Management Protocol version 3", October 2002
17. A. Bazzi, M. Diolaiti, G. Pasolini, "Measured Performance of Real Time Traffic over IEEE 802.11b/g Infrastructured Networks", VTC2005-Spring, Stockholm, Sweden, May 2005
18. A.Roversi, A.Conti, D.Dardari and O.Andrisano, "Telemeasured Performances of a DSP based CDMA Software Defined Radio", iCEER 2004International Conference on Engineering Education and Research, Czech Republic, June 27-30, 2004.
19. A. Roversi, A. Conti, D. Dardari and O. Andrisano, "A WEB-based Architecture Enabling Cooperative Telemeasurements", Tyrrhenian International Workshop on Digital Communications 2005, Sorrento, Naples.

The Dynamic Construction Of Multi-User VRML 3D Environment For Immersive Learning On The Web

Paola Pierleoni, Tommaso Di Biase, Giovanni Cancellieri, Folco Fioretti, Samuele Pasqualini

Dipartimento di Elettronica, Intelligenza Artificiale e Telecomunicazioni
Università Politecnica delle Marche, Via Brecce Bianche
60131 Ancona, Italy
p.pierleoni@univpm.it

Abstract. This paper proposes the design of virtual teaching environments using the combination of VRML, Java and JavaScript languages for the construction of highly interactive and immersive 3D worlds, whose behaviour is real-time modified by users actions. A number of users involved in the animations are present in the classroom (their presence is captured by a camera or other kind of sensors) and for the rest they are remote internet students. In both cases they are represented within a VRML world as avatars into a virtual classroom designed to be available in an immersive way by everybody. They interact with each other through a Java program that is able to process and generate events determining the behaviour of scene elements. In particular, Java application establishes the communication among users and VRML environment and updates parameters stored in a MySQL database. The database, resident into the server, is necessary to share information among the clients and to guarantee real time changes in the virtual representation.

Keywords: Virtual Reality, Immersive Learning, VRML, Java 2, EAI.

1 Introduction

Distributed virtual environments can be useful for learning and training in many fields such as distance learning, scientific simulations and for supporting collaborative work. Nevertheless, to realize computer simulation of real environments the demanded levels of interactivity and multi-user access are difficult to achieve. In this paper we describe our approach to overcome the limits of the VRML technology related to dynamic reproduction of the reality and to the subsequent users sharing of the virtual environment. The main purpose of this work is to develop an Internet-based 3D e-learning environment in which VRML is the tool to create synthetic worlds populated by avatars representing students attending the lesson, teachers and remote users connected to the system [1]. Such VRML environment is designed to provide a unique representation of the virtual community that will allow participants to interact with each other in the learning environments. The dynamic behaviour of a VRML scene is realized using Java 2, JavaScript and the EAI (External Authoring Interface) [2,3]. With these tools is possible to describe rich 3D environments that enable the user to interact. Java 2 is a platform-independent, object-oriented language, which can be used to control the movements in a VRML scene. As a general-purpose language, Java 2 can also be used to construct user interfaces. Real-time presentation of animations, although presently at the cost of rendering quality, is guaranteed by efficient VRML browsers, like Blaxxun Contact and Cortona. Current browsers may execute as built-in tools for Web browsers and as autonomous tools. In this paper we present our strategies used to achieve the integration of VRML and the Java 2 user interface, in order to develop highly interactive computer modelled animations. Each user interacts with the others through a server that is capable to store all shared information and same useful 3D object. The server side Java DataBase Connection (JDBC) is used to get or give information to MySQL database about the scene update [4,5]. We describe the VRML features and its integration with Java 2, JavaScript and the present techniques and facilities we have developed to enable the user to have a tight control over the animation. One example of interactive animations using VRML, Java 2 and JavaScript is shown. In addition we explain how a virtual scene is generated.

2 VRML overview

VRMLs technical roots reach back to the Inventor project, which started in 1989 at Silicon Graphics. The Inventor project aimed to build an object-oriented toolkit and file format for a new generation of interactive, networked 3D multimedia applications. Later that year, the Inventor team released Annotator, the first commercial product to use Web-based 3D hyperlinking technology. In the same year a request for proposals released for VRML 1.0 was presented. Eventually, the Inventor based proposal written by Silicon Graphics engineers was chosen as the working document for a VRML 1.0 specification. VRML 1.0 was released in October 1994 at the Second International Conference on the World Wide Web. This first version of VRML described 3D geometry, attributes, hierarchy, and Web hyperlinking. However, it soon became clear that VRML lacked some important features: user interactivity, animation, scripting language, networking, and multimedia integration. In January 1996, a request for proposals for VRML 2.0 was issued, and in February, six proposals were received. The Silicon Graphics proposal, Moving Worlds, received a strong majority of the votes and became the working document for the VRML 2.0 specification, published in August 1996 at Siggraph in New Orleans. At the July 1996 meeting in Kyoto, Japan, the International Organization for Standardizations (ISO) JTC1/SC24 committee circulated the August 1996 version of VRML 2.0 as a committee draft. In December 1997, ISO/IEC 14772 (VRML97) was officially published [6].

VRML files describe 3D objects and worlds using a hierarchical scene graph. Entities in the scene graph are called nodes. VRML defines 54 different node types, including geometry primitives, appearance properties, multimedia objects, animation interpolators, interaction sensors, and various types of grouping nodes. Nodes store their data in fields. VRML defines 20 different types of fields that can store everything from a single number. Nodes can contain other nodes (some types of nodes may have children) and may be contained by more than one node (they may have more than one parent). However, a node must not contain itself. This scene graph structure makes it easy to create large worlds or complicated objects from subparts. Fig. 1 illustrates a very simple scene graph where the root node contains two children.

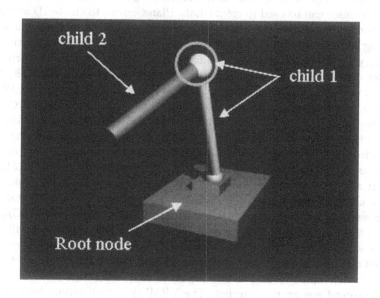

Fig. 1. Hierarchical nodes organization in a VRML scene

VRML features eight grouping nodes. These nodes define a local coordinate system for their children nodes. They create hierarchical scenes and support specialized grouping behaviours. Anchor defines its descendants as a hyperlink to a uniform resource locator (URL). Collision specifies collision-detection properties for its decedents. Group, a generic grouping node, simply clusters other nodes into a coordinate system. Inline specifies its children through a URL, enabling distributed, hierarchical file structures. Nine sensor nodes provide the built in user interaction primitives for VRML. Sensors have two categories: environmental and pointing devices. The four environmental sensors Collision, Proximity-Sensor, TimeSensor, and VisibilitySensor detect changes in the world. Collision sets the collision detection state and detects collision events between the users avatar and geometry in the scene. ProximitySensor detects when the users avatar navigates in and out of a user defined region and tracks a users position while in the region. VisibilitySensor detects the rendering visibility status of specific geometry in a scene. TimeSensor generates events time passes and serves as the basis for all animated behaviours. There are five pointing device sensors: Anchor, CylinderSensor, PlaneSensor, SphereSensor, and TouchSensor. These device sensors detect and react to user manipulation of a pointing device activated over any geometry that descended from the pointing device sensors parent node. Typically, the sensor implements user interface widgets on sibling objects. Clicking on geometry influenced by Anchor causes the browser to load a new URL or view in the existing URL. CylinderSensor, PlaneSensor, and SphereSensor define invisible geometry, representing a movable widgets shape and constrained behaviour. CylinderSensor can be used to create dials, PlaneSensors to create 1D or 2D sliders or constraints, and SphereSensors to create trackballs. TouchSensor, a generic pointing device sensor that detects pointing device motion and activation over geometry, can be used for many applications. VRML 97 includes two primitives that let a single VRML world definition span the World Wide Web. The Inline node allows the inclusion of another VRML file stored anywhere on the Web. In addition, it serves as the basic extensibility mechanism for VRML. VRML fits into the existing infrastructure of the Internet and Web. It uses existing standards wherever possible, even if those standards have some shortcomings when used with VRML. Using existing standards instead of inventing new, makes it much easier for Web developers to use their existing tools to create VRML content. It also makes it easier for somebody implementing the VRML standard, since libraries of code for popular standards already exist. VRML files may contain references to files in many other standard formats. JPEG, PNG, GIF, and MPEG files may be used as texture maps on objects. WAV and MIDI files may specify sound in the world. Files containing Java or JavaScript code may be referenced and used to implement programmed behaviour for VRML objects. Each of these independent standards was chosen to integrate with VRML, because of its widespread use on the Internet. The VRML 97 specification describes how they are used with VRML it does not attempt to define these other standards or describe how to create files in these other file formats. The definition of how VRML should be used with other standards is generally done by the organizations that define those standards. Using VRML with HTML pages, Java applets and JavaScript is very effective. Combining 2D and 3D information is often better than either 2D or 3D alone. Since 3D image visualization is normally done on personal computer monitor, three-dimensionally sensation is perceived only through a prospective ambient representation such as happens observing a photo or a film: depth concept of the environment is dummy but sufficient indeed. Using extra hardware peripherals, a lot of browsers are able to give to the final user the sensation to be in a real world creating a so called immersive virtual reality. This kind of vision is carried out creating a different picture for every eye. When brain receives a fast sequence of such images (grater than 60 Hz), it merges them in a single scene so perceiving depth. There are a lot of available techniques to create this effect and these are different for complexity and costs. In immersive systems the user is completely involved in an artificial reality by using different devices like Head Mounted Display or orientable binocular monitor (Fig. 2).

Diffusion and development limitation of these applications are strictly connected to safety problems, costs and high computational requested power. For this reasons near immersive reality is growed up a semi-immersive reality with the intent to give to the final user an idea of virtual reality, but leaving the spectator conscious of his physical reality. In his elementary release our system uses semi-immersive technologies, but is possible to extend it to completely immersive system taking advantage of VRML and 3D players facilities further than previous devices.

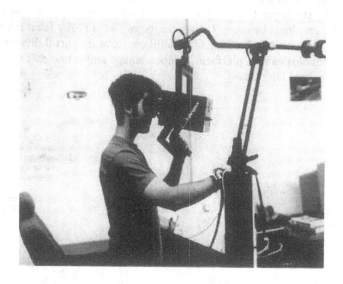

Fig. 2. Fakespace Boom System

3 The architecture of the virtual environment

We propose a multiuser collaborative and interactive client-server framework for e-learning environment. We have integrated distributed computational technologies (Java 2) with an existing language for computer graphics (VRML). On the client side, a Java applet is used to implement the graphics user interface and Java DataBase Connection (JDBC), from the Java Development Kit (JDK), is used to provide the system with a standard interface to the scene database. On the server side, a database stores data and a Java application gives management functionalities. Services can query and update the database using SQL commands. VRML, a web standard for representing 3-D shapes and interactive environment, is used by clients only. For the server, each 3-D object in database (for example avatars) is represented by a VRML format file. A service queries the database and results in a VRML scene. After the scene is transmitted to the client under a request, a VRML browser is used to view the 3-D scene inter-actively. Moreover, we address the problem of collaborative interaction. Design results of one user can be published to other members and they can see the updates in real-time (Fig. 3).

As shown, the system software architecture uses a combination of different environments based on different languages. In particular, the final result is a Java 2 applet and a VRML player in a simple HTML (HyperText Markup Language) page. The player is necessary to show and to manage 3D graphical environment on the web page or in a frame. Moreover, in the HTML page structure JavaScript parts of code are present and represent the middleware layer between the ap-plet and the graphical browser. Using languages like HTML, JavaScript, Java 2 and VRML, we have two fundamental facilities: software portability (due to the standard technologies and platform independency) and a low cost system (due to the open source languages).

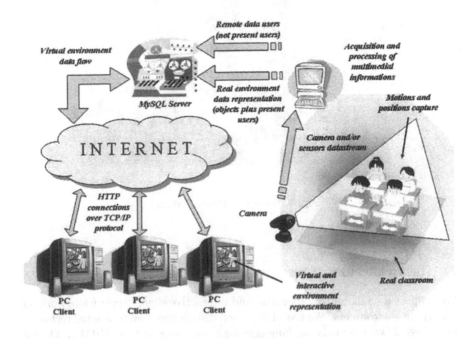

Fig. 3. Proposed system architecture

Now we want to explain how the system works. The application is based on HTML page structured in three separate software layers, communicating by specific tags. The layers are:

- a VRML player (plug-in), that is automatically loaded by used internet browser when <embed> tag is encountered in the code of HTML page. When the file to load is specified, the plug-in selection is automatically done; practically a *file.wrl* induces the internet browser to load another browser: the VRML player precisely;

- a Java applet, that is defined by a class loaded by the system resident Java virtual machine. Obviously the class is pre-compiled as Japplet and then inserts in the page through the proper <applet> tag;
- a JavaScript code, that is not compiled, but directly inserted into the HTML page where is located by <script language="JavaScript"> tag.

Indicatively the code is organized as follow:

```
        <html>
        <head>
        <title>Applet and VRML</title>
⇒      <script language="JavaScript">
            FunctionName(parameter,...) {
                Istruction;
                ...
            }
            [Other functions]
        </script>
        </head>
        <body>
⇒      <EMBED NAME="ControllName" SRC="WorldFileName.wrl"
        WIDTH=480 HEIGHT=360 align="right">
⇒      <applet code="AppletFileName.class" width = 400
        height = 61 name = "appletName" align="left" mayscript>
        Java Virtual Machine absence error string</applet>
        </body>
        </html>
```

As shown in the previous code, both applet and player are univocal addressed by **name** parameter. This is important because the parameter indeed allows interaction among the whole different layers.

To manipulate a VRML world, giving interactions with the scene graph, we use EAI (Extended Authoring Interface) that creates a bridge between VRML browser (HTML browser plus VRML player) and Java (or JavaScript). EAI has been developed to give to Java applet the possibility to access in runtime and in a direct fashion on the software representing the scene, so every VRML player was provided by VRML libraries capable to simplify the connection management and the information exchange between the browser and Java applications. These libraries were designed for Microsoft Java Virtual Machine (MS-JVM) based on Java 1.1 standard, so the new virtual machine in the system made this packages unusable. Java 2 introduction, by Sun Microsystem, has carried out many differences with previous Java 1 both in language and in development environment.

At the moment, it is not possible to use Java EAI libraries because not supported by the latest Sun-JVM releases, so access to the graphical 3D scene is more complex. An alternative solution is to implement an indirect access taking in use JavaScript as middleware between Java 2 applet and 3D world (Fig. 4). The main difference with the previous solution is due to the presence of some scripts in HTML and VRML sources. Those scripts are able to actively work on the VRML browser querying information from 3D graphical scene or modifying parameters in it. The presence of Java 2 and JavaScript at the same time provides the possibility to use APIs (Application Programming Interface) functions that could not be available with the single JavaScript employment.

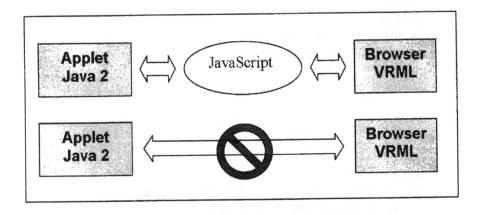

Fig. 4. Use of JavaScript as middleware

3.1 VRML environment and the world file

In order to modify a node of 3D virtual environment in an interactive way, is necessary to locate it in the nodes set univocally. To do this we must mark every node of interest with a specific name that will be used as a reference for all future data processing. In practice the marker is a tag DEF NodeName before node declaration itself.

```
⇒ DEF Sphere Transform {
    translation 0 1 0
    children [
      Shape {
        appearance Appearance {
        material Material { diffuseColor 1 0 0 }
        }
      geometry Sphere { radius 1 }
```

```
      }
   ]
}
```

The presence of this kind of tag does not change the normal ambient representation; in fact the graphic engine does not consider this tag if the file is loaded apart. Regarding to the connection of the player with the applet, is necessary to insert the following declaration in the code:

```
<EMBED NAME="World" SRC="FileName.wrl" WIDTH=480
HEIGHT=360 align="right">
```

We must underline that with the NAME tag is identified the VRML browser and through this name will be possible to interact with scripts and with different graphic sessions at the same time.

3.2 Java 2 Applet

In the multimedia context Java 2 code is inserted as a normal applet, the only difference is the call into the HTML document. Also in this case there is a similar declaration to the previous one, that is an univocal name identification realizing a communication channel between Java 2 and JavaScript.

```
<applet code="FileName.class" width = 100
height = 100 name = "appletName" mayscript>
Java Virtual Machine absence error string
</applet>
```

The code design implements the graphical interfaces normally used to allow user interaction (AWT or Swing). To guarantee the scripts access functionality into original HTLM page every applet is equipped by specific classes package called *netscape.javascript*. In particular JSObject definition provides the methods to access to the aforesaid functionalities. The following example gives an idea of the declarative section in the program code:

```
import netscape.javascript.*;

[...]

win = JSObject.getWindow(this);
JSObject doc = (JSObject)win.getMember("document");
JSObject loc = (JSObject)doc.getMember("location");

[...]

Object args [] = {"DefProx", "position_changed"};
win.call("getEvent", args);
```

The declarations chain allows connecting JavaScript object under examination (`win`) to the right window and to the exact part of the HTML code. In the last section of the Java code we can see a call to the script functionality into HTML document. The call is done by `getEvent` (script function) via `call` instance with the proper parameters stored in the string array args. In this way and in this case a query to browser information is done, but it is also possible to access to do changes.

3.3 The JavaScripts

All the scripts are in JavaScript language (Netscape group standard) and the usage of them is quite simple:

```
<script language="JavaScript">
    function getEvent(node, field) {
    currentPos = document.World.getNodeEventOut(node, field);
    }
</script>
```

As shown, the Jscript realizes `getEvent` function used by applet; this function makes a query directly to the player through `getNodeEventOut` function (EAI interface) passing the same applet parameters (node and field). It is important to notice that a similar method allows to interfere with 3D scene directly from HTLM page and externally to the applet by using proper interfaces (as buttons, text fields,) capable to work on the script.

3.4 MySQL database management

Using Java 2 application (applet) we can extend the code functionalities to allow MySQL database logging. JDBC library package gives the necessary tools to guarantee database Java 2 connectivity.

JDBC is used as a medium and not as a real tool indeed; in order to manage database accesses, queries and updates we prefer to employ a wrapper class that reduces instruction number [5]. *WrapperDataBean* class provides a mechanism that easily extends fundamental operation for database interaction, that is:

- record selection;
- record insertion;
- record update;
- record deletion.

WrapperDataBean allows executing instructions without SQL code giving an intuitive approach, more modularity and scalability. Using *WrapperDataBean* abstract class we must respect the following constrains for a correct databean realization:

- class must extend *WrapperDataBean*;

- class name and table name must coincide;
- the name of attributes must with the name of table fields;
- Java types of attributes must coincide with the types of table fields.

At the end we want to underline that through the same Java 2 applet user authentication is done to guarantee a secure database and system access.

4   Basic user-environment interaction example

In this section we describe how our system works. The application is mounted on the server side, but with few changes is possible to adapt it to work on client side (Fig. 5). Practically the client has not the whole database control panel but only the connection section and the window used to access to the virtual 3D scene. This program is capable to receive, process and generate events that control the animation behaviour. For example, a new user position may be extracted from the scene by the JavaScript through EAI interfaces, sent to the database so that all other users can update their scene (client side) querying to the same database. The combination of VRML and Java 2 is a very powerful tool for the control of animations behaviour. Interactive animations that have their behaviour defined by user actions can be modelled with VRML using a JavaScript node and its associated program, which is capable to receive parameters from another Java 2 application, responsible for the user interface. We have developed an interaction tool, that is a control panel for the animation, in which the user defines the values of parameters that control the animation in real-time. Using this application, it is possible to have a graphical interface, responsible for sending animation control parameters to the VRML file. In our first scenario, the Java-JavaScript program (consequently VRML scene) receives data from the user interface or automatically from database. The communication between all users interfaces programs are based on the client-server model where users interface is the client. The HTML file starts the Java 2 program that connects to the interface, requesting data from the server. The server then sends the requested data, the Script program processes them and returns the results to the VRML file. The communication between the VRML file and the JScript program follows the standards proposed in the VRML specification. The window that appears on the top shows the VRML 3D virtual environment state. The communication methods, based on JDBC library, simply define that the specified value will be sent to the client when it requests an update of its parameters (this update can happen at each clock event, at each animation cycle, and so on, depending on the implementation). The graphical components appear in the user interface and are sent to the client in the same order they have been created by the real scene or by the remote users. The example is platform-independent and is visualized with a simple browser equipped with a VRML player. A set of Java class libraries provide the functionality needed to write out, to render, to update and to manipulate a VRML scene.

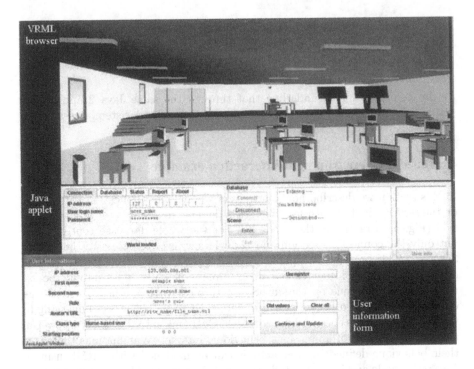

Fig. 5. User interface

5   Virtual scene generation

Simple VRML 3D scenes can be realized using directly a text editor (like notepad) and writing down code following VRML standard specifications, but more complex situations needs tools to help the virtual world design. A VRML converter is a software that allows the user to build interactive 3D objects or environments for the internet in a visual and graphical manner through the use of a user friendly interface. As the user creates objects or environments, the software writes the necessary code behind the scenes. This reduces significantly the amount of time required to build a VRML file, eliminates the need to learn VRML code and simplifies all the work process from the idea to the .wrl file creation. After the conversion any VRML file may be viewed and edited in a common text editor to improve the code or insert more complex features. Very complex attributes will require the use of the scripting language to give to a specific object an intelligent and organic behaviour; this can be done, as shown before, with few code modifications, so everything is really possible within the limitations of telecommunication technology, computer processing power and human imagination. 3D Studio Max, VIZ and Maya, among others, can be used to generate and export non-elementary environments in VRML files format [7,8,9]. The exporter in 3D Studio Max is quite a mature tool; with its dialogue box is possible to set exportation parameters as coordinate interpolations, objects position, orientation and so on. We have generated both scenes and avatars with 3D Studio Max, giving interaction, for example avatars movements, by modifying proper code nodes with Jscript instructions.

6 Conclusions

VRML enables the development of interactive animations by the use of sensor nodes and the routing of events through the nodes. The conjunction of VRML and Java 2 through the JavaScript enables the definition of more complex animations and interactions. In this paper, we describe how achieve a better interaction with the user, by the creation of graphical interfaces (Java 2 application) that manages animation using Script program. Interactive VRML animations can be used in a variety of applications such as distance learning, scientific simulation and cooperative environment. VRML scenes are connected to the external world (MySQL database) through JDBC library package giving interaction among users, updating avatars positions in the scene and so on, opening several other possibilities for the improvement of this work. Currently we are trying to achieve more interactions by using pattern detection and recognition on picture supplied by the camera positioned in the classroom. The avatars are created by means VRML language and can be stored on server side together with users coordinates and all the other data or charged from different web sites simply addressing them with the right URL.

References

1. ISO/IEC 14772-2: The Virtual Reality Modelling Language. In: Part 2: External Authoring Interface. (1997)
2. Java Specifications: http://www.sun.com/java. (2005)
3. JavaScript Guide: http://www.netscape.com. (2005)
4. Wyatt, N.: Pure java databases for deployed applications. Informix Corporation (2005)
5. Swain, M., Anderson, J.A., Korrapati, R., Swain, N.K.: Database programming using java. Proc. IEEE Southeast Con. 2002, April 5-7 (2002) 121–126
6. ISO/IEC 14772-1: The Virtual Reality Modelling Language. In: Part 1: Functional specification and UTF-8 encoding. (1997)
7. Autodesk: Autodesk 3ds max, http://www4.discreet.com/3dsmax. (2005)
8. Autodesk: Viz, http://usa.autodesk.com/adsk. (2005)
9. Alias: Maya, http://www.alias.com/glb/eng/products-services. (2005)

NeBULa (Network-Based User-Location Sensing System): A Novel And Open Location Sensing Framework Operating On A WLAN Environment

G. Ferraiuolo*, G. Massei+, L. Paura*+, A. Scarpiello+

*Dipartimento di Ingegneria Elettronica e delle Telecomunicazioni
Università di Napoli Federico II
Via Claudio 21
I-80125 Naples, Italy
{gferraiu, paura}@unina.it

+Consorzio Nazionale Interuniversitario per le Telecomunicazioni (CNIT)
Via Diocleziano 328
I-80125 Naples, Italy
{gianluca.massei, amedeo.scarpiello}@cnit.it

Abstract. Recently, the need to provide location-based services has lead to an increasing interest in accurate location sensing techniques. For indoor environments, most of proposed solutions do not generally provide accurate location estimates due to propagation problems. In this paper, we present NeBULa, a novel and open framework to locate a mobile user within an indoor Wireless LAN environment. It uses signal-strength information extracted from the wireless LAN interface and doesn't require additional and costly hardware. The performances of the proposed system have been evaluated in different operative conditions to show its effectiveness.

Keywords. Location sensing, WLAN, Scene Analysis

1   Introduction

Since location is an important need for commercial, public safety, and military applications, there is increasing interest in accurate location estimate techniques. For example, as regards commercial applications, there is an increasing need to monitoring people (as elderly and children) whereas, the location information is mandatory also, of course, for delivering location-aware services, as for example navigation services for museum visiting. The wide proliferation of wireless local area networks (W-LAN) and mobile computing devices also motivates the growing interest in location-aware systems and services. For outdoor applications the Global Positioning System and wireless Enhanced 911 services represent often satisfactory solutions. For indoor applications, such solutions do not generally provide accurate location estimates due to propagation problems. Therefore, much attention has been devoted to single out new techniques, which can cope with the hostile indoor propagation environment. Among the indoor location and tracking techniques, the Scene Analysis ones are the most popular (e.g. RADAR [1] and EKAHAU [2]). These techniques are based on recording and processing signal strength information provided by multiple base-stations (access points in the W-LAN context) positioned to provide overlapping coverage in the area of interest. So, in this paper, we propose NeBULa (Network-Based User Location sensing system): a new Scene Analysis framework, based on RADAR technique, in order to have an open platform to develop in future possible upgrading or novel location techniques, since RADAR and EKAHAU source codes are not open. To validate the proposed system a test-bed has been realized and the NeBULa system has been tested in different operative conditions. The experimental results confirm the effectiveness of our Location Sensing System. The scheme of the paper is the following: Section II briefly describes a taxonomy of location-sensing techniques, lists the most popular solutions and compares them. Section III presents the proposed location system and its alternative versions, while Section IV describes the experimental results obtained by the ad hoc realized test-bed. Finally, Section V provides conclusions and open problems.

2   Location-sensing technologies

There are several techniques that may be used to estimate the location of objects, people, or mobile terminals. They can require different infrastructures, as well as can exhibit different costs, limitations, performances and characteristics (type of information provided, accuracy, dynamic). So, the particular properties of each technique make it a suitable choice for a specific case.

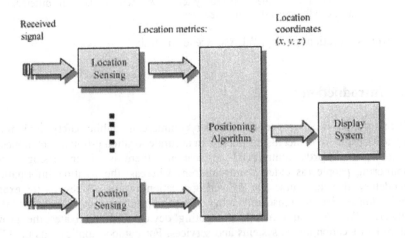

Fig. 1. Typical Location Sensing System Flowchart

Fig. 1 shows the architecture of a typical Location-Sensing System. Its main elements are:

- *Location Sensing*: elements able to measure some greatness depending on the relative positions of the mobile terminals in comparison to known points of reference;
- *Positioning Algorithm*: an algorithm that elaborates the greatness acquired by the location-sensing elements to estimate the position coordinates of the mobile terminal;
- *Display System*: a system that visualizes the estimated position. These visualization systems can report to a PC display or to a mobile unity, locally accessible in a LAN or universally accessible on the Web. The coordinates of the mobile terminal can be simply visualized on the display, or alternatively the terminal position can be shown on the map of the area concerned.

In general, the Location-Sensing Systems can be classified as follows [3]: Triangulation, Proximity, and Scene Analysis. Location system implementations may employ them individually or in combination. The *Triangulation* location sensing technique uses the geometric properties of triangles to compute object locations. Triangulation location sensing systems include GPS, the Active Bat Location System [4], the Cricket Location Support System [5], Bluesoft [6], and SpotON [7]. A *Proximity* location sensing technique entails determining when an object is proximal to a known reference point within a specific and limited range. Proximity location sensing systems include Active Badge [8], CoolTown [9], SmartFloor [10], and others. The *Scene Analysis* location sensing technique, also called *Finger-Print*, uses features of a scene observed from a particular vantage point, to draw conclusions about the location of people or objects in the scene. In particular, in a preliminary off-line phase, observed features are collected in a predefined dataset, and then an on-line monitoring of the real-time features is carried out and the observed features are compared with those stored during the off-line procedure. So, through a set of features describing the environment and an algorithm which evaluates a metric based on off-line and on-line data (*Pattern Matching Algorithm* - PMA) it is possible to estimate the object position.

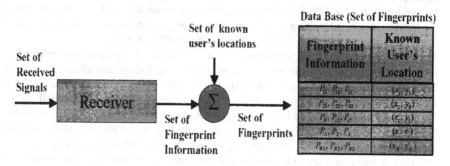

Fig. 2. Scene Analysis Location Sensing System: Off-line phase

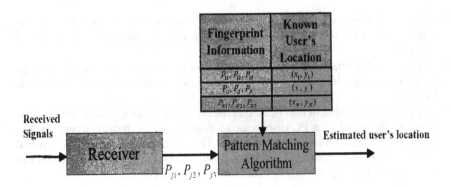

Fig. 3. Scene Analysis Location Sensing System: On-line phase

The advantage of Scene Analysis in comparison with triangulation is that the location of objects can be inferred using passive observation, it does not require terminal synchronization and attenuation models, and it suffers poorly from power losses and signal reflections. The main drawback of Scene Analysis technique is that if changes of the environment occur in a way that they alters the perceived features of the scenes, it may be necessary to repeat the off-line procedure to update the predefined dataset.

The scene itself can consist of visual images, such as frames captured by a wearable camera [11], or any other measurable physical phenomena, such as the electromagnetic signals. RADAR [1] and EKAHAU [2] location systems are examples of the latter. RADAR uses a dataset of signal strength measurements created by observing the radio transmissions of an 802.11 wireless NIC (Network Interface Card) at many positions and orientations throughout an indoor environment. The location of other 802.11 network devices can then be computed by performing table look up on the pre-built dataset. Instead, EKAHAU combines a probabilistic approach with the Scene Analysis technique: using signal strength data from various known locations, EKAHAU infers a model that can be used to make predictions about the location associated with a set of new signal strength data. It uses different physical layer technologies including WLAN, Bluetooth and GSM/GPRS.

The general guidelines which are usually adopted for the design and implementation of an efficient Location Sensing System within an indoor environment are the following: the system must not require significant infrastructure deployment and high maintenance costs only to locate users, its effectiveness must not be limited by the multiple reflections suffered by the RF signal within indoor environment, it must not require fine-grain time synchronization among mobile terminals, it must be suitable for large-scale deployment, the accuracy of system must not be limited by the AP's cell size, it must be a scalable and high-performance system even in the presence of direct sunlight (issues present in IR-based systems).

In our opinion, among the above-mentioned Location-Sensing techniques, a Scene Analysis Location system using RF signals avoids many of these issues. In fact:

- it can be implemented purely in software and is easily deployable over a standard wireless LAN without any additional hardware;

- in comparison with IR-based systems it doesn't scale poorly due to the limited IR range, while in comparison with Triangulation techniques it doesn't require terminal synchronization;
- propagation problems, such as reflection, refraction, and multipath, poorly affect its accuracy in comparison with triangulation techniques.

Since RADAR and EKAHAU source code are not open, we have implemented NeBULa (Network-Based User Location sensing system): a new Scene Analysis framework, based on RADAR technique, in order to have an open platform to develop in future possible upgrading or novel location strategies.

It is built on a WLAN 802.11b infrastructure. Access Points are located in such a way to provide overlapping coverage of the area of interest. A mobile user is equipped with a computing device (laptop or PDA) which has a wireless LAN interface capable of bidirectional communication with the access points.

3   The NeBULa System

The NeBULa system uses as features the signal strength (SS) of the beacons transmitted from APs and received at the mobile terminals. The key idea is that in an RF network the signal strength of a packet is a function of the receiver's location, so that such a property is exploited by the system to estimate the mobile's location. Then, these features are stored, together with the physical coordinates of each location, during off-line procedure (creating the *Radio Map* of the indoor environment), and processed during on-line one to determine the signal strength tuple that best matches the signal strengths measured in real time by the mobile terminal, according to a specific method.

As shown in Fig. 4, the NeBULA System is composed by three software components:

- the *NeBULa Client*, which measures the signal strength of the beacons transmitted from APs and display the user location;
- the *NeBULa Database*, which stores the Radio MAP; this is composed by the vectors $(x, y, SS1, SS2, SS3,.. , SSM)$, where M is the number of APs, x and y represent the coordinates of terminal position according to a fixed reference system, and SSj is the signal strength of the beacon transmitted from the j-th AP;
- the *NeBULa Manager*, which downloads the file from the NeBULa Client, extracts the signal strengths, creates (during the off-line phase) and manages (during the on-line phase) the Radio MAP interacting with NeBULa Database, determines the terminal position using the chosen PMA and, finally, transmits the estimated position to NeBULa Client.

We have chosen to use a MySQL database and the JAVA language for the software development, with the support of some already available API (Application Program Interface), in order to guarantee the portability of NeBULa system on Linux OS Platform.

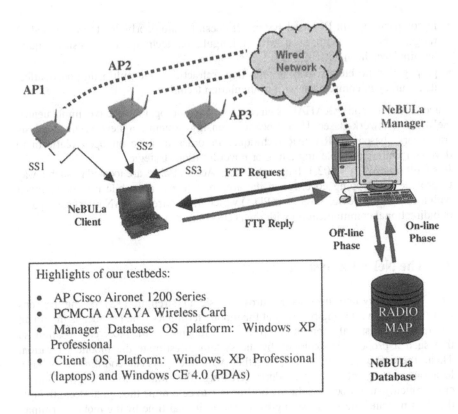

Fig. 4. Components of NeBULa System

Before proceeding with signal-strength measurements, some preliminary operations are necessary. Primarily, it is fundamental to choose in the indoor environment the points that will constitute the Radio Map, dividing the environment in cells whose centres are the Radio Map points (the cell dimension is defined through a compromise between sensibility and accuracy: the more the cells are wide, the more sensibility goes down; the more the cells are narrow, the more points are indistinguishable). Then, it is important to define the number of AP's and their positions and orientations, in accordance with the indoor environment in which the mobile terminals are located.

Finally, it is possible to proceed to off-line and on-line procedures.

During the off-line phase, the system performs the following actions in each of the Radio Map points: it measures the signal strength of AP beacons through the NeBULa Client, transmits them to NeBULa Manager, processes them and, finally, fills up the NeBULa Database.

To obtain the signal-strength levels, we have used, as NeBULa Client component, the *AVAYA wireless card software* [12] for the laptops, and *Ministumbler v0.4.0* [13] for the PDAs. Both these utilities (available for several Operating Systems) allow to know the connection quality among the mobile terminal and every Access Point (for each AP it is possible to visualize: name, MAC address, signal strength of beacons, transmission channel and signal/noise ratio) and put it into a file (*.txt* for laptops and *.ns1* for PDAs).

The NeBULa Manager is responsible for measure processing. For the file transfer we have used the FTP protocol, enabling on the mobile terminal the FTP server provided by the operating system and implementing a FTP client in the NeBULA Manager (exploiting the services offered by the *iNet Factory library* [14] of JSCAPE Company). Then, the NeBULA Manager extracts the signal strength levels from the *.txt* file (for the PDAs the *.ns1* file is preventively translated in text format) and calculate for every AP the average of the signal strength levels during the observation time.

Finally, it is necessary to create the Radio Map on the NeBULA Database. To remotely access the database, a database client is necessary to communicate to database server which information to recover, cancel or upgrade. The interaction with MySQL database server is possible through the *JBDC APIs* and *mysql-connector-java-3.0.8-stable* driver (to translate the JDBC APIs in the database server language). The mainly used JDBC APIs are the Connection and the Statement classes: the former opens the connection with the database server, while the latter sends to the database server the SQL code to execute.

In the On-line procedure, the signal-strength levels of beacons are acquired in real time and then compared with the Radio MAP data through the PMA. So, it is important to choose the algorithm having the best performance in matching the right position with the estimated one. According to RADAR system, we have used three approaches: Nearest Neighbour in Signal Space (NNSS), Multiple Nearest Neighbours in Signal Space (NNSS-AVG), and History Based Algorithm (HBA).

NNSS is the basic approach. It scans the Radio Map and calculates the Euclidean distance d among the real-time $(SS_1, .., SS_M)$ data and every Radio Map $(SS'_1,.. , SS'_M)$ tuple:

$$d = \sqrt{\sum_{j=1}^{M}\left(SS_j - SS'_j\right)^2} \qquad (1)$$

where *M* is the AP number. The NNSS purpose is to find the Radio Map tuple minimizing the distance in the signal space.

Instead, *NNSS-AVG* considers *k* nearest neighbours in signal space and calculates the average of their space coordinates. An appropriate choice of value of *k* can assure a more accurate location estimate then the NNSS PMA.

Finally, the *HBA* PMA exploits the old location information together with the physical contiguity constraint in adjacent time instants to improve the accuracy of location system resorting to memory-based processing. With HBA PMA we can overcome the ambiguity of physically distant points which are close in signal space due to undesirable propagation problems.

For every time slot Δt, the HBA PMA collects signal-strength levels from a mobile terminal, starts a NNSS-AVG search to find the *k* nearest neighbours in signal space (*k-NNSS* set), and stores the last *h* k-NNSS sets. After *h* time slot, HBA PMA singles out the most likely path of the user on the sequential trellis of depth *h* (which represents the memory length), shown in Fig. 5, through the Viterbi algorithm, and the first point of identified path is the user's estimated location in the considered time slot.

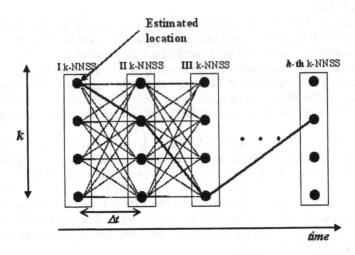

Fig. 5. HBA trellis to estimate the user's location

As branch metric we have adopted Euclidean distance d among the locations corresponding to two branch endpoints. Since the probability that users assume a speed v decreases as the speed increases, the branch metric $d=v*\Delta t$ is inversely proportional to the branch probability. So the shortest path between the vertices in the oldest and the newest set represents the most likely trajectory of the mobile user. According to the Viterbi algorithm, this procedure, however, implies that there is a time lag of h signal strength samples, and so h has to be as small as possible provided that, anyway, the memory length be effective. In Fig. 6 is shown the flow chart of NeBULa Manager.

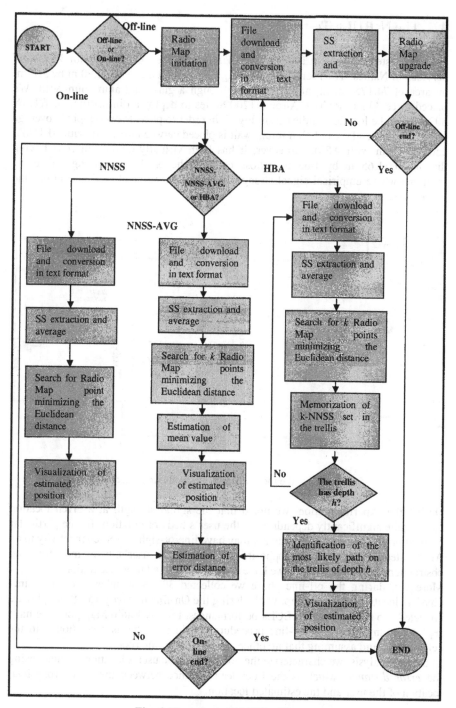

Fig. 6. Flow chart of NeBULa Manager

4   The NeBULa Test-beds

We have tested our system in a room of *Wireless Systems Laboratory for Wideband Services* in Naples. The room, shown in Fig. 7, has dimension of 12.90 m by 5.75m, an area of 74.175 sq. m, and is divided through a glass and aluminium wall. We placed three APs (AP Cisco Aironet 1200 Series to deploy an infrastructure WLAN 802.11b) at the locations indicated in Fig. 7, in order to provide overlapping coverage in the room (the glass and aluminium wall is placed between the room with AP1/AP2 and the room with AP3). Moreover, it has been virtually divided in cells having dimensions 1.65 m by 1.65 m, whose centres (the black dots in Fig. 7) denote locations where empirical signal-strength information was collected during off-line procedures.

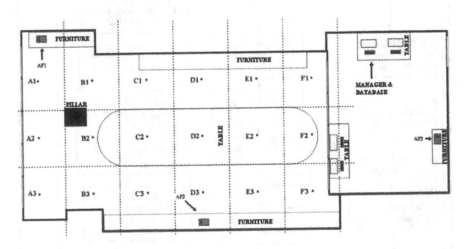

Fig. 7. Map of the room where the testbeds were conducted

During the experimentation, we noted that the signal strength at a given location varies quite significantly depending on the user's body orientation. In one particular orientation the mobile host's antenna may have line-of-sight (LoS) connectivity to an AP's antenna, while in the opposite orientation the user's body may form an obstruction. So, in the testbeds we have considered a fixed user orientation.

Moreover, during the off-line phase we collected 30 SS samples at each distinct physical location of Radio Map, while during the On-line one only 15 SS samples are collected. To measure the NeBULa performances, for any Radio Map point we have repeated many times the On-line procedure considering the user positioned in the cell's centre and assuming that no other people were in the room.

In all our analysis, we characterize the goodness of the user's location estimate using the *error distance*, which is the Euclidean distance between the actual (physical) location of the user and the estimated location.

In the first analysis, we want to assess the NeBULa performances using NNSS algorithm and different AP transmission powers. Fig. 8 shows the relevant Cumulative Distribution Function (CDF) of the error distance. Table 1 summarizes the results of Fig.8 in terms of the 25th, 50th (median), 75th percentile values of the error distance in correspondence of each transmission power.

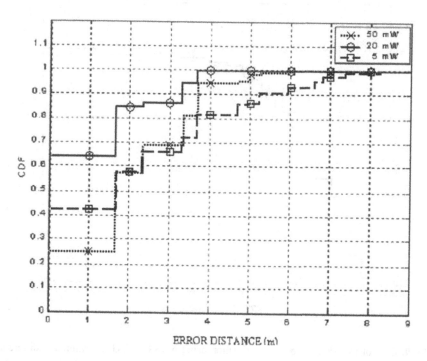

Fig. 8. CDF of the error in location estimation using NNSS algorithm and different AP tx powers

Table 1. The 25th, 50th, and 75th percentiles values of the error distance using NNSS algorithm and different AP tx powers

AP transmission powers	25th (m)	50th (m)	75th (m)
50 mW	1.65	3.70	5.95
20 mW	0	1.65	3.30
5 mW	1.65	3.30	3.71

The results show that an AP transmission power equal to *20 mW* is a good compromise to minimize the effects of transmissions problems. So in the following analysis we have assumed *20 mW* as AP transmission power.

In the second test, we want to assess the NeBULa performances using NNSS-AVG algorithm and two different values of k. Fig. 9 shows the relevant error distance CDF, and Table 2 summarizes the results in terms of the 25th, 50th (median), 75th percentile values.

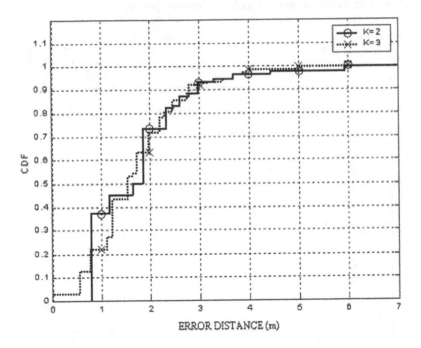

Fig. 9. CDF of the error in location estimation using NNSS - AVG algorithm and different values of k

Table 2. The 25th, 50th, and 75th percentiles values of the error distance using NNSS - AVG algorithm and different values of k

Number of neighbours (k)	25th (m)	50th (m)	75th (m)
2	1.65	2.33	2.97
3	1.55	2.20	2.80

The results show that $k=3$ is slightly preferable. In comparison with NNSS, we note better performance only with high percentile values of the error distance. In the following analysis we have assumed a number of neighbours equal to *3*.

In the third test, finally, we want to assess the NeBULa performances using HBA algorithm and three different values of the history depth h. Fig. 10 shows the relevant error distance CDF, and Table 3 summarizes the results in terms of the 50th(median), 75th and 90th percentile values.

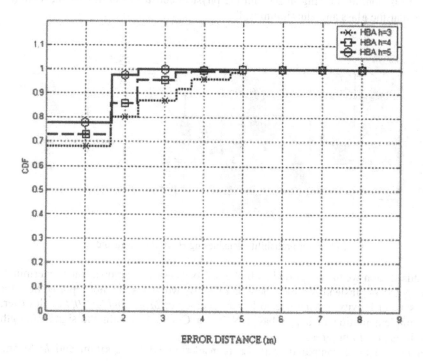

Fig. 10. CDF of the error in location estimation using HBA algorithm and different values of *h*

Table 3. The 50th, 75th, and 90th percentiles values of the error distance using HBA algorithm and different values of *h*

history depth (h)	50th (m)	75th (m)	90th (m)
3	0	1.65	3.30
4	0	1.65	2.33
5	0	0	2.33

The results show that *h=5* is preferable. In Fig.11 is shown the error probability for any point of Radio Map considering HBA algorithm and *h=5*. It is worthwhile to observe that cells with worse performance are near the sharp corners and the glass and aluminium wall, where reflection and diffraction effects are more significant

Considering that NeBULa doesn't require additional and costly hardware and is suitable for large-scale deployment, at the moment such performances are satisfactory in our opinion, also taking in account the physical constrains of the testing room (in particular the glass and aluminium wall).

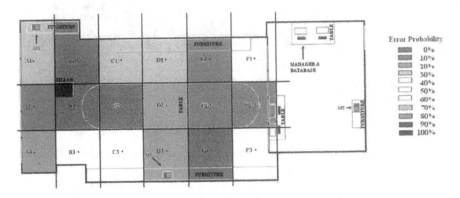

Fig. 11. Error probability considering HBA algorithm and $h=5$

In this section we have reported NeBULa test-beds using a laptop as mobile terminal. Using a PDA, instead, we have obtained slightly worse results, not reported here for the sake of brevity, maybe due to lower accuracy of *Ministumbler v0.4.0*. However, the implementation of a java-based NeBULa Client, to obtain the signal-strength levels, is now in progress.

Finally, we have compared NeBULa (considering HBA algorithm and $h=5$) and EKAHAU [2] in the same indoor environment and wireless network deployment; in this case we have considered random user orientation and physical points in the on-line phase. Table 4 summarizes the comparison in terms of the 50th, 75th and 90th percentile values.

Table 4. Performance comparison among NeBULa and EKAHAU

Location Sensing System	50th (m)	75th (m)	90th (m)
NeBULa ($h=5$)	0	1.65	3.3
EKAHAU	1.22	1.63	1.7

Considering that EKAHAU is a valid commercial product while NeBULa is a prototype still at the experimental stage, such performance comparison encourages to continue its software development.

5 Conclusions and Open Problems

In this paper, we have proposed NeBULa (Network-Based User-Location sensing system) to locate a mobile user within an indoor Wireless LAN environment. It is a new Scene Analysis system, based on RADAR technique, implemented in order to have an open platform to develop in future possible upgrading or novel location strategies (RADAR and EKAHAU source codes are not available). Its performance has been investigated through extensive testbeds. The results show the effectiveness of the proposed system in different operative conditions, considering both that it doesn't require additional and costly hardware and is suitable for large-scale deployment, and taking in account the physical constrains of the testing room (in particular the glass and aluminium wall).

Further problems are under consideration, regarding: a) the incidence of user orientation on the system performance; b) the use of many Radio Maps to minimize the error, especially to mitigate the effects due to the presence of other people in the environment, to the changes of temperature, sunlight and other environmental factors; c) the use, in alternative, of a propagation model; d) the possibility of integrate NeBULa with other Location-Sensing System like GPS to provide the absolute position.

6 Acknowledgment

This work was Partially Sponsored by *Centro Regionale di Competenza sull'ICT*

References

1. P.Bahl and V.N.Padmanabhan, "RADAR: An in-building RF-Based user location and tracking system", Proc. IEEE Infocom, March 2000
2. "EKAHAU Positioning Engine", http://www.ekahau.com/products/positioningengine/
3. J.Hightower and G.Borriello, "Location Sensing Techniques", University of Washington, Computer science and Engineering, July 2001.
4. Andy Hopper, Pete Steggles, Any Ward, and Paul Webster, "The anatomy of a context-aware application", Proc. Mobicom 1999, Seattle, August 1999.
5. N.B.Priyantha, A.Chakraborty, and H.Balakrishnan, "The cricket location-support system", Proc. Mobicom 2000, Boston, August 2000.
6. Bluesoft Incorporated. Website, 2001. http://www.bluesoft-inc.com
7. Jeffrey Hightower, Chris Vakili, Gaetano Borriello, and Roy Want, "Design and Calibration of the SpotON Ad-Hoc Location Sensing System," University of Washington, Seattle, WA, August 2001.

8. R.Want, A.Hopper, V.Falcao, and J.Gibbons, "The active badge location system", ACM Transaction on Information Systems, January 1992
9. J.Barton and T.Kindberg, "The CoolTown User Experiment", TR 2001-22, HP Laboratories, Palo Alto, Calif., 2001
10. R.J.Orr and G.D.Abowd, "The Smart Floor: a Mechanism for Natural User Identification and Tracking", Proc. 2000 Conf. Human Factors in Computing System, ACM Press, New York, 2000
11. T.Starner, B.Schiele, and A.Pentland, "Visual context awareness via wearable computing", International Symposium on Wearable Computers, IEEE Computer Society Press, Pittsburgh, PA, October 1998.
12. "Avaya Wireless PC Card – Getting Started Guide", http://support.avaya.com/elmodocs2/avayawireless/docs/pc_card/GSG_PC.pdf
13. "MiniStumbler v0.4.0 Release Notes", http://www.stumbler.net/readme/readme_Mini_0_4_0.html
14. "iNet Factory downloading", http://www.jscape.com/inetfactory/dload.html

The VIL (Virtual Immersive Learning) Test-Bed: An Innovative Approach To Distance Learning

G. Massei[+], A. Scarpiello[+], A. Vollono[+]

[+]National Inter-University Consortium for Telecommunications (CNIT)
Via Diocleziano 328, I-80125 Naples, Italy
{gianluca.massei, amedeo.scarpiello, alfonso.vollono}@cnit.it

Abstract. The Virtual Immersive Learning (VIL) test-bed intends to realize a virtual collaborative immersive environment, capable of integrating natural contexts and typical gestures, which may occur during traditional lectures, enhanced with advanced experimental sessions. The main aim of the paper is to describe the test-bed motivations, as well as the most significant strategies, both hardware and software, adopted for its implementation. The novelty of the authors' approach essentially relies on its capability of remarking and emphasizing results that are the output of VICom project, and "putting the pieces together" in a well-integrated framework. These features, along with its high portability, good flexibility, and, above all, low cost, make the proposed test-bed appropriate for educational purposes, mainly concerning measurements on telecommunication systems and virtual restoration training courses at universities, as well as research centres.

Keywords. Virtual Reality, E-learning, Tele-measurement, Tele-restoration, advanced interaction devices, natural gestures' identification, middleware

1   Introduction

VICom (*Virtual Immersive Communications*) [1] is a three-year project funded by the Italian Ministry of Education, University and Research (MIUR) and started in November 2002. The project goal is the design of a communication system architecture able to provide mobile virtual immersive services. The architecture effectiveness must be proved with two service test-beds, respectively, denoted as *Mobility in Immersive Environment* (MIE) and *Virtual Immersive Learning* (VIL). In particular, the VIL test-bed intends to realize a virtual collaborative immersive environment, capable of integrating natural contexts and typical gestures, which may occur during traditional lectures, enhanced with advanced experimental sessions. Two training courses are foreseen: the first one is oriented to virtual restoration of paintings (Fig. 1), whereas the second one concerns e-measurement applications (Fig. 2), where students are able to remotely control real devices and instrumentation in order to perform supervised measurements on telecommunications and networking systems. A 3D virtual reality application allows the real time interaction between a lecturer or instructor and a fixed number of students, who are not physically present in the same classroom. Students are grouped inside a number of well equipped classrooms, which are interconnected through the CNIT national network or other high performance networks. Each class may host a single teacher and up to four students, being able to interact with the aforementioned 3D application through projection screens, a 3D mouse or advanced interaction devices. Natural human gestures will be captured thanks to a set of coordinated cameras. Context information, such as environment status and personal information about participants in the virtual lecture, will be published by contextual 3D icons.

Fig. 1. Restoration Environment

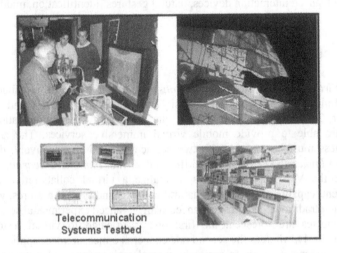

Fig. 2. E-measurement application.

Traditional approaches to Virtual Reality (VR) are based on complex and relatively expensive devices, such as head-mounted displays (HMDs), data gloves and CAVE systems [2]. Instead, the proposed approach to realize the VIL test-bed intends to emphasize results that are the output of research activities related to other work packages of the VICOM project. In particular it would exploit:

- natural gestures' identification algorithms to realize an interaction with the virtual class;
- middleware to share and manage all context information;
- a multimedia board and an embedded haptic interface to show different approaches to virtual reality applications;
- hardware/software architectures specifically designed and realized to control real instruments and devices, which may also be placed in different laboratories;
- virtual restoration tools to improve the quality of the digital reproductions.

2   The VIL Scenario

Hereafter the VIL scenario is described through a script, to define the virtual classroom environment and the interactions among lecturer and students involved in the virtual learning session.

1. The generic user reaches a VIL real classroom and logs in to the system through an accounting phase, to define the user's profile and know the seat reserved.
2. The lecturer and students enter the virtual classroom thanks to their avatars and reach their own work space.
3. The real-time lecture takes place in a virtual context-aware environment, where interactions occur in a natural way by means of Scene Analysis systems and immersive input devices.
4. Lectures are complemented with laboratory experiences, oriented to Supervised Tele-Restoration and Cooperative Tele-Measurements, exploiting specialized virtual laboratory software.

In the centres where also the MIE (Mobility in Immersive Environments) test-bed is foreseen, the users can reach the VIL classroom through the MIE virtual guide application, providing in such way the test-beds integration in a unified scenario.

3   Hardware technologies

The proposed scenario has been realized in order to be compliant with the economical resources of the VICOM project. To this aim, all useful research results from other work packages have been emphasized, rather than buying very expensive hardware available on the market.

The fulfillment of the VIL Test-Bed's goals requires the specification and the equipment acquisition of some enhanced classrooms, through which lecturer and students can take part in the immersive lecture.

Among the enhanced classrooms to be set up and interconnected, there are those at the Italian Ministry of Communications in Rome, PERCRO laboratories in Pontedera

(Pisa), TILS (Telecom Italia Learning Services, Rome), the National Resource Council (CNR) laboratories in Bologna, Rome Tor Vergata, Florence and Naples CNIT research units. Most of these centers are interconnected through a 2 Mbps proprietary CNIT satellite network (over the Skyplex network provided by Eutelsat). Since different types of enhanced classrooms are possible, each centre can choose the test-bed components to highlight. A fully equipped classroom (able to host up to four students and a lecturer) includes the hardware components explained in the following to remark all significant aspects of the test-bed.

- *Video rendering systems.* For the students' class we have select a visualization system composed by a projection screen, two linear polarisation filters, two XGA projectors and passive glasses. A LCD display is sufficient for the lecturer. Both systems must be equipped with a professional graphics workstation.
- *Audio rendering systems.* For the students' class we have chosen wireless headphones, while normal loudspeakers are sufficient for the lecturer.
- *Input devices.* Any user can interact with the GUI (*Graphical User Interface*) through input devices providing different immersion sensations. The user can choose a simple mouse, a 3D mouse with six degrees of freedom, a haptic interface, or the Multimedia Board (Fig. 3).
- *Contribution devices.* During the lectures or the laboratory experiences, audio and video interaction of any user must be allowed. For the students' class we have selected a dome camera (whose control is allowed thanks to VISCA[1] commands) and omni-directional microphones, while simple commercial devices are sufficient for the lecturer.
- *Scene Analysis systems.* The Scene Analysis systems allow the acquisition and analysis of context information. They need an accurate tuning to overcome the environment problems (room size, light, noise level, reverberation, etc...). For the VIL Test-Bed we have considered the systems explained in the following.

  - ➢ *Audio-Location System.* It allows locate the speaker's position through the phase processing of the audio signals acquired by arrays of microphones. The system output can control pan, tilt, zoom and focus of the dome camera. The system, provided by the PoliMi (Politecnico di Milano) research unit, includes array of microphones, audio mixer, computer for the processing, and possible deadening panels.
  - ➢ *Request Identification System.* It allows making a reservation for a question or intervention simply by raising a hand. The system, provided by the CNIT research unit at University of Geneva, includes dome camera and a computer for processing.
  - ➢ *Accounting System.* It allows obtaining identity, role, e-mail and other personal data of the users attending the immersive lecture. Moreover, for each user it provides the relative avatar ID and physical position in the enhanced classroom. To this aim, a computer at the entrance of the classroom is sufficient.

---

[1] VISCA (Video System Control Architecture) is a network protocol designed to interface a wide variety of video equipment to computer

- *Additional Hardware.* Other computers are necessary as ambient servers, namely to install the middleware and enable objects (such as input devices) to provide services or information.

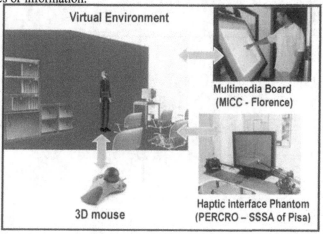

Fig. 3. Input devices for the VIL/VICOM test-bed

4   The Software Architecture

The proposed software architecture is illustrated in Fig. 4 and Fig. 5.

The *Common Experience Manager* (CEM) is certainly the main block of such architecture, as it manages both e-learning (through the *Vip-Teach Server* [3]) and experimental laboratory sessions (through the *LabNet Server* [4]). Context is captured and analyzed by the *Scene Analysis* (SA) module, throw arrays of microphones and cameras. Such information is stored in a shared memory and managed by the LIME (*Linda In a Mobile Environment*) middleware [5]. Students and lecturers are recognized thanks to SA, as they are registered during a preliminary authentication phase. The SA notifies asynchronously all information to the CEM, which assigns the appropriate role and related privileges to individual subjects.

Any student can select a synchronous or asynchronous instruction course. In the former case, the CEM manages the interaction between students and lecturer through a token-based mechanism: the lecturer is able to give entirely or to share its privileges, communicating with the CEM through an immersive GUI (shown in Fig. 6). *Interactive Inputs* (II) allow to interact with the virtual environment, while *Contribution Inputs* (CI) allow to ask questions during a lecture after being enabled by the lecturer: interventions occur by video and audio streaming. In the latter case, the *Offline Lectures Manager* (OLM) module represents the interface to enter off-line contents. To support experimental sessions focusing on e-measurements and virtual restoration, a unique Java-based interface, namely the *Environment Interface* (EI), has been designed in the proposed software architecture to interact with the *LabNet Server* (LNS) module (to set up a collaborative experience).

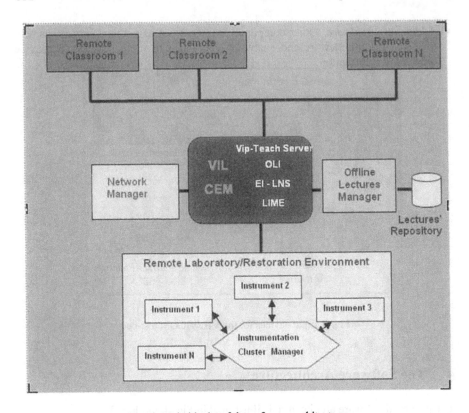

Fig. 4. Main blocks of the software architecture.

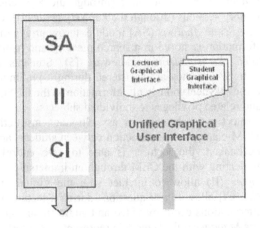

Fig. 5. Remote Classroom.

The immersive GUI is shown in Fig. 6.

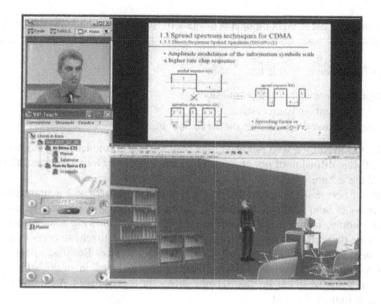

Fig. 6. Lecture Graphical User Interface

It is possible to clearly see the elements constituting the client interface of the synchronous e-learning application Vip-Teach by Lightcomm [3] (video, Powerpoint presentation, management window with the list of students present and making a reservation) and the Internet Explorer browser where 3D contents are shown. They are modeled through 3D Studio Max by Discreet [6] and controlled through eXtreme Virtual Reality (XVR) by VRmedia [7]. In particular, we expect the 3D environment, used for the presentation of context information, takes up approximately the same space reserved to slides during the lecture, while during the laboratory experiences it is important to reserve more space to the 3D interface allowing to act on digital paintings or virtual instruments.

At the moment, thanks to capabilities offered by XVR, the users are able to:

- navigate in a 3D environment;
- manage collision among 3D objects;
- select and describe objects;
- visualize waveforms on virtual instruments;
- interact with virtual objects by either a simple 2D mouse or a haptic interface;
- display a stereoscopic vision of the virtual environment.

Analyses are now in progress of the communication protocol between XVR and the LABNET architecture throw the Java-based EI, of the Tele-restoration/Tele-measurement integration, of the introduction of avatars, of the presentation of context information, and of the integration with VIP-Teach functionalities.

Besides, we are considering the possibility to additionally use the VRML standard for its characteristics of simplicity and easy integration with Java and JavaScript.

In the following, the main components of the VIL Software Architecture are analysed. Fig. 7 shows the main components of CEM: the *VIP-Teach Server* and the *LabNet Server*.

The *Vip-Teach Server* is able to:

- manage users' accounts and permissions;
- enrol the students in the lectures;
- activate the PowerPoint viewer on the remote PCs.

It can be followed by a web portal, for the management of the lectures' calendar and for the off-line diffusion of ppt presentations, and eventually by a Recorder, allowing to record the lecture.

The *LabNet Server* [4] manages the access to a generic experience, guaranteeing the interoperability and the synchronisation among the users. Particularly, owing a *Control* module, it makes the experience collaborative, allowing a super-user (the lecturer) the possibility to pass the instrumentation control (token) to users of inferior level (the students), through the VIP-Teach client interface. Besides, owing to the *Data Providing* module, the instrument data are distributed in multicast to users, which can visualize them on the 3D interface, via a java-based Adaptation Layer (previously indicated as EI).

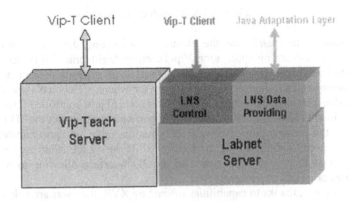

Fig. 7. Main components of CEM

Fig. 8 shows instead the main components of the Remote Classroom.

In accordance with the GUI, we have considered two main software modules: the VIP-Teach Client and the XVR-based 3D Control module (to create the object to import in the html page).

The *VIP-Teach Client* is able to:

- interact with VIP-Teach Server to manage users' accounts and permissions;
- receive the audio/video contents from own peers or transmit it to them, according to the relative roles;
- transfer the information related for token management to the LNS Control module;

- interact with LIME to publish the token holder in the context space and to extract the data (i.e., the students making a reservation for a question or intervention by raising a hand).

The *XVR-based 3D Control* module must interact instead with LIME, to present context information (i.e., user identity, students talking or in reservation), and with the LNS Data Providing module, to write and read instrument and painting data, via the java-based adaptation layer.

Fig. 8 shows that the Scene Analysis systems, the Interactive Inputs and the Contribution Inputs must interact with LIME middleware through the software module described in the following paragraph. Besides, it is evident that the contribution elements provide inputs only to the VIP-Teach client, since we do not insert the audio/video signal in the 3D environment, while the interaction devices act on both modules.

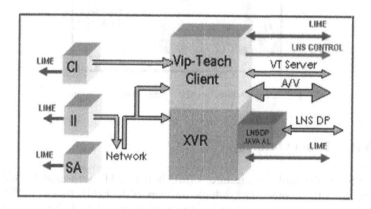

Fig. 8. Main components of Remote Classroom

Finally, the software architecture for e-measurement experiences is in shortly explained, using a top-down approach, in Fig. 9.

The SW modules involved in the architecture are explained in the following.

- The 3D GUI communicates with the remaining architecture via a Java-based interface.
- The LNS (LabNet Server) manages the access of users to the experiences.
- The Experience Manager manages the allocation of the instruments in the single experiences, the correspondence among the experience variables and actions on the instrumentation drivers.
- The Experience database contains the experience table (to list the instruments involved in each experience) and instrument table (to define the allocation state).
- The Testbed is the set of instrumentation drivers for e-measurement sessions.

In particular, to call the driver procedures, the Experience Manager uses *Remote Procedure Calls* (RPC) through SOAP (*Simple Object Access Protocol*) on TCP/IP [8]. Finally, the drivers recognize SOAP messages and translate them in reading/writing commands on the instruments involved in the experience.

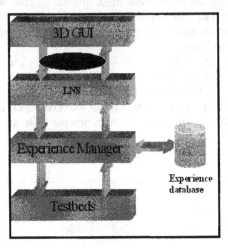

Fig. 9. SW architecture for e-measurement session.

5   VIL/VICom Ontology

In the VICom Protocol Architecture, LIME is the middleware used for hiding the distribution of the context data, which are mapped to LIME tuples. In [9] a Java interface, denominated *MIE Interface*, has been proposed to define the context-to-tuple mapping rules and therefore the format of tuples published on LIME.

To have a simple interpretation of the context data, they are written within XML statements that have to be consistent with the VICOM ontology through a software module, denominated CDM (*Context Data Manager*). The ontology defines the semantic meaning of classes and their properties, and it establishes structures and logical relationships, which, for instance, can be interpreted by an ontological motor to effect searches of semantic level. The VICom ontology is written in OWL (*Ontology Web Language*) [10].

Therefore, as shown in Fig. 10, the CDM collects own context data and publishes them in XML statements coherent with the VICom ontology, and the MIE Interface defines the way to publish instance of OWL concepts on LIME tuple space in order to build an XML statement in a distributed fashion.

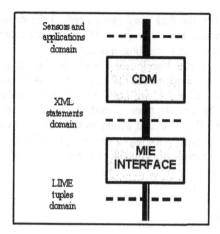

Fig. 10. LIME to application interface

This section describes at high level the Classes, the *DataType Properties* (DTP) and the *Object Properties* (OP) that we have defined for the VIL Test-bed, and the associated relationship coming out from the OWL scheme.

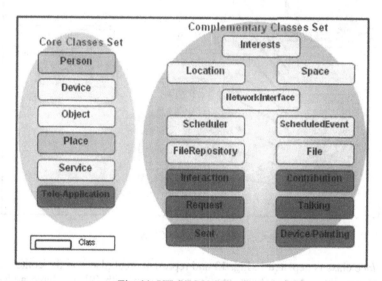

Fig. 11. VIL/VICOM Ontology

In Fig. 11, in accordance with [9], two sets of classes have been introduced in the VICom Ontology: *Core Classes* and *Complementary Classes*. Basically, the Core Classes represent the main actors in the VICom scenario. The other classes are strictly related to one or more Core Classes, and thus we call them Complementary Classes.

In comparison with [9], the dark grey classes collect the context data related only to the VIL Test-bed, while the Person and Place classes represent the MIE classes modified in accordance with e-learning applications.

In the following, we report a description of each novel or modified class.

The *Person* class is depicted in Fig. 12 and represents a real person in the context space. In comparison with [9], we have added four mandatory DTPs to keep in memory the role of the user (lecturer or student), the user's e-mail, type of course to give or to attend, and ID of related avatar in the virtual environment. Three further OPs are necessary to specify the input device type (3D mouse, haptic interface or multimedia board), the available contribution elements, and the seat occupied by the user in the real rooms.

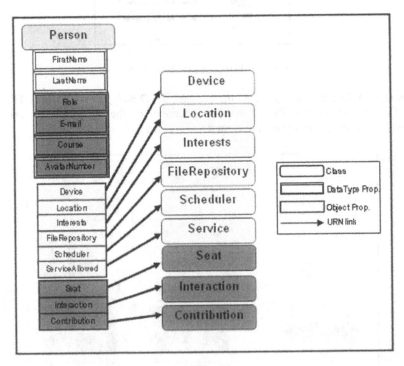

Fig. 12. Person Class

The *Place* class is depicted in Fig. 13 and represents real places in the context space, such as meeting rooms, offices, etc. In comparison with [9], we have added two DTPs: city to distinguish different cities, and address to distinguish different laboratories in the same city. It mainly springs from the presence of different enhanced classrooms for the VIL Test-Bed on the national territory, but also from the need to locate a generic measurement instrument among the possible locations (Laboratory in Naples or one of the three Laboratories in Bologna).

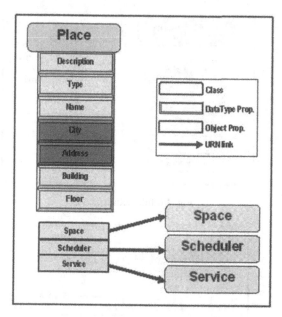

Fig. 13. Place Class

The *Interaction* and *Contribution* classes are depicted respectively in Fig. 14 and Fig. 15. They represent the input devices and the contribution elements, respectively, in the context space.

Interaction devices (simple mouse, 3D mouse, haptic interface or multimedia board) are characterized by three mandatory DTPs. Description is a string that reports general information on the device, Type is a string that indicates the group the device belongs to, Name is the device nickname and is a string. Two OPs can be present: the Place OP is used to relate the input device with a real enhanced room, while the scheduler OP is a link to the device events scheduler (some input devices such as the haptic interface are not always available).

A Contribution element (dome camera and arrays of microphones in the enhanced classroom for the students, simple camera and microphones for the lecturer and the students with dedicated hardware) is characterized by two mandatory DTPs, to specify the enhanced classroom and the IP address of the workstation on which the GUI is present.

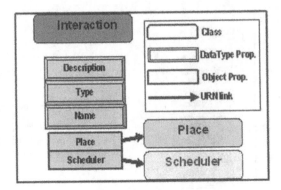

Fig. 14. Interaction Class

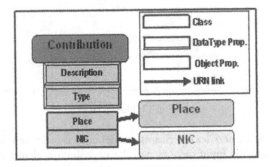

Fig. 15. Contribution Class

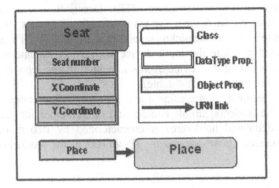

Fig. 16. Seat Class

The *Seat* class is depicted in Fig. 16 and represents the seat occupied by the user in an enhanced classroom in the context space. This context information is published by the CDM related to the accounting system. We have considered three DTPs to specify the seat ID and location (through two space coordinates in a well-known reference frame), and an OP link to the place class to fix the enhanced classroom.

The *Request* and *Talking* classes are depicted in Fig. 17 and Fig. 18, respectively. They specify the student making a reservation for a question or intervention, simply by raising a hand, and the talking one.

The Request Identification and Audio-Location systems are able to provide the location of the concerned student. Since any student is represented in the context space by an instance of Person class specifying the Seat OP link, these scene analysis systems, interacting with LIME, are able to define the concerned student and publish the relative context data.

The accounting system, instead, specifies a number of Person class properties through the own CDM.

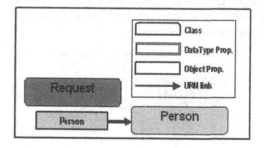

Fig. 17. Request Class

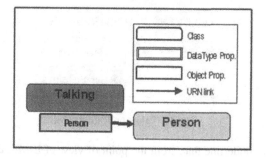

Fig. 18. Talking Class

Finally, the *Tele-Application* and *Device/Painting* classes are depicted respectively in Fig. 19 and Fig. 20. They are introduced to support the Tele-Restoration and Tele-Measurement laboratory experiences.

The Tele-Application class includes three DTPs to specify the description, the type (Tele-Restoration or Tele-Measurement) and the application availability (i.e., the measure test-bed could be unavailable due to link failure, local use, or other problems). Five OPs can be related to Tele-Application class: Place (to specify the real classroom), NIC (to register the IP address of the LNS or servers in one of the Bologna Laboratories), FileRepository (to obtain relative literature), Person (to provide the token manager, i.e., the user able to access the experience) and Device/Painting (to specify the painting or the measurement instruments involved in the experience)

The Device/Painting class represent the instruments involved in the measurement experience or the paintings to restore, in the context space. It includes four DTPs to specify the description, the type, the name and the availability (a measurement instruments might be off or locally used). Four OPs can be related to the Device/Painting class: FileRepository (to obtain relative manuals or application notes for instruments, and historical or bibliographical notes for paintings), Place (to specify the laboratory or the museum), NIC (to register the IP address of the LNS or servers in one of the Bologna Laboratories) and Person (to provide the token manager, i.e., the user able to access the experience). Such context data are necessary when the instruments are geographically distributed.

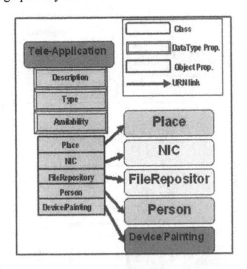

Fig. 19. Tele-Application Class

It is important to underline that, from the point of view of LIME, the VIL scenario is simpler than the MIE one, since the user mobility is limited to the enhanced classroom and the context data are limited and often unidirectional (from students to lecturer).

A LIME optimization according to the VIL scenario would consist of joining the tuples related to the physical context (Location, Request, etc...) and provide them to an "ambient server" on the site, while the others (FileRepository, Cloth-Application, etc...) could be provided through the classical peer-to-peer architecture [5].

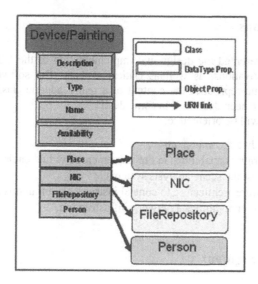

Fig. 20. Device/Painting Class

6   Figures of Merit and Performance

The VIL Test-Bed allows underline new potentialities in distance learning scenarios. However, a careful evaluation is necessary of numerous aspects connected to new applications.

The VIL Test Bed must be evaluated under a few different respects:

- *Level of integration*: "putting the pieces together" in a well-integrated framework has an inherent value, and is a factor of innovation, even though not all the parts would be based on completely novel achievements.
- *Enhancement of distance learning*: context-awareness and tele-laboratory are two significant enhancements to "traditional" distance learning. The immersivity-enhancing value of this may be evaluated with subjective tests (e.g., MOS – *Mean Opinion Score*).
- *QoS requirements*: objective evaluation may be conducted on the networking requirements and scalability of the various degrees of complexity of the platform. In particular, the tele-restoration and tele-measurement components allow the monitoring of the network traffic generated and the needs in terms of QoS.
- *LIME performance*: middleware performance may also be evaluated, in terms of response time at the application level and service discovery. However, as regards the VIL components, the role of LIME is not so critical. A technology worth investigating in future work in this area would be that of Grid computing, with the integration of real-time elements and the control of real/virtual instrumentation.

Individualization and evaluation of other legal, economical, psycho-sociological and ergonomic aspects of the new technologies are still in progress.

7   Conclusions

The paper has presented the design and implementation of the VIL test-bed and its main related motivations, as well as critical aspects. The software and hardware strategies, allowing reproduce the context of a real academic classroom in a virtual environment, have been described in some detail.
Future activities will be oriented to:

- deploy the hardware components;
- integrate e-learning capabilities in the implemented virtual environment;
- integrate and test the scene analysis systems;
- stress the software architecture to control instrumentation in this new perspective;
- test the test-bed on a multicast network infrastructure.

8   Acknowledgment

This work was funded by the Italian Ministry for Education and Scientific Research (MIUR) in the framework of the FIRB_VICOM project.

References

1. http://www.vicom-project.it
2. D. Tougaw, J. Will, "Visualizing the Future of Virtual Reality", Computing in Science & Engineering, Volume 5, Issue 4, pp. 8-11 July/August 2003.
3. http://www.lightcomm.it
4. F. Davoli, S. Vignola, S. Zappatore, "A Multimedia Laboratory Environment for Generalized Remote Measurement and Control", Proc. 2002 Tyrrhenian Internat. Workshop on Digital Communications (IWDC 02), Capri, Italy, Sept. 2002, pp. 413-418.
5. G. P. Picco, S. Cicero, G. Cortese, D. Frey, A. L. Murphy, E. Trevisani, A. Vialetti, "Software Architecture and Middleware for VICOM: Concepts, API, and Guidelines for Application Development", June 2004, www.vicom-project.it.
6. http://www.discreet.com
7. http://www.vrmedia.it
8. A.Vollono, A. Zinicola, "A New Perspective in Instrumentation Interfaces as Web Services", Proc. 2005 Tyrrhenian Internat. Workshop on Digital Communications (TIWDC 05) Sorrento, Italy, July 2005.
9. Detti, P. Loreti, "Mobility in Immersive Environment: System Architecture and Test-Bed Applications", Oct. 2004, www.vicom-project.it.
10. D. L. McGuinness, F. van Harmelen, Ed., "OWL Web Ontology Language Overview", W3C Recommendation, Feb. 2004, "http://www.w3.org/TR/2004/REC-owl-features-20040210/".

Author Index

Index